Mechanics of Solids

Mechanics

of Solids

DEPARTMENT OF MECHANICAL ENGINEERING,
IMPERIAL COLLEGE OF SCIENCE, TECHNOLOGY
AND MEDICINE,
LONDON UNIVERSITY

OXFORD
BLACKWELL SCIENTIFIC PUBLICATIONS
LONDON EDINBURGH BOSTON
MELBOURNE PARIS BERLIN VIENNA

© 1989 by
Blackwell Scientific Publications
Editorial offices:
Osney Mead, Oxford OX2 0EL
25 John Street, London WC1N 2BL
23 Ainslie Place, Edinburgh EH3 6AJ
3 Cambridge Center, Suite 208
  Cambridge, Massachusetts 02142, USA
54 University Street, Carlton
  Victoria 3053, Australia

First published 1989
Reprinted 1990

Set by Macmillan India Ltd, Bangalore 25
Printed and bound in Great Britain by
Butler & Tanner Ltd, Frome.

DISTRIBUTORS

Marston Book Services Ltd
PO Box 87
Oxford OX2 0DT
(*Orders:* Tel. (0865) 791155
      Fax. (0865) 791927
      Telex. 837515)

USA
Publishers' Business Services
PO Box 447
Brookline Village
Massachusetts 02147
(*Orders:* Tel. (617) 524–7678)

Canada
Oxford University Press
70 Wynford Drive
Don Mills
Ontario M3C 1J9
(*Orders:* Tel. (416) 441–2941)

Australia
Blackwell Scientific Publications
(Australia) Pty Ltd
54 University Street
Carlton, Victoria 3053
(*Orders:* Tel. (03) 347–0300)

British Library
Cataloguing in Publication Data

Fenner, Roger T. (Roger Theedham), 1943–
  Mechanics of solids.
  1. Solids. Mechanics
  I. Title
  531

  ISBN 0–632–02018–0
  ISBN 0–632–02019–9 (pbk.)

Library of Congress
Cataloging-in-Publication Data

Fenner, Roger T.
  Mechanics of solids/Roger T. Fenner.
  Bibliography: p.
  Includes index.
  ISBN 0–632–02018–0
  ISBN 0–632–02019–9 (pbk.)
  1. Strength of materials.
  2. Solids. I. Title.
  TA405.F423 1988
  620.1'12——dc19

Contents

Computer Programs

Preface

There are only a few basic principles to be mastered in the study of the subject variously referred to as the mechanics of solids, mechanics of materials or strength of materials. It is the application of these principles to the solution of problems, and the choice of assumptions which must be made, which present the greatest challenge. This text is intended for the first course in mechanics of solids offered to engineering students. It concentrates on developing analysis techniques from basic principles for a range of practical problems which includes simple structures, pressure vessels, beams and shafts. Many worked examples are given. The arrival of computers in general, and personal computers in particular, has revolutionized the way in which engineering problems are solved in practice, and this is being reflected in the way in which subjects such as the mechanics of solids are taught. A distinctive feature of the present book is therefore the inclusion of a number of computer techniques and programs for carrying out the analyses – not merely as appendices, but integrated into the text. The programs will also find many applications in the teaching of design.

It is not intended that the use of computer programs should replace hand calculations in the learning process, but should supplement them, and make it possible for the student to explore more complex and realistic problems of analysis and design. The approach adopted is therefore first to present the underlying theory and the traditional manual methods of solution before introducing computer techniques. The programs, which are coded in FORTRAN 77, are suitable for personal computers, but can also be run on minicomputers or mainframes. Detailed internal and external documentation is provided to aid the understanding of the programs, together with examples of their use. It is intended that students should use them to solve many of the problems which are set at the end of each chapter, and in their design work.

It is assumed that students using this book will have some experience of elementary statics (mechanics of rigid bodies), although Chapter 1 includes a review of the relevant topics. A knowledge of matrix notation for the presentation of linear algebraic equations is also assumed. Some familiarity with the solution of constant coefficient ordinary differential equations (particularly for the buckling problems considered in Chapter 8) is highly desirable, as is experience of partial differentiation and integration if Chapter 10 on more advanced applications is to be studied. Those numerical analysis techniques which are incorporated in the computer programs, particularly for the solution of simultaneous linear algebraic equations and single nonlinear algebraic equations, are described in Appendices. It is assumed that students will have sufficient knowledge of the FORTRAN programming language to be able to read and understand relatively straightforward programs.

There are many ways in which a first course on the mechanics of solids can be presented. The approach adopted here, based on the author's experience of teaching the subject, is to start with types of problems involving uniform stresses. Initially such stresses are uniaxial, as in pin-jointed structures, progressing to biaxial and even triaxial, but without shearing, as in thin-walled pressure vessels. Statically determinate situations, which require only the consideration of equilibrium conditions for the forces and stresses to be found, are treated before statically indeterminate ones. Problems involving relatively simple variations of stresses are then examined, principally the bending of beams and the torsion of shafts. Finally, an introduction to more complex situations is provided via the analysis of two-dimensional states of stress and strain, failure criteria and the differential equations of equilibrium and compatibility in two dimensions.

In addition to a review of statics, Chapter 1 introduces the concepts of stress and strain in a solid body, the influence of material properties and the principles of the mechanics of solids. These principles are those of equilibrium of forces, compatibility of strains and the stress–strain characteristics of materials, underlying themes which run through the remainder of the book. In Chapter 2, some statically determinate systems are analyzed, in particular pin-jointed structures, thin cylindrical and spherical shells, and flexible cables. A computer program is introduced for the analysis of statically determinate pin-jointed structures, using a simple form of finite element method. Stress–strain relationships for engineering materials are discussed in Chapter 3, and are used to find the deformations of statically determinate systems considered in the previous chapter. Some types of statically indeterminate systems are examined in Chapter 4, notably pin-jointed structures (by finite element computer method), liquid-filled pressure vessels, and problems involving resisted thermal expansion.

Chapters 5 and 6, which form a major part of the book, are concerned with beams and the simple theory of bending. While Chapter 5 deals with shear forces, bending moments and stresses, and the analysis of statically determinate beams (including a finite element computer method), Chapter 6 is concerned with beam deflections, leading to the analysis of statically indeterminate beams (and another finite element method). In Chapter 7, problems of torsion of circular shafts are considered. Following an introduction to problems of instability, Chapter 8 deals with the buckling of struts and columns.

In Chapter 9, attention moves away from problems involving only simple states of mainly uniaxial stresses, towards more complex situations. Transformations of stress and strain components acting on different planes at a point lead to the definition of principal and maximum shear values. A computer method is introduced for analyzing stresses or strains at a point, as is a program for determining the state of strain and stress at a point from strain gage measurements. Criteria for yielding and fracture under complex states of stress are then examined. Finally, Chapter 10 develops the principles of equilibrium and compatibility into the partial differential stress equilibrium and strain compatibility equations for problems involving general one- and two-dimensional variations of stresses and strains. These are applied to beam problems, and serve to demonstrate the levels of approximation involved in the simple theory of bending. In a one-dimensional form suitable for axisymmetric problems they are also applied to rotating disks and thick-walled cylinders used as pressure vessels, and a computer method is introduced for the determination of stresses and strains in compound thick-walled cylinders.

The coverage of topics provided by the text may well be greater than that required for particular courses. For example, not all instructors would wish to deal so fully with pin-jointed structures, although the solution of more realistic problems is so much more practical using computer techniques, and they do provide a natural introduction to finite element methods which students will meet in later courses. Also, Chapter 10, and perhaps parts of Chapter 9, may be more appropriate for a more advanced course. The introduction of computer techniques has meant that some more traditional methods have been omitted. For example, graphical methods for the analysis of pin-jointed structures are not considered. Similarly, the only manual method described for the determination of beam deflections involves integration of the moment–curvature equation: the more graphical moment–area method is not covered. Other topics which have been omitted, but which are only rarely covered to any significant depth in a first course, are energy methods and an introduction to plasticity and the analysis of elastoplastic bending and torsion problems.

Some of the computer techniques are described in the text as finite element methods. This is because they are the kinds of essentially simple techniques which led to the birth of finite element methods. It is useful to describe them in finite element terms, not least because it provides a good preparation for the introduction of more sophisticated finite element methods in later courses.

Many engineering courses have now converted entirely to SI metric units. In the United States, however, US customary (Imperial) units are still widely used, reflecting industrial practice. Consequently, in this book both sets of units are employed. In worked examples, given numerical data and the main calculated results are usually shown first in SI units, followed in parentheses by equivalent values in US customary units. The equivalence is not intended to be exact, and values are normally quoted to only two significant figures. The intention is to provide those readers less familiar with SI units with a better feel for the magnitude of the quantities involved. In many examples, detailed calculations in both sets of units are shown, side by side. The material property data presented in Appendix A are given in both SI and US customary units, together with the appropriate conversion factors.

A substantial number of problems is provided at the end of each chapter, first a set in SI units and then a set in US customary units. Where appropriate, problems are also grouped under topic headings, and within each group they are graded in difficulty. While the elementary problems involve only straight-forward application of the methods described in the text, some of the more difficult ones are more open-ended and of the design type. The necessary material properties are in most cases not given in the problems, but are provided in an appendix (Appendix A). Answers to alternate problems are listed at the end of the book, and worked solutions to all problems are contained in a separate instructor's manual.

Magnetic disk copies of all the computer programs can be obtained from the publisher. Programs may be freely copied, used, modified or translated. Although they have been carefully tested, they may contain errors, and I would appreciate being informed of any that are found. While the programs were developed specifically for teaching purposes, they may find applications in the solution of real problems. If they are used for this purpose, I will not be responsible for any errors they may contain.

Acknowledgements

The author of a textbook owes a debt of gratitude to other people, whose experience materially contributes to its formulation. In particular I gratefully acknowledge detailed advice and comments received from the following:

Professor Daniel Frederick, Department of Engineering Science and Mechanics, Virginia Polytechnic Institute and State University
Professor Carl F. Long, Thayer School of Engineering, Dartmouth College
Dr V. J. Meyers, School of Civil Engineering, Purdue University
Dr Ozer A. Arnaz, Department of Mechanical Engineering, California State University, Sacramento
General guidance and support from Dr Lee L. Lowery, Department of Civil Engineering, Texas A&M University is also much appreciated.

I also wish to acknowledge the contributions, both direct and indirect, made by many of my colleagues at Imperial College to the development of the course with its associated examples, problems and examination questions, on which the book is based. In particular I would like to thank Dr Frank Ellis, Professor Alan Swanson and Dr Simon Walker. Special thanks are also due to Dr Fusun Nadiri for her careful reading of much of the typescript, and for many helpful suggestions. Some of the examples and problems used are taken from, or based on, questions set in University of London examination papers. Permission to use this material is gratefully acknowledged. Responsibility for the solutions and answers is, however, mine alone.

Roger T. Fenner

List of Symbols

The symbols which are used throughout the book are defined in the following list. In some cases, particular symbols have different meanings in different parts of the book, although this should not cause any serious ambiguity. Other symbols or alternative definitions of the present symbols representing, for example, dimensions of components or constants of integration are introduced within the limited contexts of particular examples or pieces of analysis.

FORTRAN variable names used in the computer programs are defined in the lists associated with the programs.

A	Area (especially cross-sectional area)
$[A]$	Coefficient matrix
b	Breadth
C	Couple
C_1 to C_5	Constants in a finite element shape function
c_1, c_2	Distances from neutral surface to highest and lowest points of a beam cross section
D	Diameter
d	Depth
E	Young's modulus (modulus of elasticity)
e	Normal strain
e	Eccentricity
e_{vol}	Volumetric strain
e_Y	Yield strain
F	Force
$[F]$	Vector of applied external loads
F_r	Body force per unit volume in the r direction
f	Flexibility
$f(\ )$	Function; step or singularity function
$[f]$	Vector of loads on an element
G	Shear modulus (modulus of rigidity)
g	Acceleration due to gravity
H	Horizontal component of force
h	Height
I	Second moment of area of a beam cross section about its neutral axis
i, j	Node numbers
J	Polar second moment of area of a shaft cross section about its axis

K		Stress concentration factor; bulk modulus; radius ratio for a thick-walled cylinder
$[K]$		Overall stiffness matrix
k		Stiffness
$[k]$		Element stiffness matrix
L		Length
L_e		Effective length of the equivalent pin-ended strut
M		Moment; bending moment
m		Element number; modulus ratio
N		Number (of elements, reactions, supports etc. – indicated by appropriate subscripts); moment
n		Buckling parameter defined in equation (8.5)
P		Force
P_c		Euler critical buckling force
p		Pressure; perimeter
Q		Force; first moment of area of the region of a beam cross section above a given distance from the neutral axis
R		Reaction force at a support; radius; radius of curvature of the neutral surface of a beam
r		Radial coordinate in the cylindrical polar system; radius of gyration
S		Elastic section modulus
T		Force in a cable or member of a structure; temperature; torque
$[T]$		Vector of member forces
t		Wall thickness
U, V		Forces in the x and y directions
u, v, w		Displacements in the x, y and z directions
V		Vertical component of force; shear force; volume
$[V]$		Vector of forces and moments
W		Weight
$[W]$		Vector of applied external loads
w		Weight per unit length (of a beam or cable)
X, Y		Global Cartesian coordinates; body forces per unit volume in the x and y directions
x, y, z		Cartesian coordinates
z		Axial coordinate in the cylindrical polar system
α		Angle; coefficient of linear thermal expansion
β		Coefficient of volumetric thermal expansion
γ		Shear strain
Δ		Change of (followed by another symbol)
δ		Small increment of (followed by another symbol)
δ		Displacement; radial interference
$[\delta]$		Vector of displacements
ε		Natural strain
θ		Angular coordinate in the cylindrical polar system; angle
λ		Extension ratio; rotational stiffness
v		Poisson's ratio

ρ	Density
σ	Normal stress
σ_e	Von Mises equivalent stress
σ_H	Hydrostatic stress
σ_U	Ultimate tensile stress
σ_Y	Yield stress
τ	Shear stress
ϕ	Angle
ω	Angular velocity

1

Introduction

As engineers we are concerned with solving problems, which may involve not only the engineering sciences but also related subjects such as economics and management science. The ability to solve problems can be gained in two main ways: by practical experience of particular problems and the systematic study of underlying principles. Although both are necessary for the practising engineer, the study of principles leads more rapidly to a genuine understanding, and makes it possible to tackle new problems not previously met. For convenience, the total subject matter of engineering science is usually subdivided into a number of topics, such as solid mechanics, fluid mechanics, heat transfer, properties of materials, and so on, although there are close links between them in terms of the physical principles involved and the methods of analysis employed.

Solid mechanics as a subject is usually further subdivided into the *mechanics of rigid bodies* (sometimes just *mechanics*) and the *mechanics of (deformable) solids*, also known as *mechanics of materials* and *strength of materials*. While the mechanics of rigid bodies is concerned with the static and dynamic behavior under external forces of engineering components and systems which are treated as infinitely strong and undeformable, the mechanics of solids is more concerned with the internal forces and associated changes in geometry of the components involved. Of particular importance are the properties of the materials used, the *strength* of which will determine whether the components fail by breaking in service, and the *stiffness* of which will determine whether the amount of deformation they suffer is acceptable. The subject of mechanics of solids is therefore central to the whole activity of engineering design.

Let us consider the situation illustrated in Fig. 1.1, which shows a person standing on a ladder which has one end on a horizontal surface, and the other resting against a vertical wall. While mechanics (of rigid bodies) would be concerned with whether the ladder will slip, in mechanics of (deformable) solids we are more interested in whether the ladder is strong enough to support its human load, and whether it is stiff enough to make that load feel secure! We must expect the ladder to suffer some deflection, which will be greatest near its center. If it is correctly designed, however, this deflection will be relatively small and *recoverable*. In other words, when the load is removed, the ladder returns to its original straight form. Such behavior is referred to as *elastic* deformation. On the other hand, if it is not well designed, the ladder may suffer large deflections, with possibly gross local deformations at some position along its length, rather like a hinge being formed, and would not return to its original form when it is unloaded. Such behavior, involving large and permanent

Fig. 1.1.

changes in geometry is referred to as *plastic* deformation. Alternatively, if the material used is *brittle*, the ladder may deflect by a relatively small amount before suffering *fracture* into two pieces. In the context of ladder design, both permanent deformation and fracture must be regarded as modes of *failure*.

Finding the answers to the questions as to whether an engineering system, in this case the ladder, is strong enough and stiff enough involves two main steps. Firstly, we must model the system, and then apply appropriate physical principles to determine the forces and displacements involved. Let us now examine these steps in more detail.

1.1 Modeling of engineering systems

Choosing a suitable model for a system is a matter of making reasonable assumptions in order to simplify the real system far enough to permit it to be analyzed without an excessive amount of labor, but without at the same time simplifying it so far as to make the results of the analysis unreliable for design and other purposes. Successful modeling requires knowledge, experience and good physical insight into the way solid components behave and interact. It is usually the most important, and often the most difficult, part of a solid mechanics analysis.

The importance of knowledge and experience in modeling and analyzing engineering systems is reflected in this book by the many Examples which are described and worked through in detail in the text, together with the Problems provided at the end of each chapter. To some extent, many of these problems are artificial in the sense that the underlying assumptions are implicit, and the data presented in unambiguous forms. In problems encountered in engineering practice, the greatest difficulties often arise in deciding what assumptions are reasonable, and then obtaining sufficient data on which to base a solution. There is little to be gained, however, in attempting problems of the latter type until experience is gained solving problems of the former type.

An example of system modeling is provided by the ladder arrangement shown in Fig. 1.1. The foot of the ladder may well rest, not on a rigid surface, but on soft ground. The rounded end will therefore penetrate the surface, giving a finite area of contact, and a distribution of reaction force acting on the ladder, as shown in Fig. 1.2a. Now, unless we are particularly interested in what happens at the very end of the ladder, which we are not when considering

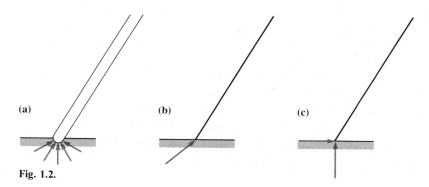

(a)   (b)   (c)

Fig. 1.2.

strength and stiffness of the ladder as a whole, we can model the ladder as a line element of negligible width subject to a single force at its lower end, as shown in Fig. 1.2b. The force is of course the *resultant* of the actual distribution of forces shown in Fig. 1.2a, acting in the appropriate direction, which in general is not along the line of the ladder. Now, instead of defining the total force acting on the foot by the resultant and its direction, we could equally well define the components of this force in two convenient perpendicular directions, such as the horizontal and vertical, along and at right angles to the ground as shown in Fig. 1.2c. In each case two, and only two, pieces of information are required to specify the force vector.

1.2 Review of statics

The types of problems we will be concerned with are either static, involving no motion, or are subject to dynamic loads which do not vary with time, and the physical principles we will need to employ include those of statics. The following notes are intended to summarize those topics in statics which are assumed to be already familiar to the reader.

1.2.1 Some definitions

1   A state of *equilibrium* is a state of no acceleration, in either the translational or rotational senses.
2   A *scalar* is a quantity having only a magnitude. Examples include mass and temperature.
3   A *vector* is a quantity having both magnitude and direction, and satisfies the parallelogram rule of vector addition. Examples include displacement and force.
4   A *force* is that interaction between bodies which tends to impart an acceleration or to deform. The interaction can occur either through direct contact of the bodies or remotely, as in the case of gravitational attraction, which is responsible for the weight of an object.
5   A *moment* is the product of the magnitude of a force and the (perpendicular) distance of its line of action from a particular point. Moment is also a vector quantity. Indeed, it is the vector product of the force and distance, and its direction is therefore normal to the plane containing the force and the point.
6   A *couple* consists of two forces equal in magnitude but opposite in direction whose lines of action are parallel but not colinear. The couple has the same moment about all points in the plane containing the two forces, equal to the product of the magnitude of either of the forces and the perpendicular distance between them. The resultant of the two forces is zero.

In order to understand the idea of a couple more clearly, consider the plane body shown in Fig. 1.3, which is subject to two forces, each of magnitude F but opposite in direction, acting in the plane of the body and applied to points A and B. The lines of action of these forces are parallel and at a perpendicular distance h apart. The resultant force acting on the body is zero. Let us consider an arbitrary point P in the plane of the body, but not necessarily within the

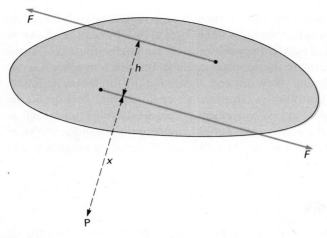

Fig. 1.3.

body itself, this point being at a perpendicular distance x from the nearer of the two forces, as shown. If we calculate moments about point P, the nearer of the two forces has a clockwise moment of magnitude Fx. As a vector, this moment is perpendicular to, and directed into, the plane of the body shown. On the other hand, the force further from P has a counterclockwise moment about it of $F(x+h)$ perpendicular to, and directed out of, the plane shown. The total moment, in the counterclockwise direction, is therefore

$$F(x+h)-Fx=Fh \tag{1.1}$$

This result is independent of the distance x, and therefore of the position of point P.

1.2.2 Laws of motion

Newton's three laws of motion, which are relevant to both dynamics and statics, can be expressed for our purposes as follows.

1   If there is no external force or moment acting on a body, then the body experiences no acceleration.
2   An external force acting on a body produces an acceleration in the direction of the force, the force being equal to the product of the mass of the body and the acceleration.
3   The force exerted by one body B_1 on another B_2 is equal in magnitude and opposite in direction to the force exerted by B_2 on B_1.

The distinction between a force which is external to a body, and one which is internal and acting between different parts of it, is an important one. The reference to external force or moment in the first law really means *resultant* external force or moment: while there may be several external forces and moments acting on a body, provided they have no resultants the body will experience no acceleration. In the context of statics, this means that the body remains at rest, and is said to be in *equilibrium*. While the second law is much more relevant to dynamics (and a similar statement can be made about moment and angular acceleration being linked by moment of inertia), we will make use of it in connection with engineering components subject to steady centrifugal loading. The third law is deceptively simple in its statement, and is of great importance to us when, for the purposes of analysis, we wish to separate different parts of a system and analyze the forces acting on each of the parts. In doing this, we often make use of *free body diagrams*.

1.2.3 Free body diagrams

The term *body* has so far been used in a general sense which perhaps implies a single physical object. Indeed, it may be, and often is, a single object. In the present context, however, it may also be any part of a physical system, consisting of a number of objects, or parts of objects, linked together. When analyzing systems, it is often necessary to subdivide them into smaller subsystems or bodies, and to consider each of these bodies, and the forces acting on them, in isolation. To do this, we draw the bodies and the forces acting on them in the form of *free body diagrams*.

Consider, for example, the system shown in Fig. 1.4a consisting of a framework supporting a mains water pipe across a road. If we only wish to find the reactions at the framework supports, then it is sufficient to consider the entire system shown in Fig. 1.4a as a free body: the only forces external to this body are the two reactions shown, together with the weights of the members of the framework, the pipe and its contents. We should note that the weight of an object, which is the gravitational force exerted on it by the earth, is always an external force. In the present example, the weights of the members of the framework are assumed to be negligibly small compared with the weight of the pipe.

If we wish to find out more about the internal forces in the system, such as the forces in the links between the framework and the pipe, then we must construct appropriate free body diagrams. Figure 1.4b shows the two free bodies formed when we separate the pipe from the framework by cutting through the links which join them. Note that the members of the framework are for convenience modeled as single lines. The forces in the links, F_1, F_2 and F_3, which were *internal* to the system as a whole, are now *external* to the framework and to the pipe when these are considered separately. While we are not yet able to find the magnitudes of these forces, we can assume that in each link the force acting on the framework is equal in magnitude and opposite in direction to the force acting on the pipe, as a consequence of Newton's third law.

We may also wish to find the forces in the members of the framework. One way of doing this (described in detail in Chapter 2) is to subdivide the framework into smaller portions by cutting through selected members. One of the free bodies formed by cutting along the dotted line marked in Fig. 1.4b is shown in Fig. 1.4c. Internal forces in the three cut members are now external forces as far as the free body is concerned, and equal and opposite forces act on the other half of the structure. Note that in this case the free body consists of three complete members of the framework, and parts of three others.

(a)

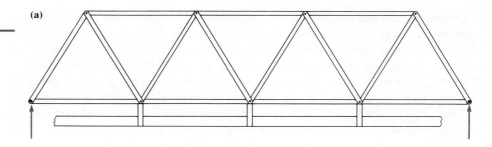

(b)

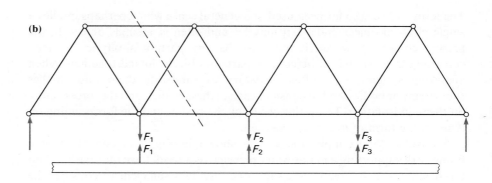

(c)

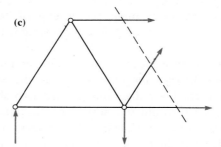

Fig. 1.4.

1.2.4 Conditions of equilibrium

If a body is in equilibrium, then we know that both the resultant force and resultant moment acting on it are zero. This allows us to establish certain conditions which must be satisfied for any body to be in equilibrium.

1   For a body acted on by only two external forces to be in equilibrium, the forces must be equal in magnitude, parallel in direction, opposite in sense and have the same line of action (Fig. 1.5a).

2   For a body acted on by three nonparallel external forces to be in equilibrium, the forces must form a closed triangle of forces (Fig. 1.5b: the

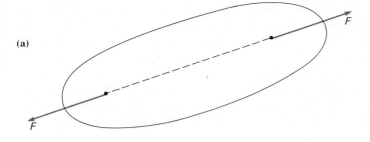

(a)

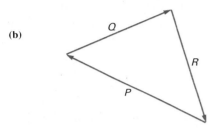

(b)

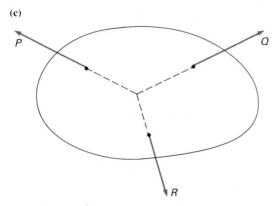

(c)

Fig. 1.5.

vector sum of the three forces is zero, the lengths of the sides of the triangle being proportional to the magnitudes of the forces), and the three lines of action must pass through a single point (Fig. 1.5c).

3   For a body subjected to any set of external forces and moments in a single plane, there are *three* independent equations of equilibrium. Equilibrium equations are formed by setting to zero either the algebraic sum of the resolved components of the forces in a particular direction in the plane concerned, or the algebraic sum of the moments of the forces about a point

in the plane. In order to be independent, the three equations must involve either (a) forces resolved in any two different directions and moments about any point, or (b) moments about any two points and forces resolved in any direction other than perpendicular to the line joining the two points, or (c) moments about any three points which are not colinear. If there are more than three unknown forces and moments, the three equilibrium equations for the body will be insufficient. This means either that additional parts of the body have to be isolated in free body diagrams and further equilibrium equations established, or that equilibrium conditions alone are insufficient to find all the forces (the problem is then *statically indeterminate*).

4   For the special case of a body acted on by three or more parallel forces, there are effectively only two independent equilibrium equations. This is because one equation, for the sum of force components in the direction perpendicular to that of the applied forces, is automatically satisfied. The two equations can sum to zero either (a) the applied forces, and moments about any point in the plane, or (b) moments about any two points in the plane, provided the line joining the points is not parallel to the forces.

1.2.5 Examples

Let us now consider some examples of statics problems, which serve particularly to illustrate the use of equilibrium equations and free body diagrams. In these and other Examples, the given data and the main calculated results are shown first in SI metric units, usually followed in parentheses by equivalent values in US customary (Imperial) units. The equivalence is not intended to be exact, and values are normally quoted to only two significant figures. In some Examples, such as the first one, calculations in both sets of units are shown, SI on the left, US customary on the right.

EXAMPLE 1.1

Part of a fence is shown in Fig. 1.6. The vertical fence post is free to rotate about a pinned joint at A, and is secured to the ground by the light stay wire BC, which is pinned to the post at B. Three horizontal wires are attached to the post, and the tension in each is known to be 300 N (67 lb). Find the tension in the stay BC and the horizontal and vertical components of the reaction force acting on the post at A.

Pin joints, such as those at A and B, are considered in detail in Section 2.1. For our present purposes, all we need to know about them is that friction forces there are negligible, and that no moments are transmitted. Our model of the problem, expressed in the form of free body diagrams for the stay wire and fence post, is therefore as shown in Fig. 1.7. The tension in the wire is shown as T, while the horizontal and vertical reaction components acting on the post at A are shown as H_A and V_A, respectively. The free body diagram for the stay wire, Fig. 1.7a, is a very simple one. With only two forces acting (since the wire is described as light, we neglect its weight), these must act along the straight wire in opposite directions.

In order to solve the problem, we first write down equations of equilibrium for the external forces acting on the fence post as shown in Fig. 1.7b.

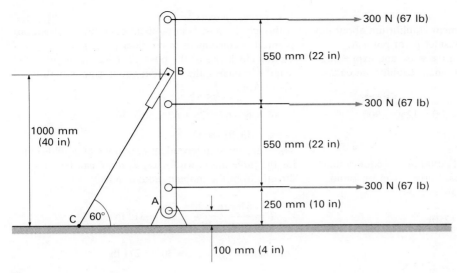

Fig. 1.6.

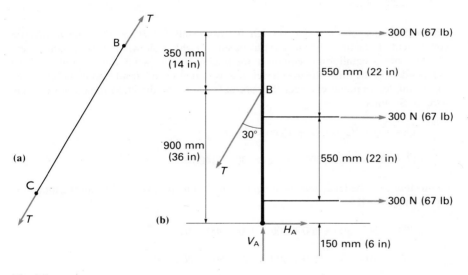

Fig. 1.7.

SI units
For equilibrium in the horizontal direction, taking forces acting from left to right as being in the positive sense

$$300 + 300 + 300 + H_A - T\sin 30° = 0$$

and for equilibrium in the vertical direction, taking upwards as the positive sense

$$V_A - T\cos 30° = 0$$

US customary units
For equilibrium in the horizontal direction, taking forces acting from left to right as being in the positive sense

$$67 + 67 + 67 + H_A - T\sin 30° = 0$$

and for equilibrium in the vertical direction, taking upwards as the positive sense

$$V_A - T\cos 30° = 0$$

Although the post is in moment equilibrium about *any* point, it is convenient to consider pivot point A. Taking clockwise as the positive sense and expressing lengths in millimeters, the moment equilibrium equation about A is

$$300 \times 150 + 300 \times 700 + 300 \times 1250 - 900$$
$$\times T \sin 30° = 0$$

These are three independent equations of equilibrium for the three unknown forces. Force T can be found directly from the moment equation as

$$T = \frac{300 \times (150 + 700 + 1250)}{450} = \underline{1400\,N}$$

and then $\qquad V_A = T \cos 30° = \underline{1212\,N}$

and $\qquad H_A = T \sin 30° - 900 = \underline{-200\,N}$

Although the post is in moment equilibrium about *any* point, it is convenient to consider pivot point A. Taking clockwise as the positive sense and expressing lengths in inches, the moment equilibrium equation about A is

$$67 \times 6 + 67 \times 28 + 67 \times 50 - 36$$
$$\times T \sin 30° = 0$$

These are three independent equations of equilibrium for the three unknown forces. Force T can be found directly from the moment equation as

$$T = \frac{67 \times (6 + 28 + 50)}{18} = \underline{313\,lb}$$

and then $\qquad V_A = T \cos 30° = \underline{271\,lb}$

and $\qquad H_A = T \sin 30° - 3 \times 67 = \underline{-44.7\,lb}$

The significance of the negative sign in this result for H_A is that the horizontal component of reaction at A acts not from left to right as shown, but from right to left.

Any other equilibrium equation we might care to write down would not be independent of the three already chosen, and would not provide additional information. For example, we might consider the sum of forces in the direction parallel to the stay wire (in SI units)

$$(3 \times 300 + H_A) \sin 30° + V_A \cos 30° - T$$

$$= (3 \times 300 - 200) \sin 30° + 1212 \cos 30° - 1400 = 0$$

confirming that the fence post is in equilibrium. Similarly, we could sum the moments about the point B

$$300 \times 350 - 300 \times 200 - 300 \times 750 - 900 \times H_A$$

$$= 300 \times 350 - 300 \times 200 - 300 \times 750 + 900 \times 200 = 0$$

Although such equations do not provide additional information, they can provide useful checks on both the equations chosen and the arithmetic.

EXAMPLE 1.2

A packing case containing a machine tool and weighing a total of 1.7 kN (380 lb) is being maneuvered with the aid of a light crowbar as shown in Fig. 1.8. One edge of the case rests on a rough floor at A, where there is sufficient friction to prevent slipping, and the crowbar provides support at point B where the case may be assumed to be smooth (frictionless). The distribution of mass within the case is such that its center of gravity is at the center point. The crowbar rests on frictionless rollers at C. Find the force which must be applied to end D of the crowbar to hold the packing case in equilibrium.

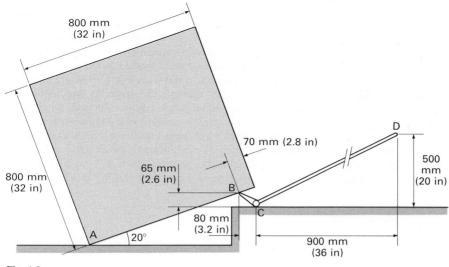

Fig. 1.8.

The designation of contacts as either smooth (with no friction) or rough (with enough friction to prevent relative motion) is not always a reasonable form of simplification. For present purposes, however, we are more interested in studying the application of equilibrium conditions than in investigating in detail a quite complex mechanics problem.

We first consider the free body diagram for the packing case, shown in Fig. 1.9a. There are only three external forces acting: the weight of 1.7 kN acting vertically

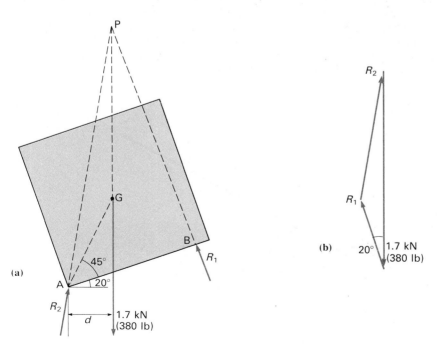

Fig. 1.9.

downwards through the center of gravity, G; the reaction force R_1 of the crowbar at B, and the reaction force R_2 of the ground at A. We know that, because the surface at B is smooth, the reaction there acts perpendicular to the surface. We cannot say the same of the reaction at A. But, because there are only three (nonparallel) forces acting, we know that their lines of action must pass through a single point: point P in Fig. 1.9a. Also, the three forces must form a closed triangle of forces, as shown in Fig. 1.9b. There are three unknowns: R_1, R_2 and the direction of R_2 (for example, the angle the direction of the force makes with the horizontal), and three equations of equilibrium from which they may be found. In this problem, however, we only require one of the unknowns, R_1, and we should therefore try to select an equation which gives this directly, without involving the other two unknowns. While no equation of equilibrium of forces achieves this, the equation of equilibrium of moments about A (the point at which R_2 acts) does. Taking the clockwise sense as positive, and expressing forces and lengths in kN and mm, respectively

$$1.7 \times d - R_1 \times 730 = 0$$

where

$$d = \text{AG} \cos 65° = \frac{800}{\sqrt{2}} \cos 65° = 239.1 \text{ mm (9.4 in)}$$

and

$$R_1 = \frac{1.7 \times 239.1}{730} = 0.5567 \text{ kN (125 lb)}$$

Now, let us consider the free body diagram for the crowbar, shown in Fig. 1.10. Since the weight of the crowbar is to be neglected, only three external forces act: R_1 due to the packing case, the supporting reaction R_3 at C, and the force F at D which we wish to find. From Newton's third law, R_1 is equal in magnitude but opposite in direction to the corresponding force in Fig. 1.9a. Because the rollers at C are frictionless, R_3 acts vertically, at right angles to the surface on which they rest. With only three (nonparallel) forces acting, we know that their lines of action must pass through a single point: point

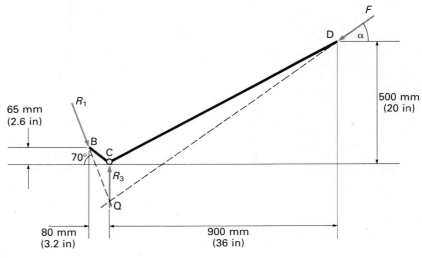

Fig. 1.10.

Q in Fig. 1.10, and they must form a closed triangle of forces. There are three unknowns: R_3, F and its direction α relative to the horizontal, and three equations of equilibrium from which they may be found. In this case we only require two of the unknowns, F and α, and should therefore try to choose two equations which give these directly, without involving R_3. These equations are for equilibrium of forces in the horizontal direction, and equilibrium of moments about C

$$R_1 \cos 70° - F \cos \alpha = 0$$

$$65 \times R_1 \cos 70° - 80 \times R_1 \sin 70° - 500 \times F \cos \alpha + 900 \times F \sin \alpha = 0$$

from which

$$F \cos \alpha = 0.5567 \cos 70° = 0.1904 \text{ kN}$$

and

$$F \sin \alpha = \frac{80 \times 0.5567 \sin 70° + 500 \times 0.1904 - 65 \times 0.5567 \cos 70°}{900}$$

$$= 0.1385 \text{ kN}$$

Therefore

$$F = \sqrt{(0.1904^2 + 0.1385^2)} = \underline{0.235 \text{ kN}} \text{ (53 lb)}$$

and

$$\alpha = \tan^{-1}\left(\frac{0.1385}{0.1904}\right) = \underline{36.0°}$$

Note that in this and most other Examples involving manual calculations, the final answers are quoted to a precision of three significant digits. The accuracy of the data and methods of analysis rarely justify any more than this. In order to help preserve this overall precision, the results of intermediate calculations are normally quoted to four significant digits.

EXAMPLE 1.3

Figure 1.11a shows part of a mechanical control system, in which torsional loading in a horizontal shaft is transmitted via the vertical arm AB to a horizontal cable attached to the arm at B. The shaft is supported in frictionless bearings adjacent to the arm at A. Find the tension in the cable and the reaction at the bearings when a torsional moment of 300 Nm (220 lb ft) is applied to the shaft.

Although this is a relatively simple example, it does serve to show how couples can arise in practice, and how they enter the equations of equilibrium. Figure 1.11b shows the model for the system in the form of a free body diagram for the arm AB. The torsional moment in the shaft is applied to the arm in the form of a couple acting at A. There are three unknown forces: tension T in the cable, and horizontal and vertical components, H_A and V_A, of the bearing reaction force acting at A. Three suitable equations for finding them are those of equilibrium of forces in the horizontal and vertical direction, and moments about A.

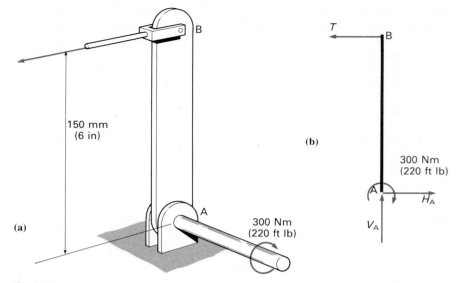

150 mm
(6 in)

(b)

300 Nm
(220 ft lb)

(a)

300 Nm
(220 ft lb)

Fig. 1.11.

SI units

$$H_A - T = 0$$

$$V_A = 0$$

$$300 - 0.15 \times T = 0$$

from which $T = H_A = 2000$ N $= 2$ kN.

US customary units

$$H_A - T = 0$$

$$V_A = 0$$

$$220 \times 12 - 6 \times T = 0$$

from which $T = H_A = 440$ lb.

Since V_A is zero, H_A is the total (resultant) reaction force at the bearings. We should note that, because there is no resultant force associated with the applied couple, the couple does not appear in either of the force equilibrium equations. Also, that in order to balance this couple in the rotational sense, another couple of equal magnitude but opposite direction is formed by the forces T and H_A.

An alternative to the equation of moment equilibrium about point A would have been a similar equation for any other convenient point, such as B (in SI units)

$$300 - 0.15 \times H_A = 0$$

This of course gives the same results. We should note that, although the couple is applied to the arm at A, it has the same moment about any other point in the plane of the problem.

EXAMPLE 1.4

Figure 1.12a shows a horizontal uniform timber beam used in the construction of a house. Its ends rest on brick walls. The weight of the beam is 650 N (150 lb), and a tool box weighing 300 N (67 lb) rests on the beam in the position shown. Find the vertical reaction forces between the beam and the walls.

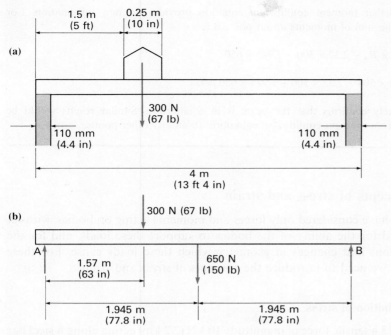

Fig. 1.12.

Like many beam problems, which are considered in detail in Chapters 5 and 6, this example involves external forces on a body which are all parallel to each other. Figure 1.12b shows the model of the system in the form of a free body diagram for the beam. The weight of the beam is taken to act at its center, while the weight of the tool box is treated as a concentrated load acting on the beam at the center of the tool box. Similarly, the supporting reactions at the wall, R_A and R_B, are assumed to act at the centers of the walls. These two are the only unknown forces, and there are only two independent equations of equilibrium available to find them. No useful information is obtained by considering equilibrium of forces in the horizontal direction, since there are no components of force in this direction.

For equilibrium of forces in the vertical direction, taking the upward direction as positive

$$R_A + R_B - 650 - 300 = 0$$

while for equilibrium of moments about A (the point of application of R_A), taking the clockwise direction as positive and working in meter, newton units for lengths and forces

$$1.57 \times 300 + 1.945 \times 650 - 3.89 \times R_B = 0$$

Hence

$$R_B = \frac{1.57 \times 300 + 1.945 \times 650}{3.89} = \underline{446\,\text{N}}\ (102\,\text{lb})$$

and

$$R_A = 650 + 300 - 446 = \underline{504\,\text{N}}\ (115\,\text{lb})$$

Even a further moment equilibrium equation provides no new information. For example, the sum of moments about point B is

$$3.89 \times R_A - 2.32 \times 300 - 1.945 \times 650$$

$$= 3.89 \times 504 - 2.32 \times 300 - 1.945 \times 650 = 0$$

which merely confirms that the beam is in equilibrium. Similar results would be obtained from moment equilibrium equations about any other points.

1.3 Concepts of stress and strain

So far we have considered only forces and moments acting on bodies, without any regard for the ability of the bodies to support these loads, and for the deformations and changes in geometry which these loads cause. For these purposes, we need to introduce the concepts of *stress* and *strain*.

1.3.1 Definition of stress

In Fig. 1.13 a tensile force of magnitude 10 kN (2.2 kip) acting along a steel bar of circular cross section with a diameter of 15 mm (0.6 in) is much more likely to cause failure than the same force acting along a bar of the same material which has a rectangular cross section 25 mm by 30 mm (1 in by 1.2 in). What is important is the force intensity, or force per unit area, which we call *stress*. In the example quoted, the cross-sectional area of the round bar is 176.7 mm² (0.2827 in²), while that of the rectangular bar is 750 mm² (1.2 in²). The mean stresses in the two bars are therefore $10 \times 10^3/176.7 \times 10^{-6} = 56.6 \times 10^6$ N/m²

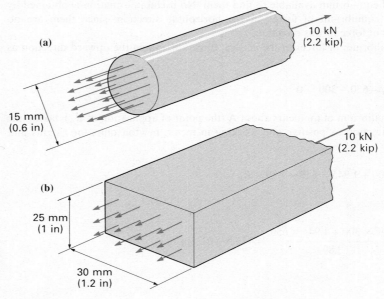

Fig. 1.13.

$(2.2/0.2827 = 7.78 \text{ ksi})$ and $10 \times 10^3/750 \times 10^{-6} = 13.3 \times 10^6 \text{ N/m}^2$ $(2.2/1.2 = 1.83 \text{ ksi})$, respectively. It is these stresses which need to be compared with the strength properties of the steel concerned. In using mean stresses, we are of course assuming that the total force carried by each bar is uniformly distributed over its cross section. While this is a good assumption for the central region of a long bar remote from its ends, it is not always so. As we shall see, stress distributions may be far from uniform, with local regions of high stress known as *stress concentrations*.

If the force carried by a component is not uniformly distributed over its cross section, then we must modify our simple definition of stress. Rather than consider the whole cross-sectional area, A, we must consider a small part of it, δA, which carries a small part, δF, of the total force, F. The definition of the local force intensity or stress acting over the chosen element of area is

$$\sigma = \frac{\delta F}{\delta A} \qquad (1.2)$$

We will often use the symbol σ (the Greek letter 's') to denote stress. Now, we must make the area δA so small compared to the total area A that the force distribution over it can be considered to be uniform, and the stress constant. Bearing in mind that real materials are not continuous media at the microscopic level, but are ultimately composed of discrete atoms, this provides a good practical definition of stress. If we do make the further assumption that materials are truly continuous media, which is reasonable for most solid mechanics problems where the scale of interest is macroscopic rather than microscopic, then the mathematical concept of local *stress at a point* is obtained as the limit

$$\sigma = \lim_{\delta A \to 0} \left(\frac{\delta F}{\delta A} \right) \qquad (1.3)$$

Since stress is a force intensity or force per unit area, the natural unit of stress in the SI system is newton per square meter, or N/m^2. This compound unit is sometimes referred to as a pascal or Pa. In the context of engineering components and structures, a newton is a small force, while a square meter is a large area. Consequently, one N/m^2 or Pa is a very small stress. Indeed, most metallic materials are capable of carrying stresses of the order of millions of newtons per square meter, and a much more convenient unit of stress is a MN/m^2 $(= 10^6 \text{ N/m}^2)$ or MPa. In the US customary (Imperial) system, the unit of stress is pound per square inch, which is abbreviated to psi, or ksi (10^3 psi).

1.3.2 Normal and shear stresses

We have defined stress as force per unit area and used Fig. 1.13 to illustrate the distribution of force over the cross sections of long bars. In these examples, not only are the stresses uniform over the cross-sectional areas, but the forces, and hence the stresses, are normal to the areas concerned. The resulting forces per unit area are known as *normal* (or *direct*) *stresses*. Normal stresses can be either *tensile* (as in Fig. 1.13), with the force acting out of the area to which it is applied, or *compressive*, with the force acting into the area. For most engineering purposes, a sign convention is adopted which defines tensile normal stresses as positive, and compressive normal stresses as negative.

The normal stresses shown in Fig. 1.13 are *uniaxial*, in that they act in one direction only, the direction along the length of each bar. Such a simple *state of stress* is only relatively rarely encountered in practice. More common are the *biaxial* and *triaxial* states illustrated in Fig. 1.14, where there are stresses in two and three mutually perpendicular directions, respectively. An example of a biaxial state of stress is provided by a thin flat rectangular sheet of material loaded (by different amounts) on opposite pairs of edges. The general case of a triaxial state of stress arises in many different ways, and later chapters of this book provide a number of examples.

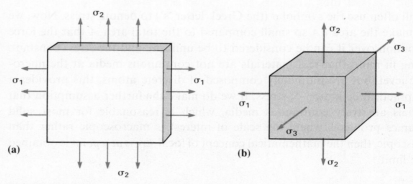

(a)

(b)

Fig. 1.14.

Let us consider now the situation shown in Fig. 1.15 where the cross-sectional area of a block of material is subject to a distribution of forces which are parallel, rather than normal, to the area concerned. Such forces are associated with a shearing of the material, and are referred to as *shear forces*. The resulting force intensities are known as *shear stresses*, the mean shear stress being $\tau = F/A$, where F is the total force and A the area over which it acts. If the

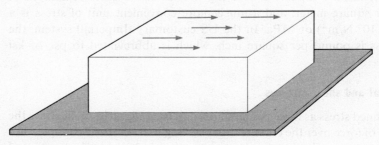

Fig. 1.15.

$$\tau = \lim_{\delta A \to 0} \left(\frac{\delta F}{\delta A} \right) \qquad (1.4)$$

where δF is the magnitude of the force acting over the small element of area δA. We will use the symbol τ (the Greek letter 't', suggesting tangential) to denote shear stress. While it will become necessary in due course to define a sign convention for shear stresses, we can defer this until we have to deal with relatively complicated problems. Also, in such problems it will not be possible to consider the state of stress at a point as being that due to either a single normal or shear stress, but as a complex combination of normal and shear stresses in two or three dimensions.

EXAMPLE 1.5

In order to demonstrate the way in which normal and shear stresses arise in practice, and to show how mean values of these stresses can be calculated, let us consider the pinned joint between two steel bars shown in Fig. 1.16. Figure 1.16a shows a general view of the joint, while Figs 1.16b and 1.16c show the main dimensions. The round bar which is marked with the letter A ends in a clevis marked B, into which the rectangular bar marked C fits. Both components are drilled to take the cylindrical pin marked D which joins them, and tensile forces of 90 kN (20 kip) are applied to their ends.

Firstly, we may calculate the normal stresses in the bars remote from the joint. In bar A

SI units

$$\sigma_A = \frac{(90 \times 10^3 \text{ N})}{\left(\frac{\pi}{4} \times 0.045^2 \text{ m}^2 \right)} = 56.6 \times 10^6 \text{ N/m}^2$$

$$= 56.6 \text{ MN/m}^2$$

which is a tensile stress, and therefore positive. Similarly, in bar C

$$\sigma_C = \frac{(90 \times 10^3 \text{ N})}{(0.09 \text{ m}) \times (0.035 \text{ m})} = 28.6 \times 10^6 \text{ N/m}^2$$

$$= 28.6 \text{ MN/m}^2$$

which is also tensile. Other normal stresses in the bars which need to be checked are those in the joint region level with the centerline of the pin, where the reductions of area due to the holes drilled for the pins are

US customary units

$$\sigma_A = \frac{(20 \times 10^3 \text{ lb})}{\left(\frac{\pi}{4} \times 1.8^2 \text{ in}^2 \right)} = 7860 \text{ psi}$$

$$= 7.86 \text{ ksi}$$

which is a tensile stress, and therefore positive. Similarly, in bar C

$$\sigma_C = \frac{(20 \times 10^3 \text{ lb})}{(3.6 \text{ in}) \times (1.4 \text{ in})} = 3970 \text{ psi}$$

$$= 3.97 \text{ ksi}$$

which is also tensile. Other normal stresses in the bars which need to be checked are those in the joint region level with the centerline of the pin, where the reductions of area due to the holes drilled for the pins are

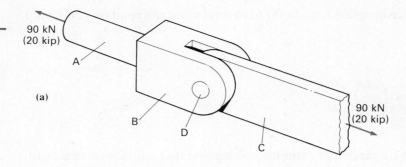

(a)

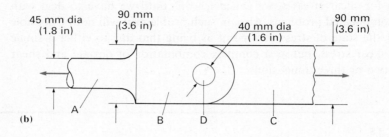

(b)

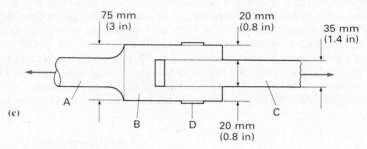

(c)

Fig. 1.16.

SI units

greatest. Figures 1.17a and b show the relevant cross sections through B and C. The mean tensile stresses at these sections may be calculated using the dimensions shown in Fig. 1.16 (together with the fact that the total load taken by each of the sections must equal the applied force of 90 kN) as

$$\sigma_1 = \frac{(90 \times 10^3 \text{ N})}{4 \times (0.025 \text{ m}) \times (0.020 \text{ m})}$$

$$= 45.0 \times 10^6 \text{ N/m}^2 = 45.0 \text{ MN/m}^2$$

$$\sigma_2 = \frac{(90 \times 10^3 \text{ N})}{2 \times (0.025 \text{ m}) \times (0.035 \text{ m})}$$

$$= 51.4 \times 10^6 \text{ N/m}^2 = 51.4 \text{ MN/m}^2$$

US customary units

greatest. Figures 1.17a and b show the relevant cross sections through B and C. The mean tensile stresses at these sections may be calculated using the dimensions shown in Fig. 1.16 (together with the fact that the total load taken by each of the sections must equal the applied force of 20 kip) as

$$\sigma_1 = \frac{(20 \times 10^3 \text{ lb})}{4 \times (1 \text{ in}) \times (0.8 \text{ in})} = 6250 \text{ psi}$$

$$= 6.25 \text{ ksi}$$

$$\sigma_2 = \frac{(20 \times 10^3 \text{ lb})}{2 \times (1 \text{ in}) \times (1.4 \text{ in})} = 7140 \text{ psi}$$

$$= 7.14 \text{ ksi}$$

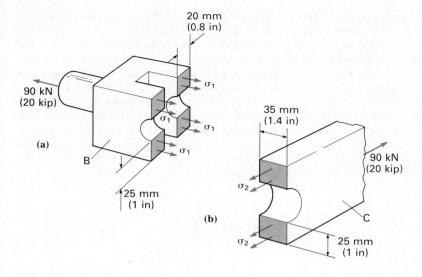

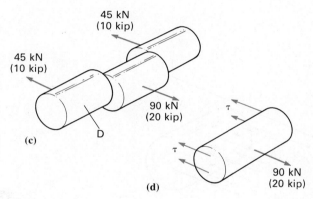

Fig. 1.17.

Now, the cylindrical pin D is loaded in such a way that there is a tendency for its central region (inside the hole drilled in the end of bar C) to be sheared away from its ends, as illustrated in Fig. 1.17c. This action is like that produced by hedge-cutting 'shears'. Figure 1.17d shows the central region of the pin separated from the ends along the planes which carry the greatest shear forces. The shear stresses on these planes must be such as to balance the applied force, and the mean value of shear stress is

$$\tau = \frac{(90 \times 10^3 \text{ N})}{2 \times \left(\dfrac{\pi}{4} \times 0.040^2 \text{ m}^2\right)} = 35.8 \times 10^6 \text{ N/m}^2$$

$$= 35.8 \text{ MN/m}^2$$

Now, the cylindrical pin D is loaded in such a way that there is a tendency for its central region (inside the hole drilled in the end of bar C) to be sheared away from its ends, as illustrated in Fig. 1.17c. This action is like that produced by hedge-cutting 'shears'. Figure 1.17d shows the central region of the pin separated from the ends along the planes which carry the greatest shear forces. The shear stresses on these planes must be such as to balance the applied force, and the mean value of shear stress is

$$\tau = \frac{(20 \times 10^3 \text{ lb})}{2 \times \left(\dfrac{\pi}{4} \times 1.6^2 \text{ in}^2\right)} = 4970 \text{ psi}$$

$$= 4.97 \text{ ksi}$$

The way in which the 90 kN (20 kip) force is applied to the pin is not that implied by Fig. 1.17d, which shows a point force somehow applied at its center. In practice, the force is transmitted by means of a distribution of compressive stress on one side of the pin, as shown in Fig. 1.18, with an identical distribution on the inner surface of the hole with which the pin is in contact. While Fig. 1.18a shows the end of bar C, Fig. 1.18b shows the central region of the pin previously shown in Fig. 1.17d, both subject to compressive stress σ_3. Such a stress is often referred to as a *bearing stress* or *contact stress*. This stress is applied to at most half of the circumferences of the pin and hole, and in practice is far from uniform in magnitude. Nevertheless, it is possible to define a nominal value as the mean force intensity over the maximum cross-sectional area of the pin or hole at right angles to the applied force. This cross-sectional area is taken as the product of length and diameter of the cylinder concerned. In this case

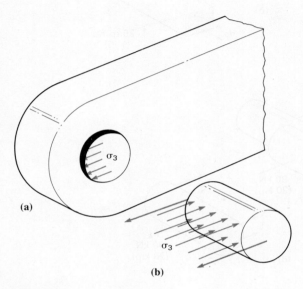

(a)

(b)

Fig. 1.18.

SI units

$$\sigma_3 = -\frac{(90 \times 10^3 \text{ N})}{(0.040 \text{ m}) \times (0.035 \text{ m})}$$

$$= -64.3 \times 10^6 \text{ N/m}^2 = -64.3 \text{ MN/m}^2$$

the negative sign implying compression. Similarly, the nominal bearing stress over the contact surface between the ends of the pin and the holes through B is

$$\sigma_4 = \frac{(90 \times 10^3 \text{ N})}{2 \times (0.040 \text{ m}) \times (0.020 \text{ m})}$$

$$= -56.3 \times 10^6 \text{ N/m}^2 = -56.3 \text{ MN/m}^2$$

US customary units

$$\sigma_3 = -\frac{(20 \times 10^3 \text{ lb})}{(1.6 \text{ in}) \times (1.4 \text{ in})}$$

$$= -8930 \text{ psi} = -8.93 \text{ ksi}$$

the negative sign implying compression. Similarly, the nominal bearing stress over the contact surface between the ends of the pin and the holes through B is

$$\sigma_4 = -\frac{(20 \times 10^3 \text{ lb})}{2 \times (1.6 \text{ in}) \times (0.8 \text{ in})} = -7810 \text{ psi}$$

$$= -7.81 \text{ ksi}$$

Reviewing the calculated values of tensile and shear stresses in the components of the joint, the largest tensile value is 56.6 MN/m² in bar A, which is only slightly larger than the 51.4 MN/m² calculated for the end of bar C, level with the pin. The largest shear stress, in the pin, is 35.8 MN/m².

Reviewing the calculated values of tensile and shear stresses in the components of the joint, the largest tensile value is 7.86 ksi in bar A, which is only slightly larger than the 7.14 ksi calculated for the end of bar C, level with the pin. The largest shear stress, in the pin, is 4.97 ksi.

On the basis of the largest calculated stress, we would predict that if failure occurs it would be in bar A, possibly remote from the joint. This, however, is too simple a view, and at the very least we should know how the strength in shear for the material concerned compares with that in tension. In other words, we need a criterion for failure under conditions other than simple uniaxial tension. This topic is dealt with in Chapter 9. Also, we have calculated simple average values of stresses over the cross sections concerned. While this is reasonable for σ_A and σ_C, the tensile stresses in the rods away from the joint, it is not satisfactory for σ_1 and σ_2, the stresses in the joined members close to the pin, nor for τ, the shear stress in the pin. As we shall see in the next section, these stresses are not uniformly distributed. Consequently, the greatest tensile stress is likely to occur in bar C, at the cross section shown in Fig. 1.17b, despite the fact that this bar is subject to a relatively low stress remote from the joint.

1.3.3 Stress concentrations

In the above example, the effect of drilling a hole in the end of member C to take the pin is to increase significantly the mean stress at the cross section through the center of the hole, due to the reduction of area at this section. In addition to this increase, there is a further elevation of the maximum stress due to the nonuniformity of the stress distribution. Figure 1.19 shows the form of distribution obtained, with the tensile stresses at the edges of the hole substantially higher than those at the outer edges of the bar. In other words, there is a *stress concentration* at the hole, and it is customary to define a *stress concentration factor*, K, as the ratio of local maximum stress to mean stress

$$K = \frac{\sigma_{max}}{\sigma_{mean}} \tag{1.5}$$

In situations such as the present one, the stress concentration factor is of the order of 2 or more.

The two most common causes of stress concentrations are holes and sudden changes of section. Figure 1.20a shows a flat strip of width $2b$ with a relatively small central hole of radius a, subjected to tensile loading at its ends. Figure 1.20b shows the form of stress distribution over section XX, through the center of the hole. The stress concentration is severe, but its effect is very local to the hole. The stress concentration factor depends only on the ratio a/b, and for $a/b \to 0$ it is exactly 3. For larger holes, the value may be determined either

23

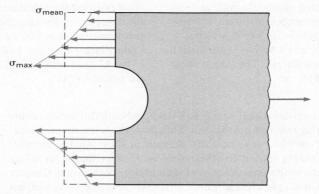

Fig. 1.19.

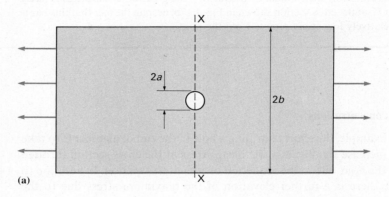

(a)

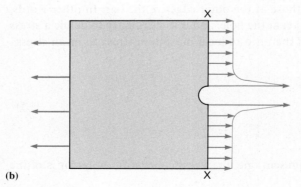

(b)

Fig. 1.20.

experimentally (by photoelastic techniques) or computationally (by finite element or other methods), and results for this and many other similar geometries have been published (in, for example, *Stress Concentration Factors*, by R. E. Petersen, published by John Wiley). The fact that the presence of holes in a component leads to large local stress concentrations is unfortunate because holes are often essential, not least for making pinned, bolted or riveted connections.

Figure 1.21a shows an example of a sudden change of section, where the width of a flat strip under tension reduces abruptly from $2b$ to $2a$. Very high stresses are induced at the re-entrant corners, but these can be reduced with the aid of radiused *fillets* (fillets of radius r are shown). Nevertheless, there are still substantial stress concentrations: Fig. 1.21b shows the form of stress distribution over section YY. The magnitude of the stress concentration factor depends on the geometric parameters a/b and r/a, and published results are again available.

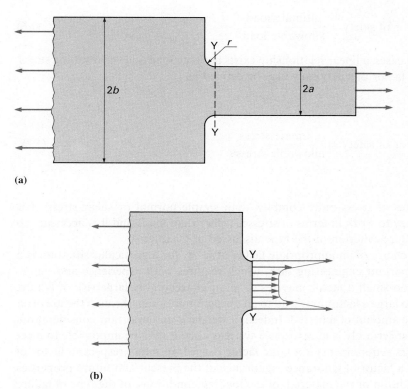

(a)

(b)

Fig. 1.21.

Stress concentrations in engineering components are important not only because they are responsible for stresses much higher than we would expect from simple calculations, but also because they tend to encourage the formation of cracks, which can lead to catastrophic failure.

1.3.4 Allowable stresses

In engineering practice, the calculation of stresses is not an end in itself. The values obtained must be related to the strength properties of the materials concerned, either to help in the assessment of the safety of an existing structure or piece of equipment, or to assist in the design of new systems. The *ultimate tensile strength* of a material can be measured as the maximum stress in simple uniaxial tension that it can withstand. Similarly, an *ultimate shear strength* can also be measured, although this is less widely used. At least for simple states of normal or shear stress, an *ultimate load* for a component can then be defined as the product of ultimate stress and cross-sectional area. Engineering structures and components must be designed and used in such a way that the actual loads they experience are substantially lower than the ultimate loads they are theoretically capable of sustaining. The maximum permitted actual load is referred as the *allowable load*, and the ratio of ultimate to allowable load as the *factor of safety*

$$\text{Factor of safety} = \frac{\text{ultimate load}}{\text{allowable load}} \tag{1.6}$$

In many cases, a linear relationship exists between load and the resulting stress, and the factor of safety may also be defined as

$$\text{Factor of safety} = \frac{\text{ultimate stress}}{\text{allowable stress}} \tag{1.7}$$

For states of stress more complex than simple normal or shear stress, it is customary to work in terms of stresses rather than loads and it is necessary to use a failure criterion of the type discussed in Chapter 9.

The choice of an appropriate factor of safety for a particular situation is a very important engineering task, which requires both experience and judgement. Too small a factor may result in an unacceptably large risk of failure, while too large a factor may lead to an uneconomic design through the use of an excessive amount of material. Indeed, if weight is an important consideration, as it most certainly is in aerospace systems, then it may be impossible to meet the design requirements if a large factor is used. In many respects a factor of safety is a factor of ignorance – ignorance of the present and future properties and condition of the material, of the loading conditions, of the type of failure which may occur, and to some extent of the accuracy of the methods of analysis used to calculate stresses. Considerations which are inherent in the choice of a factor of safety are as follows.

1 *Material properties* The chemical composition, heat treatment, physical condition and mechanical properties of materials produced by manufacturing processes are all subject to variations, and residual stresses may be introduced. Also, over the period for which the component or structure is designed to be used, there may be changes, such as corrosion, which adversely affect the material properties.

2 *Loading conditions* Precise details of the loads which will be experienced in service are often not available, and the values used for design purposes

are only estimates. Also, loads are rarely constant, but change with time. A load which varies repeatedly from, say, zero up to some maximum value and down to zero again is more likely to cause failure than if the maximum load is applied steadily, due to the phenomenon of *fatigue*. In this context, loads which produce high tensile stresses are more damaging than those which cause compressive stresses of the same magnitudes.

3   *Type of failure*   A failure which occurs with little or no warning is generally much more dangerous than one which can be anticipated. For example, a brittle material suffers only small deformations before fracture, while a ductile material first undergoes relatively large plastic deformations.

4   *Accuracy of analysis*   All methods of analysis involve some approximations, either in modeling the system or in solving the equations which are derived from the model. With the advent of cheap accessible digital computers, however, it is now possible to carry out much more thorough and accurate analyses than used to be the case. Also, we may sometimes be able to choose approximations which are known to be *conservative*, and err on the safe side. For example, an analysis which overestimates rather than underestimates the actual stresses can lead to a smaller factor of safety being used.

5   *Consequences of failure*   In many engineering systems, some of the components are primary load bearers, the failure of which would be catastrophic, while others are secondary and could fail with only minor consequences. The latter can be designed with smaller factors of safety than the former.

Having chosen a factor of safety, and knowing the ultimate stress for the material concerned, we can calculate an allowable stress from equation (1.7). It is then this figure we use in our analysis of the particular component or structure as the stress level which must not be exceeded.

1.3.5 Definitions of strain

Figure 1.22 shows a straight uniform bar in simple uniaxial tension. The force applied is F, and the cross-sectional area of the bar A, giving a uniform tensile stress of $\sigma = F/A$. Now, the effect of this stress is to deform the bar in such a way as to increase its length from its initial value of L_0 to a new length of $L_0 + \Delta L$. There are also changes in the dimensions of the bar cross section, but these need not concern us at this stage, and will be discussed in Chapter 3. We can

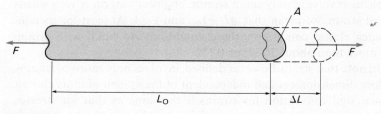

Fig. 1.22.

anticipate that the magnitude of the change in length, ΔL, will depend on the initial length of the bar. Indeed, we can anticipate that the change in length will be directly proportional to the initial length. For example, had we started with a bar of length $2L_0$, the change in length would have been $2\Delta L$. In other words, a possible measure of the amount of deformation which occurs, and which is independent of the length of bar chosen, is the ratio between the change in length and the original length. This quantity is known as *engineer's strain*, and frequently just *strain*, and is given the symbol e

$$e = \frac{\Delta L}{L_0} = \frac{L - L_0}{L_0} \tag{1.8}$$

It is a normal (or direct) strain, as opposed to a shear strain, in the same way that the associated stress σ is a normal stress.

While there is almost universal agreement on the definition of stress as force per unit area, this is not so in the case of strain. For example, if the change in length is substantial in relation to the initial length, it may be convenient to define change in strain as the ratio between the change in length and the current length,

$$\Delta\varepsilon = \frac{\Delta L}{L} \tag{1.9}$$

This leads to a definition of the total strain (*natural strain*) as

$$\varepsilon = \sum \Delta\varepsilon = \int_{L_0}^{L} \frac{dL}{L} = \ln\left(\frac{L}{L_0}\right) = \ln(1 + e) \tag{1.10}$$

Also, in some applications, notably the analysis of rubber components which often suffer large strains, an *extension ratio* may be used

$$\lambda = \frac{L}{L_0} = 1 + e \tag{1.11}$$

Whichever definition is employed, and there are a number of others in use, they are all related to each other by relatively simple functions, and therefore provide essentially the same information.

For problems involving only small strains, engineer's strain is very widely used. By small strain, we mean that $\Delta L \ll L_0$, and $e \ll 1$. At least for metallic materials under elastic conditions, these conditions do hold, with strains typically no more than about 10^{-3}, or 0.1%.

We should note that strain, however defined, involves only ratios of lengths, and is therefore dimensionless and independent of the system of units chosen. Also, the usual sign convention for strains is the same as that for stresses, namely tensile positive and compressive negative. In other words, the change in length, ΔL, is taken as positive for an increase in length and negative for a decrease in length.

Figure 1.23 shows a rectangular block of material of height d_0, the bottom surface of which is fixed, and the top surface of area A is subject to a total shear force of magnitude F, giving a mean shear stress of $\tau = F/A$. Now, suppose that the displacement of the top surface relative to the fixed bottom surface in the direction of the applied shear force is Δs. Following the approach adopted for normal strain, we can define a shear strain as the ratio between a displacement and an initial length. In this case, we must choose a displacement and a length which are not in the same direction but at right angles to each other, to give *shear strain*, for which we use the symbol γ (the Greek letter 'g')

$$\gamma = \frac{\Delta s}{d_0} \tag{1.12}$$

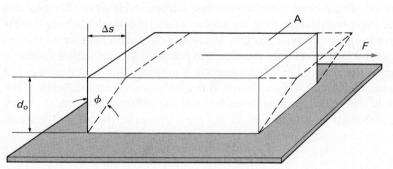

Fig. 1.23.

Now, if the angle ϕ is the angle through which the vertical sides of the block are rotated by the action of the applied shear force, then the shear strain is also given by

$$\gamma = \tan \phi \approx \phi \tag{1.13}$$

the approximation of shear strain by the angle ϕ (expressed in radians) being valid under small strain conditions.

1.4 Influence of material properties

We have already seen that the properties of materials have a considerable influence on the design of engineering components and structures. In particular, the strength properties determine the level of stresses which can be tolerated: we looked at the specification of maximum allowable stresses via factors of safety in Section 1.3.4. On the other hand, the stiffness properties determine the deformations and displacements under load, by providing the link between stresses and strains. More details of this link for typical engineering materials under elastic conditions are described in Chapter 3.

Strength and stiffness properties of materials have to be measured experimentally. The most widely used test for this purpose is the *simple tension test*, in which a uniform bar of the material is subjected to a uniaxial tensile load along its length. Figure 1.24a shows a typical specimen of circular cross section used for this purpose: the center of the specimen is of constant diameter, but the ends are larger to allow them to be gripped in the testing machine, either by friction grips or the ends may be threaded. Note the use of rounded fillets at the changes of section to reduce the levels of stress concentration there. The strain in the specimen is obtained either from the measured extension over a known length of the bar or from *strain gages* (which are described in Chapter 9). Figure 1.25 shows typical results of such a test. The nominal stress σ is the ratio between the measured load on the specimen and its initial cross-sectional area. Figure 1.25a shows a stress–strain curve for mild steel (low carbon steel). At low stresses and strains the curve rises steeply from the origin O up to the point Y, where there is an abrupt change in behavior. At first, strain increases with little or no change in stress, after which stress does increase, but relatively slowly, until it reaches a maximum at the point U and subsequently decreases. This final decrease is a consequence of defining stress in terms of the initial cross-sectional area of the specimen: in this region *necking* (local nonuniform reduction in area as shown in Fig. 1.24b) leads to a reduction in the load, although the local true stress

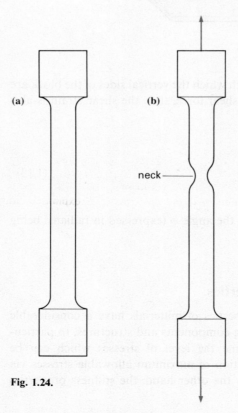

(a)   (b)

neck

Fig. 1.24.

(load divided by current area) is still increasing. Final failure of the specimen occurs by fracture at the neck and corresponds to point X on the curve. We take the value of stress at the point U as the *ultimate tensile strength* of the material, the corresponding load being the *ultimate load* the specimen can take (see Section 1.3.4). The large overall strain and large reduction of cross-sectional area at the neck distinguish mild steel as a *ductile* material.

If the tensile test illustrated in Fig. 1.25a is halted at any point between O and Y, and the load removed, the response of the material returns along the curve to the unstressed and unstrained state at O. In other words, the material is *elastic* in this region. The relationship between stress and strain is also a linear one, although it is the reversible nature of the process rather than its linearity which makes the behavior elastic. While most metals exhibit essentially linear elastic behavior, many nonmetals are significantly nonlinear in their elastic regions. If the test is halted at some point beyond Y and the load removed, rather than returning to the initial state, the response of the material would be to follow a path essentially parallel to the line OY down to zero stress, leaving a permanent strain in the material. In other words, the material is plastic beyond the point Y, which is defined as the *yield point*. We refer to the corresponding stress and strain as the *yield stress* and *yield strain*. Unlike many other materials, steel has a relatively sharp yield point which can be readily identified on its stress–strain curve. Typical numerical values in Fig. 1.25a would be a yield stress of 400 MN/m² (60 ksi), yield strain 0.2%, ultimate tensile strength 650 MN/m² (95 ksi), and strain at fracture of several per cent.

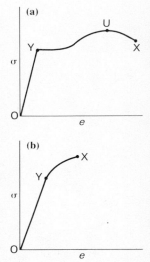

Fig. 1.25.

Figure 1.25b shows another type of stress–strain curve, which differs in two important respects from that for mild steel. Firstly, the yield point, Y, is difficult to distinguish. Secondly, failure occurs after only a small amount of plastic deformation and necking: the material is *brittle*. Materials of this type include cast iron, concrete and many high-strength alloy steels.

Although less commonly used as a test, it is also possible to subject a specimen of material to shearing rather than tension. This can be done by twisting a hollow tube of the material about its axis (a situation which is examined in Chapter 7). Figure 1.26 shows a typical shear stress–strain curve, which again is elastic and linear from the origin up to the yield point, Y, and then plastic and nonlinear up to the failure point X. We should be aware of the fact that neither the yield stress nor the yield strain in shear are equal to the corresponding values in tension. The reasons for this are explained in Chapter 9, where criteria for yielding under different states of stress are discussed.

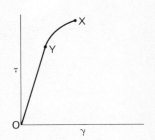

Fig. 1.26.

So far we have considered only some of the properties of engineering materials which we will need for analysis purposes. Typical numerical values for all such properties are, however, listed in Appendix A. These data will be used not only in Examples which are worked through in detail in the text, but will also be needed in the Problems provided at the ends of the chapters. To some extent therefore, we are simulating the conditions under which practising engineers work: material properties relevant to a particular problem are not provided explicitly as part of the problem specification, but must be selected from an appropriate source of data.

1.5 Principles of mechanics of solids

The three physical principles governing the mechanics of deformable solids are as follows.

1 *Equilibrium of forces* In Section 1.2, we reviewed the subject of statics, whose main concern is with establishing equations of equilibrium for the forces acting on engineering systems, and forces are directly related to stresses.

2 *Compatibility of strains* The *geometry of deformation* imposed on a particular system directly influences the displacements it experiences, and hence the strains.

3 *Stress–strain characteristics of the material* or materials concerned. As we have seen in Section 1.4, stress and strain are directly linked by the physical properties of materials.

The complete solution of any problem requires the use of all three of these principles.

The only one of the three we have not yet considered is the second one, concerned with compatibility of strains. An example of what is involved in this principle is provided by Fig. 1.27a, which shows an elevator cage at rest and supported by two long vertical cables. At their upper ends, the cables pass over a rigid horizontal winding drum, and there is no slipping of the cables relative

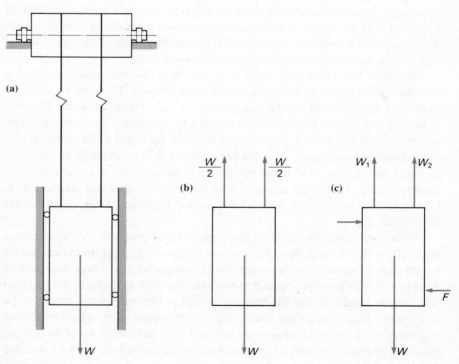

Fig. 1.27.

to the drum: the effective lengths of the cables are the same. The cage itself is constrained by rollers running on tracks so that it can only move in the vertical direction. As the cage cannot rotate, and can be regarded as rigid, the geometry of deformation imposed on the two cables is that their ends attached to the cage must be displaced downwards by the same amount, irrespective of the load which each one carries. Since the cables are of the same length, it follows that the strain in each must be the same.

Now, let us suppose that the two cables are identical in terms of both cross-sectional area and material properties, and that they are attached to the cage at points which are at the same horizontal distance from the vertical center line of the cage. Due to the symmetry of the system under these conditions, we may deduce that each cable is subject to the same tension, and Fig. 1.27b shows the free body diagram for the cage. If the total weight of the cage and its contents is W, then for equilibrium of forces in the vertical direction the tension in each cable is $W/2$. In this case, we have been able to determine the forces from the conditions of static equilibrium alone, and we describe such problems as *statically determinate*. If we also wish to find the downward vertical displacement of the cage due to its weight, however, we would have to involve the stress–strain characteristics of the cable material. Further examples of statically determinate problems are provided by Examples 1.1 to 1.4.

Figure 1.27c shows the free body diagram for the cage when the cables differ, either in their cross-sectional areas or in the materials used in their construction. Not only are the tensions W_1 and W_2 in the cables no longer identical, but lateral forces, F, must be applied to the cage via the rollers to keep it in rotational equilibrium. Although we can deduce from the equation of equilibrium of forces in the vertical direction that $W = W_1 + W_2$, we can make no further progress towards solving the problem until we introduce the compatibility condition that both cables extend by the same amount, together with the stress–strain characteristics of the cable materials. The problem is now *statically indeterminate*, and detailed consideration of such problems is deferred until Chapter 4.

1.6 Use of computers

Traditional methods of analysis for solid mechanics problems were intended for use in conjunction with hand calculations. The amount of algebraic and arithmetic manipulation involved were therefore necessarily limited if solutions were to be obtained with an acceptable amount of effort. To this end it was often necessary to introduce so many approximations that the reliability of the results was seriously impaired, and very large factors of safety had to be used. With the advent of high-speed digital computers capable of carrying out arithmetic operations much more reliably and cheaply than by manual methods, this situation has changed, and is continuing to change. At first the only computers were mainframe machines which were physically large and relatively inaccessible to engineers. But now personal computers are available which are as powerful as the mainframes of two decades ago, at a very small fraction of the cost. Using such machines, we can solve a wide range of problems to a much greater level of detail than we would attempt to do by hand.

There are, however, hazards associated with using computers. Firstly, there is a tendency to believe that because a result is produced with a large number of significant figures it must be not only correct, but correct to the number of figures displayed. There is always scope for simple checks to be made to confirm at least the order of magnitude of the computed solution. Secondly, it is important for us to understand what a computer program is doing, and the assumptions inherent in the physical model of the real system and the method used to analyze it. Therefore, particularly at the learning stage, there is no substitute for solving at least some relatively simple problems by hand before tackling more difficult ones with the aid of a computer.

The approach adopted in this book is therefore to first present the underlying theory and, where appropriate, the more traditional manual methods of solution, before introducing computer techniques. Although the methods of analysis have much in common, there are frequently significant differences of approach. In manual methods we often seek to minimize the amount of algebra and arithmetic involved by careful choice of equations involving only those unknowns which are actually required in the particular problem. On the other hand, using a computer which can manipulate data very rapidly, we are more interested in methods which can proceed without intervention from the analyst, even if this results in more information than we really need. In other words, we seek methods which are straightforward to program for automatic calculation.

All the programs presented in this book are coded in FORTRAN 77. Despite some deficiencies as a programming language, FORTRAN is still the most commonly used for engineering calculations. It is also very portable between machines supplied by different manufacturers, and between mainframe, mini and personal computers. All the programs were initially developed and run on a personal computer (an Apple Macintosh). Detailed internal documentation in the form of comments in the coding are provided, together with detailed definitions of all the program variable names, and examples of the use of the programs in solving problems.

While programming languages are generally well standardized in the way they provide for numerical calculations, which helps to make programs portable, this is unfortunately much less true of the ways in which different computer systems allow the user to present results in graphical or pictorial form. Consequently, in the interests of portability, no attempt has been made here to provide the programs with either graphical input or output which could have made them easier to use, and the results easier to understand.

With even relatively simple computer programs for engineering calculations it is possible to begin to undertake tasks which would have been impractical using manual methods. We can use computer simulations of engineering systems to explore the effects of changing the physical parameters, such as the dimensions of the members of a structure. We can also investigate the sensitivities of the responses such as the displacements and stresses to such changes, with a view to optimizing the design of the system. Computer simulations are also very valuable for studying systems which are either too expensive, dangerous or time-consuming to explore experimentally using real hardware.

Problems

SI UNITS – STATICS

1.1 Three cables are attached to an eyebolt which is fixed to a vertical wall, and all three cables lie in the same vertical plane. The tension in one cable is 4 kN vertically upwards, in the second is 8 kN horizontally away from the wall, and in the third is 6 kN downwards and away from the wall at 60° to the horizontal. Find the magnitude and direction of the resultant force transmitted to the wall.

1.2 A uniform ladder of length 5 m and weight 150 N is placed with one end on a rough horizontal surface and the other resting against a smooth vertical wall. The ladder makes an angle of 75° with the horizontal. Find the reaction forces acting on the ladder at its ends when a person of mass 65 kg stands on a rung which is 3 m from the foot of the ladder, measured along its length.

1.3 The simple crane structure shown in Fig. P1.3 consists of a light jib AB of length 4 m which is freely pivoted at A and supported at 60° to the horizontal by a cable CD. The cable is inclined at 30° to the horizontal and attached to the jib at point D which is 3 m from A measured along AB. A rope passing over a pulley wheel at B is used to lift a weight W of 2 kN, the other end of the rope being inclined at 15° to the vertical. Assuming there is no friction at the pulley, find the tension in cable CD, and the reaction forces supporting the structure at A and C.

1.4 The structure shown in Fig. P1.4 consists of the three light bars AD, BD and CD which are pinned together at point D, and to a vertical wall at points A, B and C. Assuming there is no friction at the joints, determine the reaction forces supporting the structure at points A, B and C when a vertical load of 10 kN is applied at point D.

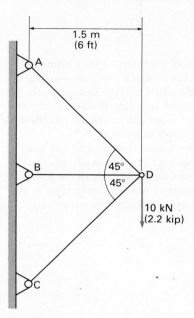

Fig. P1.4.

1.5 A ship has two propellers whose axes of rotation are parallel to its longitudinal axis and 10 m apart. Each propeller develops a thrust of 400 kN, one ahead and the other astern, while turning. The ship is being assisted by one tug on each side pressing laterally, one of the tugs being 60 m from the bows and the other 100 m. Find the force which each tug must exert on the ship to counteract the effect of the propellers.

1.6 Figure P1.6 shows an elevator cage of total weight W supported by two cables which are 1.5 m apart. Although these are attached to the cage at points which are symmetrically placed with respect to the center of gravity, they do not carry equal loads but 45% and 55% of W, respectively, as shown. As a

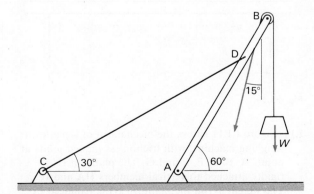

Fig. P1.3.

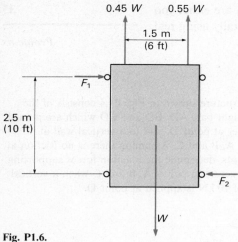

0.45 W 0.55 W

1.5 m
(6 ft)

F_1

2.5 m
(10 ft)

F_2

W

Fig. P1.6.

result of this lack of symmetry of the loading, lateral reaction forces F_1 and F_2 as shown are set up between the guide rollers and the vertical tracks (not shown) on which they run. If the vertical distance between the upper and lower sets of rollers is 2.5 m, find these forces in terms of W.

1.7 Figure P1.7 shows a piston, connecting rod and crankshaft arrangement. A gas pressure of $p = 2 \text{ MN/m}^2$ is just sufficient to make the piston move in the position shown. Assuming there is no friction at the joints or between the piston and

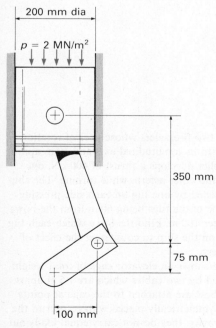

200 mm dia

$p = 2 \text{ MN/m}^2$

350 mm

75 mm

100 mm

Fig. P1.7.

cylinder in which it slides, and the weights of the components can be neglected, calculate the force in the connecting rod and the torsional moment transmitted to the shaft.

1.8 A uniform horizontal beam 5 m long with a weight of 400 N is supported at its ends. If weights of 120 N, 160 N and 200 N are placed on the beam at distances of 2 m, 3 m and 4 m, respectively, from one end, find the reaction at each of the supports.

1.9 A horizontal beam having a mass per unit length of 15 kg/m is rigidly built in to a vertical wall at one end and protrudes 1.5 m from the wall, as shown in Fig. P1.9. If a vertical concentrated load of 250 N is applied to the beam 0.5 m from its free end, find the moment M applied to the beam by the wall.

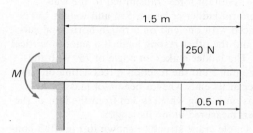

1.5 m

250 N

M

0.5 m

Fig. P1.9.

1.10 Figure P1.10 shows a horizontal beam rigidly built in at each end to vertical walls which are 3 m apart. If a vertical load of 7 kN is applied at a point 1 m from one wall, find the supporting reaction forces and moments acting on the ends of the beam.

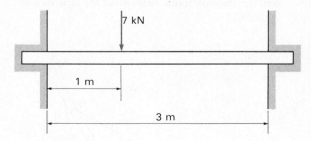

7 kN

1 m

3 m

Fig. P1.10.

1.11 Figure P1.11 shows the mechanism of a platform weighing machine, with frictionless pinned joints at points A, B, C, E, F and G. The platform HI is rigidly attached to vertical members HA and IE. The weight W on the platform HI is balanced by force Q at point D. Show that the magnitude of Q

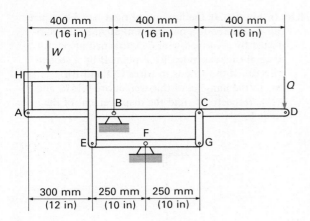

Fig. P1.11.

is independent of the position of load W on the platform, and find this magnitude in terms of W. The weights of the components may be neglected.

1.12 Figure P1.12 shows a hollow circular steel cylinder resting on a smooth horizontal surface. The two solid steel spheres, which may be assumed to be smooth, are placed inside the cylinder. Find the minimum length, L, of cylinder required if it is not to tip over.

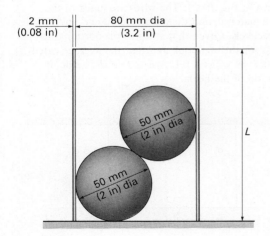

Fig. P1.12.

1.13 A smooth cylindrical drum weighing 800 N is being maneuvered up a step with the aid of a light crowbar as shown in Fig. P1.13. Find the force F which must be applied to the end of the crowbar at 30° to the horizontal to just start the drum to move up the step.

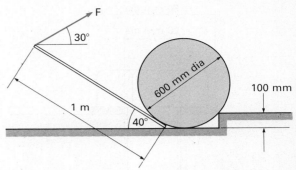

Fig. P1.13.

SI UNITS – STRESSES AND STRAINS

1.14 Figure P1.14 shows a circular shaft with a diameter of 50 mm over the section marked A, stepping down to 25 mm over the section marked B, and subject to an axial tensile force of 50 kN. A washer of outer diameter 60 mm just fits over section B and it rests against the step in the shaft. If there is no pressure (shown as p in the figure) acting on the washer, find the mean stresses in sections A and B of the shaft.

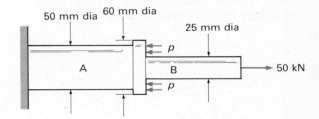

Fig. P1.14.

1.15 If, in Fig. P1.14, the uniform pressure over the surface of the washer is $p = 5$ MN/m², again find the mean stresses in sections A and B of the shaft.

1.16 For the pin-joint arrangement shown in Fig. 1.16, find the ratio between the diameter of member A and that of pin D if both components are to fail at the same value of axial load on the joint. The failure stress in simple shear may be assumed to be half that in simple tension.

1.17 Figure P1.17 shows part of a joint in a timber framework where a steel bolt passes through a hole in piece of timber. The steel washer which is placed between the bolt head and timber has inner and outer diameters of 20 mm and 50 mm, respectively. If the ultimate compressive strength for the timber

37

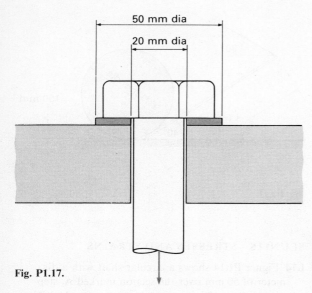

Fig. P1.17.

is 5 MN/m^2 in the direction parallel to the bolt, estimate the maximum possible tensile force in the bolt.

1.18 Given that the ultimate shear strength of an alloy steel is 250 MN/m^2, calculate the force required to punch a 50 mm diameter hole in a 10 mm thick sheet of the material, as shown in Fig. P1.18, and determine the mean stress in the punch.

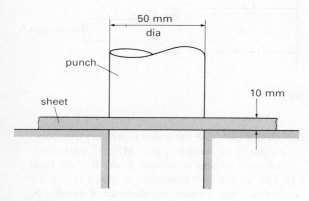

Fig. P1.18.

1.19 A tubular steel member of the structure of an off-shore oil drilling rig has an outer diameter of 2.8 m, a wall thickness of 22 mm, and is subject to a tensile force of 14 MN. If the yield strength of the steel is 400 MN/m^2, determine the factor of safety against yielding of (i) the plain tube, and (ii) the tube with a small radial hole drilled in it.

1.20 In Fig. P1.20, the horizontal rigid bar AB, which can rotate about a pinned joint at A, is also supported by a vertical wire CD attached at point C. A weight is suspended from point B by a second wire BE. If the strains in wires CD and BE due to the hanging of this weight are 0.15% and 0.1%, respectively, find the displacement of the point E.

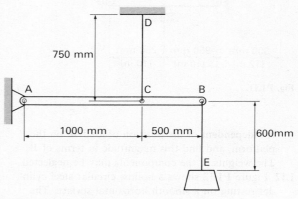

Fig. P1.20.

1.21 A uniform heavy rigid beam of length L is suspended on two wires of the same initial length h, as shown in Fig. P1.21. The wires are made of the same material, and their diameters are d and $1.5d$, respectively. Given that strain is proportional to stress for this material, and that the average extension of the wires is 0.1% of L, find the angle of rotation of the beam.

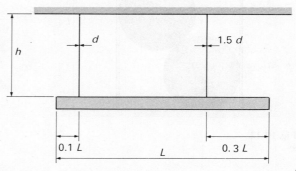

Fig. P1.21.

1.22 In Problem 1.21, find what ratio of the diameters of the wires would be required to avoid any rotation of the beam.

1.23 Figure P1.23 shows part of the mechanical control system for an aircraft. Crank ABC is free to rotate about a pinned joint at B, and the tension in the wire attached at A is 1.4 kN. If the crank is in equilibrium, find the tension in the wire attached at C. Also, if the pin at B is 5 mm in diameter, and is made from a material with an ultimate shear strength of 150 MN/m², estimate the safety factor against failure of this pin.

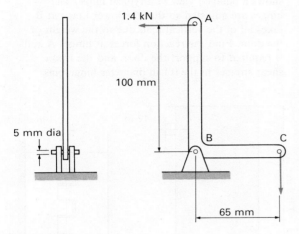

Fig. P1.23.

US CUSTOMARY UNITS – STATICS

1.24 A uniform ladder of length 15 ft and weight 40 lb is placed with one end on a rough horizontal surface and the other resting against a smooth vertical wall. The ladder makes an angle of 75° with the horizontal. Find the reaction forces acting on the ladder at its ends when a person weighing 150 lb stands on a rung which is 9 ft from the foot of the ladder, measured along its length.

1.25 Solve Problem 1.4 (using the dimension and load shown in parentheses in Fig. P1.4).

1.26 Solve Problem 1.6 (using the dimensions shown in parentheses in Fig. P1.6).

1.27 A uniform horizontal beam 20 ft long with a weight of 200 lb is supported at its ends. If weights of 100 lb, 150 lb and 200 lb are placed on the beam at distances of 3 ft, 9 ft and 16 ft, respectively, from one end, find the reaction at each of the supports.

1.28 Figure P1.28 shows a horizontal beam supported at its ends and subject to two concentrated loads and a load uniformly distributed over the length of beam between them. Neglecting the weight of the beam, calculate the vertical support reactions at its ends.

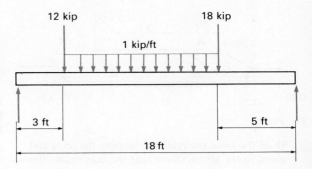

Fig. P1.28.

1.29 Solve Problem 1.11 (using the dimensions shown in parentheses in Fig. P1.11).

1.30 Solve Problem 1.12 (using the dimensions shown in parentheses in Fig. P1.12).

US CUSTOMARY UNITS – STRESSES AND STRAINS

1.31 Solve Problem 1.16.

1.32 Figure P1.32 shows a riveted joint between two pieces of steel plate. If the tensile force applied to the ends of the plates is 25 kip, and each of the rivets carries the same proportion of this force, calculate: (i) the mean stress in the plates away from the joint, (ii) the mean stress in the plates at a cross section through the centers of a line of rivets, (iii) the mean shear stress in the rivets, and (iv) the nominal bearing stress at the rivets.

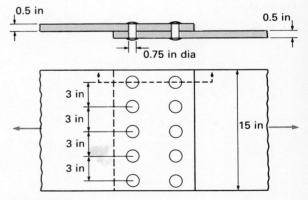

Fig. P1.32.

1.33 Figure P1.33 shows a joint made between two rectangular strips of a thermoplastic material, A and B, by means of two rectangular pieces, C and

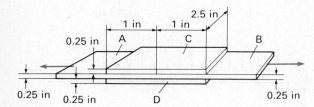

Fig. P1.33.

D, of the same material overlapping the joint and glued to the strips. Given that the maximum allowable shear stress in the adhesive layers is 100 psi, and neglecting any tensile strength due to adhesive between the ends of A and B, calculate the maximum allowable tensile force that can be applied to the ends of the strips away from the joint.

1.34 A tubular steel member of the structure of an off-shore oil drilling rig has an outer diameter of 10 ft, a wall thickness of 0.875 in, and is subject to a tensile force of 3500 kip. If the yield strength of the steel is 60 ksi, determine the factor of safety against yielding of (i) the plain tube, and (ii) the tube with a small radial hole drilled in it.

1.35 The composite block shown in Fig. P1.35 consists of 0.6 in thick layers of rubber separated by 0.4 in

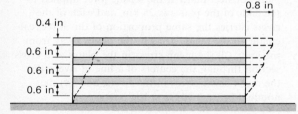

Fig. P1.35.

thick sheets of steel. The block is subjected to shear loading which causes the top sheet of steel to be displaced 0.8 in relative to the bottom one. Assuming that steel is very much stiffer in shear than rubber, calculate the mean shear strain in the rubber layers.

1.36 Solve Problem 1.21.

1.37 Solve Problem 1.22.

1.38 Figure P1.38a shows a uniform door weighing 170 lb supported on two hinges, while Fig. P1.38b shows a detailed view of a typical hinge. The hinges are adjusted so that the lower hinge at B takes all of the vertical load due to the weight of the door. Find the reaction forces at hinges A and B required to support the door, and the mean shear stresses in the 0.1 in diameter hinge pins.

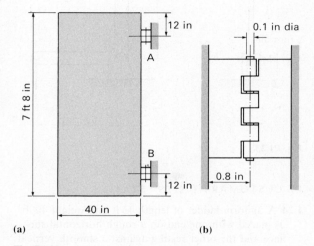

(a)   **(b)**

Fig. P1.38.

2

Statically Determinate Systems

In Chapter 1 we saw that an engineering structure or system of components can be described as statically determinate if the forces in the members can be calculated from a consideration of equilibrium requirements alone. In other words, we do not have to take into account either the properties of the materials concerned or any imposed geometric conditions, although we would certainly have to do so if we wished to calculate also the strains and displacements which result. Statically determinate systems therefore form a particularly straightforward class of problems which it is appropriate for us to consider first. They will, however, serve the very important function of further illustrating the use of both equilibrium equations and free body diagrams.

The statically determinate systems we will consider are pin-jointed structures, uniformly loaded thin shells in the form of either cylinders or spheres, and flexible cables. All of these have important applications in engineering practice.

2.1 Pin-jointed structures

The very simple but nevertheless typical structure shown in Fig. 2.1 serves to support a hoist for lifting heavy pieces of equipment. It consists of a horizontal boom, whose ends are indicated by the letters B and C, which is supported by an inclined bar with ends labeled A and B. At A and C the members of the structure, namely the boom BC and the bar AB, are pin-jointed to a massive vertical wall, while a downward vertical load is applied to the pin joint at B.

The pin joints at A, B and C may take many different forms, depending upon, among other things, the cross-sectional shapes of the members. For example, if these are flat strips of steel of rectangular cross section, the joints at A and C could be as shown enlarged in Fig. 2.2. The end of the member is sandwiched between two short plates firmly attached or built into the wall. Both plates and the end of the member are drilled to take a cylindrical pin which is long enough to protrude at each end. The pin is retained in position by, for example, a small split pin at each end, as shown. Alternatively, the ends of the pin could be threaded to take retaining nuts. The pin joint at B may take the form shown in Fig. 2.3, where the vertical load is applied to a clevis which sandwiches both members AB and BC, and a single pin is passed through the whole assembly.

Some other types of pin joint are shown in Fig. 2.4. A common feature of such joints is that the members are not clamped tightly together, and that the pins used need not fit the holes drilled for them very closely. This means that there is little or no resistance to relative rotation of the members. Only

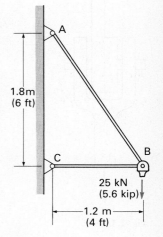

Fig. 2.1.

41

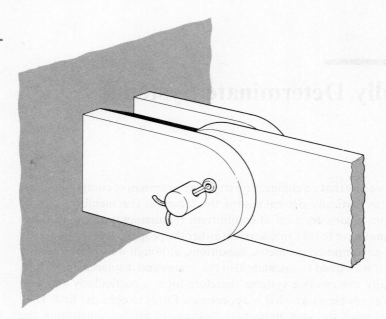

Fig. 2.2.

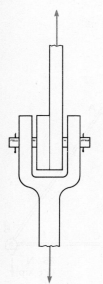

Fig. 2.3.

frictional forces acting at the contact surfaces between the pins and members can provide such resistance, but with pin diameters very small compared with the lengths of the members these act at such small distances from the centers of rotation that the moments involved are negligible. In other words, we may assume that at a pin joint no moments are applied to the members.

Although simple pin joints are used in engineering structures, they have the disadvantage of weakening the members. Drilling holes for the pins reduces the cross-sectional area of a member, and, as we saw in Chapter 1, the presence of holes causes stress concentrations which further increase the risk of failure. One way round this difficulty is to use members with their ends enlarged in the regions of the pin holes, as shown in Fig. 2.5. While such components are sometimes appropriate, they can no longer be simply made by cutting and drilling lengths of material of uniform cross section, and are therefore more expensive to produce. Consequently, it is common practice to reinforce the joints in other ways. Consider, for example, the type of structure shown in Fig. 2.6, which is used in bridge construction. The seven joints marked by the letters A to G are connected by eleven members, all of the same length. Joint C, which is typical, is shown in more detail in Fig. 2.7. A reinforcing plate is used to connect the four members which meet at this point. Incidentally, the members are not flat strips but have cross sections in the form of an 'L'. Each member is connected to the plate by two rivets which serve to clamp the components together. Alternatively, nuts and bolts can be used for the same purpose, or the members may be welded to the plates. Any of these arrangements offers considerably more resistance to relative rotation of the members than a simple pin joint. Nevertheless, provided the lengths of the members are large compared to their cross-sectional dimensions, the pin joint assumption of negligible moments applied to the members is still a good one.

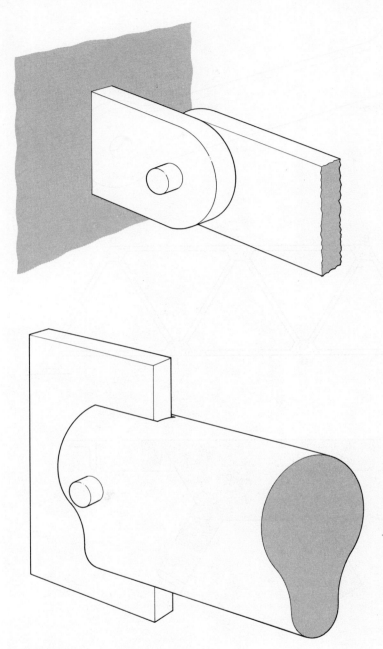

Fig. 2.4.

Further assumptions which we will apply in order to analyze pin-jointed structures are that external forces are applied only at the pin joints, and that the weights of the members are negligible compared with these applied forces. While the latter assumption is a very reasonable one for structures of moderate size, the weights of members do become significant in very large structures. The effect of both assumptions, combined with that of negligible moments applied

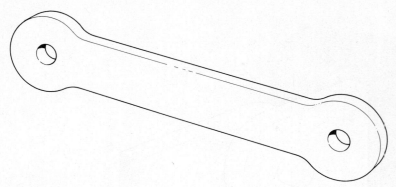

Fig. 2.5.

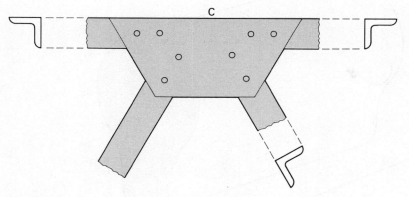

Fig. 2.6.

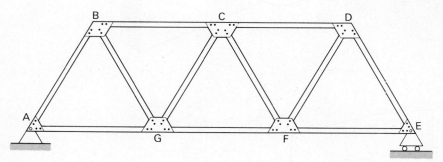

Fig. 2.7.

at the pin joints, is that the only force transmitted by a member is a tensile or compressive force along its length.

The structures shown in Figs 2.1 and 2.6 are both essentially two-dimensional, in that in each case the members of the structure lie in a single plane in space, although the members themselves have finite dimensions in the out-of-plane direction. Also, a bridge formed from a structure of the sort shown in Fig. 2.6 would normally be composed of at least two such frameworks laid parallel to each other with lateral links between them, including a deck for

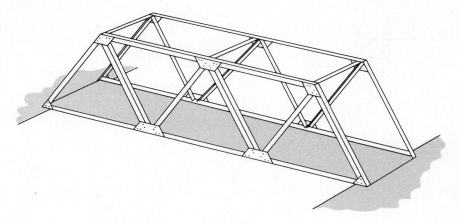

Fig. 2.8.

carrying traffic as in Fig. 2.8. Incidentally, the deck would be connected to the side frames at the joints between the lower horizontal members, so that our assumption of forces applied only at the joints would still apply. Such a three-dimensional structure is often referred to as a *space frame*. In this particular case, however, the side frames are relatively independent of each other, and may be treated as plane pin-jointed structures. Initially, we will be concentrating our attention on such plane structures, before considering how the methods we develop can be extended to space frames.

2.1.1 Equilibrium analysis of plane pin-jointed structures

In order to determine the forces in the members of a statically determinate pin-jointed structure, we need only employ equations of equilibrium of forces. If we need to find the forces in all the members, then we can effectively isolate each of the joints in turn, treating it as a free body with external forces applied, and establish the conditions for these forces to be in equilibrium. On the other hand, if we only need forces in a few of the members, a more selective approach known as the *method of sections* is more convenient to use. This method is described in Section 2.1.2.

Let us proceed with the aid of two examples: the two-member hoist support shown in Fig. 2.1, and the bridge framework shown in Fig. 2.6.

EXAMPLE 2.1

The geometry of the plane pin-jointed structure shown in Fig. 2.1 is such that the vertical distance between the centers of the pins at A and C is 1.8 m (6 ft), while the horizontal distance between centers at B and C is 1.2 m (4 ft). The magnitude of the downward vertical load at joint B is 25 kN (5.6 kip). It is convenient before trying to analyze the structure to first represent it by a simplified model of the real framework, but a model which retains all the important features. Such a model is shown in Fig. 2.9. The members AB and BC are represented by single lines, while the pin joints at A, B and C are shown as small circles. The angle which AB makes with BC, which is horizontal, is $\phi = \tan^{-1}$

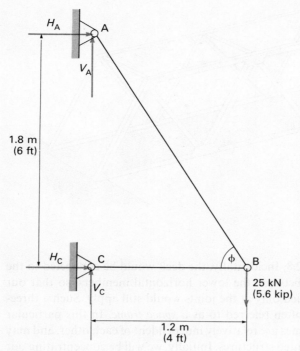

Fig. 2.9.

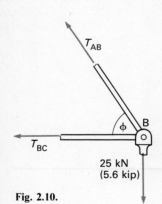

Fig. 2.10.

$(1.8/1.2) = 56.31°$. There are supporting reactions acting on the structure at the pin joints A and C where the structure is attached to a vertical wall. These can be represented by either horizontal and vertical components at each point, H_A and V_A at A, and H_B and V_B at B as shown, or by a single resultant force and its direction at each point. The effect is much the same, since we require two pieces of information to specify the loading at each point, but for present purposes the component form is more convenient to use.

If we isolate first the joint at B by sectioning through members AB and BC as shown in Fig. 2.10, a free body is created which is subject to only three external forces: the vertical load of 25 kN (5.6 kip) and the forces in AB and BC. No moments are applied, and the weights of the members themselves are ignored. The symbol T is chosen to represent forces (tensions) in the members, and it is convenient to employ a subscript notation such as the one shown, where the subscript letters refer to the joints at the ends of the member concerned. In other words, T_{AB} and T_{BC} represent the forces in members AB and BC. The order in which the two subscripts is shown is not important, so that T_{AB} and T_{BA} are identical. It is, however, important to show the forces in the members as tensile, and therefore acting away from the joint, even though we can anticipate that one of them (in member BC) will be compressive. This will automatically give results according to the sign convention of tensile positive, compressive negative.

Since the lines of action of the three forces applied to the free body pass through a single point, the center of the pin at B, no moment equilibrium equation will give more information than is provided by force equilibrium equations in two directions. Now, it is generally convenient to take these directions either as horizontal and vertical, or as parallel and perpendicular to some direction which is appropriate in the particular physical problem. In the present case, with one member horizontal, the horizontal and vertical directions are suitable. Even so, we should stop to consider which direction to take first. There are two unknown forces, T_{AB} and T_{BC}, both of which have components in the horizontal direction, but only one of which, T_{AB}, has a component in the vertical direction. Therefore, while the equation of equilibrium of forces in the horizontal direction involves both unknowns, that in the vertical direction involves only one, and should be considered first. Taking forces in the upward vertical direction as positive, this equation is

SI units

$$T_{AB} \sin 56.31° - 25 = 0$$

from which

$$T_{AB} = + \frac{25}{0.8321} = \underline{+ 30.05 \text{ kN}}$$

where the positive sign indicates that the force in AB is tensile. Similarly, the equation of equilibrium of forces in the horizontal direction, taking the positive sense as being from left to right, is

$$- T_{AB} \cos 56.31° - T_{BC} = 0$$

and, since T_{AB} has already been found, we can obtain the second unknown T_{BC} as

$$T_{BC} = - T_{AB} \cos 56.31° = \underline{- 16.67 \text{ kN}}$$

where the negative sign implies a compressive force in BC, as expected.

Figure 2.11 shows the forces acting on the joints at A and C, respectively. Once again, the forces in the members AB and BC are shown as tensile, acting away from the joints. The horizontal and vertical components of the reaction forces on the structure at A and B are in the directions already chosen in Fig. 2.9. For

US customary units

$$T_{AB} \sin 56.31° - 5.6 = 0$$

from which

$$T_{AB} = + \frac{5.6}{0.8321} = \underline{+ 6.730 \text{ kip}}$$

where the positive sign indicates that the force in AB is tensile. Similarly, the equation of equilibrium of forces in the horizontal direction, taking the positive sense as being from left to right, is

$$- T_{AB} \cos 56.31° - T_{BC} = 0$$

and, since T_{AB} has already been found, we can obtain the second unknown T_{BC} as

$$T_{BC} = - T_{AB} \cos 56.31° = \underline{- 3.733 \text{ kip}}$$

where the negative sign implies a compressive force in BC, as expected.

Figure 2.11 shows the forces acting on the joints at A and C, respectively. Once again, the forces in the members AB and BC are shown as tensile, acting away from the joints. The horizontal and vertical components of the reaction forces on the structure at A and B are in the directions already chosen in Fig. 2.9. For

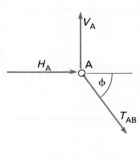

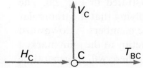

Fig. 2.11.

SI units

equilibrium of the forces acting in the horizontal and vertical directions on joint A

$$V_A - T_{AB} \sin 56.31° = 0$$

from which $V_A = 30.05 \sin 56.31° = \underline{25\ kN}$

and $\quad T_{AB} \cos 56.31° + H_A = 0$

giving $\quad H_A = -30.05 \cos 56.31° = \underline{-16.67\ kN}$

The negative sign on this result for H_A merely means that our original assumption that the reaction force on the structure at joint A would be from left to right (away from the wall) was wrong, and it actually acts towards the wall. Similarly, for equilibrium of the forces acting in the horizontal and vertical directions on joint B

$$T_{BC} + H_C = 0$$

from which $\quad H_C = -T_{BC} = \underline{+16.67\ kN}$

and $\quad\quad V_C = 0$

Having determined all the unknown forces in the members and the external reactions at the supports by considering equilibrium of the individual joints, we can now check that the entire structure is in equilibrium. For equilibrium of the forces shown in Fig. 2.9 in the vertical and horizontal directions

$$V_A + V_C - 25 = 0$$

$$H_A + H_C = 0$$

and for equilibrium of moments about, say, point A

$$1.2 \times 25 - 1.8 \times H_C = 0$$

conditions which are satisfied by the numerical values already calculated.

US customary units

equilibrium of the forces acting in the horizontal and vertical directions on joint A

$$V_A - T_{AB} \sin 56.31° = 0$$

from which $\quad V_A = 6.730 \sin 56.31° = \underline{5.6\ kip}$

and $\quad T_{AB} \cos 56.31° + H_A = 0$

giving $\quad H_A = -6.730 \cos 56.31° = \underline{-3.733\ kip}$

The negative sign on this result for H_A merely means that our original assumption that the reaction force on the structure at joint A would be from left to right (away from the wall) was wrong, and it actually acts towards the wall. Similarly, for equilibrium of the forces acting in the horizontal and vertical directions on joint B

$$T_{BC} + H_C = 0$$

from which $\quad H_C = -T_{BC} = \underline{+3.733\ kip}$

and $\quad\quad V_C = 0$

Having determined all the unknown forces in the members and the external reactions at the supports by considering equilibrium of the individual joints, we can now check that the entire structure is in equilibrium. For equilibrium of the forces shown in Fig. 2.9 in the vertical and horizontal directions

$$V_A + V_C - 5.6 = 0$$

$$H_A + H_C = 0$$

and for equilibrium of moments about, say, point A

$$4 \times 5.6 - 6 \times H_C = 0$$

conditions which are satisfied by the numerical values already calculated.

EXAMPLE 2.2

Figure 2.12 shows the simplified model of the structure illustrated in Fig. 2.6. The effective length of each of the eleven members is 4 m (13 ft), implying that the triangular shapes formed are all equilateral, with angles of 60° between the members. A downward vertical force of magnitude 108 kN (24 kip) is applied to the structure at the joint marked G. Since the framework is pin-jointed to a fixed support at A, there will in general be both horizontal and vertical components of reaction acting on the structure there, which are shown as H_A and V_A. On the other hand, at the other support at E, the structure is

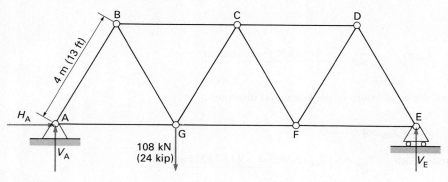

Fig. 2.12.

pinned to a roller which is free to move horizontally, and which is assumed to be frictionless. At such a roller support, there is only one reaction component, which is normal to the direction in which movement can occur, in this case a vertical reaction V_E. We should note that, had there been another fixed pin joint at E, the structure would have been statically indeterminate. This is because the horizontal reactions at A and E would then be determined, not only by equilibrium requirements, but also by the fact that the points concerned remain the same distance apart when the structure is loaded. We will return to the criteria for a structure to be statically determinate after the present example.

Before starting to analyze the structure joint by joint, it is convenient to find first the support reactions by considering the equilibrium of the structure as a whole. For equilibrium of forces in the horizontal and vertical directions, we conclude that

$$H_A = 0 \tag{2.1}$$

$$V_A + V_E - 108 = 0 \tag{2.2}$$

We may also write one further equation, for moment equilibrium about a convenient point, such as the point E. Taking moments about E to be positive in the clockwise direction

$$12\,V_A - 8 \times 108 = 0 \tag{2.3}$$

from which

$$V_A = \frac{8 \times 108}{12} = \underline{72\ \text{kN}}$$

and, substituting this value into equation (2.2), the other vertical reaction may be found as

$$V_E = 108 - 72 = \underline{36\ \text{kN}}$$

Figure 2.13 shows the forces acting at the point A, and for equilibrium in the vertical direction we may write

$$T_{AB} \sin 60° + 72 = 0$$

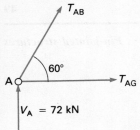

Fig. 2.13.

or

$$T_{AB} = -\frac{72}{\sin 60°} = -83.14 \text{ kN (compressive)}$$

and, for equilibrium in the horizontal direction

$$T_{AG} + T_{AB} \cos 60° = 0$$

from which $\quad T_{AG} = -T_{AB} \cos 60° = +41.57 \text{ kN (tensile)}$

Note that, as in the previous Example, we choose to consider first the direction (the vertical direction) in which only one of the two unknown forces (T_{AB} and T_{AG}) has a component. Now, in order to find further forces in the members, we can choose to consider next either joint B or joint G. The order of selecting joints is important. We know that we can only write two equilibrium equations at each joint, and therefore only find two unknown forces. Having previously found the forces in members AB and AG, there are still two unknowns at point B (forces in BC and BG), but *three* at point G (forces in BG, GC and GF). Point B is therefore the one to select, and Fig. 2.14 shows the forces applied to it. For equilibrium in the vertical direction

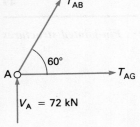

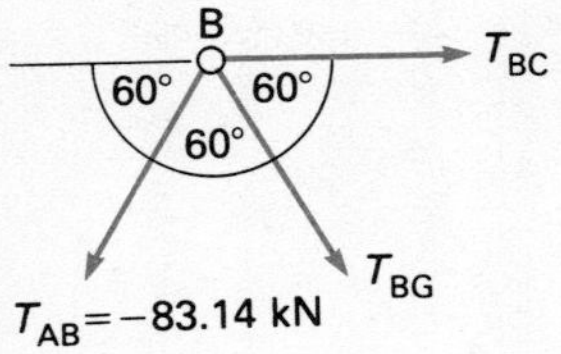

Fig. 2.14.

$$-T_{AB} \sin 60° - T_{BG} \sin 60° = 0$$

$$T_{BG} = -T_{AB} = +83.14 \text{ kN (tensile)}$$

and in the horizontal direction

$$-T_{AB} \cos 60° + T_{BG} \cos 60° + T_{BC} = 0$$

$$T_{BC} = -83.14 \cos 60° - 83.14 \cos 60° = -83.14 \text{ kN (compressive)}$$

Now we can proceed to point G, and Fig. 2.15 shows the forces acting on it. For equilibrium of these forces in the vertical direction

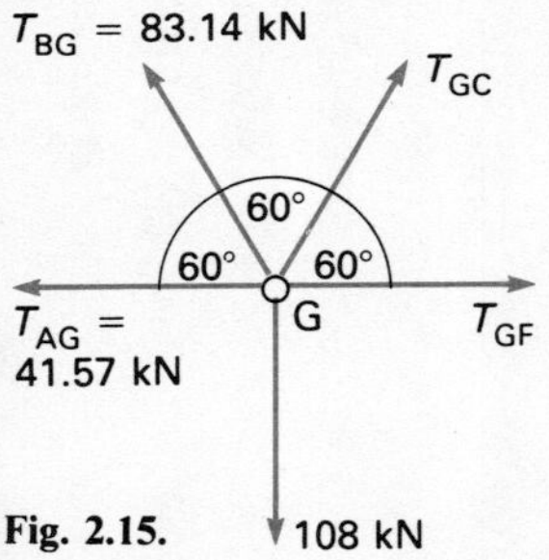

$$T_{BG} \sin 60° + T_{GC} \sin 60° - 108 = 0$$

$$T_{GC} = -83.14 + \frac{108}{0.8660} = +41.57 \text{ kN (tensile)}$$

and for equilibrium of forces in the horizontal direction

$$-T_{AG} - T_{BG} \cos 60° + T_{GC} \cos 60° + T_{GF} = 0$$

$$T_{GF} = 41.57 + 83.14 \times 0.5 - 41.57 \times 0.5 = +62.35 \text{ kN (tensile)}$$

Fig. 2.15.

We next proceed to examine point C, shown in Fig. 2.16, because there are only two unknown forces there. For equilibrium in the vertical direction

$$-T_{GC} \sin 60° - T_{CF} \sin 60° = 0$$

$$T_{CF} = -T_{GC} = -41.57 \text{ kN (compressive)}$$

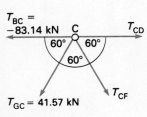

Fig. 2.16.

50

and for equilibrium in the horizontal direction

$$-T_{BC} - T_{GC} \cos 60° + T_{CF} \cos 60° + T_{CD} = 0$$

$$T_{CD} = -83.14 + 41.57 \times 0.5 + 41.57 \times 0.5 = -41.57 \text{ kN (compressive)}$$

Proceeding to point F, shown in Fig. 2.17, equilibrium in the vertical direction requires that

$$T_{CF} \sin 60° + T_{FD} \sin 60° = 0$$

$$T_{FD} = -T_{CF} = +41.57 \text{ kN (tensile)}$$

and for equilibrium in the horizontal direction

$$-T_{GF} - T_{CF} \cos 60° + T_{FD} \cos 60° + T_{FE} = 0$$

$$T_{FE} = 62.35 - 41.57 \times 0.5 - 41.57 \times 0.5 = +20.78 \text{ kN (tensile)}$$

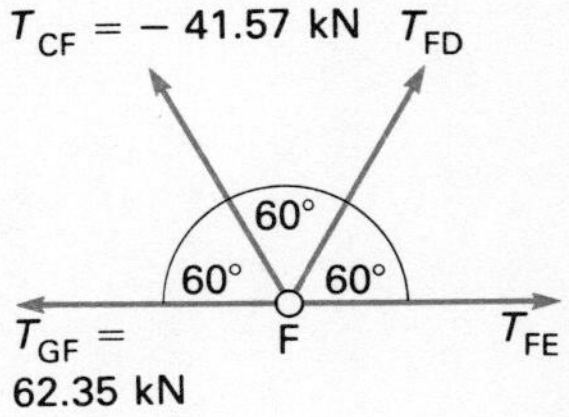

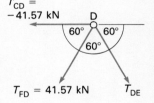

Fig. 2.17.

When we now examine the forces acting at joint D, shown in Fig. 2.18, we find that there is only one unknown force, in member DE. This can be found by considering equilibrium of forces in the vertical direction

$$-T_{FD} \sin 60° - T_{DE} \sin 60° = 0$$

$$T_{DE} = -T_{FD} = -41.57 \text{ kN (compressive)}$$

We should, however, also consider the resultant force on D in the horizontal direction, which is

$$-T_{CD} - T_{FD} \cos 60° + T_{DE} \cos 60°$$

$$= 41.57 - 41.57 \times 0.5 - 41.57 \times 0.5 = 0$$

In other words, joint D is in equilibrium. Similarly, when we consider the final joint at E, shown in Fig. 2.19, there are no unknown forces. The resultant force in the horizontal direction is

$$-T_{FE} - T_{DE} \cos 60° = -20.78 + 41.57 \times 0.5 = 0$$

and the resultant force in the vertical direction is

$$T_{DE} \sin 60° + V_E = -41.57 \times 0.8660 + 36 = 0$$

confirming that joint E is in equilibrium.

Fig. 2.18.

While these last three equilibrium equations serve the very useful purpose of confirming that our other calculations are correct, it is worth reflecting on why there are such equations providing essentially duplicate information. It is because, before we started to analyze the structure joint by joint, we employed three equations of equilibrium for the structure as a whole, namely equations (2.1), (2.2) and (2.3), which provided no more information than can be obtained from considering the joints individually. Our reason for starting with them was, however, a very practical one, in that it allowed us to find all the unknown forces as we passed from joint to joint along

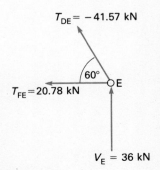

Fig. 2.19.

the structure, rather than generating a rather large set of simultaneous equations (14 in total, 2 for each joint), which would then have to be solved. In contrast, in the simple structure considered in Example 2.1, this strategy was not necessary, and the overall equilibrium equations were used to check the results.

Having considered two particular examples of the analysis of plane pin-jointed structures, we can establish both a general procedure for such problems, and the conditions which determine whether a given structure is statically determinate. For each of the joints in the structure, we can write equations of equilibrium of forces in two directions. In a general procedure it is convenient to choose these directions as the horizontal or x direction, and the vertical or y direction. We can choose to number the members of a structure in some convenient but essentially arbitrary order. For example, in Fig. 2.12, member AB can be regarded as the first, BC as the second, CD as the third, and so on for all eleven elements. Consequently, T_1 is the force in AB, T_2 the force in BC, and so on. We can also choose to number the joints in some equally arbitrary order, such as A first, then B, C, D and so on. Let us define the orientation of each member at a particular joint in terms of an angle θ between the member (strictly the line joining the centers of the pin joints at its ends) and the positive x axis, taking the joint in question as the local origin. In Fig. 2.12, for example, $\theta_1 = 60°$ at joint 1 and 240° at joint 2, $\theta_2 = 0°$ at joint 2 and 180° at joint 3, and so on. If the letter m is used as a counter for the members, and letter i as the counter for the joints, the equations of equilibrium for joint i are

$$\Sigma T_m \cos \theta_m + F_x^{(i)} + R_x^{(i)} = 0 \tag{2.4}$$

$$\Sigma T_m \sin \theta_m + F_y^{(i)} + R_y^{(i)} = 0 \tag{2.5}$$

where $F_x^{(i)}$ and $F_y^{(i)}$ are the components in the positive x and y directions, respectively, of any *external* force applied at joint i (such as the 108 kN force in the negative y direction at joint G in Fig. 2.12). Also, $R_x^{(i)}$ and $R_y^{(i)}$ are the components in the positive x and y directions, respectively, of the reaction force applied at joint i, when joint i is a support point for the structure (for example, reactions H_A and V_A at joint A, and V_E at joint E in Fig. 2.12). The summations of the components of the member forces in equations (2.4) and (2.5) are carried out over only those members which are joined at point i. The final set of simultaneous linear algebraic equations are then solved for the unknown forces in the members and reaction forces at the supports. As we have already seen in Example 2.2, this is not necessarily the most convenient approach if we are going to solve the equations by hand, but it has the merit of being quite general and therefore suitable for automatic solution using a computer. A computer solution technique is described in Section 2.1.3.

The fact that we have reduced the analysis of statically determinate plane pin-jointed structures to the solution of a set of linear algebraic equations allows us to establish criteria for examining in advance whether a particular structure is statically determinate. Firstly, we can only hope to obtain solutions if the number of equations and the number of unknowns are compatible. Suppose that the total number of joints in the structure is N_j, the total number of members is N_m, and the total number of unknown reaction force components at the supports is N_r. As we have already seen, a fixed pin-jointed support must be treated as involving two reaction force components (even though we can see by inspection that one of them will be zero, as in the case of H_A in Fig. 2.12, for example). On the other hand, a rolling or sliding support in which friction is ignored involves only one reaction force, in the direction normal to the direction of movement. Figure 2.20 illustrates the conventional representations of these two types of support. At the rolling support, the structure is shown pin-jointed to a roller unit which is free to follow the supporting surface, which is not necessarily horizontal. We can refer to the two types of support as pinned and rolling, respectively.

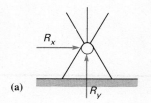

The total number of equilibrium equations available, two per joint, is $2N_j$, while the total number of unknowns is $(N_m + N_r)$. If the numbers of unknowns and equations are equal, with

$$N_m + N_r = 2N_j \qquad (2.6)$$

it is reasonable to expect that we can obtain solutions, implying that the structure is statically determinate. Indeed, this is certainly a necessary condition for a statically determinate structure, but it is not always a sufficient condition. What is also important is the way in which the members of the structure are arranged. This is perhaps best explained with the aid of some examples.

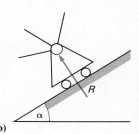

(b)
Fig. 2.20.

If the structure is such that

$$N_m + N_r > 2N_j \qquad (2.7)$$

the number of unknowns exceeds the number of equations, and solutions cannot be obtained. In other words, the structure is statically indeterminate, and involves either members or support reactions which are *redundant* in terms of forming a framework which can exist as an anchored load-carrying structure. The number of such redundancies is given by the difference between the number of unknowns and the number of equations, $(N_m + N_r - 2N_j)$. Methods for analyzing statically indeterminate structures are described in Chapter 4. The other possibility to be considered is

$$N_m + N_r < 2N_j \qquad (2.8)$$

which means that the number of equations exceeds the number of unknowns. As we shall see in the following examples, this implies that we have a *mechanism*, which is incapable of supporting static loads, either because the members are free to move relative to each other or because the structure itself is not anchored to its surroundings.

Consider the simple structure shown in Fig. 2.9, and already analyzed in Example 2.1. There are two members (AB and BC), three joints (A, B and C) and four reaction forces (two at each of the pinned supports A and C). With $N_m = 2$,

$N_r = 4$, and $2N_j = 6$, equation (2.6) is satisfied, and the structure is, as we have already demonstrated by detailed analysis, statically determinate.

Turning to the more complex structure shown in Fig. 2.12, and previously analyzed in Example 2.2, the number of members is $N_m = 11$, number of reactions, $N_r = 3$ (two at the pinned support at A, and one at the roller support at E), and the number of joints, $N_j = 7$. Equation (2.6) is satisfied, and, at least in this case, the structure is statically determinate. It is interesting to note that, as already anticipated in Example 2.2, if there had been a pinned support at E, thereby increasing N_r from three to four, the structure would have been statically indeterminate.

EXAMPLE 2.3

Figure 2.21 shows in model form four pin-jointed structures. Determine whether these are statically determinate, statically indeterminate, or mechanisms. All joints are shown as circles: elsewhere members cross without touching.

(a) Number of members, $N_m = 3$, number of reactions, $N_r = 6$ (two at each of the pinned supports), and the number of joints, $N_j = 4$. Therefore, $N_m + N_r - 2N_j = 1$, and the structure is statically indeterminate, with one redundancy. Any one of the three members could be removed without causing the framework to collapse.

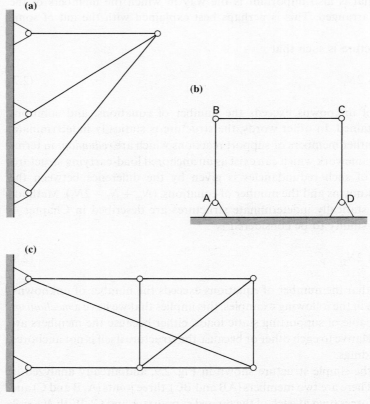

Fig. 2.21a–c

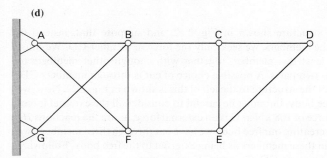

(d)

Fig. 2.21d

(b) Number of members, $N_m = 3$, number of reactions, $N_r = 4$ (two at each of the pinned supports), and the number of joints, $N_j = 4$. Therefore, $N_m + N_r < 2N_j$, and the structure is a mechanism. Members AB and CD are free to rotate together (linked by member BC) about joints A and D, respectively.

(c) Number of members, $N_m = 8$, number of reactions, $N_r = 4$, and the number of joints, $N_j = 6$. Therefore, $N_m + N_r = 2N_j$, and the structure is statically determinate. Despite the rather unconventional (and not very practical) arrangement of the members, the structure is not free to collapse as a mechanism, nor may we remove any of the members without causing it to collapse.

(d) Number of members, $N_m = 10$, number of reactions, $N_r = 4$, and the number of joints, $N_j = 7$. Therefore, $N_m + N_r = 2N_j$, but the structure is not statically determinate. We can identify three quite distinct regions of the structure: (i) the members linking points C, D and E, which do form a statically determinate structure, (ii) the members linking points A, B, F and G, which form a statically indeterminate structure, and (iii) the members linking points B, C, E and F, which form a mechanism (similar to structure (b)). The members have not been arranged in a practically useful way: if either one of the diagonal members AF or BG is moved to link either B and E or C and F, then the structure would become statically determinate (the result of the formal test would be unchanged).

2.1.2 Method of sections for plane pin-jointed structures

So far we have concerned ourselves with a method for analyzing plane pin-jointed structures which involves progressively finding all the unknown forces in the members and reactions at the supports. This is appropriate if we really do need all these forces, and forms the basis of the computer method described in the next subsection. If we only need to find the forces in certain members, then to calculate all the unknowns by hand is very laborious. In that case we can use the *method of sections* to be much more selective about the forces we calculate. The general method we have already considered involves the analysis of the equilibrium of isolated sections of the structure, namely the small regions around each of the joints, obtained by effectively cutting through all of the members meeting at the particular joint. We now extend this idea to larger portions of the structure. The choice of which members to cut through is dependent on both the particular structure concerned and the forces required, and the method is perhaps most readily understood with the aid of an example. As a general principle, however, we must arrange to cut through the members whose internal forces we wish to determine.

EXAMPLE 2.4

Consider once again the structure shown in Fig. 2.12, and suppose that, instead of wanting the forces in all the members, we seek only the force in member CD. We must therefore cut through at least this member, together with enough other members to separate the structure into two parts. A possible choice of cut is through members CD, CF and GF, and the part of the structure to the left of this is shown in Fig. 2.22. Now, we may treat this part as a free body, but must be careful to consider all the external forces acting on it. The vertical force of 108 kN at G is an external force, as are the reactions H_A and V_A at joint A. Since in creating our free body we have cut through three members, we must also treat the forces in these members as being external to the free body. Following our earlier practice, these forces in CD, CF and GF are shown as tensile, and therefore acting away from the cut ends of the members. We should note that forces in members AB, BC, CG, GA, GB and GC are internal forces as far as the chosen free body is concerned. Of the six external forces, only one is known initially. We may, however, find the reaction forces at A with the aid of overall equilibrium equations (2.1) to (2.3). The three remaining unknown forces in the cut members can be found from three equations of equilibrium for the free body. With only one of the forces, T_{CD}, being required, however, we can try to choose an equation which allows just this one unknown to be found. Now, an equation for equilibrium of forces in any direction will involve either or both of the other unknowns, T_{CF} and T_{GF}. We should therefore try to choose an equation of moment equilibrium which does not. There is only one such equation, which is the one obtained by taking moments about the point of intersection of T_{CF} and T_{GF}, namely the point F. The fact that F is a point outside the free body is not important. The equation is

$$V_A \times 8 - 108 \times 4 + T_{CD} \times 4 \sin 60° = 0$$

giving

$$T_{CD} = \frac{108 - 72 \times 2}{\sin 60°} = -41.57 \text{ kN (compressive)}$$

which is identical to the result obtained earlier.

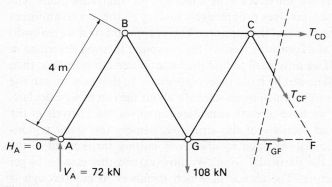

Fig. 2.22.

2.1.3 A computer method for plane pin-jointed structures

A highly selective method of analysis such as the method of sections, which requires intelligent application to particular problems, is not well suited to computer use. Once we have supplied the relevant geometric and loading data to a computer program, however, it should be a trivial matter for the machine to compute all the unknown forces, from which we can then select the items of interest.

The method established in Section 2.1.1, which is based on equilibrium equations for each and every node of the structure, is amenable to programming, especially in the generalized form expressed in equations (2.4) and (2.5). All that is required is a method for solving numerically a set of simultaneous linear algebraic equations. Techniques for doing this fall into two broad categories, namely direct and indirect (or iterative) methods. Although iterative methods have many important applications in engineering analysis, for present purposes direct methods are more appropriate. The most straightforward and widely used of these is *Gaussian elimination*, which is described in detail, together with a computer program subroutine for its implementation, in Appendix B.

Our aim then is to organize the information necessary to define a particular pin-jointed structure in a way which allows it to be supplied as data to a computer program. We need to define the geometry of the structure in terms of the positions of the joints and the connections between them by means of members, the external forces applied to the joints, and the positions and types (pinned or roller) of supports. Having done that, the program can then generate the coefficients of the linear equilibrium equations. The approach we will use is based on *finite element* methodology. Indeed, the present form of analysis is a rudimentary type of finite element method. Finite element methods were first developed for structural analysis, particularly of statically indeterminate frameworks, although they are today also applied to a very wide range of continuum problems in solid mechanics, fluid mechanics, heat transfer and many other branches of engineering science. In essence, these methods involve first dividing the structure or continuum into smaller components or *elements*, the behavior of each one of which can be represented in a relatively simple way. Elements are linked together at a limited number of *nodes* or *nodal points*. The complexity of the overall system is approximated by using a large number of elements, and using a computer to solve the resulting large set of algebraic equations. While, in continuum problems, the choice of elements and nodes relies at least partly on the experience of the analyst, in the case of pin-jointed structures the choice is a clear-cut one: the members form the elements, and the joints are the nodes. With this finite element background in mind, we will adopt this terminology in the computer program. A computer program for statically indeterminate plane pin-jointed structures, which draws even more heavily on finite element concepts is introduced in Chapter 4.

Figure 2.23 shows a typical element of a plane pin-jointed structure, which we may number as element *m*. It lies in the plane of Cartesian coordinates X and Y. We may regard these as *global* coordinates in the sense that they are used to define positions of nodes for the structure as a whole, and not just the

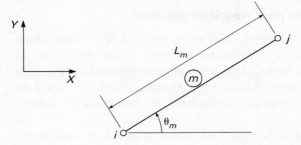

Fig. 2.23.

particular element we are currently considering. The element links two nodes, which we may number as i and j, whose positions are defined in terms of coordinates (X_i, Y_i) and (X_j, Y_j), respectively. We arbitrarily select i as the first node and j as the second node of the element. Using the first node as the local origin, we then define the orientation of the element in terms of the angle θ_m between the element and the direction of the positive X axis, as shown.

Rearranging equations (2.4) and (2.5) to have unknowns on the left and known quantities on the right-hand sides of the equals signs, we obtain as the equations for typical node i

$$\Sigma T_m \cos \theta_m + R_x^{(i)} = -F_x^{(i)} \tag{2.9}$$

$$\Sigma T_m \sin \theta_m + R_y^{(i)} = -F_y^{(i)} \tag{2.10}$$

When these are repeated for every node in the structure, we obtain a set of linear equations which can be expressed in matrix form as

$$[A][T] = [F] \tag{2.11}$$

where the column vector $[T]$ contains the unknowns, the square matrix $[A]$ contains known coefficients such as unity and sines and cosines of the element angles, while column vector $[F]$ contains the components of the applied external forces. We now choose to arrange the unknowns in $[T]$ in any convenient order such as T_1, T_2, T_3 and so on up to the total number of elements involved, followed by the reaction force components.

Program SDPINJ (standing for **S**tatically **D**eterminate **PIN-J**ointed structure analysis) provides a FORTRAN implementation of the above analysis.

```
      PROGRAM  SDPINJ
C
C  PROGRAM TO FIND THE FORCES IN THE MEMBERS OF A STATICALLY DETERMINATE
C  PLANE PIN-JOINTED STRUCTURE.
C
      DIMENSION  X(50),Y(50),NPI(100),NPJ(100),A(100,101),T(100),FX(50),
     1           FY(50),NODE(50),ANG(50)
      NNPMAX=50
      NELMAX=100
      PI=4.*ATAN(1.)
      OPEN(5,FILE='DATA')
      OPEN(6,FILE='RESULTS')
      WRITE(6,61)
   61 FORMAT('STATICALLY DETERMINATE PLANE PIN-JOINTED STRUCTURE')
```

```
C
C   INPUT AND TEST THE NUMBER OF NODES (JOINTS).
        READ(5,*) NNP
        IF(NNP.GT.NNPMAX) THEN
            WRITE(6,62) NNP
  62        FORMAT(/ 'NUMBER OF NODES = ',I4,' TOO LARGE - STOP')
            STOP
        END IF
C
C   INPUT AND OUTPUT THE CARTESIAN COORDINATES OF THE NODES.
        READ(5,*) (X(I),Y(I),I=1,NNP)
        WRITE(6,63) (I,X(I),Y(I),I=1,NNP)
  63    FORMAT(/ 'COORDINATES OF THE NODES'
       1 / 2('NODE        X              Y          ') / (2(I3,2E12.4,4X)))
C
C   INPUT AND TEST THE NUMBER OF ELEMENTS (MEMBERS).
        READ(5,*) NEL
        IF(NEL.GT.NELMAX) THEN
            WRITE(6,64) NEL
  64        FORMAT(/ 'NUMBER OF ELEMENTS = ',I4,' TOO LARGE - STOP')
            STOP
        END IF
C
C   INPUT AND OUTPUT THE NODES CONNECTED BY THE ELEMENTS.
        READ(5,*) (NPI(M),NPJ(M),M=1,NEL)
        WRITE(6,65) (M,NPI(M),NPJ(M),M=1,NEL)
  65    FORMAT(/ 'NODES CONNECTED BY THE ELEMENTS'
       1        / 4('  M    I    J    ') / (4(I4,2I5,4X)))
C
C   INITIALIZE THE EQUATION COEFFICIENTS AND EXTERNAL FORCES ON THE NODES.
        NEQN=2*NNP
        DO 1 IROW=1,NEQN
        DO 1 ICOL=1,NEQN
  1     A(IROW,ICOL)=0.
        DO 2 I=1,NNP
        FX(I)=0.
  2     FY(I)=0.
C
C   INPUT AND TEST THE NUMBER OF NODES AT WHICH EXTERNAL FORCES ARE APPLIED.
        READ(5,*) NNPF
        IF(NNPF.GT.NNPMAX) THEN
            WRITE(6,66) NNPF
  66        FORMAT(/ 'NUMBER OF LOADED NODES = ',I4,' TOO LARGE - STOP')
            STOP
        END IF
C
C   INPUT AND OUTPUT THE FORCE COMPONENTS AT THE LOADED NODES.
        IF(NNPF.GT.0) THEN
            READ(5,*) (NODE(K),FX(NODE(K)),FY(NODE(K)),K=1,NNPF)
            WRITE(6,67) (NODE(K),FX(NODE(K)),FY(NODE(K)),K=1,NNPF)
  67        FORMAT(/ 'FORCE COMPONENTS AT THE LOADED NODES'
       1        / '  NODE',12X,'FX',13X,'FY' / (I5,5X,2E15.4))
        END IF
C
C   INPUT AND TEST THE NUMBERS OF NODES AT WHICH THE STRUCTURE IS PINNED
C   TO ITS SURROUNDINGS (NNPP), AND SUPPORTED ON ROLLERS (NNPR).
        READ(5,*) NNPP,NNPR
        NNPT=NNPP+NNPR
        IF(NNPT.GT.NNPMAX) THEN
            WRITE(6,68) NNPT
  68        FORMAT(/ 'NUMBER OF NODES AT WHICH STRUCTURE SUPPORTED = ',I4,
       1            ' TOO LARGE - STOP')
            STOP
        END IF
C
C   INPUT AND OUTPUT THE NODES AT WHICH THE STRUCTURE IS PINNED.
        IF(NNPP.GT.0) THEN
            READ(5,*) (NODE(K),K=1,NNPP)
            WRITE(6,69) (NODE(K),K=1,NNPP)
  69        FORMAT(/ 'NODES FORMING PINNED SUPPORTS : ',10I4 / (18I4))
        END IF
```

```
C
C   INPUT AND OUTPUT THE NODES AT WHICH STRUCTURE RESTS ON ROLLERS,
C   TOGETHER WITH THE ANGLES (IN DEGREES MEASURED COUNTERCLOCKWISE FROM
C   THE POSITIVE X-AXIS) AT WHICH THESE ROLLERS ARE FREE TO MOVE.
        IF(NNPR.GT.0) THEN
            READ(5,*) (NODE(NNPP+K),ANG(NNPP+K),K=1,NNPR)
            WRITE(6,70) (NODE(NNPP+K),ANG(NNPP+K),K=1,NNPR)
 70     FORMAT(/ 'NODES AND ANGLES OF ROLLER SUPPORTS'
     1          /  '  NODE          ANG (DEG) ' / (I5,F17.2))
        END IF
C
C   COMPARE THE NUMBER OF UNKNOWNS WITH THE NUMBER OF EQUATIONS.
        NUNK=NEL+2*NNPP+NNPR
        IF(NUNK.GT.NEQN) THEN
            WRITE(6,71)
 71     FORMAT(/ 'THE STRUCTURE IS STATICALLY INDETERMINATE - STOP')
            STOP
        END IF
        IF(NUNK.LT.NEQN) THEN
            WRITE(6,72)
 72     FORMAT(/ 'THE STRUCTURE IS A MECHANISM - STOP')
            STOP
        END IF
C
C   SET UP THE LINEAR EQUATIONS, FIRST CONSIDERING EACH ELEMENT IN TURN.
        DO 3 M=1,NEL
        I=NPI(M)
        J=NPJ(M)
        ELENGT=SQRT((X(J)-X(I))**2+(Y(J)-Y(I))**2)
        COSINE=(X(J)-X(I))/ELENGT
        SINE=(Y(J)-Y(I))/ELENGT
        A(2*I-1,M)=COSINE
        A(2*I,M)=SINE
        A(2*J-1,M)=-COSINE
 3      A(2*J,M)=-SINE
C
C   THEN THE PINNED SUPPORTS.
        IF(NNPP.GT.0) THEN
            DO 4 K=1,NNPP
            I=NODE(K)
            A(2*I-1,NEL+2*K-1)=1.
 4          A(2*I,NEL+2*K)=1.
        END IF
C
C   THEN THE ROLLER SUPPORTS.
        IF(NNPR.GT.0) THEN
            DO 5 K=NNPP+1,NNPT
            I=NODE(K)
            ANGR=ANG(K)*PI/180.
            A(2*I-1,NEL+NNPP+K)=-SIN(ANGR)
 5          A(2*I,NEL+NNPP+K)=COS(ANGR)
        END IF
C
C   FINALLY THE EXTERNAL FORCES APPLIED TO THE NODES.
        DO 6 I=1,NNP
        A(2*I-1,NEQN+1)=-FX(I)
 6      A(2*I,NEQN+1)=-FY(I)
C
C   SOLVE THE LINEAR ALGEBRAIC EQUATIONS BY GAUSSIAN ELIMINATION.
        CALL  SOLVE(A,T,NEQN,2*NNPMAX,2*NNPMAX+1,IFLAG)
C
C   A UNIT VALUE OF THE ILL-CONDITIONING FLAG IMPLIES THAT THE STRUCTURE
C   IS NOT STATICALLY DETERMINATE.
        IF(IFLAG.EQ.1) THEN
            WRITE(6,73)
 73     FORMAT(/ 'STRUCTURE IS NOT STATICALLY DETERMINATE - STOP')
            STOP
        END IF
C
C   OUTPUT THE COMPUTED ELEMENT FORCES AND REACTIONS AT THE SUPPORTS.
        WRITE(6,74) (M,T(M),M=1,NEL)
 74     FORMAT(/ 'COMPUTED FORCES IN THE ELEMENTS (TENSILE POSITIVE)'
     1          / 3('    ELEMENT    FORCE  ') / (3(I8,E15.4)))
```

```
C
C  OUTPUT THE COMPUTED REACTION COMPONENTS AT THE PINNED SUPPORTS.
      IF(NNPP.GT.0) THEN
         WRITE(6,75) (NODE(K),T(NEL+2*K-1),T(NEL+2*K),K=1,NNPP)
   75    FORMAT(/ 'COMPUTED REACTION COMPONENTS AT THE PINNED SUPPORTS'
     1         /  '  NODE     R (X DIRN)     R (Y DIRN)' / (I5,2E15.4))
      END IF
C
C  OUTPUT THE COMPUTED REACTION FORCES AT THE ROLLER SUPPORTS.
      IF(NNPR.GT.0) THEN
         WRITE(6,76) (NODE(NNPP+K),T(NEL+2*NNPP+K),K=1,NNPR)
   76    FORMAT(/ 'COMPUTED REACTION FORCES AT THE ROLLER SUPPORTS'
     1         /  '  NODE      REACTION ' / (I5,E15.4))
      END IF
      STOP
      END
```

The following list provides the definitions of the FORTRAN variables used, arranged in alphabetical order.

Variable	Type	Definition
A	real	Coefficient matrix $[A]$ extended to include vector $[B]$ as its last column
ANG	real	Array storing angles (in degrees) at which roller supports are free to move
ANGR	real	Angle in radians of roller support
COSINE	real	Cosine of element angle θ_m at first node i
ELENGT	real	Element length, L_m (used in equations (2.12) below)
FX, FY	real	Arrays storing the external force components in the global coordinate directions at the nodes
I	integer	Nodal point number (sometimes first node of an element)
ICOL	integer	Column number in array A
IFLAG	integer	Flag for a singular or very ill-conditioned coefficient matrix (normally returned as zero by SOLVE, one if the ill-conditioning test is satisfied)
IROW	integer	Row number in array A
J	integer	Nodal point number (second node of an element)
K	integer	A counter (used in READ and WRITE statements)
M	integer	Element number
NEL	integer	Number of elements
NELMAX	integer	Maximum number of elements permitted by the array dimensions
NEQN	integer	Number of equations to be solved
NNP	integer	Number of nodal points
NNPF	integer	Number of nodes at which external forces are applied
NNPMAX	integer	Maximum number of nodal points permitted by the array dimensions
NNPP	integer	Number of nodes at which structure is pinned to its surroundings
NNPR	integer	Number of nodes at which structure is supported on rollers
NNPT	integer	Total number of nodes at which structure is supported
NODE	integer	Array storing node numbers
NPI	integer	Array storing the numbers of the first nodes of the elements
NPJ	integer	Array storing the numbers of the second nodes of the elements
NUNK	integer	Total number of unknowns
PI	real	The mathematical constant π
SINE	real	Sine of element angle θ_m at first node i
T	real	Array storing the vector of unknowns $[T]$
X	real	Array storing X global coordinates of the nodes
Y	real	Array storing Y global coordinates of the nodes

The arrays used in SDPINJ are dimensioned in such a way as to allow a structure with up to 50 nodes and 100 elements to be analyzed, implying that up to 100 equations (twice the number of nodes) may have to be solved. Note that array A, which stores the coefficient matrix $[A]$ extended to include vector $[B]$

is allowed to have up to 101 columns. The array dimensions can of course be changed, provided the values of NNPMAX and NELMAX defined in the next two statements are also adjusted.

The program reads information from a file named DATA, and writes onto a file named RESULTS. All READ statements call for free format input data. After writing out a heading, the program first reads the number of nodal points in the structure to be analyzed, and then tests that this does not exceed the maximum allowed by the array dimensions. We should note that the checking of input data in the program is comparatively rudimentary, and certainly not up to the standards which would be required if the program were to be used as a 'black box', without access to the source coding. Unfortunately, thorough checking of data demands a great deal of programming, to the extent that the amount of coding involved would be excessive for present purposes. Since the source coding is available, however, we can adopt a compromise. Some of the more obvious and potentially disastrous data errors are trapped, and in all cases input data are immediately written out to allow them to be checked.

After reading in the global coordinates of the nodes, X and Y for node 1, then node 2 and so on, the program reads in and tests the number of elements, followed by the numbers of the first and second nodes of each of the elements in turn. All coefficients of matrix $[A]$ and the external force components at the nodes are then set to zero in preparation for the assembly process to come. The number of nodes at which external forces are applied is read in, followed by the numbers of the nodes concerned together with the components in the X and Y directions of the corresponding forces. Next, the numbers of nodes at which the structure is supported on fixed pin joints and rollers are read in, followed by the numbers of the nodes forming pinned supports, then the numbers of the nodes resting on rollers together with the angles at which these rollers are free to move.

The total numbers of unknowns can now be calculated, one force in each element, two reaction components at each pinned support, and one at each roller support. Equations (2.7) and (2.8) are then used to determine whether the structure is statically indeterminate or a mechanism, respectively, in either case no further calculations being possible.

A convenient way to assemble the coefficients of the linear equations is by considering each of the elements in turn, rather than the nodes. Our typical element, shown in Fig. 2.23 and numbered m, has first and second nodes i and j, respectively, which in the program are recovered from arrays NPI and NPJ and stored in variables I and J. If L_m (ELENGT) is the length of the element, which is the distance between nodes i and j, then the cosine and sine of the element angle θ_m may be defined as

$$\cos \theta_m = \frac{X_j - X_i}{L_m}, \qquad \sin \theta_m = \frac{Y_j - Y_i}{L_m} \qquad (2.12)$$

If we arrange the equilibrium equations for the nodes in the order in which the nodes are numbered, and take the equation for the X direction before that for the Y direction in each case, then the coefficient in the mth column of $[A]$

(multiplying the unknown T_m) and the $(2i-1)$th row (the equation in the X direction for node i) is, according to equation (2.9), $\cos \theta_m$. Similarly, the coefficient in the same column but the next or $2i$th row (the equation in the Y direction for node i) is, according to equation (2.10), $\sin \theta_m$. Now, element m also contributes to the equilibrium equations at node j, where its angle relative to the positive X axis is not θ_m but $\theta_m + 180°$. Since $\cos(\theta_m + 180°) = -\cos \theta_m$, and $\sin(\theta_m + 180°) = -\sin \theta_m$, there are further contributions of $-\cos \theta_m$ and $-\sin \theta_m$ to the $(2j-1)$th and $2j$th rows (equations for node j) and mth column of $[A]$.

There are two unknown reaction force components (in the X and Y directions) at each of the pinned supports, which are arranged in vector $[T]$ immediately after the forces in the elements, in the order in which these supports are defined. A value of unity is therefore entered in each of the $(2i-1)$th and $2i$th rows of matrix $[A]$, in the appropriate columns. There is also one unknown reaction force at each roller support, in the direction normal to the direction of roller motion as shown in Fig. 2.24. These roller reaction forces are arranged in vector $[T]$ immediately after the reactions at the pinned supports, in the order in which the roller supports are defined. From Fig. 2.24, we can see that the components of the unknown reaction force R in the positive X and Y directions at the node concerned, i, are

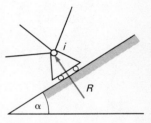

$$R_x^{(i)} = -R \sin \alpha, \qquad R_y^{(i)} = +R \cos \alpha \tag{2.13}$$

Fig. 2.24.

where α is the angle, measured counterclockwise from the positive X axis, at which the roller is free to move (which is stored by the program in array ANG). Values of $-\sin \alpha$ and $\cos \alpha$ are therefore entered in the $(2i-1)$th and $2i$th rows of matrix $[A]$, respectively, in the appropriate columns. Finally, the components of any externally applied forces are entered as required by equations (2.9) and (2.10) in the right-hand side vector $[B]$, which is now the last column of $[A]$.

The equations can now be solved, using subroutine SOLVE which is described in detail in Appendix B. If a unit value of IFLAG is returned to SDPINJ, this means that SOLVE has detected a singular or very ill-conditioned coefficient matrix, the most likely reason for which is that, although equation (2.6) is satisfied, the structure is not statically determinate, due to poor arrangement of the elements (such as in Example 2.3d). The program writes out a message to this effect and terminates execution. Otherwise, the computed forces in the elements are written out, followed by the reactions at the pinned and roller supports.

We note that, except for roller support angles which must be supplied in degrees, no units are specified for either the input data or the computed results. This is because we may use any convenient units: all force data must be supplied to the program in the same units, and the results will be presented in these units. We should also note that the dimensions of the structure (coordinates of the nodal points) are only used to define sines and cosines of element angles, and do not otherwise affect the results. Any convenient unit of length may be used.

We will now examine how program SDPINJ can be used to analyze pin-joined structures with the aid of two problems, the first of which has already been solved in Examples 2.2 and 2.4.

EXAMPLE 2.5

Consider again the structure shown in Fig. 2.12, which is redrawn in Fig. 2.25 with the nodes and elements shown numbered in convenient, but arbitrary orders. Element numbers are shown ringed, while node numbers are unringed. It is convenient to define the origin for global coordinates X and Y at node 1 as shown. Figure 2.26 shows the file of input data used to set up this problem for solution by program SDPINJ.

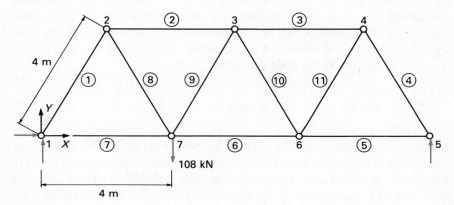

Fig. 2.25.

```
7                                                    (Number of nodes)
0. 0.,2. 3.4641,6. 3.4641,10. 3.4641,12. 0.,8. 0.,4. 0.   (Node coordinates)
11                                                   (Number of elements)
1 2,2 3,3 4,4 5,5 6,6 7,7 1,2 7,7 3,3 6,6 4     (Nodes of the elements)
1                                (Number of nodes subjected to external forces)
7 0. -108.                       (Number of loaded node, and force components)
1 1                              (Numbers of pinned and roller supports)
1                                (Node number of the pinned support)
5 0.                             (Node number and angle of the roller support)
```

Fig. 2.26.

The file has been annotated to indicate what each line of data represents. Note that, when entering data for the external force at node 7, the components are zero (in the X direction) and -108 (kN), since the applied force is directed downwards, in the negative Y direction. The file RESULTS produced by the program is shown in Fig. 2.27, the unit of force being kN as for the input data. As we would expect, the computed forces in the elements and reactions at the supports are identical to those obtained by manual methods in Example 2.2.

COORDINATES OF THE NODES

NODE	X	Y	NODE	X	Y
1	0.0000E+00	0.0000E+00	2	0.2000E+01	0.3464E+01
3	0.6000E+01	0.3464E+01	4	0.1000E+02	0.3464E+01
5	0.1200E+02	0.0000E+00	6	0.8000E+01	0.0000E+00
7	0.4000E+01	0.0000E+00			

NODES CONNECTED BY THE ELEMENTS

M	I	J	M	I	J	M	I	J	M	I	J
1	1	2	2	2	3	3	3	4	4	4	5
5	5	6	6	6	7	7	7	1	8	2	7
9	7	3	10	3	6	11	6	4			

FORCE COMPONENTS AT THE LOADED NODES

NODE	FX	FY
7	0.0000E+00	−0.1080E+03

NODES FORMING PINNED SUPPORTS :   1

NODES AND ANGLES OF ROLLER SUPPORTS

NODE	ANG (DEG)
5	.00

COMPUTED FORCES IN THE ELEMENTS (TENSILE POSITIVE)

ELEMENT	FORCE	ELEMENT	FORCE	ELEMENT	FORCE
1	−0.8314E+02	2	−0.8314E+02	3	−0.4157E+02
4	−0.4157E+02	5	0.2078E+02	6	0.6235E+02
7	0.4157E+02	8	0.8314E+02	9	0.4157E+02
10	−0.4157E+02	11	0.4157E+02			

COMPUTED REACTION COMPONENTS AT THE PINNED SUPPORTS

NODE	R (X DIRN)	R (Y DIRN)
1	0.0000E+00	0.7200E+02

COMPUTED REACTION FORCES AT THE ROLLER SUPPORTS

NODE	REACTION
5	0.3600E+02

Fig. 2.27.

EXAMPLE 2.6

Figure 2.28 shows in model form a plane pin-jointed structure which is part of the framework for a crane. The structure is subjected to a downward force of 165 kN (37 kip) at 20° to the vertical at the joint labeled as node 12, and is supported at node 1 on a fixed pin joint and node 2 on a roller free to move at 30° to the horizontal. Use program SDPINJ to find the forces in the members, and hence determine the member carrying the greatest force. Confirm this greatest force using the method of sections.

Firstly, we need to choose a global coordinate system to define the geometry of the structure, and the usual X horizontal and Y vertical, with an origin arbitrarily located at node 1, is convenient. Next, we can number nodes and elements in any order, such as the one shown. Figure 2.29 shows the file of input data used to define the problem. Note that the force applied at node 12 is entered in component form as $-165 \sin 20°$ kN and $-165 \cos 20°$ kN in the coordinate directions. Figure 2.30 shows the results obtained, from which we can see that the most heavily loaded member is element 12, with a compressive force of 465.2 kN (105 kip). Note that the forces in elements 5, 9 and 19 are all zero. This is because one end of each of these members is connected to a node at which the other elements are at right angles to it. For example, node 4: for this joint to be in equilibrium, the force in element 5 must be zero. This does not mean, however, that these members are redundant, simply that they are inactive under the particular loading applied to the structure. To remove any one of them would convert the framework into a mechanism.

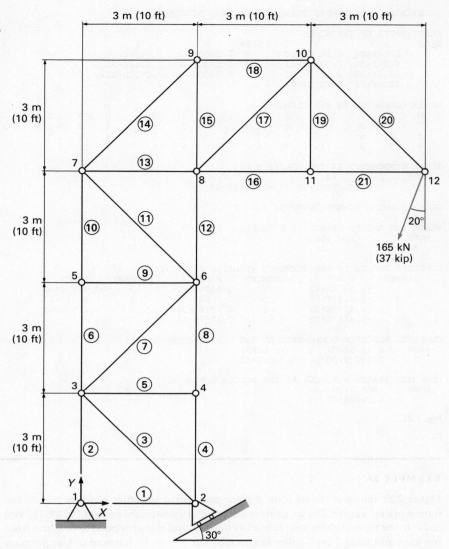

Fig. 2.28.

```
12
0. 0.,3. 0.,0. 3.,3. 3.,0. 6.,3. 6.,0. 9.,3. 9.,3. 12.,6. 12.,6. 9.,9. 9.
21
1 2,1 3,2 3,2 4,3 4,3 5,3 6,4 6,5 6,5 7,6 7,6 8,7 8,7 9,8 9,8 11,8 10,9 10,
10 11,10 12,11 12
1
12 -56.43 -155.05
1 1
1
2 30.
```

Fig. 2.29.

COORDINATES OF THE NODES

NODE	X	Y	NODE	X	Y
1	0.0000E+00	0.0000E+00	2	0.3000E+01	0.0000E+00
3	0.0000E+00	0.3000E+01	4	0.3000E+01	0.3000E+01
5	0.0000E+00	0.6000E+01	6	0.3000E+01	0.6000E+01
7	0.0000E+00	0.9000E+01	8	0.3000E+01	0.9000E+01
9	0.3000E+01	0.1200E+02	10	0.6000E+01	0.1200E+02
11	0.6000E+01	0.9000E+01	12	0.9000E+01	0.9000E+01

NODES CONNECTED BY THE ELEMENTS

M	I	J	M	I	J	M	I	J	M	I	J
1	1	2	2	1	3	3	2	3	4	2	4
5	3	4	6	3	5	7	3	6	8	4	6
9	5	6	10	5	7	11	6	7	12	6	8
13	7	8	14	7	9	15	8	9	16	8	11
17	8	10	18	9	10	19	10	11	20	10	12
21	11	12									

FORCE COMPONENTS AT THE LOADED NODES

NODE	FX	FY
12	-0.5643E+02	-0.1551E+03

NODES FORMING PINNED SUPPORTS :   1

NODES AND ANGLES OF ROLLER SUPPORTS

NODE	ANG (DEG)
2	30.00

COMPUTED FORCES IN THE ELEMENTS (TENSILE POSITIVE)

ELEMENT	FORCE	ELEMENT	FORCE	ELEMENT	FORCE
1	-0.2272E+03	2	0.1408E+03	3	0.7980E+02
4	-0.3523E+03	5	0.0000E+00	6	0.2537E+03
7	-0.7980E+02	8	-0.3523E+03	9	0.0000E+00
10	0.2537E+03	11	0.7980E+02	12	-0.4652E+03
13	-0.3665E+03	14	0.4385E+03	15	-0.3101E+03
16	-0.2115E+03	17	-0.2193E+03	18	0.3101E+03
19	0.0000E+00	20	0.2193E+03	21	-0.2115E+03

COMPUTED REACTION COMPONENTS AT THE PINNED SUPPORTS

NODE	R (X DIRN)	R (Y DIRN)
1	0.2272E+03	-0.1408E+03

COMPUTED REACTION FORCES AT THE ROLLER SUPPORTS

NODE	REACTION
2	0.3416E+03

Fig. 2.30.

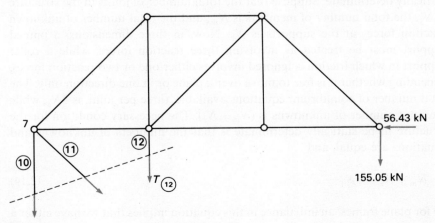

Fig. 2.31.

In order to check the force in element 12 using the method of sections, we can cut the structure through elements 10, 11 and 12, as shown in Fig. 2.31. Since the forces in elements 10 and 11 both pass through node 7, it is appropriate to consider moment equilibrium about this point of the part of the structure shown. The equation is

$$9 \times 165 \cos 20° + 3 \times T_{12} = 0$$

$$T_{12} = -3 \times 165 \cos 20° = -465.2 \text{ kN (compressive)}$$

which agrees with the computed result.

Having computed the forces in the members of a pin-jointed structure, as we have done in these two Examples, such information can be used for design purposes. For instance, the choice of cross-sectional areas of the members is strongly influenced by the magnitudes of the forces they must carry, as is the design of the joints. For members in compression, the possibility of buckling, a mode of failure which is discussed in Chapter 8, must also be considered.

2.1.4 Analysis of space frames

While we have so far only considered two-dimensional pin-jointed structures, the same fundamental methods are equally applicable to three-dimensional space frames. In other words, we can isolate each of the joints in turn, treating it as a free body with external forces applied, and establish the conditions for these forces to be in equilibrium. In three dimensions, however, we must consider equilibrium in three directions rather than two. The method of sections can also be applied if we are interested in finding only a few of the forces in the members, and computer methods are also applicable. Indeed, the complexities of three-dimensional framework geometry and the greater number of members involved in even relatively simple space frames makes the use of manual methods very tedious and prone to error.

In three dimensions, we must modify slightly our test for a structure to be statically determinate. Suppose that the total number of joints in the structure is N_j, the total number of members is N_m, and the total number of unknown reaction forces at the supports is N_r. Now, in three dimensions, a pinned support must be treated as involving three reaction forces, while a roller support in which friction is ignored involves either one or two reaction forces, depending whether it is free to move over a plane or in one direction only. The total number of equilibrium equations available, three per joint, is $3N_j$, while the total number of unknowns is $(N_m + N_r)$. The necessary condition for the structure to be statically determinate is that the numbers of unknowns and equations are equal, and

$$N_m + N_r = 3N_j \tag{2.14}$$

As for plane frames, an imbalance in this equation implies that we have either a statically indeterminate structure or a mechanism.

Thin-walled shells in the form of either cylinders of circular cross section or spheres are widely used in engineering for such purposes as pipes and vessels containing liquids or gases under pressure, examples being steam boilers, heat exchangers and chemical reactors. Provided such shells are thin, and are subjected to uniform pressure loading, then they may be treated as statically determinate systems and the stresses in them calculated using simple formulae. The restriction to uniform loading usually means that we must ignore the weights of both the shells themselves and any fluids they contain. Another practical example of a shell subject to uniform loading, though not due to internal pressure, is that of a thin ring or cylinder rotated at high speed about its axis. In this case, the loading is produced by centrifugal effects.

While thin shell problems are statically determinate, and we can calculate stresses directly, in many cases of practical interest we are also interested in the strains produced. These are considered in Chapter 3. Meanwhile, the range of useful shell problems we can solve is necessarily limited.

2.2.1 Pressurized thin-walled cylinder

Let us consider a long cylinder of circular cross-section, with an internal radius of R_i and a constant wall thickness of t, as shown in Fig. 2.32a, which is subject to a difference in hydrostatic pressure of p between its inner and outer surfaces. In many cases, p is the gage pressure within the cylinder, taking the external ambient pressure to be zero on the gage. By thin-walled we mean that the thickness t is very much smaller than the radius R_i, and we may quantify this by stating that the ratio t/R_i of thickness to radius should be less than about 0.1. An appropriate coordinate system to use to describe such a system is the cylindrical polar one, r, θ, z, shown, where z lies along the axis of the cylinder, r is radial to it and θ is the angular coordinate about the axis. The small piece of the wall of the cylinder which is shown shaded in Fig. 2.32a is shown in isolation in Fig. 2.32b. In extracting this piece, we have cut the wall along planes perpendicular and parallel to the axis of the cylinder, and the normal stress components acting on the cut surfaces are the longitudinal or *axial stress*, σ_z, in the axial direction, and the circumferential or *hoop stress*, σ_θ, in the circumferential direction as shown. The term hoop stress derives from the fact that the metal hoops which hold the wooden staves of a barrel experience this type of circumferential loading. Collectively, the hoop and axial stress components are sometimes referred to as the *membrane stresses*, because they act in the plane of the thin cylinder wall, which can be regarded as a membrane.

Before we try to derive expressions for the stresses in the wall of the cylinder, it is useful to reflect on the way in which such a component fails in service when subjected to an excessively high internal pressure. While it might fail by bursting along a path following the circumference of the cylinder, under normal circumstances it fails by bursting along a path parallel to the axis. This suggests that the hoop stress is significantly higher than the axial stress.

By showing a relatively simple biaxial system of uniform normal stress components acting on the small piece of cylinder wall in Fig. 2.32b, we are

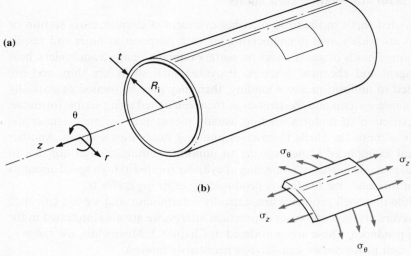

Fig. 2.32.

making two assumptions. Firstly, that there are no shear stresses acting in the wall, which we shall now establish as a valid simplification involving no approximation. Secondly, that the normal axial and hoop stresses do not vary through the wall thickness, which is a good approximation for a thin-walled cylinder. We will consider the more general case of a thick-walled cylinder in Chapter 10.

Let us consider the cross-sectional view of the cylinder, looking along its axis, shown in Fig. 2.33. The cylinder has been cut into two semicircular pieces by the horizontal plane passing through its axis. It is convenient for us to refer to the upper and lower halves as (a) and (b), respectively. Now, where we have cut the cylinder wall, there are stress components which in general give rise to resultant forces normal and parallel to the cut surfaces, also a moment. Since the cut surfaces are horizontal, it is convenient to denote the vertical forces normal to them as V', which we show as tensile on both parts (a) and (b). Now, we have not yet specified the axial length over which this force acts, and for our immediate purposes this length is not important. For completeness, however, let us assume that it acts over unit axial length. Similarly, let us denote the horizontal shear forces per unit axial length on the cut surfaces by H', and assume that they act outwards on the upper half (a). Now, they must act in opposite directions on the two sides of this body for it to be in equilibrium in the horizontal direction. Also, shear forces of the same magnitude must act inwards on the lower half (b), in order to ensure that Newton's third law is satisfied (see Section 1.2.2). Finally, let us denote the moments per unit axial length acting on the cut surfaces by M', and assume their directions are such that they tend to open both the upper and lower halves: Newton's third law is again satisfied. Since the upper and lower parts (a) and (b) are identical halves of the same completely symmetrical cylinder, we should be able to take (b) and rotate it through 180° either about the axis of the cylinder or about the horizontal diameter and obtain (a) with exactly the same loads acting on it. When we carry

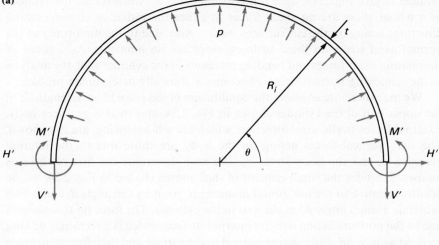

(a)

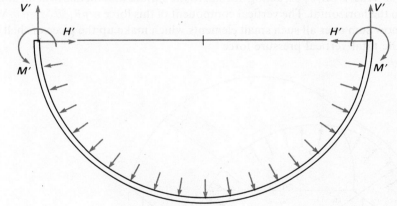

(b)

Fig. 2.33.

out this rotation, although the vertical forces V' and the moments M' are identical to those acting on (a), this is not true of the shear forces H', which are in opposite directions. We must therefore conclude that $H' = 0$. This provides an example of a general principle: *there are no shear stresses (and hence forces) on a plane of symmetry through a body.* The symmetry required for this to be true applies not only to the geometry of the body but also to the loads acting on it, and is sometimes referred to as *mirror symmetry.* We will come across further examples elsewhere.

Note that, in arriving at the conclusion that $H' = 0$, we have made no reference to the relative thickness of the cylinder, and it is equally applicable to thick-walled cylinders. The assumption which is only applicable to thin-walled cylinders, however, is that the hoop and axial stresses are constant through the wall thickness. The effect of this in terms of the loads shown in Fig. 2.33 is that the force V', which is the resultant of the normal hoop stress acting over the cut surface of one wall of the cylinder, acts at the mid thickness of the wall. As we shall see in Section 5.4, this also means that there is no moment, and $M' = 0$. Considering now the cross-sectional plane normal to the axis cutting the cylinder to give Fig. 2.33, since this is also a plane of symmetry for the cylinder as a whole, there are no shear forces in either the radial or circumferential directions acting on the cut surfaces shown. Also, due to the uniformity of the normal axial stress on these surfaces, there are no moments. As a result of eliminating shear forces and bending moments in the cylinder wall, the analysis of the remaining normal stresses becomes a statically determinate problem.

We may therefore consider the equilibrium of the piece of axial length δz of the upper half of the cylinder shown in Fig. 2.34, and treat it as a free body. External forces in the axial direction, which are self-balancing, are not shown. The only vertical forces acting on the body are those due to the internal pressure, p, and the hoop stress σ_θ. We may determine the first of these as follows. Consider the small element of shell shown shaded in Fig. 2.34, whose position relative to the horizontal diameter is given by the angle θ, and which subtends a small angle $\delta\theta$ at the axis of the cylinder. The force on this element due to the pressure acting over its internal surface, which is a rectangle δz long by $R_i\,\delta\theta$ wide, is $pR_i\,\delta\theta\,\delta z$, acting normal to the surface and therefore at an angle of θ to the horizontal. The vertical component of this force is $pR_i\,\delta\theta\,\delta z \sin\theta$. We may now sum over all such small elements which make up the piece of shell to find the total vertical pressure force

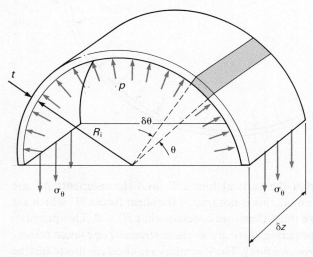

Fig. 2.34.

$$P = \int_0^\pi pR_i\,\delta z \sin\theta\,\mathrm{d}\theta = pR_i\,\delta z \int_0^\pi \sin\theta\,\mathrm{d}\theta = 2pR_i\,\delta z \qquad (2.15)$$

This result could have been obtained in a much more straightforward way, without the need to integrate over the internal surface of the cylinder. Suppose that, instead of taking our free body as just the piece of cylinder wall as in Fig. 2.34, we also include the fluid it contains (which we have previously assumed to have negligible weight) to create the free body shown in Fig. 2.35. The internal pressure p now acts over the area of the horizontal plane of symmetry through the axis of the cylinder which is contained within the

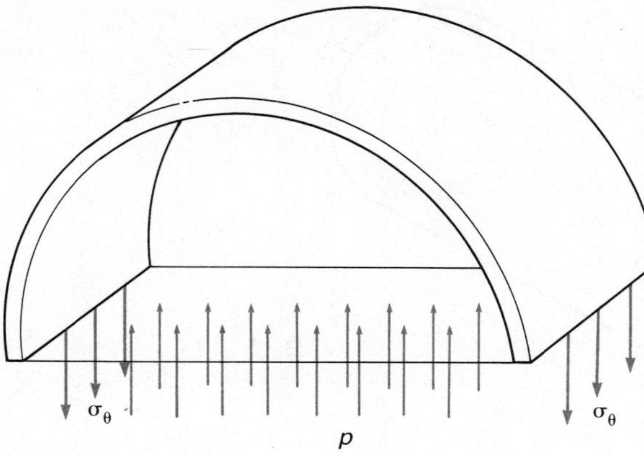

Fig. 2.35.

cylinder. This rectangular area has dimensions $2R_i$ (the internal diameter of the cylinder) by δz, giving a total vertical pressure force of $P = 2pR_i\delta z$ as in equation (2.15).

The other vertical forces acting on the free body shown in either Fig. 2.34 or Fig. 2.35 are those associated with the uniform hoop stress, σ_θ, which acts over the two cut surfaces of the cylinder wall, each of which has an area of $t\delta z$. The total force is therefore $2\sigma_\theta t\delta z$, which must balance the total pressure force, P, for the free body to be in equilibrium. Therefore

$$2\sigma_\theta t\delta z = 2pR_i\delta z$$

$$\sigma_\theta = \frac{pR_i}{t} \tag{2.16}$$

The form of result for axial stress σ_z depends on the type of end constraints applied to the cylinder. The most common case, generally referred to as *closed-ended*, is where the ends are closed by some form of end caps, which can be flat, hemispherical or some other curved shape. The important feature is that the end caps are attached to the cylinder, so that the axial force exerted by the internal pressure on these caps is transferred to the cylinder walls. We must bear in mind, however, that the present formulae for hoop and axial stresses are not applicable in the regions of the cylinder close to the end caps, where more complicated states of stress exist. Figure 2.36 shows part of a closed cylinder cut by a plane at right angles to its axis. We may treat this as a free body, subject to axial forces which must be in equilibrium. The total axial pressure force is that due to internal pressure acting over the internal area of the cylinder in the cross-sectional plane, or $p\pi R_i^2$. The axial force associated with the axial stress acting over the cross-sectional area of the wall is $\sigma_z\pi(R_o^2 - R_i^2)$, where R_o is the outer radius of the cylinder, and $R_o = R_i + t$. Therefore

$$p\pi R_i^2 = \sigma_z\pi(R_o^2 - R_i^2) = \sigma_z\pi(R_o - R_i)(R_o + R_i) = 2\sigma_z\pi t R_m$$

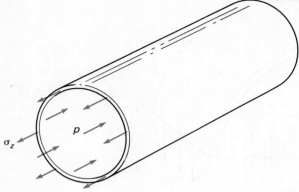

Fig. 2.36.

where $R_m = (R_o + R_i)/2$ is the mean radius of the cylinder, and

$$\sigma_z = \frac{pR_i^2}{2tR_m} \tag{2.17}$$

Since the mean and internal radii are almost identical, it is common practice to approximate this result by

$$\sigma_z = \frac{pR_i}{2t} \tag{2.18}$$

Indeed, such an approximation is essential if we are given merely the radius (or diameter) of the cylinder, without being told whether it is the inner, mean or outer dimension. Since $R_m > R_i$, equation (2.18) is a *conservative* approximation in that it slightly overestimates the axial stress. Comparing equations (2.16) and (2.18), it is clear that $\sigma_\theta = 2\sigma_z$, and our initial expectation that the hoop stress is greater than the axial stress is confirmed. We should also note that both the hoop and axial stresses are larger than the internal pressure by a factor of the order of R/t, which we know to be large for a thin shell. This is important because, in addition to the hoop and radial stresses, we also have stresses in the radial direction, normal to the surface of the cylinder. Figure 2.37 shows a cross section through the wall in the plane normal to the axis of the cylinder, with a small shaded element of material subject to hoop and radial stresses, σ_θ and σ_r. Now, the radial stress varies through the wall thickness, from $-p$ at the inner surface to zero at the outer surface. In other words, the magnitude of the radial stress is of the order of p, and therefore very small compared with the hoop and axial stresses. We will make use of this to ignore radial stresses when we come to consider strains and displacements in thin-walled cylinders in Chapter 4.

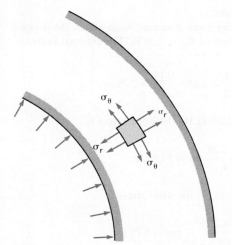

Fig. 2.37.

Another type of end constraint on the cylinder, generally referred to as *open-ended*, is that obtained when none of the pressure force on the end closures is transferred to the cylinder walls. An example of this is provided by the system shown in Fig. 2.38, where the cylinder is closed by pistons. Ignoring frictional forces between the pistons and cylinder, the axial forces on the pistons are balanced by external forces in the piston rods, and in the cylinder walls

$$\sigma_z = 0 \qquad\qquad\qquad\qquad\qquad\qquad (2.19)$$

Fig. 2.38.

In general, the axial stress in a cylinder must be based on whatever end forces are applied to it.

EXAMPLE 2.7

A closed cylinder having a wall thickness of 3.5 mm (0.14 in) and a mean diameter of 120 mm (4.7 in) is subjected to an internal pressure of 5 MN/m² (730 psi). Calculate the stresses in the cylinder wall.

SI units

We are given the mean diameter, from which the mean radius is obtained as $R_m = 60$ mm. The internal radius is therefore

$$R_i = R_m - \frac{t}{2} = 60 - \frac{3.5}{2} = 58.25 \text{ mm}$$

and from equation (2.16) the hoop stress is

$$\sigma_\theta = \frac{5 \times 58.25}{3.5} = \underline{83.2 \text{ MN/m}^2}$$

From equation (2.17) the axial stress is

$$\sigma_z = \frac{5 \times 58.25^2}{2 \times 3.5 \times 60} = \underline{40.4 \text{ MN/m}^2}$$

while from equation (2.18) it is

$$\sigma_z = \frac{5 \times 58.25}{2 \times 3.5} = \underline{41.6 \text{ MN/m}^2}$$

·US customary units

We are given the mean diameter, from which the mean radius is obtained as $R_m = 2.35$ in. The internal radius is therefore

$$R_i = R_m - \frac{t}{2} = 2.35 - \frac{0.14}{2} = 2.28 \text{ in}$$

and from equation (2.16) the hoop stress is

$$\sigma_\theta = \frac{730 \times 2.28}{0.14} \text{ psi} = \underline{11.9 \text{ ksi}}$$

From equation (2.17) the axial stress is

$$\sigma_z = \frac{730 \times 2.28^2}{2 \times 0.14 \times 2.35} \text{ psi} = \underline{5.77 \text{ ksi}}$$

while from equation (2.18) it is

$$\sigma_z = \frac{730 \times 2.28}{2 \times 0.14} \text{ psi} = \underline{5.94 \text{ ksi}}$$

a somewhat higher figure. A typical radial stress is $p/2$, or 2.5 MN/m² (365 psi). Note that it does not matter what length units we use for the cylinder radii and thickness (m or mm, for example), provided they are consistent, since these dimensions appear in the formulae only as ratios.

EXAMPLE 2.8

A cylindrical pressure vessel of length 3.2 m (10 ft 6 in), diameter 0.84 m (33 in) and wall thickness 4.5 mm (0.18 in) contains compressed air at a gage pressure of 1.2 MN/m² (170 psi). The vessel is closed by flat plates welded to the ends of the cylinder, and a compressive force of 300 kN (67 kip) is applied to these plates. Calculate the hoop and axial stresses in the cylinder wall remote from the end plates.

Figure 2.39a shows the vessel and applied loading, while Fig. 2.39b shows the free body formed by cutting through the vessel close to its left-hand end. The radius of the cylinder, R, is 0.42 m (we are not told whether this is the internal, mean or outer

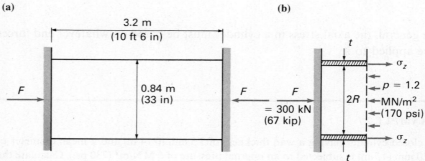

Fig. 2.39.

76

dimension), and the internal pressure is 1.2 MN/m². Firstly, we may calculate the hoop stress in the usual way, using equation (2.16)

$$\sigma_\theta = \frac{1.2 \times 0.42}{0.0045} = \underline{112 \text{ MN/m}^2} \text{ (16 ksi)}$$

In order to find the axial stress, we must consider the equilibrium of axial forces acting on the free body shown in Fig. 2.39b

$$F + 2\pi Rt\sigma_z - p\pi R^2 = 0$$

where F is the applied axial compressive force of 300 kN (0.3 MN). Hence

$$\sigma_z = \frac{pR}{2t} - \frac{F}{2\pi Rt}$$

$$\sigma_z = \frac{1.2 \times 0.42}{2 \times 0.0045} - \frac{0.3}{2\pi \times 0.42 \times 0.0045} = \underline{30.7 \text{ MN/m}^2} \text{ (4.5 psi)}$$

In this problem we have to be more careful in our choice of units, since the second term on the right-hand side of the equation for axial stress involves a force divided by a product of lengths. We have chosen to work in meganewtons for forces and meters for lengths, to give stresses in MN/m². This is a good example of a type of end constraint which gives an axial stress in the range between the closed and open-ended conditions. Had the force F been tensile, the axial stress would have been outside this range. Note that the length of the cylinder is not used in the calculations, although it does serve to indicate that it is a relatively long cylinder, in which the calculated values of hoop and axial stress are reasonable approximations for much of the cylinder wall.

2.2.2 Pressurized thin-walled sphere

Let us consider now a thin-walled spherical vessel with an internal radius of R_i and a constant wall thickness of t, as shown in Fig. 2.40, which is subject to a difference in hydrostatic pressure of p between its inner and outer surfaces. By thin-walled we again mean that $t \ll R_i$, more specifically that $t/R_i < 0.1$, and we

(a)   (b)

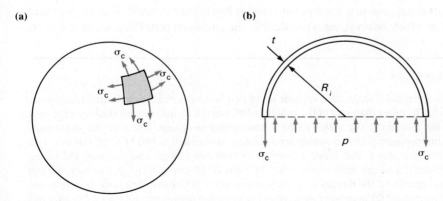

Fig. 2.40.

again assume that the membrane stresses are uniform through the wall thickness. Now, in the case of a sphere, we have symmetry of geometry and loading about each and every plane through its center. Therefore, it is not possible to separate the membrane stresses into hoop and axial as for a cylinder. Indeed, the membrane stress which we may now call the circumferential stress, σ_c, is the same in all directions tangential to the shell surface at any position over this surface: Fig. 2.40a shows this stress acting on all sides of a typical small element of the shell wall. We may determine σ_c by considering the equilibrium of the free body shown in Fig. 2.40b, formed by cutting the sphere along any plane through its center. Symmetry and the uniformity of the circumferential stress through the shell wall ensure that there are no shear forces or bending moments acting on the cut edges of the wall, making the problem statically determinate. For equilibrium of external forces acting on the free body

$$p\pi R_i^2 = 2\sigma_c \pi t R_m$$

where $R_m = R_i + t/2$ is the mean radius of the sphere, and

$$\sigma_c = \frac{pR_i^2}{2tR_m} \tag{2.20}$$

Equating the mean and internal radii, we obtain a conservative approximation to this result as

$$\sigma_c = \frac{pR_i}{2t} \tag{2.21}$$

Note that equations (2.20) and (2.21) are identical to (2.17) and (2.18) for the axial stress in a closed cylinder. This means that, for a given radius, thickness, and internal pressure, a spherical vessel is less highly stressed than a cylindrical one, in which hoop stresses are double the circumferential stresses in a sphere.

EXAMPLE 2.9

A thin spherical storage tank is made up of two hemispherical shells bolted together at flanges. The mean diameter of the sphere is 400 mm (16 in) and its wall thickness is 6 mm (0.24 in). Assuming the vessel to be a homogeneous sphere, calculate the maximum working pressure if the allowable tensile stress in the shell is 180 MN/m^2 (26 ksi). The flanges have outer and inner diameters of 490 mm (19 in) and 394 mm (15.76 in), respectively, and are held together by 24 bolts of 18 mm (11/16 in) diameter, with a gasket separating the flanges over their whole area. When the tank is at its maximum working pressure, the compressive stress in the gasket (assumed uniform) is 2.5 MN/m^2 (360 psi). Find the tensile stress in the bolts under these conditions.

SI units

To find the maximum working pressure we apply equation (2.20), using a mean radius of 200 mm, inner radius 197 mm, shell thickness 6 mm and a circumferential stress equal to the maximum allowable of 180 MN/m². Hence

$$p = \frac{2tR_m\sigma_c}{R_i^2} = \frac{2 \times 6 \times 200 \times 180}{197^2} = \underline{11.13 \text{ MN/m}^2}$$

Now, to find the stress in the bolts, we need to revert to the method we used to determine the circumferential stress in a homogeneous sphere. Figure 2.41a

US customary units

To find the maximum working pressure we apply equation (2.20), using a mean radius of 8 in, inner radius 7.88 in, shell thickness 0.24 in and a circumferential stress equal to the maximum allowable of 26 ksi. Hence

$$p = \frac{2tR_m\sigma_c}{R_i^2} = \frac{2 \times 0.24 \times 8 \times 26}{7.88^2} = \underline{1.608 \text{ ksi}}$$

Now, to find the stress in the bolts, we need to revert to the method we used to determine the circumferential stress in a homogeneous sphere. Figure 2.41a

(a)

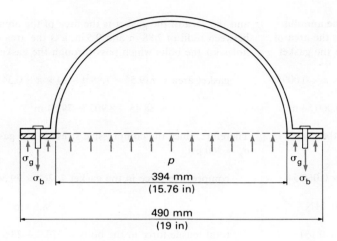

(b)

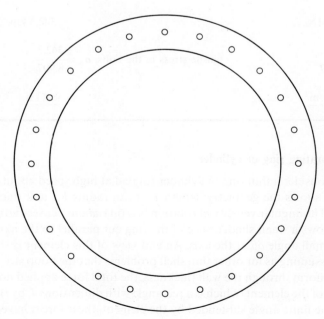

Fig. 2.41.

shows the free body obtained when we cut the system through the mid-surface of the gasket, which is a plane of symmetry, while Fig. 2.41b shows the plan view of the gasket. For the half of the body to be in equilibrium, the force due to the internal pressure acting over the internal area of the sphere projected onto the plane of symmetry, together with the compressive force in the gasket, must be balanced by the total tensile force in the bolts. Now

$$\text{pressure force} = p\pi R_i^2 = 11.13 \times \pi \times 0.197^2$$

$$= 1.357 \text{ MN}$$

and the area of the gasket is the area of the annulus between radii of 197 mm and 245 mm, less the area of the holes for the bolts which pass through the gasket

$$\text{gasket area} = \pi(0.245^2 - 0.197^2) - 24 \times \pi \times 0.009^2$$

$$= 0.066\,65 - 0.006\,107 = 0.060\,54 \text{ m}^2$$

Therefore, since the compressive stress in the gasket is $\sigma_g = 2.5 \text{ MN/m}^2$,

$$\text{compressive force in the gasket} = 2.5 \times 0.060\,54$$

$$= 0.151 \text{ MN}$$

$$\text{total tensile force in the bolts} = 1.357 + 0.151$$

$$= 1.508 \text{ MN}$$

$$\text{tensile stress in the bolts, } \sigma_b = \frac{1.508}{0.006\,107}$$

$$= \underline{247 \text{ MN/m}^2}$$

shows the free body obtained when we cut the system through the mid-surface of the gasket, which is a plane of symmetry, while Fig. 2.41b shows the plan view of the gasket. For the half of the body to be in equilibrium, the force due to the internal pressure acting over the internal area of the sphere projected onto the plane of symmetry, together with the compressive force in the gasket, must be balanced by the total tensile force in the bolts. Now

$$\text{pressure force} = p\pi R_i^2 = 1.608 \times \pi \times 7.88^2$$

$$= 313.7 \text{ kip}$$

and the area of the gasket is the area of the annulus between radii of 7.88 in and 9.5 in, less the area of the holes for the bolts which pass through the gasket

$$\text{gasket area} = \pi(9.5^2 - 7.88^2) - 24 \times \pi \times 0.3437^2$$

$$= 88.45 - 8.907 = 79.54 \text{ in}^2$$

Therefore, since the compressive stress in the gasket is $\sigma_g = 360 \text{ psi}$,

$$\text{compressive force in the gasket} = 360 \times 79.54 \text{ lb}$$

$$= 28.6 \text{ kip}$$

$$\text{total tensile force in the bolts} = 313.7 + 28.6$$

$$= 342.3 \text{ kip}$$

$$\text{tensile stress in the bolts, } \sigma_b = \frac{342.3}{8.907}$$

$$= \underline{38.4 \text{ ksi}}$$

2.2.3 Thin rotating ring or cylinder

Let us now consider a thin ring or cylinder rotated at high speed about its axis. Figure 2.42a shows the geometry: length L, mean radius R_m and thickness t, with $t \ll R_m$. The angular velocity of rotation is ω (in *radians* per second). Figure 2.42a also shows a small shaded slice of the ring cut parallel to the axis which subtends a small angle $\delta\theta$ at the axis. An end view of this element is shown in Fig. 2.42b. Assuming as for other thin shell problems that the hoop stress, σ_θ, in the ring is uniform through the wall thickness, the total force applied normal to each cut face of the element (which is a rectangle with dimensions L by t) is $\sigma_\theta L t$. Because of the finite angle subtended by the element, these forces have inward radial components of $\sigma_\theta L t \sin(\delta\theta/2)$. Now, the center of gravity of the element is (a)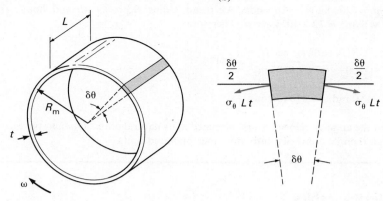

(b)

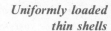

Fig. 2.42.

at a distance of R_m from the axis of rotation, at which position the centripetal acceleration is $\omega^2 R_m$. Therefore, since the volume of the element is $Lt\,R_m\,\delta\theta$, its mass is $\rho Lt\,R_m\,\delta\theta$, where ρ is the density of the material of the ring. Applying Newton's second law of motion to the element in the radial direction

$$2\sigma_\theta Lt \sin\left(\frac{\delta\theta}{2}\right) = \rho Lt\,R_m\,\delta\theta\,\omega^2 R_m$$

and, making the approximation $\sin(\delta\theta/2) = \delta\theta/2$, which is valid for a small angle, we obtain

$$\sigma_\theta = \rho R_m^2 \omega^2 \tag{2.22}$$

Note that this result is independent of both the length and thickness of the ring, and is equally applicable to short thin rings and long thin cylinders. At the same time, the ring is effectively open-ended in that there is no axial force applied to it, and the axial stress is zero.

It is worth noting that equation (2.22) for hoop stress can also be written as $\sigma_\theta = \rho V^2$, where $V = R_m\omega$ is the mean peripheral velocity of the ring. Therefore, for a given material, with a particular density and maximum allowable tensile stress, there is a maximum peripheral velocity which is independent of the size of the ring. This is why, for example, conventional rubber tyres cannot be used for attempts on the world land speed record.

EXAMPLE 2.10

The cast iron flywheel on the drive for a rolling mill has a rim of mean diameter 1.9 m (75 in). If the maximum allowable tensile stress in the rim is 12 MN/m² (1700 psi), find the maximum rotational speed at which the flywheel can run. Neglect the effects of the flywheel spokes and hub, and treat the rim as a thin ring.

In this straightforward example, we can apply equation (2.22) directly, using density $\rho = 7200\,\text{kg/m}^3$ (Appendix A), mean radius $R_m = 0.95\,\text{m}$ and hoop stress $\sigma_\theta = 12\,\text{MN/m}^2 = 12 \times 10^6\,\text{kg/ms}^2$. Therefore

$$\omega^2 = \frac{(12 \times 10^6\,\text{kg/ms}^2)}{(7200\,\text{kg/m}^3) \times (0.95\,\text{m})^2} = 1847\,\text{s}^{-2}$$

$$\omega = 43.0\,\text{rad/s} = \underline{410\,\text{rev/min}}$$

Note that the angular velocity is first obtained in its natural units of *radians* per second, which can then be converted into any other preferred units.

2.3 Flexible cables

There are many engineering examples of the use of flexible cables subject to lateral loads. Perhaps the most spectacular of these are suspension bridges and cable cars used for transport in mountainous regions, but overhead electrical power transmission lines are more common. The cables used for such applications are rarely solid in cross section, but are of stranded construction which increases their flexibility. In any case, the ratios of length to thickness of cables are usually so great that the bending moments that can be sustained by even a solid cable are very small compared with the other loadings applied to it. Consequently, it is reasonable for us to assume perfect flexibility, with no bending moments at any position along a cable. This also means that the tension in the cable always acts along the cable, in other words in the direction of the local tangent to the curve followed by the cable. The curved shape of a suspended cable is due to the weight of the cable itself, which is one of the main sources of loading. We will also assume that the cable suffers only small strains, meaning that any extension of the cable due to the loads applied produces negligible changes in the geometry of the system. With these simplifications, cable problems become statically determinate.

A further simplification which is often a reasonable one to apply is that the cable displays only a small amount of *dip*. In other words, over the length of an essentially horizontal cable, the variation in its height is small compared with the *span* or horizontal distance between the support points. By small we mean that the ratio of maximum dip (below the supports) to span is less than about 0.1. Consequently, it is possible to assume that the weight of the cable is uniformly distributed in the horizontal direction, rather than along its true shape. This is just another way of saying that the actual length of the cable is approximately equal to its horizontal length for the purposes of calculating its weight. As we shall see, this greatly simplifies the analysis of cable problems.

2.3.1 Simplified analysis for cables with small dip

We can develop the principles of analysis for a cable with small dip in terms of the typical cable system shown in Fig. 2.43. The cable, which has a uniformly distributed weight of w per unit length, is suspended between two fixed points, A and B, which are a distance L apart horizontally and a distance h apart vertically, and $h \ll L$ (the dip is deliberately exaggerated in the illustration). While this system is typical of real cables to the extent that the support points do not have to be at the same height, it is simplified in that there are no concentrated loads applied to the cable. We will consider the addition of such loads later.

Let us assume that horizontal reaction components H_A and H_B and vertical reaction components V_A and V_B act on the cable in the directions shown at the support points A and B. The flexibility of the cable means that we do not have to include moments at these points. For equilibrium of forces acting on the cable in the horizontal direction

$$H_A = H_B \tag{2.23}$$

while in the vertical direction

$$V_A + V_B - wL = 0 \tag{2.24}$$

since the total weight of the cable is assumed to be the weight per unit length multiplied by the *horizontal* length. We may also consider moment equilibrium about one convenient point, such as B

$$H_A h + V_A L - \frac{wL^2}{2} = 0 \tag{2.25}$$

The resultant force due to the distributed weight of the cable, wL, acts vertically downwards at the mid-span position, a horizontal distance of $L/2$ from B. With four unknown support reaction components and only three equilibrium equations for the cable, the problem is not yet soluble, and further information is required. In physical terms, we are free to vary the horizontal component of tension in the cable, decreasing the maximum dip as the tension is increased, increasing it as the tension is reduced. The additional information we require may be given in a number of different ways. For example, the coordinates of one point on the cable might be specified, or the maximum dip, or either the horizontal component of tension or the maximum tension in the cable. Let us assume that we know the maximum dip to be d below support point A. The lowest point of the cable is marked as C in Fig. 2.43, and occurs at an

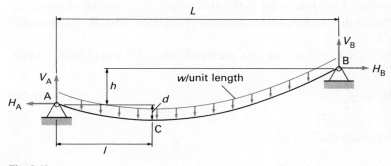

Fig. 2.43.

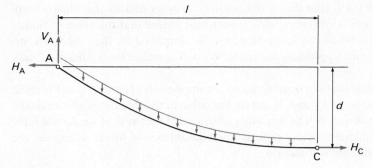

Fig. 2.44.

(unknown) horizontal distance l from A. In order to proceed, we now cut the cable through the point C and consider one of the two sections so formed, such as AC shown in Fig. 2.44, as a free body. At C, the tension in the cable must be horizontal because the tangent to the cable there is horizontal, and we need only show a horizontal force component, H_C. For equilibrium of forces acting on section AC in the horizontal direction

$$H_A = H_C \tag{2.26}$$

With only vertical loading applied to the cable, the horizontal component of tension must remain constant along its length. For equilibrium in the vertical direction

$$V_A - wl = 0 \tag{2.27}$$

again considering the weight of the cable to be horizontally distributed. Moment equilibrium can be considered about any point, although points A or C, with external forces acting through each, are the most convenient. In this case, point A is the better choice, because two forces pass through it, as compared with one through C. The equation is

$$-H_C d + \frac{wl^2}{2} = 0 \tag{2.28}$$

since the resultant force due to the distributed weight of section AC of the cable, wl, acts vertically downwards at the middle of this section, a horizontal distance of $l/2$ from A.

To solve these equations, we first use equations (2.26) and (2.28) to obtain

$$H_A = H_C = \frac{wl^2}{2d} \tag{2.29}$$

and from equation (2.27)

$$V_A = wl \tag{2.30}$$

$$\frac{wl^2h}{2d} + wlL - \frac{wL^2}{2} = 0 \qquad (2.31)$$

which may be simplified to

$$hl^2 + 2dLl - L^2d = 0 \qquad (2.32)$$

giving a quadratic equation for the unknown length l. Provided h is not zero, the roots of this equation are

$$l = \frac{L}{h}[-d \pm \sqrt{(d^2 + hd)}] \qquad (2.33)$$

On the other hand, if h is zero, which means that the support points are at the same height, then equation (2.32) has only one root

$$l = \frac{L}{2} \qquad (2.34)$$

and the maximum dip occurs at the center of the cable, as we would expect.

Of the two roots given by equation (2.33), one is always positive, and the other negative. While the positive one, which cannot exceed $L/2$ in magnitude, corresponds to the situation shown in Fig. 2.43, the negative one still has physical significance, and is associated with the situation shown in Fig. 2.45, where the cable tension is so high that there is no minimum point within the length of the cable. The shape of the cable is such that, if the curve were extended to the left of support A, a lowest point would eventually be reached. Therefore, to avoid ambiguity, we would need to be told whether the point of maximum dip occurs between points A and B. Having found l, we can now go back to equations (2.29) and (2.30) to find the horizontal component of tension in the cable, and the vertical reaction at A, and hence the vertical reaction at B from equation (2.24). Since $l < L/2$, it follows that $V_A < wL/2$, $V_B > wL/2$, hence $V_B > V_A$, and the maximum tension in the cable occurs at point B, and is given by

$$T_B = \sqrt{(H_B^2 + V_B^2)} \qquad (2.35)$$

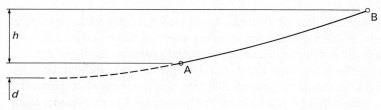

Fig. 2.45.

The assertion that this is the maximum tension perhaps requires some justification. The minimum tension in the cable occurs at its lowest point, and is equal to the horizontal component of tension which exists throughout the cable. As we move away from the lowest point, the vertical component of tension increases progressively due to the weight of the cable. Indeed, the rate of increase of this component with horizontal distance from the lowest point is equal to the weight per unit length of the cable. Consequently, the maximum tension must occur at one of the supports, the one which is further from the lowest point, and therefore the higher of the two. If concentrated vertical loads are added to the cable, while it remains true that the maximum tension occurs at one of the supports, it is not necessarily true that this support is the higher of the two.

Having referred to the shape of the cable on several occasions, it would be useful to establish what form this takes. We can do this by cutting the cable both at the lowest point C and at some arbitrary point X at a horizontal distance x from C where the height of the cable is y above C, as shown in Fig. 2.46. For equilibrium of moments acting on section CX about point X

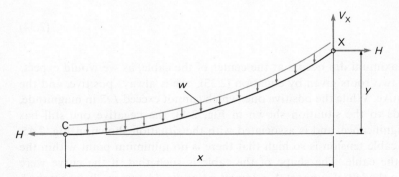

Fig. 2.46.

$$Hy - \frac{wx^2}{2} = 0 \qquad (2.36)$$

where H is the horizontal component of tension throughout the cable, and therefore

$$y = \frac{wx^2}{2H} \qquad (2.37)$$

which is a quadratic equation for y in terms of x. The shape of the cable is a parabola.

Although we have derived algebraic formulae for a cable with supports at different levels, which are useful for studying some common features of cable behavior, these are not applicable to more general problems involving concentrated loads on the cable. This, combined with the fact that we may be given data defining a problem in several different ways, means that particular problems are generally best solved from first principles. Let us consider some practical examples.

EXAMPLE 2.11   **87**

A cable weighing 8 kN/m (550 lb/ft) has a span of 60 m (200 ft) with one support 2 m (6.6 ft) higher than the other, and is subjected to a horizontal force of 1000 kN (220 kip) at each end. Find the position and magnitude of the greatest dip, and the maximum tension in the cable.

This is a relatively simple example, with no concentrated loads on the cable, and with the horizontal component of tension given. The system is illustrated in Fig. 2.47.

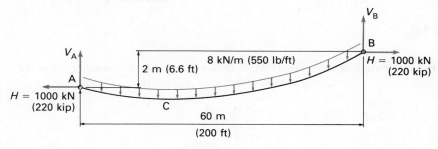

Fig. 2.47.

SI units

For moment equilibrium of the cable about support point B

$$H \times 2 + V_A \times 60 - \frac{8 \times 60^2}{2} = 0$$

from which

$$V_A = \frac{4 \times 60^2 - 1000 \times 2}{60} = 206.7 \text{ kN}$$

and for equilibrium of forces in the vertical direction

$$V_A + V_B - 8 \times 60 = 0$$

$$V_B = 8 \times 60 - 206.7 = 273.3 \text{ kN}$$

Note that the vertical component of tension is greater at the higher of the two support points, also that with H given the reactions at the supports can be found by considering only the equilibrium of the cable as a whole. The maximum tension in the cable, which occurs at B, is

$$T_B = \sqrt{(1000^2 + 273.3^2)} = \underline{1040 \text{ kN}}$$

In order to locate the lowest point, C, of the cable in terms of horizontal distance *l* from, and dip *d* below, support A, we need to cut the cable at point C: free

US customary units

For moment equilibrium of the cable about support point B

$$H \times 6.6 + V_A \times 200 - \frac{0.55 \times 200^2}{2} = 0$$

from which

$$V_A = \frac{0.275 \times 200^2 - 220 \times 6.6}{200} = 47.74 \text{ kip}$$

and for equilibrium of forces in the vertical direction

$$V_A + V_B - 0.55 \times 200 = 0$$

$$V_B = 0.55 \times 200 - 47.74 = 62.26 \text{ kip}$$

Note that the vertical component of tension is greater at the higher of the two support points, also that with H given the reactions at the supports can be found by considering only the equilibrium of the cable as a whole. The maximum tension in the cable, which occurs at B, is

$$T_B = \sqrt{(220^2 + 62.26^2)} = \underline{229 \text{ kip}}$$

In order to locate the lowest point, C, of the cable in terms of horizontal distance *l* from, and dip *d* below, support A, we need to cut the cable at point C: free

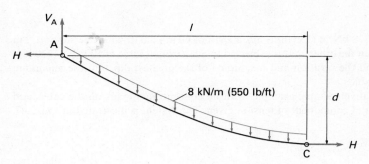

Fig. 2.48.

body AC is shown in Fig. 2.48. For equilibrium in the vertical direction

$$V_A - 8l = 0$$

$$l = \frac{206.7}{8} = 25.84 \text{ m}$$

and the lowest point is 4.16 m to the left of the center of the cable. For moment equilibrium of AC about A

$$\frac{8l^2}{2} - Hd = 0$$

$$d = \frac{8 \times 25.84^2}{2 \times 1000} = 2.67 \text{ m}$$

body AC is shown in Fig. 2.48. For equilibrium in the vertical direction

$$V_A - 0.55l = 0$$

$$l = \frac{47.74}{0.55} = 86.80 \text{ ft}$$

and the lowest point is 13.2 ft to the left of the center of the cable. For moment equilibrium of AC about A

$$\frac{0.55l^2}{2} - Hd = 0$$

$$d = \frac{0.55 \times 86.8^2}{2 \times 220} = 9.42 \text{ ft}$$

With such a small maximum dip, the small dip approximation is a reasonable one. Incidentally, had we been given the value of d (rather than H), we could have obtained the value of l from equation (2.33).

ΣXAMPLE 2.12

A flexible cable weighing 4 kN/m (270 lb/ft) is attached to two fixed points which are 120 m (390 ft) apart horizontally and 3 m (9.8 ft) apart vertically. Two concentrated vertical loads each of 150 kN (34 kip) are applied to the cable at horizontal distances of 20 m (66 ft) from each of its ends. If the maximum dip of the cable is 5 m (16 ft) below its highest point, find the maximum tension in the cable.

This is a less straightforward example, involving concentrated loads, and in which the maximum dip (but not its position) is given. The system is illustrated in Fig. 2.49. For equilibrium of forces on the cable in the vertical direction

$$V_A + V_B - 150 - 150 - 4 \times 120 = 0$$

$$V_A + V_B = 780 \tag{2.38}$$

and for moment equilibrium about support point B

$$H \times 3 + V_A \times 120 - 150 \times 100 - 150 \times 20 - \frac{4 \times 120^2}{2} = 0$$

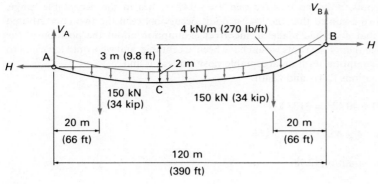

Fig. 2.49.

$$H + V_A \times 40 = 15\,600 \tag{2.39}$$

To make further progress, we must now cut the cable at its lowest point, C. Although we know the dip at this point, we do not know its position. In particular, we do not know where it occurs in relation to the concentrated loads, although in view of their positions relatively close to the ends of the cable it is reasonable to assume that it occurs at a location between them. Figure 2.50 shows the forces acting on section AC of the cable. For equilibrium in the vertical direction

$$V_A - 150 - 4l = 0$$

$$V_A = 150 + 4l \tag{2.40}$$

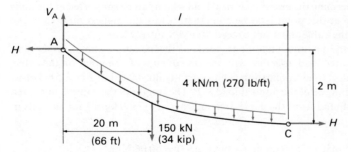

Fig. 2.50.

and for moment equilibrium about point A

$$- H \times 2 + 150 \times 20 + \frac{4l^2}{2} = 0$$

$$H = 1500 + l^2 \tag{2.41}$$

Substituting expressions (2.40) and (2.41) into equation (2.39), we obtain

$$l^2 + 160l - 8100 = 0$$

$$l = -80 \pm \sqrt{(80^2 + 8100)} = -80 \pm 120.42 \text{ m}$$

Of the two roots, only the positive one, $l = 40.42$ m, lies in the acceptable range, $20 < l < 100$ (we assumed that the lowest point occurs between the two concentrated loads). Note that if we had made an incorrect assumption about the position of the lowest point, neither of the roots would have been acceptable, and we would have had to modify the assumption until an acceptable position was found.

Using equations (2.41) and (2.40)

$$H = 1500 + 40.42^2 = 3133 \text{ kN}$$

$$V_A = 150 + 4 \times 40.42 = 311.7 \text{ kN}$$

and then from equation (2.38)

$$V_B = 780 - 311.7 = 468.3 \text{ kN}$$

The maximum tension in the cable occurs at point B, and is given by

$$T_B = \sqrt{(3133^2 + 468.3^2)} = \underline{3170 \text{ kN}} \quad (710 \text{ kip})$$

With concentrated loads applied to the cable, its shape is no longer that of a single parabola. Nevertheless, the portions of the cable between points at which concentrated loads are applied (including the support points) do take the shapes of parts of parabolas.

EXAMPLE 2.13

The ends of a flexible steel cable are attached to supports which are 60 m (200 ft) apart horizontally, and 2 m (6.6 ft) apart vertically. The cross-sectional area of the cable is 650 mm^2 (1 in^2). Determine the greatest vertical load which can be supported at the mid-span point if the dip at this point is to be 5 m (16 ft) below the highest point, and the maximum stress in the cable must not exceed 200 MN/m^2 (29 ksi).

In this example, the maximum tension is given (via the maximum stress), rather than the applied concentrated load, together with the coordinates of one point on the cable, which is also a point of load application. The system is illustrated in Fig. 2.51. Let us assume that the magnitude of the applied load is F. The weight of the cable is not given directly, but can be found from the cable area and density (7850 kg/m^3 for steel, from Appendix A)

$$w = (650 \times 10^{-6} \text{ m}^2) \times (7850 \text{ kg/m}^3) \times (9.81 \text{ m/s}^2) = 50.06 \text{ N/m}$$

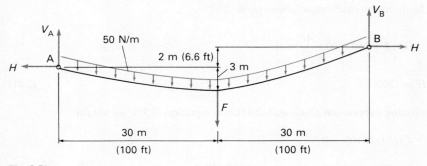

Fig. 2.51.

Also, the maximum tension is found from the maximum stress as

maximum tension $= 650 \times 10^{-6} \times 200 \times 10^6$ N $= 130$ kN

Now, with the concentrated load at the center of the cable, we can anticipate that the maximum tension will occur at the highest point, at support B. We therefore write an equilibrium equation for the whole cable which does not involve the vertical component of reaction at A, by considering moment equilibrium about A, as follows (using kN as the unit of force)

$$H \times 2 - V_B \times 60 + F \times 30 + \frac{50.06 \times 10^{-3} \times 60^2}{2} = 0 \qquad (2.42)$$

To make further progress, we must now cut the cable at its center point, C, where we know the dip. Although it is not directly relevant to the solution of the problem, this point is also the lowest point of the cable. In view of the concentrated load at C, however, we are certainly not entitled to assume that the only component of tension there is the horizontal one (the slope of the cable on either side of the concentrated load cannot be zero if it is to be supported). An equation of vertical equilibrium for either section AC or BC is therefore not useful in that it introduces an additional unknown (the vertical component of tension at C), but the equation of moment equilibrium about C for section BC does provide useful information

$$H \times 5 - V_B \times 30 + \frac{50.06 \times 10^{-3} \times 30^2}{2} = 0$$

$$H = V_B \times 6 - 4.505 \qquad (2.43)$$

Now $\quad \sqrt{(V_B^2 + H^2)} = 130$

$$V_B^2 + (V_B \times 6 - 4.505)^2 = 130^2$$

$$37V_B^2 - 54.06\,V_B - 16\,880 = 0$$

$$V_B = \frac{54.06 \pm \sqrt{(54.06^2 + 4 \times 37 \times 16\,880)}}{2 \times 37}$$

Since only the positive root gives a relevant solution in the present context

$V_B = 22.1$ kN

and from equation (2.43)

$H = 128.1$ kN

Finally, the required force F may be found from equation (2.42) as

$$F = \frac{22.1 \times 60 - 128.1 \times 2 - 90.11}{30} = \underline{32.7 \text{ kN}} \ (7.4 \text{ kip})$$

2.3.2 Introduction to catenaries

We have seen that if we make the assumption of small dip for a uniform flexible cable, so that we can treat the weight of the cable as being uniformly distributed in the horizontal direction, then the shape of the cable is parabolic. If we do not make this simplification, then a more complex shape is obtained, which is known as a *catenary*. It is instructive to examine this shape, if only to justify the small dip approximation.

Figure 2.52 shows a small element of flexible cable loaded only by its own weight. At the left-hand end of this element, marked A, the horizontal and vertical components of tension in the cable are H and V, respectively, while at the right-hand end, marked B, they are H and $V + \delta V$. The horizontal distance between the centers of the cut faces at A and B is δx, while the vertical distance between them is δy, x and y being the horizontal and vertical coordinates, respectively. The length of the element is given by the approximation

$$\delta s \approx \sqrt{[(\delta x)^2 + (\delta y)^2]} = \delta x \sqrt{\left[1 + \left(\frac{\delta y}{\delta x}\right)^2\right]}$$

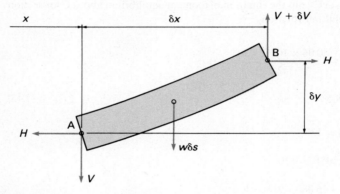

Fig. 2.52.

For equilibrium of forces on the element in the vertical direction

$$V + \delta V - V - w\delta s = 0$$

$$\delta V = w\delta s = w\delta x \sqrt{\left[1 + \left(\frac{\delta y}{\delta x}\right)^2\right]}$$

and, as $\delta x \to 0$, we obtain the exact differential relationship

$$\frac{\mathrm{d}V}{\mathrm{d}x} = w \sqrt{1 + \left(\frac{\mathrm{d}y}{\mathrm{d}x}\right)^2} \tag{2.44}$$

Now, since the resultant tension in the cable always acts along the local tangent to the curve followed by the cable, the gradient at A is given by

$$\frac{dy}{dx} = \frac{V}{H}$$

and the gradient at B by

$$\frac{dy}{dx} = \frac{V + \delta V}{H}$$

The difference between these two gradients is $\delta V/H$, and the second derivative of y with respect to x is given by

$$\frac{d^2 y}{dx^2} = \lim_{\delta x \to 0} \left(\frac{1}{H} \frac{\delta V}{\delta x} \right) = \frac{1}{H} \frac{dV}{dx} \qquad (2.45)$$

Combining equations (2.44) and (2.45)

$$\frac{d^2 y}{dx^2} = \frac{w}{H} \sqrt{1 + \left(\frac{dy}{dx} \right)^2} \qquad (2.46)$$

If we take the origin for the x and y coordinates to be at the lowest point of the cable, the solution to this second order ordinary differential equation is

$$y = \frac{H}{w} \left(\cosh\left(\frac{wx}{H} \right) - 1 \right) \qquad (2.47)$$

and the shape of the catenary is given by the hyperbolic cosine function. An important quantity in this equation is $\beta = wx/H$, because if $\beta \ll 1$ then we can use the approximation

$$\cosh\left(\frac{wx}{H} \right) \approx 1 + \frac{1}{2} \left(\frac{wx}{H} \right)^2$$

to reduce equation (2.47) to

$$y = \frac{wx^2}{2H}$$

which is identical to equation (2.37), which we obtained using the small dip approximation. Hence, we can write

$$\frac{y}{x} = \frac{\beta}{2} \qquad (2.48)$$

as compared with the more accurate expression given by equation (2.47) which becomes

$$\frac{y}{x} = \frac{(\cosh \beta - 1)}{\beta} \qquad (2.49)$$

We stated that the small dip approximation was a reasonable one provided that the ratio of maximum dip to span of the cable was not more than about 0.1. Now, this ratio is of the order of $y/2x$ since the coordinate x is measured from the lowest point of the cable and the span associated with a particular maximum dip y is of the order of $2x$. From equation (2.48), β is therefore about 0.4, and equations (2.47) and (2.48) give

$$\frac{y}{x} = 0.2027 \qquad \text{and} \qquad \frac{y}{x} = 0.2$$

which are in close agreement. In other words, what we are comparing are the approximate and true values of the maximum dip (depth y of the lowest point below the support points) for a cable with a given uniform load and horizontal component of tension suspended from two points at the same height and a distance of $2x$ apart. We conclude that the small dip approximation is a reasonable one, at least up to a dip to span ratio of $1:10$.

Problems

SI METRIC UNITS – PIN-JOINTED STRUCTURES

2.1 The dimensions of the plane pin-jointed structure shown in Fig. P2.1 are such that the angles between the members, and between the members and the vertical wall to which the structure is attached, are either 45° or 90°. If the structure is subjected to vertical loads of W and $2W$ at the joints marked D and C, respectively, calculate the forces in all the

bars, indicating whether these forces are tensile or compressive.

2.2 Use program SDPINJ to solve Problem 2.1.

2.3 The members of the framework shown in Fig. P2.3, which is used as a roof truss, are pin-jointed together at the points marked A, B, C, D and E. All the members except AB and BC are of the same length. The structure is supported at joints A and C, with C being free to move horizontally. Downward vertical loads of magnitude $5P$ and P are applied at joints B and E, respectively. Calculate the forces in all the members in terms of P.

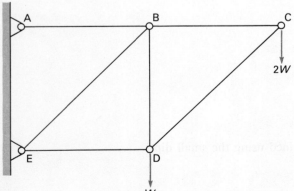

Fig. P2.1.

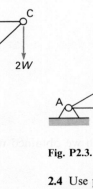

Fig. P2.3.

2.4 Use program SDPINJ to solve Problem 2.3.

2.5 The pin-jointed bridge structure shown in Fig. P2.5 is loaded by a vertical force of 250 kN which acts

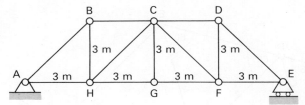

Fig. P2.5.

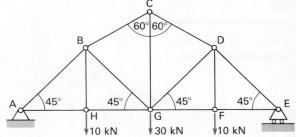

Fig. P2.9.

at each of the points F, G and H in turn. For each of these load positions, find the forces in members CD, CF and GF.

2.6 Use program SDPINJ to find the forces in all the members of the framework analyzed in Problem 2.5, when the 250 kN load is applied at point G.

2.7 The members of the roof truss framework shown in Fig. P2.7 pin-jointed together. Calculate the reaction forces at the support points, together with the forces in all the members of the structure, in terms of the applied force, F.

2.8 Use program SDPINJ to solve Problem 2.7.

2.9 Calculate the forces in all the members of the pin-jointed structure shown in Fig. P2.9.

2.10 Use program SDPINJ to solve Problem 2.9.

2.11 In the pin-jointed structure shown in Fig. P2.11, equal and opposite forces of magnitude 17 kN are applied to the joints marked A and E along the line joining these points. Use the method of sections to find the forces in members BC and BF.

2.12 Use program SDPINJ to find the forces in all the members of the structure analyzed in Problem 2.11.

2.13 The pin-jointed bridge girder framework shown in Fig. P2.13 is used at an angle of 15° to the horizontal. While the joint marked A is a fixed pin joint, point F is free to slide at 30° to the horizontal. A vertical load of 100 kN is applied at each of points G, H and I in turn. Use program SDPINJ

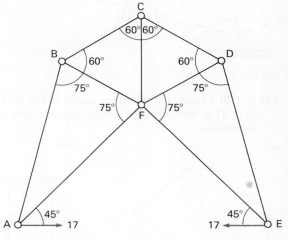

Fig. P2.11.

to find the greatest absolute force in each of the members for these three loading conditions.

2.14 Use the method of sections to check the results of Problem 2.13 for members CD, DH and GH.

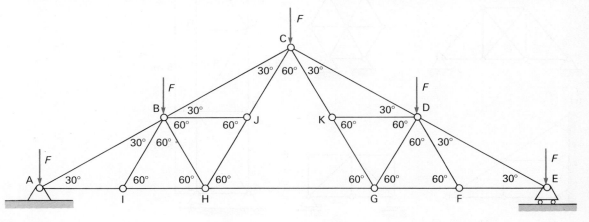

Fig. P2.7.

95

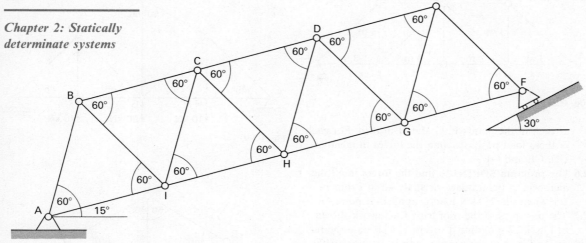

Fig. P2.13.

2.15 to 2.20 Classify the structures shown in Fig. P2.15
(a) to (f) as statically determinate, indeterminate or
mechanism.

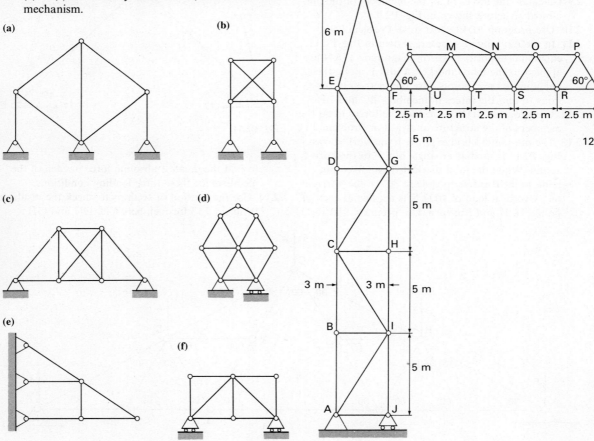

Fig. P2.15.

Fig. P2.27.

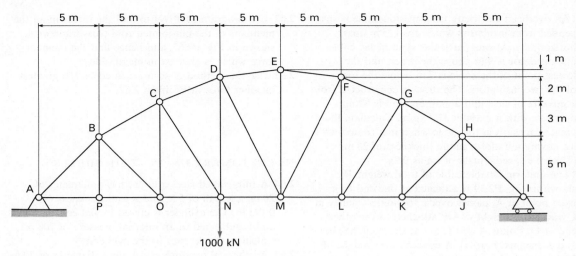

Fig. P2.29.

2.21 to 2.26 Confirm the results of Problems 2.15 to 2.20 using program SDPINJ.

2.27 Use program SDPINJ to find the forces in all the members of the pin-jointed crane structure shown in Fig. P2.27.

2.28 Use the method of sections to check the results of Problem 2.27 for members KN, FL and FU.

2.29 Use program SDPINJ to find the forces in all the members of the pin-jointed bridge structure shown in Fig. P2.29.

2.30 Use the method of sections to check the results of Problem 2.29 for members DE, DM and NM.

SI METRIC UNITS – THIN SHELLS

2.31 A thin-walled steel cylinder has a diameter of 0.5 m, a length of 2.5 m, and a wall thickness of 3 mm. The cylinder is closed by flat end plates and is subjected to an internal pressure of 1.5 MN/m². Calculate the stresses in the cylinder.

2.32 Part of a high-speed centrifuge takes the form of a circular brass cylinder with closed ends whose mean diameter is 85 mm, length 240 mm and wall thickness 3 mm. Under operating conditions, the cylinder is rotated at 10 000 rev/min about its axis and a gage pressure of 0.5 MN/m² is applied to its external surfaces. Calculate the stresses caused in the wall of the cylinder.

2.33 A spherical pressure vessel has a diameter of 700 mm and a wall thickness of 15 mm. If the maximum allowable circumferential stress is 180 MN/m², calculate the maximum working pressure of the vessel.

2.34 For an attempt on the world land speed record for wheeled vehicles, solid tyres made of a material having a density of 300 kg/m³ and a maximum allowable normal stress of 50 MN/m² are to be used. Determine the maximum ground speed that can be achieved and whether there is an optimum wheel diameter. The radial width of the tyre can be assumed to be much smaller than its diameter.

SI METRIC UNITS – FLEXIBLE CABLES

2.35 The supports at a particular section of a cable-car system are 100 m apart and at the same level. The solid steel cable used weighs 54 N/m and is circular in cross section. The cable car has a mass of 500 kg, and when it is midway between the supports the maximum tensile stress in the cable is 120 MN/m². Calculate the dip at this central point.

2.36 A cable weighing 5 kN/m is suspended from two points A and B which are 80 m apart, with B 2 m higher than A. The cable also carries a weight of 50 kN at point C, which is 20 m from A. The maximum dip in the cable is observed to be 2 m below the level of A, and occurs at a point D lying between C and B. Find the horizontal and vertical components of the reactions at A and B and the maximum tension in the cable.

2.37 A flexible cable weighing 3 kN/m is attached to two fixed points which are 100 m apart horizontally and 3 m apart vertically. A concentrated vertical load of 150 kN is applied to the cable at its center, which is at a vertical distance of 7 m below the highest point of the cable. Find the maximum tension in the cable.

97

2.38 An overhead electricity transmission cable is suspended from insulators which are 125 m apart horizontally. At one particular span of the cable, one insulator is 9 m above the other, and the lowest point of the cable is 7 m below the lower of the two insulators. The cable is constructed from a mixture of steel and aluminum and is 32 mm diameter with a mass of 3.75 kg/m. Calculate the greatest tension in the cable when it is coated with ice having a uniform radial thickness of 28 mm. (Take the specific mass of ice as 0.9).

2.39 The uniform flexible cable of total weight W shown in Fig. P2.39 is attached at one end to a fixed point at A, passes over a frictionless pulley at C and has a weight of $6W$ attached to the other end at D. Points A and C are at the same height and a distance L apart. A concentrated weight of $2W$ is attached to the cable at point B, which is a horizontal distance $L/10$ from A. The diameter of the pulley at C, the distance CD and the maximum dip in the cable are small compared with L. Find expressions for: (i) the horizontal and vertical positions of the lowest point of the cable, (ii) the greatest tension in the cable, (iii) the resultant force acting on the pulley at C.

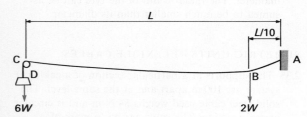

Fig. P2.39.

US CUSTOMARY UNITS – PIN-JOINTED STRUCTURES

2.40 to 2.43 Solve Problems 2.1 to 2.4.
2.44 and 2.45 Solve Problems 2.7 and 2.8.
2.46 Solve Problem 2.11 with forces of 17 kip applied to the joints marked A and E along the line joining these points as in Fig. P2.11.
2.47 Use program SDPINJ to find the forces in all the members of the structure analyzed in Problem 2.46.
2.48 Solve Problem 2.13 with a vertical load of 100 kip applied at each of points G, H and I (Fig. P2.13) in turn.
2.49 Use the method of sections to check the results of Problem 2.48 for members CD, DH and GH.
2.50 to 2.61 Solve Problems 2.15 to 2.26.

2.62 Use program SDPINJ to find the forces in all the members of the pin-jointed roof truss framework shown in Fig. P2.62, and hence find the member force which is greatest in magnitude.
2.63 Use the method of sections to check the greatest member force in Problem 2.62.

US CUSTOMARY UNITS – THIN SHELLS

2.64 A thin-walled steel cylinder has a diameter of 20 in, a length of 5 ft, and a wall thickness of 0.125 in. The cylinder is closed by flat end plates and is subjected to an internal pressure of 200 psi. Calculate the stresses in the cylinder.
2.65 A spherical pressure vessel has a diameter of 30 in and a wall thickness of 0.625 in. If the maximum allowable circumferential stress is 25 ksi, calculate the maximum working pressure of the vessel.
2.66 The pressure vessel shown in Fig. P2.66 consists of a cylinder of internal diameter 6.5 in, wall thickness 0.15 in and length 18 in. The ends of the cylinder are closed by rigid flat plates which are held in position by four rods passing between them and secured by nuts on the outside. The diameter of the rods is 0.75 in. Calculate the stresses in the cylinder when the vessel is subjected to an internal gage pressure of 360 psi, and the tensile stress in the rods is 10 ksi.

US CUSTOMARY UNITS – FLEXIBLE CABLES

2.67 A flexible cable weighing 30 lb/ft is attached to two points A and B which are 330 ft apart horizontally, B being 15 ft above A. A concentrated vertical load of 750 lb is applied to the cable at a point C which is 6 ft below and 65 ft horizontally from A. A second concentrated load of 500 lb is applied to the cable at a point D which is 100 ft horizontally from B. Find the position and magnitude of the greatest dip in the cable.
2.68 A flexible cable weighing 90 lb/ft is attached to two fixed points, A and B, which are 100 ft apart horizontally and at the same height. A concentrated vertical load of 4.5 kip is applied to the cable at a horizontal distance of 30 ft from A, and another concentrated vertical load of 3.5 kip is applied to the cable at horizontal distance of 35 ft from B. If the maximum dip of the cable (which is observed to occur at a point between the two ap-

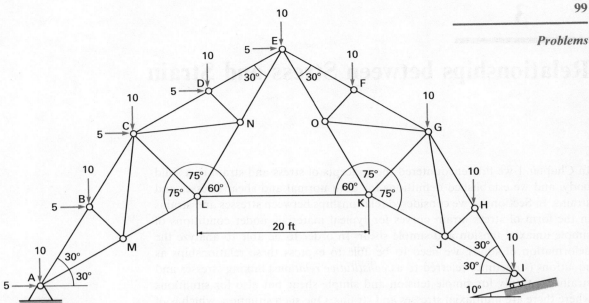

Fig. P2.62.

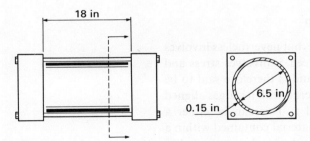

Fig. P2.66.

plied loads) is 5 ft below A and B, find the magnitude and position of the maximum tension in the cable.

2.69 A uniform steel cable is suspended between two supports 300 ft apart with a difference in level *h*. The horizontal component of tension is 6 kip. The maximum dip occurs 75 ft from the lower of the two supports, and the tensile stress in the cable at this point is 12 ksi. Determine the diameter of the cable, the maximum dip below the level of the lower support, and the value of *h*.

2.70 Solve Problem 2.39.

3

Relationships between Stress and Strain

In Chapter 1 we first encountered the concepts of stress and strain in a solid body, and we established definitions for both normal and shear stresses and strains. In Section 1.4 we considered relationships between stresses and strains in the form of stress–strain curves for typical materials under conditions of simple uniaxial tension and simple shear. In order to be able to analyze the deformation of solids, we need to be able to express these relationships as equations (sometimes referred to as *constitutive relations*) linking stresses and strains, not only for simple tension and simple shear but also for situations where there are multiaxial stresses and strains. One such situation, which is of particular importance because it is sometimes used experimentally to determine material properties, is that of hydrostatic loading and deformation.

3.1 Hydrostatic stress and volumetric strain

A hydrostatic state of stress and strain is a simple one, but nevertheless involves stresses and strains in more than one direction. Indeed, it involves stress and strain components in all three directions in space, and is therefore said to be *triaxial*. Figure 3.1 shows a rectangular block of material with its edges aligned with the three Cartesian coordinate axes x, y and z. This can be either a complete brick-shaped body or a small element of material contained within a body of arbitrary size and shape. Let us suppose that the block is subjected to normal stresses acting on all its faces. Because of the alignment with the coordinate axes, it is convenient to denote these stresses by the symbol σ with

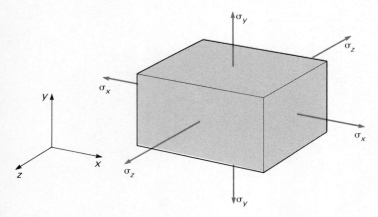

Fig. 3.1.

100

subscript x, y or z to indicate the direction of a particular component. For example, σ_x is the stress in the direction parallel to the x axis, acting on the faces of the block which are perpendicular to this direction. We should note that the forces associated with the stresses on opposite faces are self-balancing, ensuring that the block is in equilibrium: the weight of the block is ignored.

A *hydrostatic* state of stress is one in which the normal stress components in all directions are identical. Therefore, in the Cartesian coordinate system

$$\sigma_x = \sigma_y = \sigma_z = \sigma_H \qquad (3.1)$$

where σ_H is the hydrostatic stress. The term hydrostatic is borrowed from fluid mechanics, where it is used to describe the pressure in a fluid at rest. While pressure in a fluid is a compressive stress, hydrostatic stress in a solid need not be compressive.

The type of strain associated with a hydrostatic state of stress is a change of volume, and we can define a volumetric strain as the ratio between the change in volume and the original volume. Volumetric strains in solids are small. Suppose that in the unstrained state the uniform rectangular block of material shown in Fig. 3.2 has sides of lengths L_x, L_y and L_z parallel to the x, y and z coordinate axes and is deformed in such a way that these lengths increase by ΔL_x, ΔL_y and ΔL_z, respectively. In accordance with equation (1.8), the normal strains in the three coordinate directions are therefore

$$e_x = \frac{\Delta L_x}{L_x}, \quad e_y = \frac{\Delta L_y}{L_y}, \quad e_z = \frac{\Delta L_z}{L_z}$$

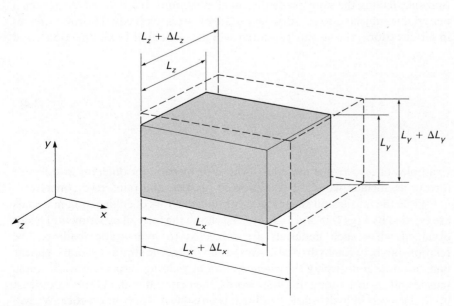

Fig. 3.2.

and the volume of the deformed block is

$$V = (L_x + \Delta L_x)(L_y + \Delta L_y)(L_z + \Delta L_z) = L_x L_y L_z (1 + e_x)(1 + e_y)(1 + e_z)$$

compared with an initial volume of $V_0 = L_x L_y L_z$. Consequently, the volumetric strain is

$$e_{vol} = \frac{V - V_0}{V_0}$$

$$= (1 + e_x)(1 + e_y)(1 + e_z) - 1 \tag{3.2}$$

$$= e_x + e_y + e_z + e_x e_y + e_y e_z + e_z e_x + e_x e_y e_z$$

Now, provided the strains are small, $e_x \ll 1$, $e_y \ll 1$ and $e_z \ll 1$, which implies that products of strains such as $e_x e_y$ and $e_x e_y e_z$ are very small compared with the strains themselves, and we may simplify this expression to

$$e_{vol} = e_x + e_y + e_z \tag{3.3}$$

This result holds for any state of strain. Volumetric strain is sometimes referred to as *dilatation*.

Many engineering materials, including most metals, can be regarded as *isotropic*, having the same properties in all directions. If a body made of such a material is subjected to a hydrostatic state of stress, with equal normal stresses in all directions, the strains produced will also be equal in all directions, and

$$e_x = e_y = e_z = \frac{e_{vol}}{3} \tag{3.4}$$

Examples of *anisotropic* materials, whose properties are different in different directions, include wood, fiber reinforced plastics, and reinforced concrete.

While the tensile and shear stress–strain behaviors of engineering materials are typified by Figs 1.25 and 1.26, Fig. 3.3 shows the form of experimental result obtained when such materials are subjected to hydrostatic loading. The relationship between hydrostatic stress and volumetric strain is usually a linear one, but does not display the phenomenon of yielding associated with simple tensile and shear loading. In other words, the material remains elastic, at least up to the levels of hydrostatic loading it is possible to apply in practice. We will examine the reasons for this in Chapter 9.

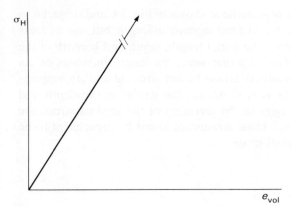

Fig. 3.3.

3.2 Elastic stress–strain equations

For the moment we will confine our attention to the elastic behavior of isotropic materials. For many engineering materials, this also means that we can assume that the strains involved are small (in the sense discussed in Section 1.3.5) and that stresses and strains are directly proportional (Figs 1.25, 1.26 and 3.3). Relatively simple equations linking stresses and strains can therefore be derived.

3.2.1 Normal loading

For simple uniaxial tension, the relationship between stress and strain in the elastic region can be expressed as

$$\sigma = Ee \tag{3.5}$$

where the constant E is the slope of the stress–strain curve (Fig. 1.25) in this region. This equation is often referred to as *Hooke's Law* (after the English scientist Robert Hooke, 1635–1703, who first demonstrated the proportionality between load and deflection for elastic springs). The constant E is known as the *modulus of elasticity* for the material. It is also commonly referred to as the *Young's modulus of elasticity* or just *Young's modulus* (after another English scientist Thomas Young, 1773–1829). Typical numerical values of Young's modulus for engineering materials are given in Appendix A. Since strain e is a dimensionless ratio of lengths, the units of E are those of stress. Although Young's modulus is usually measured only in simple tension, for most common materials the same value is obtained in compression. We will therefore assume that equation (3.5) is applicable to both positive and negative stresses and strains. It is very important to appreciate that it is only applicable to uniaxial loading.

If we take a rectangular bar of material as shown in Fig. 3.4, and subject it to an axial stress σ, then not only does the bar increase in length, but also its cross section becomes smaller. Suppose the initial length, depth and breadth of the bar are L_0, d_0 and b_0 as shown, and that while the length increases by an amount ΔL, the depth and breadth decrease by amounts Δd and Δb, respectively. So, while the axial strain is $e_L = \Delta L/L_0$, the strains in the depth and breadth directions, at right angles to the direction of the applied stress, are $e_d = -\Delta d/d_0$ and $e_b = -\Delta b/b_0$. These strains are found by experiment to be directly proportional to the axial strain

$$e_d = e_b = -v e_L$$

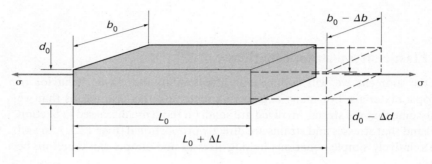

Fig. 3.4.

where the constant v (the Greek letter 'n') is known as *Poisson's ratio* (after the French mathematician Simeon Poisson, 1781–1840). This constant ratio between lateral and axial strain is, like Young's modulus, an elastic property of the particular material concerned, and typical values are given in Appendix A. Most metals have values of Poisson's ratio around 0.3. Materials such as glass and concrete have positive values lower than this, while plastics and rubbers have appreciably higher values, in the case of rubber approaching 0.5. The range of values encountered in practice is from zero to 0.5. The value of exactly 0.5 has a special significance which we will examine in Section 3.2.4.

We should note that a state of uniaxial stress such as that shown in Fig. 3.4 gives rise to a triaxial state of strain. If we now consider a triaxial state of stress, such as that shown in Fig. 3.1, we must also expect to create a triaxial state of strain. The rectangular block shown can be either a brick-shaped body or a small element of a much larger body of arbitrary size and shape. Due to the stress σ_x acting alone (that is, with $\sigma_y = \sigma_z = 0$), we know that the strains are $+\sigma_x/E$, $-v\sigma_x/E$, and $-v\sigma_x E$ in the x, y and z directions, respectively. Similarly, due to the stress σ_y acting alone, the strains are $-v\sigma_y/E$, $+\sigma_y/E$, and $-v\sigma_y/E$ in the x, y and z directions, and due to the stress σ_z acting alone, the strains are $-v\sigma_z/E$, $-v\sigma_z/E$, and $+\sigma_z/E$ in the x, y and z directions. Therefore, if all three stresses act, the total strains in the coordinate directions are

$$e_x = \frac{1}{E}[\sigma_x - v(\sigma_y + \sigma_z)] \qquad\qquad (3.6a)$$

$$e_y = \frac{1}{E}[\sigma_y - v(\sigma_z + \sigma_x)] \qquad\qquad (3.6b)$$

$$e_z = \frac{1}{E}[\sigma_z - v(\sigma_x + \sigma_y)] \qquad\qquad (3.6c)$$

This set of equations is obtained by adding or superimposing the strains associated with the three stress components. It is often referred to as *generalized Hooke's Law*. We should note that it is only in the simple case of uniaxial stress in, for example, the x direction (so that $\sigma_y = \sigma_z = 0$) that equation (3.6a) reduces to the form of equation (3.5)

$$\sigma_x = E e_x$$

The superposition of strains associated with the three different stress components provides an example of the *principle of superposition*. This states that *the total effect of a combination of loads applied to a body is the sum of the effects of the individual loads applied separately, provided*: (i) *that these effects are directly proportional to the loads which produce them*, (ii) *that the strains produced are small*. It is a principle we will also use in other situations.

EXAMPLE 3.1

The mild steel bolt shown in Fig. 3.5 has a diameter of 12 mm (0.5 in) and a thread with a pitch of 1 mm (0.04 in). It is used to hold together two castings which may be assumed to be rigid, and which have a combined thickness of 200 mm (8 in). After the nut has been adjusted to bring the parts just into contact, it is then tightened by one eighth of a turn. Find the force induced in the bolt, and the change in its diameter.

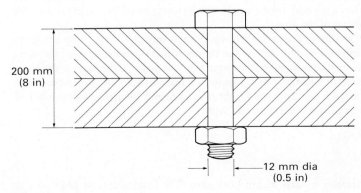

Fig. 3.5.

This is a straightforward example of uniaxial loading, but with strain specified rather than load. In the absence of the castings, an eighth turn of the nut would reduce the distance between the bolt head and nut by an eighth of the thread pitch, or 0.125 mm (0.005 in). The rigidity of the castings prevents this reduction taking place, and the 200 mm (8 in) length of the bolt is increased by 0.125 mm (0.005 in) (the length of bolt drawn into the nut by its rotation), implying a positive strain of 0.125/200 (or 0.005/8) which is 0.625×10^{-3}. For uniaxial tensile loading equation (3.5) is applicable.

SI units

Given the value of Young's modulus for mild steel as 207 GN/m² (Appendix A), the axial stress in the bolt is

$$\sigma = 207 \times 10^{9} \times 0.625 \times 10^{-3} = 129.4 \text{ MN/m}^{2}$$

and the force in the 12 mm diameter bolt is

$$\text{force} = 129.4 \times 10^{6} \times \frac{\pi \times 0.012^{2}}{4} \text{ N} = \underline{14.6 \text{ kN}}$$

US customary units

Given the value of Young's modulus for mild steel as 30×10^{6} psi (Appendix A), the axial stress in the bolt is

$$\sigma = 30 \times 10^{6} \times 0.625 \times 10^{-3} \text{ psi} = 18.75 \text{ ksi}$$

and the force in the 0.5 in diameter bolt is

$$\text{force} = 18.75 \times \frac{\pi \times 0.5^{2}}{4} = \underline{3.68 \text{ kip}}$$

Now, the strain in any lateral direction normal to the axis of the bolt is equal to the axial strain multiplied by Poisson's ratio (0.3 for mild steel, from Appendix A), with a change in sign

$$\text{lateral strain} = -0.3 \times 0.625 \times 10^{-3}$$

$$= -0.1875 \times 10^{-3}$$

and

$$\text{change in diameter} = -0.1875 \times 10^{-3} \times 12$$

$$= \underline{-2.25 \times 10^{-3} \text{ mm}}$$

the negative sign implying a reduction.

Now, the strain in any lateral direction normal to the axis of the bolt is equal to the axial strain multiplied by Poisson's ratio (0.3 for mild steel, from Appendix A), with a change in sign

$$\text{lateral strain} = -0.3 \times 0.625 \times 10^{-3}$$

$$= -0.1875 \times 10^{-3}$$

and

$$\text{change in diameter} = -0.1875 \times 10^{-3} \times 0.5$$

$$= \underline{-0.0937 \times 10^{-3} \text{ in}}$$

the negative sign implying a reduction.

EXAMPLE 3.2

An elastic bar of length L, cross-sectional area A, and weight W is suspended vertically from one end and has a vertical force F applied to its lower end. Find an expression for the increase in volume of the bar.

This is another example of uniaxial loading, but this time the stress varies along the bar (due to its distributed weight). Let us use the coordinate y to define position along the bar measured from its lower end, as shown in Fig. 3.6. At $y = 0$ the force in the bar is just the applied force F, while at $y = L$ this has increased to $F + W$. Since the cross-sectional area of the bar is constant, the weight W is uniformly distributed along the bar, and the force and axial stress vary linearly with y

$$\sigma_{y} = \frac{1}{A}\left(F + \frac{Wy}{L}\right)$$

while the lateral stresses σ_{x} and σ_{z} are both zero. The axial strain at position y is therefore $e_{y} = \sigma_{y}/E$, and the lateral strains $e_{x} = e_{z} = -v\sigma_{y}/E$, E and v being the Young's modulus and Poisson's ratio of the material of the bar. Using equation (3.3), the volumetric strain at position y is given by

$$e_{\text{vol}} = e_x + e_y + e_z = \frac{(1 - 2v)}{E} \sigma_y$$

The change in volume of the small element of the bar of height δy and initial volume $A\delta y$ at position y shown in Fig. 3.6 is therefore

$$\Delta(A\delta y) = A\delta y \frac{(1 - 2v)}{E} \sigma_y$$

and the change in volume of the whole bar is found by summing such changes over all the elements of the bar. Consequently

$$\Delta V = \int_0^L \frac{A}{E}(1 - 2v)\sigma_y \, \mathrm{d}y = \int_0^L \frac{1}{E}(1 - 2v)\left(F + \frac{Wy}{L}\right)\mathrm{d}y$$

$$\Delta V = \frac{L}{E}(1 - 2v)\left(F + \frac{W}{2}\right)$$

which is the required result.

In this particular example, with a linear variation of stress (and strain) along the bar, we could have arrived at the same result by a simpler route. We could have argued that the average axial stress along the bar is

$$\sigma_{\text{ave}} = \frac{1}{A}\left(F + \frac{W}{2}\right)$$

and the corresponding average volumetric strain is

$$e_{\text{vol}} = \frac{1}{AE}(1 - 2v)\left(F + \frac{W}{2}\right)$$

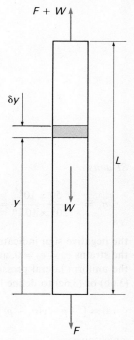

Fig. 3.6.

When multiplied by the volume of the bar, LA, this gives the same expression for the change of volume, ΔV. Such simple averaging is only permissible with linear variations of stress and strain with distance along the bar.

EXAMPLE 3.3

A rectangular block of brass is 100 mm (4 in) long with a cross section 20 mm by 20 mm (0.8 in by 0.8 in). It is compressed by a force of 50 kN (11 kip) applied parallel to its 100 mm (4 in) dimension, while all lateral strain is prevented by the application of uniform lateral pressure to all four long faces. Find the change in length and the apparent modulus of elasticity.

This example involves triaxial stresses, but uniaxial strain, and is illustrated in Fig. 3.7. Cartesian coordinates x, y and z are chosen to be parallel with the edges of the block, with x parallel to the long edges. In the absence of more detailed information, we must assume that the stresses are uniform over each of the faces of the block. The information we are given defines one of the normal stresses

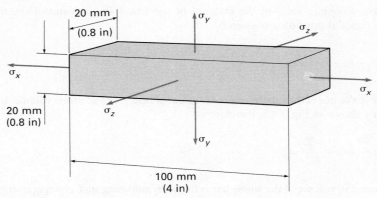

Fig. 3.7.

SI units

$$\sigma_x = -\frac{50 \times 10^3}{20^2 \times 10^{-6}} = -125 \times 10^6 \text{ N/m}^2$$

the negative sign indicating compression, also two of the strains $e_y = e_z = 0$, and tells us that $\sigma_y = \sigma_z = -p$, the uniform lateral pressure. Using either equation (3.6b) or (3.6c) to define lateral strain

$$0 = [-p - v(\sigma_x - p)]$$

which gives

$$p = -\frac{v\sigma_x}{1 - v}$$

Taking the values of Young's modulus and Poisson's ratio for brass as 103 GN/m² and 0.35, respectively (Appendix A)

$$p = \frac{0.35 \times 125 \times 10^6}{1 - 0.35} = 67.31 \times 10^6 \text{ N/m}^2$$

Also, using equation (3.6a)

$$e_x = \frac{1}{E}[\sigma_x - v(-p - p)]$$

$$= -\frac{(125 - 0.35 \times 2 \times 67.31) \times 10^6}{103 \times 10^9}$$

$$= -7.561 \times 10^{-4}$$

US customary units

$$\sigma_x = -\frac{11}{0.8^2} = -17.19 \text{ ksi}$$

the negative sign indicating compression, also two of the strains $e_y = e_z = 0$, and tells us that $\sigma_y = \sigma_z = -p$, the uniform lateral pressure. Using either equation (3.6b) or (3.6c) to define lateral strain

$$0 = [-p - v(\sigma_x - p)]$$

which gives

$$p = -\frac{v\sigma_x}{1 - v}$$

Taking the values of Young's modulus and Poisson's ratio for brass as 15×10^3 ksi and 0.35, respectively (Appendix A)

$$p = \frac{0.35 \times 17.19}{1 - 0.35} = 9.256 \text{ ksi}$$

Also, using equation (3.6a)

$$e_x = \frac{1}{E}[\sigma_x - v(-p - p)]$$

$$= -\frac{17.19 - 0.35 \times 2 \times 9.256}{15 \times 10^3}$$

$$= -7.141 \times 10^{-4}$$

giving a change of length of

$$\Delta L = e_x L = -7.561 \times 10^{-4} \times 100$$

$$= \underline{-0.0756 \text{ mm}}$$

the negative sign indicating a reduction. The apparent modulus of elasticity is the ratio between the applied stress and the resulting strain in the x direction

$$E' = \frac{-125 \times 10^6}{-7.561 \times 10^{-4}} \text{ N/m}^2 = \underline{165 \text{ GN/m}^2}$$

giving a change of length of

$$\Delta L = e_x L = -7.141 \times 10^{-4} \times 4$$

$$= \underline{-2.86 \times 10^{-3} \text{ in}}$$

the negative sign indicating a reduction. The apparent modulus of elasticity is the ratio between the applied stress and the resulting strain in the x direction

$$E' = \frac{-17.19}{-7.141 \times 10^{-4}} = \underline{24.1 \times 10^3 \text{ ksi}}$$

3.2.2 Shear loading

For simple shear, the relationship between stress and strain in the elastic region can be expressed as

$$\tau = G\gamma \qquad (3.7)$$

where the constant G is the slope of the stress–strain curve (Fig. 1.26) in this region. This equation is referred to as *Hooke's Law for shearing*. The property G is the *modulus of rigidity* or *shear modulus (of elasticity)* for the material, and has the units of stress. Typical numerical values of shear modulus for engineering materials are given in Appendix A. Unlike equation (3.5) for simple tension, equation (3.7) for simple shear also applies when more than one stress component (shear or normal) is acting.

3.2.3 Hydrostatic loading

For hydrostatic loading conditions, the relationship between hydrostatic stress and volumetric strain can be expressed as

$$\sigma_H = K e_{vol} \qquad (3.8)$$

where the constant K is the slope of the stress–strain curve (Fig. 3.3), and is the *bulk modulus (of elasticity)* for the material. As for the other elastic moduli, K has the units of stress. Typical numerical values of bulk modulus for engineering materials are given in Appendix A.

EXAMPLE 3.4

A solid aluminum alloy cone, with a base radius, R, of 50 mm (2 in) and height, h, of 125 mm (5 in) is immersed in a liquid whose hydrostatic pressure is then raised to 100 MN/m² (15 ksi). Determine the effect on the shape of the cone, and the changes in the radius, height and volume of the cone.

Since a hydrostatic state of stress in a uniform isotropic body produces the same strain in every direction (equation (3.4)), the shape of the cone is not affected, only its size. We may take the bulk modulus for aluminum alloy as 57.5 GN/m² (Appendix A), and using equation (3.8) find the volumetric strain as

$$e_{vol} = -\frac{100 \times 10^6}{57.5 \times 10^9} = -1.739 \times 10^{-3}$$

the negative signs indicating compressive hydrostatic stress and reduction in volume. Now, since the volume of the cone is

$$V = \frac{\pi R^2 h}{3} = \frac{\pi \times 50^2 \times 125}{3} = 327.2 \times 10^3 \text{ mm}^3$$

the change in volume is

$$\Delta V = -1.739 \times 10^{-3} \times 327.2 \times 10^3 = \underline{-569 \text{ mm}^3} \ (-0.035 \text{ in}^3)$$

Also, from equation (3.4), the normal strain in any direction is

$$e = -\frac{1.739 \times 10^{-3}}{3} = -5.797 \times 10^{-4}$$

Therefore, the change in base radius of the cone is

$$\Delta R = -5.797 \times 10^{-4} \times 50 = \underline{-0.0290 \text{ mm}} \ (-1.1 \times 10^{-3} \text{ in})$$

and the change in height is

$$\Delta h = -5.797 \times 10^{-4} \times 125 = \underline{-0.0725 \text{ mm}} \ (-2.9 \times 10^{-3} \text{ in})$$

3.2.4 Relationships between the elastic constants

In the course of considering the elastic behavior of isotropic materials under normal, shear and hydrostatic loading, we have introduced a total of four elastic constants, namely the moduli E, G and K, together with Poisson's ratio, v. It turns out that not all of these are independent of the others. In fact, given any two of them, the other two can be found.

Let us first establish a relationship between E, K and v. In equation (3.3) we obtained a generally applicable expression for volumetric strain, as the sum of the normal strain components in three mutually perpendicular coordinate directions. In equations (3.6), these strains were defined in terms of the normal stresses, and we may therefore conclude that

$$e_{vol} = e_x + e_y + e_z = \frac{(1-2v)}{E}(\sigma_x + \sigma_y + \sigma_z) \qquad (3.9)$$

which is true for any system of applied stresses. Now, equation (3.8) relates volumetric strain to hydrostatic stress, σ_H, under conditions where $\sigma_H = \sigma_x = \sigma_y = \sigma_z$ (equation (3.1)), and from which we obtain the result

$$e_{vol} = \frac{\sigma_H}{K} = \frac{(1-2v)}{E}(3\sigma_H)$$

or

$$E = 3K(1-2v) \qquad (3.10)$$

In addition to providing this useful link between three of the elastic constants, equation (3.9) also gives other information about material behavior. In particular, if the Poisson's ratio, v, is exactly equal to 0.5, then $(1-2v) = 0$ and the volumetric strain is zero, irrespective of the stresses applied. In other words, the material is *incompressible*. Conversely, if v is not exactly equal to 0.5, then there is always some change in volume associated with elastic deformation (unless $(\sigma_x + \sigma_y + \sigma_z) = 0$). Incidentally, in the context of equation (3.10), a value of v of exactly 0.5 implies an infinite value of K, rather than a zero value of E.

In addition to equation (3.10), there is a similar relationship between E, G and v, which takes the form

$$E = 2G(1+v) \qquad (3.11)$$

although we will not be in a position to prove this result until Chapter 9. Under conditions of simple tension and simple shear, all real materials tend to experience displacements in the directions of the applied forces, and under hydrostatic tensile loading they tend to increase in volume. In other words, the three elastic moduli, E, G and K, cannot be negative which, according to equations (3.10) and (3.11), means that

$$-1 \leqslant v \leqslant 0.5 \qquad (3.12)$$

As already indicated, in practice no real materials have negative values of Poisson's ratio.

EXAMPLE 3.5

Given that the shear and bulk moduli of elasticity of a linearly elastic solid material are 15 and 145 GN/m², respectively, determine the Young's modulus and Poisson's ratio.

This is a very straightforward example demonstrating that if two of the elastic constants are known, the other two can be found. Eliminating E from equations (3.10) and (3.11)

$$3K(1 - 2v) = 2G(1 + v)$$

from which

$$v(2G + 6K) = 3K - 2G$$

$$v = \frac{3 \times 145 - 2 \times 15}{2 \times 15 + 6 \times 145} = \underline{0.45}$$

and

$$E = 2G(1 + v) = 2 \times 15 \times (1 + 0.45) = \underline{43.5 \text{ GN/m}^2}$$

EXAMPLE 3.6

Rubber bearing pads used to support bridge structures need to be much stiffer in compression than they are in shear. Assume that such a pad can be treated as a thin elastic layer sandwiched between, and bonded to, rigid plates and that the effect of bonding is to prevent normal strains in the plane of the layer. Determine the ratio between the apparent Young's modulus for compression normal to the layer and the apparent shear modulus for shearing of one plate relative to the other, where Poisson's ratio for the rubber is 0.49. Compare this ratio of apparent moduli with the ratio between the true moduli.

This example demonstrates some of the interesting consequences of using a nearly incompressible material in a situation where deformation is very restricted. Figure 3.8 shows the rubber pad, with the top plate removed, with coordinates x and z lying in the plane of the pad, and y normal to the pad. We are told that the effect of the bonding of the pad to the plates is that $e_x = e_z = 0$, and using equations (3.6) we deduce that

$$\sigma_x = v(\sigma_y + \sigma_z) \quad \text{and} \quad \sigma_z = v(\sigma_x + \sigma_y)$$

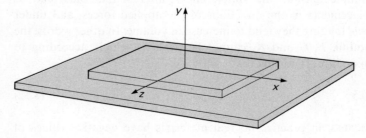

Fig. 3.8.

$$\sigma_x = \sigma_z = \frac{v\sigma_y}{1-v}$$

Then, from equation (3.6b)

$$e_y = \frac{1}{E}\left[\sigma_y - \frac{2v^2\sigma_y}{(1-v)}\right] = \frac{\sigma_y}{E}\frac{(1-2v)(1+v)}{(1-v)}$$

and the apparent Young's modulus is the ratio between σ_y and e_y

$$E' = \frac{\sigma_y}{e_y} = \frac{E(1-v)}{(1-2v)(1+v)}$$

Now, since shear deformation is unaffected by the bonding of the pad to the plates, the true and apparent shear moduli are identical, and are related to the true Young's modulus by equation (3.11)

$$G = G' = \frac{E}{2(1+v)}$$

Therefore, the ratio between the apparent Young's and shear moduli is

$$\frac{E'}{G'} = \frac{2(1-v)}{(1-2v)} = \frac{2 \times 0.51}{0.02} = \underline{51}$$

while the ratio between the true values is

$$\frac{E}{G} = 2(1+v) = 2 \times 1.49 = \underline{2.98}$$

The normal stiffness of the pad is dramatically increased by the bonding and the nearly incompressible nature of rubber.

3.2.5 Thermal strains

The stress–strain equations for linear elasticity we have considered so far assume that the state of strain in a material is due entirely to the stresses. Changes in temperature can also produce strains, without necessarily introducing stresses. The thermal expansion of a material is defined in terms of either its *coefficient of linear thermal expansion*, α, (which is often referred to as the *coefficient of thermal expansion*) or *coefficient of volumetric thermal expansion*, β. It is conventional to quote linear values for solids and volumetric values for fluids, although as we shall see, there is a simple relationship between them. Typical numerical values for common engineering materials are given in Appendix A. For metals, α is of the order of $10^{-5}\,°C^{-1}$, while for liquids β is of the order of $10^{-4}\,°C^{-1}$.

By using a single value for the coefficient of linear expansion of a material we are assuming that the material is isotropic and has the same thermal expansion properties in all directions. What we mean by the coefficient of linear expansion, α, is that for a unit rise in temperature the normal strain produced in all directions is equal to α, and that this strain is proportional to temperature rise, *provided the body concerned is free to expand*. In other words, if the temperature rise, or difference between the final and initial temperatures, is ΔT (and a temperature decrease is represented by a negative ΔT), the strains in the three coordinate directions are given by

$$e_x = \alpha \Delta T, \qquad e_y = \alpha \Delta T, \qquad e_z = \alpha \Delta T \tag{3.13}$$

In this simple case, strains have no stresses associated with them. Since strain is dimensionless, α must have the units of reciprocal temperature. Similarly, what we mean by coefficient of volumetric expansion, β, is that for unit rise in temperature the volumetric strain produced in a body free to expand is equal to β, and

$$e_{\text{vol}} = \beta \Delta T \tag{3.14}$$

Using equation (3.3) to define volumetric strain

$$e_{\text{vol}} = e_x + e_y + e_z = 3\alpha \Delta T = \beta \Delta T$$

$$\text{and} \quad 3\alpha = \beta \tag{3.15}$$

which provides the simple link between the two coefficients.

If we have both applied stresses and a temperature change, thermal strains given by equation (3.13) may be added to those given by the generalized Hooke's Law equations (3.6) to give the total strains as

$$e_x = \frac{1}{E}[\sigma_x - v(\sigma_y + \sigma_z)] + \alpha \Delta T \tag{3.16a}$$

$$e_y = \frac{1}{E}[\sigma_y - v(\sigma_z + \sigma_x)] + \alpha \Delta T \tag{3.16b}$$

$$e_z = \frac{1}{E}[\sigma_z - v(\sigma_x + \sigma_y)] + \alpha \Delta T \tag{3.16c}$$

While the normal strains in a body are affected by changes in temperature, shear strains are not. To see why this should be so, we go back to the fundamental definition of shear strain we adopted in Section 1.3.5, namely the relative angular displacement of two sides of a block or element of material such as that shown in Fig. 1.23. If the temperature of the block changes then its size certainly changes, but its shape, and therefore the angle between adjacent sides, does not.

In using equations such as (3.16) to link stresses, strains and change in temperature, we are assuming that the values of the elastic constants E and v are not significantly affected by the change in temperature. Although such material properties, particularly Young's modulus, are mildly dependent on temperature the effect is only likely to be significant if large changes in temperature, of the order of several hundred degrees Celsius, are involved.

EXAMPLE 3.7

Figure 3.9 shows a solid brass bar of length $L = 100$ mm (4 in) and uniform diameter $D = 15$ mm (0.6 in) which just fits between two rigid surfaces. If the temperature of the bar is raised by 20°C (36°F), find the stresses induced in it.

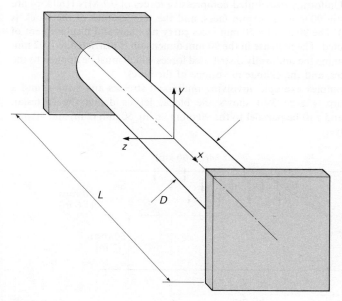

Fig. 3.9.

Provided we ignore local effects at the ends of the bar, where frictional forces between the bar and rigid surfaces may create relatively complicated states of stress, we can treat this as a problem of uniaxial stress induced by the prevention of thermal expansion. Taking x as the coordinate along the axis of the bar as shown, the normal stresses σ_y and σ_z in the directions y and z at right angles to the axis are zero. If the surfaces between which the cylinder is located are truly rigid, and not affected by the change in temperature, then the strain e_x in the bar in the x direction is zero, and from equation (3.16a)

$$\sigma_x = -E\alpha\Delta T$$

It is perhaps useful to consider an alternative, but entirely equivalent, way in which we can arrive at the same result. Suppose the bar is free to expand in all directions when its temperature is increased. The thermal strain in every direction is then $\alpha \Delta T$, and because there are no stresses these are the only strains. Now, in order to impose the effect of the rigid surfaces, the bar must be compressed in the axial direction to return it to its original length. In other words, we must apply a uniaxial compressive stress σ_x to create a corresponding strain of $e_x = -\alpha \Delta T$, and therefore $\sigma_x = -E\alpha \Delta T$.

We can evaluate this stress using values for the properties of brass given in Appendix A: $E = 103 \text{ GN/m}^2$, and $\alpha = 19 \times 10^{-6}\,°\text{C}^{-1}$, and a temperature rise of $\Delta T = 20\,°\text{C}$. Hence

$$\sigma_x = -103 \times 10^9 \times 19 \times 10^{-6} \times 20 \text{ N/m}^2 = \underline{-39.1 \text{ MN/m}^2}\ (-5.7 \text{ ksi})$$

It is interesting to note that this result does not depend on either the length or diameter of the cylinder.

EXAMPLE 3.8

A prismatic steel block of rectangular cross section 40 mm by 50 mm (1.6 in by 2 in) is 90 mm (3.6 in) long. Uniformly distributed compressive forces of 0.7 MN (160 kip) are applied to the opposite 90 mm by 40 mm faces, and the temperature of the block is raised by 15°C (27°F). The 90 mm by 50 mm faces carry no load, and displacement of these faces is unrestricted. The increase in the 90 mm dimension is restricted to 0.02 mm $(0.8 \times 10^{-3}$ in). Determine the uniformly distributed forces which must be applied to the 40 mm by 50 mm faces, and the change in volume of the block.

This is a more complex example, involving multiaxial stresses and strains, and a change in temperature. Figure 3.10 shows the block: let us define the cartesian coordinate axes x, y and z to be parallel to the 90 mm (3.6 in), 50 mm (2 in) and 40 mm (1.6 in) sides, respectively.

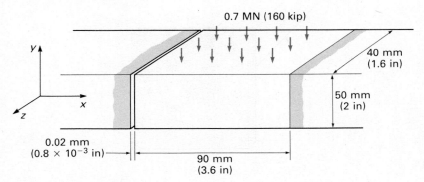

Fig. 3.10.

SI units

We are told that the 90 mm by 40 mm faces (the top and bottom surfaces of the block as shown) are uniformly loaded with compressive forces of 0.7 MN. Consequently, we can calculate the stress in the direction normal to these surfaces as

$$\sigma_y = -\frac{0.7 \times 10^6}{0.09 \times 0.04} = -194.4 \times 10^6 \text{ N/m}^2$$

US customary units

We are told that the 3.6 in by 1.6 in faces (the top and bottom surfaces of the block as shown) are uniformly loaded with compressive forces of 160 kip. Consequently, we can calculate the stress in the direction normal to these surfaces as

$$\sigma_y = -\frac{160}{3.6 \times 1.6} = -27.78 \text{ ksi}$$

We are also told that the 90 mm by 50 mm faces are not loaded, implying that $\sigma_z = 0$. In the x direction we are given only the maximum permitted displacement of 0.02 mm, and hence maximum strain. Let us assume that this is the actual strain (which we can confirm as correct if it leads to a compressive stress in the x direction), so that

$$e_x = \frac{0.02}{90} = 2.222 \times 10^{-4}$$

Now, this strain is composed of both elastic and thermal effects, according to equation (3.16a). Taking the required properties for steel from Appendix A as $E = 207 \text{ GN/m}^2$, $v = 0.3$ and $\alpha = 11 \times 10^{-6} \,^{\circ}\text{C}^{-1}$, this equation becomes

$$2.222 \times 10^{-4} = \frac{[\sigma_x - 0.3 \times (-194.4 \times 10^6 + 0)]}{207 \times 10^9}$$
$$+ 11 \times 10^{-6} \times 15$$

from which

$$\sigma_x = (2.222 \times 10^{-4} - 1.65 \times 10^{-4}) \times 207 \times 10^9$$
$$- 0.3 \times 194.4 \times 10^6$$
$$= -46.48 \times 10^6 \text{ N/m}^2 \text{ (compressive)}$$

and the required forces on the opposite 40 mm by 50 mm faces are

$$-46.48 \times 10^6 \times 0.04 \times 0.05 = -92\,960 \text{ N}$$
$$= -93.0 \text{ kN} \quad \text{(compressive)}$$

The fact that σ_x is compressive justifies the assumption that the length of the block is limited by the maximum permitted displacement in the x direction. Now, in order to find the change in volume, equation (3.3) for the volumetric strain (which is valid for any state of strain, however caused) can be used. We already know e_x, and can find the other two normal strains, from equations (3.16b) and (3.16c), as

$$e_y = \frac{[-194.4 \times 10^6 - 0.3 \times (0 - 46.48 \times 10^6)]}{207 \times 10^9}$$
$$+ 11 \times 10^{-6} \times 15$$
$$= -7.068 \times 10^{-4}$$

We are also told that the 3.6 in by 2 in faces are not loaded, implying that $\sigma_z = 0$. In the x direction we are given only the maximum permitted displacement of 0.8×10^{-3} in, and hence maximum strain. Let us assume that this is the actual strain (which we can confirm as correct if it leads to a compressive stress in the x direction), so that

$$e_x = \frac{0.8 \times 10^{-3}}{3.6} = 2.222 \times 10^{-4}$$

Now, this strain is composed of both elastic and thermal effects, according to equation (3.16a). Taking the required properties for steel from Appendix A as $E = 30 \times 10^3 \text{ ksi}$, $v = 0.3$ and $\alpha = 6.5 \times 10^{-6} \,^{\circ}\text{F}^{-1}$, this equation becomes

$$2.222 \times 10^{-4} = \frac{[\sigma_x - 0.3 \times (-27.78 + 0)]}{30 \times 10^3}$$
$$+ 6.5 \times 10^{-6} \times 27$$

from which

$$\sigma_x = (2.222 \times 10^{-4} - 1.755 \times 10^{-4}) \times 30 \times 10^3$$
$$- 0.3 \times 27.78$$
$$= -6.933 \text{ ksi (compressive)}$$

and the required forces on the opposite 1.6 in by 2 in faces are

$$-6.933 \times 1.6 \times 2$$
$$= -22.2 \text{ kip} \quad \text{(compressive)}$$

The fact that σ_x is compressive justifies the assumption that the length of the block is limited by the maximum permitted displacement in the x direction. Now, in order to find the change in volume, equation (3.3) for the volumetric strain (which is valid for any state of strain, however caused) can be used. We already know e_x, and can find the other two normal strains, from equations (3.16b) and (3.16c), as

$$e_y = \frac{[-27.78 - 0.3 \times (0 - 6.933)]}{30 \times 10^3}$$
$$+ 6.5 \times 10^{-6} \times 27$$
$$= -6.812 \times 10^{-4}$$

117

SI units

$$e_z = \frac{[0 - 0.3 \times (-194.4 \times 10^6 - 46.48 \times 10^6)]}{207 \times 10^9}$$

$$+ 11 \times 10^{-6} \times 15$$

$$= + 5.141 \times 10^{-4}$$

The volumetric strain is therefore

$$e_{\text{vol}} = 2.222 \times 10^{-4} - 7.068 \times 10^{-4} + 5.141 \times 10^{-4}$$

$$= + 2.950 \times 10^{-5}$$

and the change (increase) in volume is found by multiplying this strain by the original volume

$$\Delta V = 2.950 \times 10^{-5} \times 90 \times 50 \times 40$$

$$= 5.31 \text{ mm}^3$$

US customary units

$$e_z = \frac{[0 - 0.3 \times (-27.78 - 6.933)]}{30 \times 10^3}$$

$$+ 6.5 \times 10^{-6} \times 27$$

$$= + 5.226 \times 10^{-4}$$

The volumetric strain is therefore

$$e_{\text{vol}} = 2.222 \times 10^{-4} - 6.812 \times 10^{-4} + 5.226 \times 10^{-4}$$

$$= + 6.360 \times 10^{-5}$$

and the change (increase) in volume is found by multiplying this strain by the original volume

$$\Delta V = 6.360 \times 10^{-5} \times 3.6 \times 2 \times 1.6$$

$$= 0.732 \times 10^{-3} \text{ in}^3$$

We note that working in the two different sets of units, with only approximately the same block dimensions, loading and material properties, two very different values are obtained for the volumetric strain. This is because the volumetric strain is the sum of larger positive and negative linear strains of similar magnitudes, which makes it very sensitive to small changes in the data.

3.3 Other stress–strain relationships

As we have already seen in Section 1.4, engineering materials which are linearly elastic at low stresses and strains do not continue this behavior under more severe loading. Above a certain stress, known as the yield stress, the relationship between stress and strain under either normal or shear loading becomes nonlinear, as shown in Figs. 1.25 and 1.26. Curves such as these can often be approximated by fitting appropriate forms of mathematical functions, although these are generally awkward to use in any subsequent analysis. Alternatively, the plastic part of a stress–strain relationship can be represented by the coordinates of a number of discrete points on the curve, intermediate values of stress and strain being determined by interpolation. This approach is inconvenient for manual calculations, but can be very useful when solving problems numerically with the aid of a computer.

Figure 1.25a shows a typical tensile stress–strain curve for mild steel, a particular feature of which is that just beyond the yield point of the material the curve is virtually horizontal over a substantial range of strain, certainly to several times the yield strain. This gives rise to the simplified form of stress–strain curve for mild steel shown in Fig. 3.11a, where the plastic part of

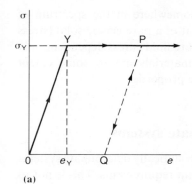

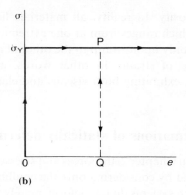

Fig. 3.11.

the curve is horizontal: similar behavior is obtained in compression. The relationship between stress and strain can be expressed mathematically as

$$\frac{d\sigma}{de} = E \quad \text{for} \quad -\sigma_Y \leqslant \sigma \leqslant +\sigma_Y$$

$$\sigma = \sigma_Y \qquad \frac{de}{dt} \geqslant 0 \qquad\qquad (3.17)$$

$$\sigma = -\sigma_Y \qquad \frac{de}{dt} \leqslant 0$$

where σ_Y is the yield stress, and t is time. Under plastic conditions, with $\sigma = \pm\sigma_Y$, the strain may change with time. If with the material at, say, point P on the plastic part of the curve the load is removed, then the stress–strain response follows the straight line PQ, which is parallel to the initial elastic line OY. In some problems, where the strains are large compared with the strain at point Y, it is reasonable to make the further assumption that the elastic part of the stress–strain curve can be ignored, as shown in Fig. 3.11b, and $\sigma = \pm\sigma_Y$ is assumed in the absence of unloading. While stress–strain behavior of the type shown in Fig. 3.11b is referred to as *perfectly plastic*, that shown in Fig. 3.11a is described as *elastic–perfectly plastic*.

So far we have assumed that the stresses and strains in elastic solids are independent of time. While this is reasonable for most metallic materials at normal ambient temperature, it is not always so. Materials such as lead, concrete, glass and plastics exhibit significant time-dependent behavior at such temperatures, and all materials do so at sufficiently high temperatures. To take an extreme example, the alloys used to make turbine blades for aircraft engines, which are required to operate at temperatures of the order of 1000°C, suffer *creep* or continuous extension with time under tensile loading, and it is this which can severely limit the useful working life of a blade.

We normally associate time-dependent behavior more with fluids than with solids, the stress in a fluid being related to the *rate* or time derivative of strain via its viscosity. In reality, all materials lie somewhere in the spectrum of behavior which ranges from at one extreme that of a pure elastic solid (stress dependent only upon strain) to the other of a pure fluid (stress dependent only on the rate of strain). In other words, all materials are to some extent *viscoelastic*, exhibiting both viscous and elastic properties.

3.4 Deformations of statically determinate systems

As we saw in Chapter 2, the forces and stresses in statically determinate systems can be found by considering only the equilibrium requirements. This is not to say that such systems do not deform, merely that the strains and displacements produced do not affect the forces and stresses. If we wish to find the deformations, we can now do so with the aid of the stress–strain equations.

3.4.1 Pin-jointed structures

We saw in Section 2.1 that a typical member of a pin-jointed structure is relatively long and thin, and may therefore be assumed to be subject only to a tensile or compressive force along its length, resulting in a state of uniaxial stress. Let us assume that the cross-sectional area of the member is constant, and of magnitude A. If the force in the member is T (with the usual tensile positive sign convention), the axial stress is $\sigma = T/A$, and from Hooke's Law the strain in the absence of any change in temperature is

$$e = \frac{\sigma}{E} = \frac{T}{AE} \tag{3.18}$$

where E is the Young's modulus of the material. Taking the length of the member as L, measured between the centers of the pins at its ends, and ignoring local deformations at the pin holes, the change in length is

$$\Delta L = eL = \frac{TL}{AE} \tag{3.19}$$

Using this result, together with the previously calculated forces in all the members of the structure, we can calculate the changes in lengths of the members. Up to this point in the analysis, we have used only two of the three physical principles of the mechanics of solids laid down in Section 1.5, namely *equilibrium of forces* and *stress–strain characteristics of the materials.* In order to find the deformed shape of the structure, which can be expressed in terms of the displacements of its joints, we must also satisfy *compatibility of strains* (*geometry of deformation*).

Consider, for example, the simple pin-jointed structure analyzed in Example 2.1, and illustrated in Figs 2.1 and 2.9. Figure 3.12 shows the model for this structure with the calculated forces in the two members and their lengths.

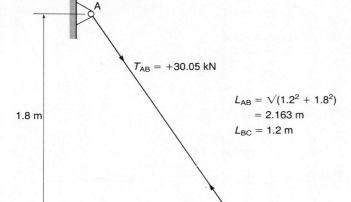

$T_{AB} = +30.05$ kN

$L_{AB} = \sqrt{(1.2^2 + 1.8^2)}$
$\quad\quad = 2.163$ m
$L_{BC} = 1.2$ m

1.8 m

C $\quad T_{BC} = -16.67$ kN B

1.2 m

25 kN

Fig. 3.12.

Assuming the cross-sectional area of each of the (steel) members is 325 mm², their changes in length can be calculated as

$$\Delta L_{AB} = \frac{+30.05 \times 10^3 \times 2.163}{325 \times 10^{-6} \times 207 \times 10^9} \text{ m} = +0.9662 \text{ mm}$$

$$\Delta L_{BC} = \frac{-16.67 \times 10^3 \times 1.2}{325 \times 10^{-6} \times 207 \times 10^9} \text{ m} = -0.2973 \text{ mm}$$

What we are also interested in, however, is the displacement of point B. To find this displacement, we must take into account the compatibility requirement that, while members AB and BC change in length by the amounts calculated, they must move in such a way as to remain straight and to remain pinned together at B. Figure 3.13 shows how this occurs, with the changes in length greatly magnified so that we can see what is happening. If we imagine the members to be uncoupled at B, then in the absence of rotations of either member, end B of AB would be displaced to the point labelled B_1, while end B of BC would be displaced to B_2 (Fig. 3.13a). The two members must now be rotated, about their fixed ends A and C, respectively, until B_1 and B_2 coincide, at B_3. In other words, B_3 is the point of intersection of the arc of radius AB_1, center A, with the arc of radius CB_2, center C. Now, the changes of length and therefore angles of rotation of the members are very much exaggerated in the illustration. In reality they are so small that there is negligible loss of accuracy in replacing the arcs by the tangents to these arcs at points B_1 and B_2. These

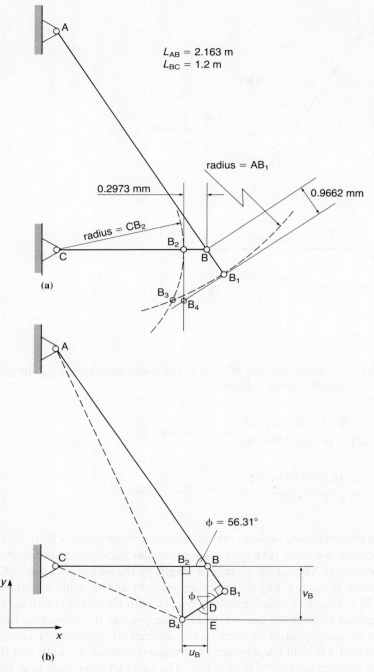

$L_{AB} = 2.163$ m
$L_{BC} = 1.2$ m

radius = AB_1

0.2973 mm

0.9662 mm

radius = CB_2

B_2

B

B_1

B_3

B_4

(a)

$\phi = 56.31°$

C

B_2

B

B_1

ϕ

D

v_B

B_4

E

y

x

u_B

(b)

Fig. 3.13.

tangents meet at the point B_4, and Fig. 3.13b shows a redrawn view of the assumed geometry changes, with dotted lines indicating the deformed shape of the structure. From this illustration we can find the horizontal and vertical components of the displacement of point B to position B_4. Suppose we denote these components by u_B in the x (horizontal) and v_B in the y (vertical) directions, as shown. Now

$$u_B = BB_2 = \Delta L_{BC} = -0.2973 \text{ mm}$$

with the negative sign indicating that the displacement is in the negative x direction. Also

$$v_B = BD + DE$$

where the vertical through point B cuts the line B_1B_4 at D, and the horizontal through B_4 at E. In terms of the known changes in length

$$|v_B| = |\Delta L_{AB} \operatorname{cosec} \phi| + |\Delta L_{BC} \cot \phi|$$

$$= 0.9662 \times 1.2018 + 0.2973 \times 0.6667 = 1.3594 \text{ mm}$$

and since this displacement component is in the downward vertical direction, $v_B = -1.3594$ mm. Having found the horizontal and vertical components of the displacements of B, we can of course determine the magnitude and direction of the resultant displacement there.

It is instructive to compare these approximate values of u_B and v_B with the exact values. The latter can be obtained by considering the geometry of the deformed triangular shape of the structure shown in Fig. 3.14. Since we are interested in very small changes of lengths and angles, we must work to high precision (eight significant figures) in our calculations. The deformed length of member BC is

$$a = 1.2 - 0.000\,297\,3 = 1.199\,702\,7 \text{ m}$$

while the distance between fixed points A and C remains at $b = 1.8$ m, and the deformed length of member AB is

$$c = \sqrt{(1.2^2 + 1.8^2)} + 0.000\,966\,2 = 2.164\,297\,0 \text{ m}$$

The angle between member BC and the vertical line AC, which was originally a right angle, is increased to $90° + \varepsilon$. Applying the cosine rule to the triangle

$$c^2 = a^2 + b^2 - 2ab \cos(90° + \varepsilon)$$

or

$$\cos(90° + \varepsilon) = \frac{a^2 + b^2 - c^2}{2ab}$$

from which $\quad \varepsilon = 0.064\,94°$

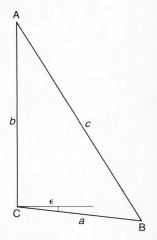

Fig. 3.14.

The exact value of the horizontal displacement of point B is

$$u_B = a \cos \varepsilon - 1.2 = -2.981 \times 10^{-4} \, \text{m} = -0.2981 \, \text{mm}$$

and the exact value of the vertical displacement is

$$v_B = -a \sin \varepsilon = -1.3597 \, \text{mm}$$

Clearly, the errors in the approximate values of -0.2973 mm and -1.3594 mm (some 0.3% and 0.02%, respectively) are negligibly small. Another point to notice is the very small change in angle, ε, of member BC of less than $0.07°$. Since the angles of the members play an important role in the equilibrium analysis of pin-jointed structures (Section 2.1.1), we may conclude that we were perfectly justified in the present problem in using the *undeformed* geometry of the structure to calculate the forces in the members. This is also true of the large majority of engineering structures, and is a consequence of the small strains associated with the elastic deformation of metallic materials.

An important difference between the approximate and exact methods described above for calculating displacements of joints of pin-jointed structures is that while the exact method involves the use of a nonlinear relationship (the cosine rule) between the lengths and changes in length of the members, the approximate method uses only linear equations linking the deformations. This is of particular significance when we come to automate the calculation of displacements in such structures, in Chapter 4, using an approach which relies on creating a set of linear algebraic equations for solution by computer. In the above simple example, involving only two members and one joint which is not fixed, the amount of calculation involved in finding displacements of the structure is significant, particularly in relation to that involved in finding the forces in the members. With structures of much greater complexity, the effort involved becomes prohibitive and a computer method of solution is highly desirable. We will defer consideration of such a method until Chapter 4, when we will be able to treat both statically determinate and statically indeterminate pin-jointed structures.

3.4.2 Pressurized thin-walled cylinder

In Section 2.2.1 we used equilibrium considerations alone to find expressions for the stresses in the wall of a thin-walled cylinder subject to internal hydrostatic pressure. From equations (2.16) and (2.18), the membrane stresses in the wall of a cylinder with unrestrained closed ends in the hoop and axial directions are

$$\sigma_\theta = \frac{pD}{2t} \quad \text{and} \quad \sigma_z = \frac{pD}{4t} \tag{3.20}$$

where D is the diameter of the cylinder (previously we worked in terms of radius), t is its radial thickness and p is the internal pressure. If we are given sufficient information to be able to distinguish between inner, outer and mean diameter, then the most appropriate value to use for D is that of the inner diameter.

Figure 3.15a shows the geometry and the cylindrical polar coordinates, while Fig. 3.15b shows a small piece of the cylinder wall with the hoop and axial stresses acting on it. In Section 2.2.1 we concluded that the radial stress, σ_r, which acts normal to the curved plane of this piece of cylinder, is negligibly small compared with the membrane stresses.

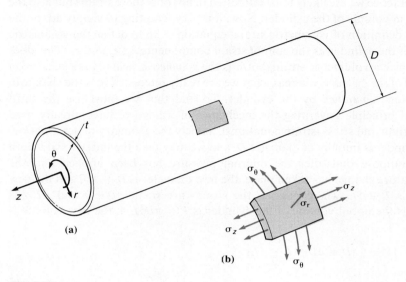

Fig. 3.15.

Having used equilibrium considerations to find the stresses, we can now use stress–strain equations in the form of Hooke's Law with thermal strains, equations (3.16), to find the strains in the cylinder wall. Although these stress–strain equations were expressed in terms of stress and strain components in a Cartesian coordinate system, they are equally applicable to components in any *orthogonal* (mutually perpendicular) coordinate system, including cylindrical polar coordinates. Since the radial stress is zero

$$e_\theta = \frac{1}{E}[\sigma_\theta - v\sigma_z] + \alpha \Delta T$$

$$= \frac{pD}{4Et}(2 - v) + \alpha \Delta T \tag{3.21}$$

and

$$e_z = \frac{1}{E}[\sigma_z - v\sigma_\theta] + \alpha \Delta T$$

$$= \frac{pD}{4Et}(1 - 2v) + \alpha \Delta T \tag{3.22}$$

also

$$e_r = \frac{1}{E}[-v(\sigma_\theta + \sigma_z)] + \alpha\Delta T$$

$$= -\frac{3vpD}{4Et} + \alpha\Delta T \tag{3.23}$$

In practice we are likely to be interested in not only these strains but also the change in volume of the cylinder. Now, it is very tempting to simply adapt the general definition of volumetric strain, equation (3.3), to define the volumetric strain of the cylinder as the sum of strain components e_θ, e_z and e_r. This does indeed give a volumetric strain, but it is the *volumetric strain in the material of the wall of the cylinder*, whereas what we are really interested in is the change in the volume *contained* by the cylinder. To find this, we must use the third physical principle governing the mechanics of solids (we have already used equilibrium and stress–strain equations), namely the geometry of deformation. The cylinder is initially of diameter D, and we may take the initial axial length as L. Suppose that after the internal pressure has been applied and the temperature change has taken place, the new diameter is $D + \Delta D$, and the new length is $L + \Delta L$, where because the strains are small $\Delta D \ll D$ and $\Delta L \ll L$. Whereas the initial volume of the cylinder is $V = \pi D^2 L/4$, the new volume is

$$V + \Delta V = \frac{\pi}{4}(D + \Delta D)^2 (L + \Delta L)$$

and

$$\Delta V = \frac{\pi D^2 L}{4}\left[\left(1 + \frac{\Delta D}{D}\right)^2 \left(1 + \frac{\Delta L}{L}\right) - 1\right]$$

and the volumetric strain of the contents of the cylinder is

$$e_{vol} = \frac{\Delta V}{V} = \left(1 + 2\left(\frac{\Delta D}{D}\right) + \left(\frac{\Delta D}{D}\right)^2\right)\left(1 + \frac{\Delta L}{L}\right) - 1$$

Retaining only terms involving $\Delta D/D$ and $\Delta L/L$, but not products of these quantities (which are negligibly small), this expression becomes

$$e_{vol} = 2\left(\frac{\Delta D}{D}\right) + \frac{\Delta L}{L}$$

Now, while $\Delta L/L$ is the axial strain, e_z, in the cylinder wall, the identity in terms of the strains of $\Delta D/D$ is perhaps not so clear. Since it is the ratio between the change in diameter and the initial diameter, which is the same as the ratio between the change in radius and the initial radius, we might be tempted to equate it to the radial strain, e_r. But e_r is the strain in the radial direction *in the cylinder wall*, not in the volume contained by the cylinder. In other words, e_r is the ratio between the *change in thickness and the initial thickness* of the wall. The term $\Delta D/D$ can, however, be related to the hoop strain, e_θ, in the circumferential direction, which can be defined as the ratio between the change in circumference and the initial circumference

$$e_\theta = \frac{\pi(D + \Delta D) - \pi D}{\pi D} = \frac{\Delta D}{D} \qquad (3.24)$$

Therefore, the volumetric strain of the contents of the cylinder is

$$e_{\text{vol}} = 2e_\theta + e_z \qquad (3.25)$$

and, using equations (3.21) and (3.22) to define the hoop and radial strains, this becomes

$$e_{\text{vol}} = \frac{pD}{4Et}(5 - 4v) + 3\alpha\Delta T \qquad (3.26)$$

In order to find the change in volume, we multiply this strain by the initial volume, $V = \pi D^2 L/4$.

EXAMPLE 3.9

A long thin-walled cylindrical steel tank has a diameter of 800 mm (32 in) and a wall thickness of 5 mm (0.2 in). The ends of the tank are closed by flat plates which are free to move apart axially, and when it is pressurized its diameter is observed to increase by 0.48 mm (0.019 in). Find the pressure in the tank.

This example involves the straightforward application of equations (3.21) and (3.24). The hoop strain is

$$e_\theta = \frac{0.48}{800} = 6.00 \times 10^{-4}$$

and, in the absence of any change in temperature, the internal pressure is related to the hoop strain by

$$p = \frac{4Ete_\theta}{D(2 - v)}$$

Substituting in the numerical values (including the elastic constants for steel from Appendix A)

$$p = \frac{4 \times 207 \times 10^9 \times 0.005 \times 6.00 \times 10^{-4}}{0.800 \times (2 - 0.3)}$$

$$= 1.83 \times 10^6 \text{ N/m}^2 = \underline{1.83 \text{ MN/m}^2} \text{ (270 psi)}$$

EXAMPLE 3.10

Repeat Example 3.9 assuming that the pressure in the tank is applied by means of freely-moving pistons, as shown in Fig. 2.38.

While the hoop strain is unchanged at 6.00×10^{-4}, the fact that the cylinder is now open-ended with no axial stress means that equation (3.21) is no longer valid. The state of stress is now uniaxial (in the hoop direction), and with no change in temperature

$$\sigma_\theta = \frac{pD}{2t} = Ee_\theta$$

from which

$$p = \frac{2Ete_\theta}{D} = \frac{2 \times 207 \times 10^9 \times 0.005 \times 6.00 \times 10^{-4}}{0.800}$$

$$= 1.55 \times 10^6 \text{ N/m}^2 = \underline{1.55 \text{ MN/m}^2} \text{ (220 psi)}$$

EXAMPLE 3.11

The copper pipe used in a domestic heating system has a diameter of 15 mm (0.6 in) and a wall thickness of 0.5 mm (0.02 in). The total length of piping used is 60 m (200 ft). Under operating conditions, the temperature of the system rises by 60°C (110 °F) and the pressure by 0.5 MN/m² (70 psi) above ambient conditions. Calculate the stresses in the pipe wall, its change in thickness, and the change in internal volume of the system.

The change in temperature alone has no effect on the stresses which, since the piping is effectively closed-ended but free to expand axially, are given by equation (3.20) as

SI units

$$\sigma_\theta = \frac{0.5 \times 10^6 \times 0.015}{2 \times 0.0005} \text{ N/m}^2 = \underline{7.50 \text{ MN/m}^2}$$

and

$$\sigma_z = \frac{\sigma_\theta}{2} = \underline{3.75 \text{ MN/m}^2}$$

The properties of copper may be obtained from Appendix A as $E = 120$ GN/m², $v = 0.33$ and $\alpha = 16 \times 10^{-6} \, °\text{C}^{-1}$. The radial strain in the pipe wall is given by equation (3.23) as

$$e_r = -\frac{3 \times 0.33 \times 0.5 \times 10^6 \times 0.015}{4 \times 120 \times 10^9 \times 0.0005} + 16 \times 10^{-6} \times 60$$

$$= -3.094 \times 10^{-5} + 9.6 \times 10^{-4} = 9.291 \times 10^{-4}$$

the thermal strain being the dominant component. The change in wall thickness is therefore

$$\Delta t = e_r t = 9.291 \times 10^{-4} \times 0.5 = \underline{+4.65 \times 10^{-4} \text{ mm}}$$

US customary units

$$\sigma_\theta = \frac{70 \times 0.6}{2 \times 0.02} = \underline{1050 \text{ psi}}$$

and

$$\sigma_z = \frac{\sigma_\theta}{2} = \underline{525 \text{ psi}}$$

The properties of copper may be obtained from Appendix A as $E = 17 \times 10^6$ psi, $v = 0.33$ and $\alpha = 8.9 \times 10^{-6} \, °\text{F}^{-1}$. The radial strain in the pipe wall is given by equation (3.23) as

$$e_r = -\frac{3 \times 0.33 \times 70 \times 0.6}{4 \times 17 \times 10^6 \times 0.02} + 8.9 \times 10^{-6} \times 110$$

$$= -3.057 \times 10^{-5} + 9.79 \times 10^{-4} = 9.484 \times 10^{-4}$$

the thermal strain being the dominant component. The change in wall thickness is therefore

$$\Delta t = e_r t = 9.484 \times 10^{-4} \times 0.02 = \underline{+1.90 \times 10^{-5} \text{ in}}$$

The volumetric strain can be found with the aid of equation (3.26) as

$$e_{\text{vol}} = \frac{0.5 \times 10^6 \times 0.015 \times (5 - 4 \times 0.33)}{4 \times 120 \times 10^9 \times 0.0005}$$

$$+ 3 \times 16 \times 10^{-6} \times 60$$

$$= 1.15 \times 10^{-4} + 2.88 \times 10^{-3} = 2.995 \times 10^{-3}$$

Since the initial volume of the piping system is

$$V = \frac{\pi}{4} \times 0.015^2 \times 60 = 0.010\,60 \text{ m}^3$$

the change in volume is

$$\Delta V = 0.010\,60 \times 2.995 \times 10^{-3}$$

$$= \underline{3.18 \times 10^{-5} \text{ m}^3 \ (31.8 \text{ cm}^3)}$$

The volumetric strain can be found with the aid of equation (3.26) as

$$e_{\text{vol}} = \frac{70 \times 0.6 \times (5 - 4 \times 0.33)}{4 \times 17 \times 10^6 \times 0.02}$$

$$+ 3 \times 8.9 \times 10^{-6} \times 110$$

$$= 1.136 \times 10^{-4} + 2.937 \times 10^{-3} = 3.051 \times 10^{-3}$$

Since the initial volume of the piping system is

$$V = \frac{\pi}{4} \times 0.6^2 \times 200 \times 12 = 678.6 \text{ in}^3$$

the change in volume is

$$\Delta V = 678.6 \times 3.051 \times 10^{-3}$$

$$= \underline{2.07 \text{ in}^3}$$

3.4.3 Pressurized thin-walled sphere

In Section 2.2.2 we used equilibrium considerations to find an expression for the circumferential stress in the wall of a thin-walled sphere subject to internal hydrostatic pressure. Equation (2.21) gives a good approximation to this stress as

$$\sigma_c = \frac{pD}{4t} \tag{3.27}$$

where D is the (inner) diameter of the cylinder, t is its radial thickness and p is the internal pressure.

Figure 3.16a shows the geometry, while Fig. 3.16b shows a small piece of the wall of the sphere with the circumferential stress acting on each of its edges. The radial stress, σ_r, which acts normal to the curved plane of this piece of cylinder, is negligibly small compared with the circumferential stress, and the state of stress in the wall is one of biaxial tension with equal stresses in any two perpendicular directions, and indeed in any direction locally tangential to the sphere. Using stress–strain equations (3.16), we can find the strains in the wall of the sphere as

$$e_c = \frac{1}{E}[\sigma_c - v\sigma_c] + \alpha \Delta T$$

$$= \frac{pD}{4Et}(1 - v) + \alpha \Delta T \tag{3.28}$$

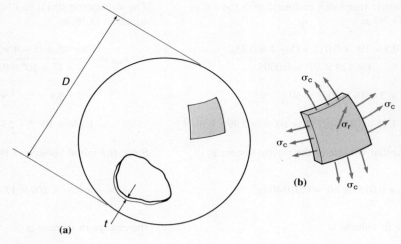

(a)

(b)

Fig. 3.16.

and

$$e_r = \frac{1}{E}[-v(\sigma_c + \sigma_c)] + \alpha \Delta T$$

$$= -\frac{vpD}{2Et} + \alpha \Delta T \qquad (3.29)$$

where, as in the case of a thin-walled cylinder, the radial strain, e_r, is only useful for finding the change in thickness of the wall of the sphere.

In order to find the change in the internal volume of the sphere, we can consider its diameter changing from D to $D + \Delta D$, producing a change in volume from $V = \pi D^3/6$ to

$$V + \Delta V = \frac{\pi}{6}(D + \Delta D)^3$$

and a volumetric strain of the contents of the sphere of

$$e_{vol} = \frac{\Delta V}{V} = \left(1 + \frac{\Delta D}{D}\right)^3 - 1 \approx 3\left(\frac{\Delta D}{D}\right)$$

Now, the strain, e_c, in any circumferential direction around the sphere is the ratio between the change in circumference and the initial circumference, and hence equal to the ratio $\Delta D/D$ between the change in diameter and the initial diameter (as in equation (3.24)). The volumetric strain is therefore

$$e_{vol} = 3e_c \qquad (3.30)$$

and, using equation (3.28) to define the circumferential strain, this becomes

$$e_{vol} = \frac{3pD}{4Et}(1 - v) + 3\alpha\Delta T \qquad (3.31)$$

To find the change in volume, we multiply this expression by the initial volume $V = \pi D^3/6$.

EXAMPLE 3.12

A spherical aluminum vessel of 350 mm (14 in) diameter and 4 mm (0.16 in) thick is just full of an incompressible liquid under ambient conditions. If a further 5×10^4 mm^3 (50 cm^3 or 3 in^3) of the liquid is pumped into the vessel, find the resulting internal pressure and change in diameter.

The initial volume of the vessel is

$$V = \frac{\pi D^3}{6} = \frac{\pi \times 350^3}{6} = 22.45 \times 10^6 \text{ mm}^3$$

and, since this volume is increased by 5×10^4 mm^3, the volumetric strain of the contents of the vessel is

$$e_{vol} = \frac{5 \times 10^4}{22.45 \times 10^6} = 2.227 \times 10^{-3}$$

The elastic properties of aluminum are given in Appendix A as $E = 70$ GN/m^2, $v = 0.34$. Equation (3.31) can therefore be used to obtain the internal pressure as

$$p = \frac{4 \times 70 \times 10^9 \times 0.004 \times 2.227 \times 10^{-3}}{3 \times 0.350 \times (1 - 0.34)}$$

$$= 3.60 \times 10^6 \text{ N/m}^2 = \underline{3.60 \text{ MN/m}^2} \text{ (520 psi)}$$

Also, from equation (3.30), the circumferential strain is one third of the volumetric strain

$$e_c = \frac{2.227 \times 10^{-3}}{3} = 0.7423 \times 10^{-3}$$

and the change in diameter is therefore

$$\Delta D = 0.7423 \times 10^{-3} \times 350 = \underline{0.260 \text{ mm}} \text{ (0.010 in)}$$

In practice, the compressibility of the liquid contained will have a significant effect on this result. Example 4.8 repeats this problem with water as the liquid.

EXAMPLE 3.13

A tank for storing compressed air is made from a thin-walled cylinder of diameter D, length $2D$ and thickness t, with hemispherical ends made from the same material. If the radial expansion of the cylinder and hemispheres at their junctions is to be the same when the vessel is subjected to an internal pressure p, calculate the necessary uniform thickness of the hemispheres, given that the Poisson's ratio of the vessel material is 1/3. Also, find an expression for the change in internal volume of the vessel.

This example usefully combines the analyses of thin-walled cylinders and spheres. Figure 3.17 shows the geometry of the vessel: the thickness of the hemispherical ends is t'. Since in practice the cylinder and hemispheres must be welded together, it is desirable that these separate components tend to deform by the same amount under the same internal pressure. If this is not the case, and they tend to expand radially by different amounts, then when they are welded together and internal pressure applied local bending will occur. Although we wish to equate radial expansions, this does *not* imply equal *radial strains* (which would only ensure the same relative changes in wall *thickness*), but rather equal *circumferential (hoop) strains*.

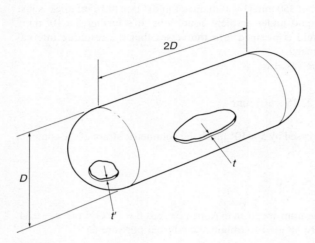

Fig. 3.17.

From equation (3.21), the hoop strain in the cylindrical portion of the vessel is

$$e_\theta = \frac{pD}{4Et}(2 - v)$$

where E and v are the Young's modulus and Poisson's ratio of the vessel material. Similarly, the circumferential strain in the hemispherical ends is given by equation (3.28) as

$$e_c = \frac{pD}{4Et'}(1 - v)$$

For these two expressions to be equal

$$\frac{t'}{t} = \frac{1 - v}{2 - v} = \frac{1 - \frac{1}{3}}{2 - \frac{1}{3}} = 0.4$$

and the thickness of the hemispherical ends should be only 40% of that of the cylinder.

Now, in terms of volumes, the vessel is composed of a cylinder of diameter D, length $2D$, and hence volume $V_c = \pi D^3/2$, together with a sphere of diameter D, and volume $V_s = \pi D^3/6$. The volumetric strain for the cylinder is given by equation (3.26) as

$$e_{vol} = \frac{pD}{4Et}(5 - 4v)$$

and its change in volume is therefore

$$\Delta V_c = \frac{\pi p D^4}{8Et}(5 - 4v)$$

Similarly, the volumetric strain for the sphere is given by equation (3.31) as

$$e_{vol} = \frac{3pD}{4Et'}(1 - v)$$

and its change in volume is

$$\Delta V_s = \frac{\pi p D^4}{8Et'}(1 - v)$$

With $v = 1/3$ and $t' = 0.4\,t$, the total change in volume of the vessel is

$$\Delta V = \Delta V_c + \Delta V_s = \frac{2\pi p D^4}{3Et}$$

Problems

SI METRIC UNITS –
STRESS–STRAIN EQUATIONS

3.1 An aluminum alloy tube of length 600 mm and circular cross section with an inner diameter of 32 mm and an external diameter of 36 mm is subjected to an axial tensile force of 25 kN. Find the change in length of the tube.

3.2 Find the change in length of the shaft in Problem 1.14, assuming it is made of aluminum. Sections A and B are 400 mm and 300 mm long, respectively.

3.3 Find the change in length of the shaft in Problem 1.15, assuming it is made of brass. Sections A and B are 350 mm and 200 mm long, respectively.

3.4 In Problem 1.20, the weight supported at point E is 2.5 kN. Find the diameters of the steel wires CD and BE.

3.5 An open-ended piece of aluminum alloy tube with a diameter of 200 mm is dropped into the sea and comes to rest at a depth where the pressure is 45 MN/m². Find the change in the diameter of the tube.

3.6 A steel bar of length 750 mm and circular cross section whose diameter varies linearly from 12 mm at one end to 17 mm at the other is subjected to an axial tensile force of 20 kN. Find the change in length of the bar.

3.7 A uniform cable of length L, made from a material of density ρ and Young's modulus E, is suspended vertically from one end. Find an expression for the increase in length of the cable due to its own weight.

3.8 A thin sheet of brass lying in the x, y plane is stretched until the uniform strains in the x and y directions are $e_x = 1.2 \times 10^{-4}$ and $e_y = 2.0 \times 10^{-4}$. Assuming there is no stress normal to the plane of the sheet ($\sigma_z = 0$), find the stresses in the sheet.

3.9 Repeat Problem 3.8 assuming that the piece of brass is not a thin sheet but is very thick in the z direction, and that there is no strain in this direction.

3.10 The sheet of Problem 3.8 is held rigidly at its edges in its stretched state and then subjected to a uniform temperature decrease of 20 °C. Calculate the resulting stresses in the sheet.

3.11 The Young's modulus and shear modulus of a material are 150 GN/m² and 67 GN/m², respectively. Find the values of Poisson's ratio and the bulk modulus of the material.

3.12 The steel rod shown in Fig. P3.12 passes through a hydraulic cylinder. The sealing rings may be assumed to be frictionless. Find the change in overall length of the rod and the change in diameter at its center when the pressure in the oil in the cylinder is raised to 50 MN/m².

3.13 A solid plug of elastic material (Fig. P3.13) with a Young's modulus E and Poisson's ratio v is contained within a rigid cylinder of diameter D, and in the stress-free state has a length of L. If the cylinder is closed by a rigid piston to which a compres-

sive force F is applied, find an expression for the reduction in length of the plug.

3.14 Invert equations (3.16) to obtain stresses in terms of strains and temperature change.

3.15 A square steel plate with sides of length 500 mm and a thickness of 20 mm has a circular hole of 30 mm diameter drilled at its center. If the edges of the plate are unrestrained, determine the shape and size of the hole when the temperature of the plate is increased uniformly by 100 °C.

3.16 In Example 3.1, in addition to the tightening of the nut, the temperature of the bolt and nut is reduced by 20 °C. Find the final force in the bolt, and the total reduction in its diameter.

3.17 A prismatic steel block of rectangular cross sec-

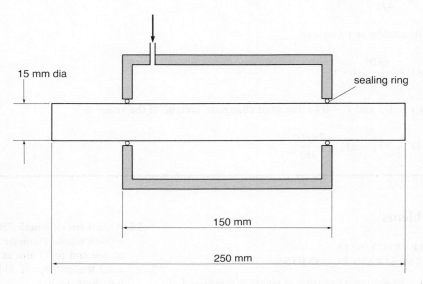

15 mm dia

sealing ring

150 mm

250 mm

Fig. P3.12.

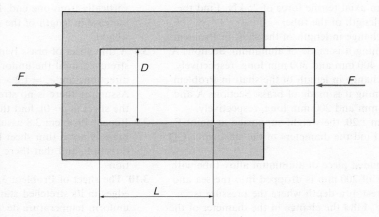

D

F F

L

Fig. P3.13.

134

tion 60 mm by 80 mm is 100 mm long. Uniformly distributed compressive forces of 1.2 MN are applied to the opposite 100 mm by 60 mm faces. The 100 mm by 80 mm faces carry no load and displacement of these faces is unrestricted. The increase in length of the 100 mm dimension is restricted to 0.015 mm. Determine the uniformly distributed forces which must be applied to the 60 mm by 80 mm faces, and the change in volume of the block.

SI METRIC UNITS – STATICALLY DETERMINATE SYSTEMS

3.18 In Problem 1.3, if the jib AB can be regarded as rigid, and the steel cable CD has a cross-sectional area of 50 mm², find the horizontal and vertical components of displacement of point B.

3.19 The pin-jointed steel framework shown in Fig. P3.19 consists of two members, AB and AC, each of which has a rectangular 40 mm by 15 mm cross section. A load of 40 kN is applied to the framework at A. Find the displacement of this point in the direction of the applied load.

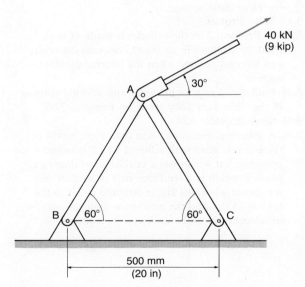

Fig. P3.19.

3.20 Calculate the changes in outer diameter and wall thickness of the tube in Problem 3.1.

3.21 A thin-walled cylindrical steel tank with its ends closed by flat plates (which are free to move apart axially) is subjected to internal hydrostatic press-

ure. If the initial length of the cylinder is three times its diameter, find the ratio between the changes in length and diameter.

3.22 Find the changes in length, internal diameter and wall thickness of the cylinder in Problem 2.32.

3.23 Find the change in diameter and internal volume of the aluminum alloy vessel in Problem 2.33.

3.24 An evacuated thin-walled spherical vessel is placed inside a larger spherical vessel made from the same material, and the gap between the two vessels is then just filled with liquid. The inner vessel is of radius R_1 and thickness t_1, while the outer is of radius R_2 and thickness t_2. More liquid is then added to the gap, increasing the pressure there by p. Find an expression for the change in volume of the gap.

3.25 A solid steel block, 500 mm in diameter and 300 mm thick, rests on rigid supports as shown in Fig. P3.25. Below and coaxial with the block is a thin-walled closed steel cylinder with an internal diameter of 100 mm, a length of 200 mm, a wall thickness of 2 mm and rigid ends. The cylinder rests on a rigid surface and when it is unpressurized there is a gap of 0.01 mm between the upper end of the cylinder and the underside of the block. Find (i) the internal pressure at which the end of the cylinder will just contact the block, and (ii) the change in volume, compared with the unpressurized condition, when the pressure is increased to twice the value found in (i).

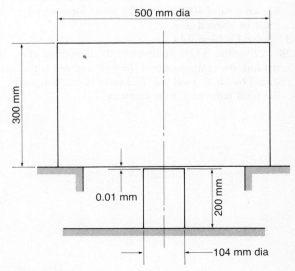

Fig. P3.25.

3.26 A 200 ft length of steel wire is to carry a tensile load of 340 lb. Find the minimum acceptable wire diameter if the maximum allowable stress in the wire is 30 ksi, and the length of the wire must not increase by more than 2 in when there is no change in temperature.

3.27 A solid piece of cast iron is dropped into the sea and comes to rest on the sea bottom at a depth where the hydrostatic pressure of the sea water is 10 ksi. Find the percentage change in density of the material.

3.28 A steel bar of length 30 in and circular cross section whose diameter varies linearly from 0.5 in at one end to 0.75 in at the other is subjected to an axial tensile force of 4.5 kip. Find the change in length of the bar.

3.29 to 3.31 Solve Problems 3.7 to 3.9.

3.32 Solve Problem 3.10 for a temperature decrease of 36°F.

3.33 The Young's modulus and shear modulus of a material are 20×10^6 psi and 8.8×10^6 psi, respectively. Find the values of Poisson's ratio and the bulk modulus of the material.

3.34 A rectangular block of concrete has sides of length 20, 40 and 120 in. If the block is immersed in water at a pressure of 200 psi, find the changes in the lengths of the sides, also the change in volume. The concrete may be assumed to be impermeable to water.

3.35 Solve Problem 3.13.

3.36 In Problem 1.35 the incompressible rubber has a Young's modulus of 7 ksi. If the top sheet of steel has an area of 400 in², find the shear force which must be applied to it.

3.37 Solve Problem 3.14.

3.38 In Example 3.1, in addition to the tightening of the nut, the temperature of the bolt and nut is reduced by 36°F. Find the final force in the bolt, and the total reduction in its diameter.

3.39 A brass bar just fits into the gap between two rigid surfaces at ambient temperature. If the maximum allowable compressive stress in the bar is 12 ksi, find the maximum permissible increase in temperature of the bar.

3.40 A phosphor bronze sleeve is to be secured in place on a steel shaft by having an interference fit between them. While the diameter of the solid shaft is 6 in, the sleeve is machined to an inner diameter of 5.996 in and outer diameter of 6.6 in. In order to assemble the arrangement, either heating or cooling is applied. Assuming that a diametral clearance of 0.002 in is required for ease of assembly, calculate the uniform changes in temperature required if (i) the sleeve alone is heated, (ii) the shaft alone is cooled, or (iii) both the sleeve and shaft are heated.

3.41 Solve Problem 3.19, using the dimensions and load shown in parentheses in Fig. P3.19. Each of the members AB and AC has a rectangular 1.6 in by 0.6 in cross section.

3.42 Solve Problem 3.21.

3.43 In Problem 2.66 the cylinder is made of steel. Find the changes in its length, external diameter, and internal volume when the internal pressure is applied.

3.44 Find the change in diameter and internal volume of the aluminum alloy vessel im Problem 2.65.

3.45 Solve Problem 3.24.

3.46 A spherical steel tank 48 in diameter and 0.4 in thick is just filled with a liquid which is incompressible, but which has a coefficient of linear expansion which is three times that of steel. If the maximum allowable circumferential stress in the tank is 45 ksi, find the maximum uniform increase in temperature of the tank and its contents.

4

Statically Indeterminate Systems

Let us review the position we have reached. In Chapter 1 (Section 1.5), we established the three physical principles governing the mechanics of deformable solids, namely equilibrium of forces, compatibility of strains (geometry of deformation), and material stress–strain characteristics. While the complete solution of any problem, including the determination of all the forces, stresses, displacements and strains, requires the use of all three of these, in some cases we can find just the forces and stresses in a system without introducing either the compatibility of strains or the material properties. Such systems we have classified as statically determinate, and in Chapter 2 we analyzed some common practical problems of this type. If we wish to find displacements and strains in a system, be it statically determinate or statically indeterminate, then we must consider strain compatibility and stress–strain relationships. In Chapter 3 we established these relationships for linearly elastic materials, and used them to find the deformations of statically determinate systems.

We are now in a position to analyze statically indeterminate systems. First we will consider plane pin-jointed structures, followed by a range of other types of statically indeterminate problems, such as thin-walled vessels filled with compressible fluids, and resisted thermal expansion.

4.1 Pin-jointed structures

As we have already seen in Section 2.1 for the equilibrium analysis of statically determinate frameworks, pin-jointed structure problems are very suitable for solution by computer methods. This is because, although a large number of calculations are involved, they are individually straightforward and repetitive, and therefore convenient to automate. We saw in Section 2.1 how relatively simple statically determinate structures give rise to sizeable sets of equilibrium equations. In Section 3.4.1 we saw that if deformations are also required, as they certainly are for statically indeterminate structures, the calculations for even very simple frameworks become tedious if they have to be performed manually. We are therefore seeking a method suitable for use on a computer. In the past, attempts to establish such methods have led to the development of the *finite element* technique. Although this technique is now very widely used, not only in solid mechanics but also in fluid mechanics, heat transfer and in many other branches of engineering science, its origins lie in structural analysis. It is therefore appropriate for us to develop our analysis as a finite element method, and to describe it using finite element terminology.

4.1.1 Finite element analysis

We start by dividing the plane pin-jointed structure into its individual members or *elements*, the behavior of each one of which can be analyzed in a relatively simple way. The behavior of the overall structure can then be established by linking the elements together at the appropriate joints or *nodal points* (*nodes*). Figure 4.1 shows a typical element, which we may number as element *m* (element numbers are ringed to distinguish them from node numbers). It lies in the plane of Cartesian coordinates *X* and *Y*. We may regard these as *global* coordinates in the sense that they are used to define positions of nodes for the structure as a whole, and not just the particular element we are currently considering. The element links two nodes, which we may number as *i* and *j*, whose positions are defined in terms of coordinates (X_i, Y_i) and (X_j, Y_j), respectively. We arbitrarily select *i* as the first node and *j* as the second node of the element. Using the first node as the local origin, we then define the orientation of the element in terms of the angle θ_m between the element and the direction of the positive *X* axis, as shown. The length (measured between the centers of the joints) and cross-sectional area of the element are L_m and A_m, respectively, and the Young's modulus of its material is E_m. In general, all three of these may be different for different elements of the structure.

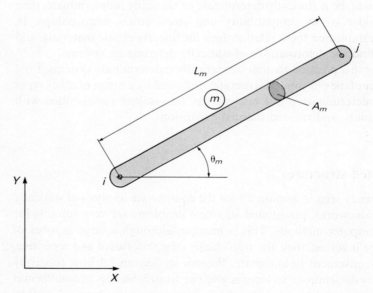

Fig. 4.1.

Now, when the structure is loaded, the nodes are displaced and the elements are strained. Let us take the displacements in the *X* and *Y* directions of node *i* to be u_i and v_i, respectively, and those of node *j* to be u_j and v_j, as shown in Fig. 4.2. Similarly, we can take the external forces in the *X* and *Y* directions applied to the element at node *i* to be U_i and V_i, respectively, and those at node *j* to be U_j and V_j, as shown in Fig. 4.3, which is a free body diagram for the element (body forces such as weight are assumed to be negligible). These external forces are

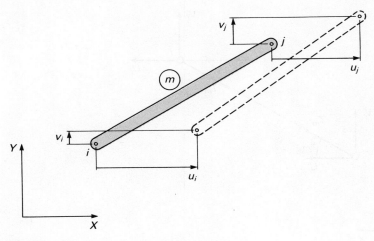

Fig. 4.2.

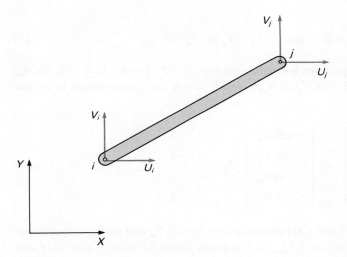

Fig. 4.3.

due to the other elements connected at the particular nodes and to any applied forces there which are external to the structure. Figure 4.4, which now shows the element modeled as a single line joining the nodes, also shows the element effectively sectioned through its center, with the internal force in the element, T_m, acting in accordance with the usual tensile positive sign convention, away from each of the nodes.

We can start to develop the analysis by considering the equilibrium of forces applied to each of the nodes of the typical element. For the forces acting on node i in Fig. 4.4 to be in equilibrium in the X and Y directions

$$U_i + T_m \cos \theta_m = 0 \quad \text{and} \quad V_i + T_m \sin \theta_m = 0 \tag{4.1}$$

Similarly, for the forces acting on node j to be in equilibrium in the X and Y directions

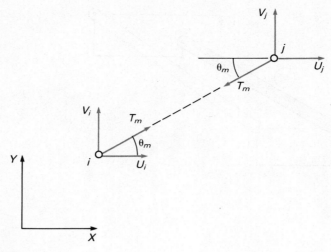

Fig. 4.4.

$$U_j - T_m \cos \theta_m = 0 \quad \text{and} \quad V_j - T_m \sin \theta_m = 0 \tag{4.2}$$

From these four equations we deduce that $U_i = -T_m \cos \theta_m$, $V_i = -T_m \sin \theta_m$, $U_j = T_m \cos \theta_m$ and $V_j = T_m \sin \theta_m$, a result which can be expressed in matrix form as

$$[f]_m = \begin{bmatrix} U_i \\ V_i \\ U_j \\ V_j \end{bmatrix} = \begin{bmatrix} -\cos \theta_m \\ -\sin \theta_m \\ \cos \theta_m \\ \sin \theta_m \end{bmatrix} T_m = \begin{bmatrix} -c \\ -s \\ c \\ s \end{bmatrix} T_m \tag{4.3}$$

where the symbols c and s are abbreviations for $\cos \theta_m$ and $\sin \theta_m$, respectively. The *element force vector*, $[f]_m$, is a column *vector* (a matrix with only one column) containing the four components of the external forces applied to the element at its nodes, arranged in an appropriate order. While we could choose this order arbitrarily, it is convenient to take the components applied to the first node (node i) first, followed by those applied to the second node (node j), in each case taking the component in the X global coordinate direction before that in the Y direction. Although we have used equilibrium conditions to arrive at equations (4.3), they are really only definitions of the components in the global coordinate directions at the nodes of the axial force in the element.

We need the forces in this form to be able to establish equilibrium conditions for the structure as a whole later in the analysis. Note that, in deriving equations (4.3), we are assuming that the angle of the element, θ_m, does not change significantly when the structure deforms.

We next choose the variables we wish to have as the primary unknowns in the analysis; in other words, those variables we intend to solve for in the first instance, and from which other unknowns may then be obtained as required. In the analysis of statically determinate structures (Section 2.1), we solved for the forces in the elements and the reactions at the supports, but in that case we had no alternative. Now we can choose either forces or displacements, and in fact we choose the latter, in particular the displacement components (in the global coordinate directions) at the nodal points of the structure. As the analysis is developed, we shall see that this choice results in a natural balance between the number of unknowns and the number of equations for finding them: there are two displacement components at each node, and two equations of equilibrium of forces at each node.

In a finite element analysis of a more complicated problem, having chosen nodal point displacements as the unknowns, we would have to choose suitable approximations to describe the variations of the displacements and strains within the elements. In this case no approximation is necessary: provided the cross-sectional area of each element is constant, the axial strain along it is also constant. Referring to Fig. 4.2, the displacement of node i of the typical element along the axis of the element towards node j is $(u_i \cos \theta_m + v_i \sin \theta_m)$, while that of node j in the same direction is $(u_j \cos \theta_m + v_j \sin \theta_m)$. Consequently, the axial strain in the element is

$$e_m = [(u_j \cos \theta_m + v_j \sin \theta_m) - (u_i \cos \theta_m + v_i \sin \theta_m)]/L_m$$

or, in matrix form

$$e_m = \frac{1}{L_m}[-c \ -s \ c \ s] \begin{bmatrix} u_i \\ v_i \\ u_j \\ v_j \end{bmatrix} = \frac{1}{L_m}[-c \ -s \ c \ s][\delta]_m \qquad (4.4)$$

where $[\delta]_m$ is the *element displacement vector*, which contains the four nodal point displacement components, u_i, v_i, u_j and v_j, arranged in the same order as for the element force components in equation (4.3). Once again, we are assuming that the angle of the element, θ_m, does not change significantly when the structure deforms.

In equation (4.3), nodal point forces are related to the element axial force, while in (4.4) axial strain is related to the nodal point displacements. Provided we can link the axial force to the axial strain, we will have a relationship between nodal point forces and nodal point displacements. This link is provided by the stress–strain characteristics of the element material. Using equation (3.18), which is derived from Hooke's Law in the absence of any temperature change for an elastic member under uniaxial load

$$T_m = E_m A_m e_m$$

which results in

$$[f]_m = [k]_m [\delta]_m \qquad (4.5)$$

where

$$[k]_m = \frac{E_m A_m}{L_m} \begin{bmatrix} -c \\ -s \\ c \\ s \end{bmatrix} [-c \ -s \ c \ s] = \frac{E_m A_m}{L_m} \begin{bmatrix} c^2 & cs & -c^2 & -cs \\ cs & s^2 & -cs & -s^2 \\ -c^2 & -cs & c^2 & cs \\ -cs & -s^2 & cs & s^2 \end{bmatrix}$$

(4.6)

is the *element stiffness matrix*. In a linearly elastic system, the force applied at a point in a particular direction is directly proportional to the resulting displacement in the same direction there, the constant of proportionality being the *stiffness*. The axial stiffness (ratio of axial force to the change of length) of our typical element is $T_m/\Delta L_m$, or $E_m A_m/L_m$, where ΔL_m is its change of length. This axial stiffness appears as a common factor in equation (4.6). A force applied at a node of the element, not in the direction along its axis but at some angle to this direction, also produces a change of length, and hence displacements, proportional to the axial stiffness, with the constant of proportionality depending on the angle concerned. Equations (4.5) and (4.6) serve to summarize the relationships between the four element force components and the corresponding displacements, via stiffnesses which involve trigonometric functions of the element angle. Consider, for example, the case of an element parallel to the X axis, with $\theta_m = 0$, s = 0, c = 1

$$\begin{bmatrix} U_i \\ V_i \\ U_j \\ V_j \end{bmatrix} = \frac{E_m A_m}{L_m} \begin{bmatrix} 1 & 0 & -1 & 0 \\ 0 & 0 & 0 & 0 \\ -1 & 0 & 1 & 0 \\ 0 & 0 & 0 & 0 \end{bmatrix} \begin{bmatrix} u_i \\ v_i \\ u_j \\ v_j \end{bmatrix}$$

which means that

$$U_i = -U_j = \frac{E_m A_m}{L_m} \Delta L_m \quad \text{where} \quad \Delta L_m = (u_i - u_j)$$

and $V_i = V_j = 0$

which is the expected result for simple axial loading.

It is worth noting that all of the rows of the element stiffness matrix in equations (4.6) are proportional to each other, and so are all of the columns. This means that the determinant of the matrix is zero, and the matrix is *singular* (a condition which is discussed in Appendix B).

Having established the form of relationship between the externally applied forces and the corresponding displacements for a typical element, we can apply this relationship to each of the elements in the structure. We now need to gather together all the resulting element stiffness information to establish the behavior of the whole structure. To do this, we employ two of the three principles of the mechanics of solids, namely the geometry of deformation, and equilibrium of forces acting at the nodes (we have already used stress–strain relationships to determine element stiffnesses). The geometry of deformation condition which must be satisfied is that, when the structure deforms, the displacements of all the elements connected together at a particular node must be the same. This is demonstrated by Fig. 4.5 which shows a typical structure in both its un-deformed state (solid lines) and much-exaggerated deformed state (dotted lines). In terms of the symbols we have chosen to use, displacement components u_i and v_i at node i are the same for all elements which are pinned together there.

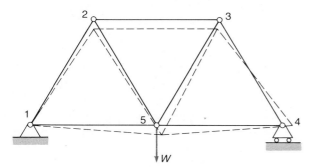

Fig. 4.5.

We now write equilibrium equations for the forces acting at each of the nodes of the structure. Figure 4.6a shows a typical node i which connects a number of elements together. It is subject to forces which are external to the structure (in the sense that in Fig. 4.5 the load W at node 5 is an external load), which may be expressed as known force components $F_x^{(i)}$ and $F_y^{(i)}$ in the positive global coordinate directions, as shown. For most nodes in a typical structure, one or both of these force components is zero. At this stage, we exclude any consideration of nodes which are support points for the structure, involving external reaction forces which are as yet unknown. The externally applied forces at node i are transmitted to the elements connected there, and must be in equilibrium with them. While we could consider the forces acting along each of the elements, for present purposes it is more appropriate to consider the components of these element forces in the global coordinate directions. Figure 4.6b shows the ends of the elements detached from the node, with these components acting on them. For equilibrium of forces in each of the global coordinate directions, the external force applied to the node must be equal to the sum of the forces transmitted to the elements, and

$$\Sigma\, U_i^{(m)} = F_x^{(i)} \tag{4.7a}$$

$$\Sigma\, V_i^{(m)} = F_y^{(i)} \tag{4.7b}$$

where $U_i^{(m)}$ and $V_i^{(m)}$ are the forces at node i acting on element m, and the summations are carried out over elements $m\,(m_1, m_2, m_3$ and m_4 in Fig. 4.6b) which are connected together at node i.

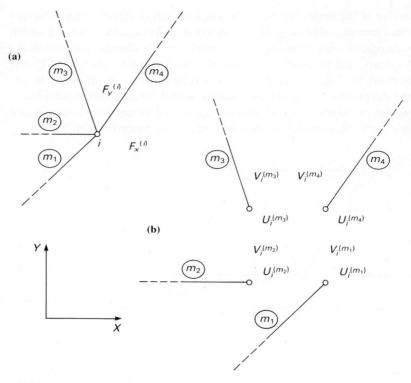

Fig. 4.6.

Forces $U_i^{(m)}$ and $V_i^{(m)}$ applied to typical element m are terms in the element force vector $[f]_m$ which, according to equations (4.5) and (4.6), can be expressed as sums of products of stiffnesses and displacements. While the stiffnesses are known functions of the element geometry and physical properties, the displacements concerned are the four as yet unknown displacement components of the nodes of the element. When the summations required by equations (4.7) are complete, we can therefore expect to have a pair of algebraic equations which are linear in certain unknown displacements. These are the displacement components of node i itself, together with those of the adjacent nodes which are directly connected to node i by elements of the structure. For example, in Fig. 4.5 the equilibrium equations for node 5 will involve the displacements at all the nodes of the structure, while those for, say, node 4 will involve only the displacements of nodes 3, 4 and 5.

When the same process of assembling equilibrium equations is complete for all the nodes of the structure, we have a set of $2n$ linear algebraic equations (where n is the number of nodes in the structure) for the $2n$ displacement components (two per node), which can be expressed in the form

$$[K][\delta] = [F] \tag{4.8}$$

The column vector $[\delta]$ contains the unknown displacements, the square *overall stiffness matrix* $[K]$ contains known stiffnesses, while the column vector $[F]$

contains the known components of the applied external forces. Although we are free to arrange the variables in $[\delta]$ and $[F]$ in any order, it is convenient to choose an order similar to that used for element displacements and forces, namely to take the components of the first node (node 1) first, followed by those of the second node (node 2), and so on for all the other nodes of the structure, in each case taking the component in the X global coordinate direction before that in the Y direction. Therefore

$$[\delta] = \begin{bmatrix} u_1 \\ v_1 \\ u_2 \\ v_2 \\ \cdots \\ u_n \\ v_n \end{bmatrix} \quad \text{and} \quad [F] = \begin{bmatrix} F_x^{(1)} \\ F_y^{(1)} \\ F_x^{(2)} \\ F_y^{(2)} \\ \cdots \\ F_x^{(n)} \\ F_y^{(n)} \end{bmatrix} \tag{4.9}$$

with corresponding displacement and force components occupying the same positions within their respective vectors. The order of arrangement is still arbitrary to the extent that we are free to number the nodes in any convenient order.

The two equilibrium equations for the pth node of the structure are the $(2p - 1)$th and $2p$th equations in (4.8) for the X and Y directions, respectively. The $(2p - 1)$th and $2p$th rows of stiffness matrix $[K]$ therefore contain the stiffness coefficients linking forces at node p with displacements at adjacent nodes. In these rows, the stiffnesses associated with displacements of node q, say, are located in the $(2q - 1)$th and $2q$th columns of $[K]$. If p and q are not directly linked by an element, the corresponding overall stiffnesses (4 in total) are all zero. If p and q are directly linked by an element, the corresponding overall stiffnesses are obtained from the stiffness matrix for that element. If p and q refer to the same node, the corresponding overall stiffnesses (which are sometimes referred to as *self stiffnesses*) are found by summing stiffness contributions from all the elements which are linked together at that node. This process of assembling overall stiffnesses from element stiffnesses should become clearer after we have examined the computer program for carrying out the solution process automatically, and after we have studied Example 4.1, which works through a simple problem in detail.

At first sight we may be tempted to think that we can solve equations (4.8) as they stand. We do, after all, have $2n$ linear algebraic equations involving $2n$ unknowns. If we attempted to solve them, however, we would find that the overall stiffness matrix is, like the individual element stiffness matrices, singular. The physical significance of a singular stiffness matrix is that we have not yet taken into account the conditions which must be applied at the support points of the structure, conditions which are really part of the imposed geometry of deformation. Without these conditions, the structure as a whole is free to move as a rigid body (it is a mechanism). This rigid body movement, which could take the form of translation or rotation of the structure, or a combination of the two, could occur in the absence of applied forces. In Section 2.1 we imposed support conditions on statically determinate structures by including as unknowns reaction forces in appropriate directions at the relevant nodes. Such reaction forces certainly exist, for both statically determinate and statically indeterminate structures, but since we are now choosing to use displacements as the unknowns it is more appropriate to express the conditions at the supports in terms of these displacements. There are at least two types of support to be considered, namely pinned and roller, which were introduced in Fig. 2.20, and are also shown in Fig. 4.7. At a pinned support, Fig. 4.7a, there can be no movement as the structure is loaded, and both displacement components of the nodal point, here numbered i, are therefore zero. This condition is readily implemented in equations (4.8) by replacing the $(2i - 1)$th and $2i$th equations (which are the equilibrium equations for node i) by

$$u_i = 0 \quad \text{and} \quad v_i = 0 \tag{4.10}$$

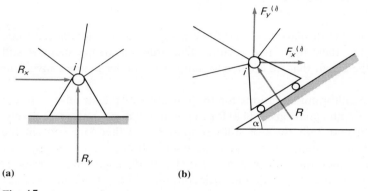

(a)   (b)

Fig. 4.7.

respectively. On the other hand, at a roller support with no friction, Fig. 4.7b, the node is constrained to move freely in a fixed direction, which we can define as being at an angle α to the positive X axis. The external forces acting at the node are the known components $F_x^{(i)}$ and $F_y^{(i)}$, together with an unknown reaction force, R, at right angles to the direction of movement, as shown. In the direction of movement, equilibrium of the external forces with those transmitted to the elements requires that

$$\Sigma U_i^{(m)} \cos \alpha + \Sigma V_i^{(m)} \sin \alpha = F_x^{(i)} \cos \alpha + F_y^{(i)} \sin \alpha \tag{4.11}$$

with the summations being carried out over the elements connected at the node. In the direction perpendicular to this there is no displacement, and

$$u_i \sin \alpha - v_i \cos \alpha = 0 \tag{4.12}$$

Equations (4.11) and (4.12) become the $(2i - 1)$th and $2i$th equations of the set (4.8). Another possible type of support condition is that associated with an elastic support, where the displacement at the relevant node is directly proportional to the reaction force there.

Equations (4.8) are now ready to be solved and, once solved, we have the displacements of all the nodal points. From these can be found the forces and stresses in the elements, via the strains defined by equation (4.4), giving a full solution to the problem. This finite element method is equally applicable to statically determinate and statically indeterminate plane pin-jointed structures, and can be extended into three dimensions to deal with space frames.

In Section 3.4.1 we considered two methods, exact and approximate, for calculating nodal point displacements of a simple statically determinate structure. It is worth examining how the present finite element technique relates to these two approaches. In the exact method we used the nonlinear cosine rule for the geometry of a triangle to find the very small change in one of its angles. On the other hand, in the approximate method, the calculation was linearized by effectively assuming no change in the angles of the elements in order to determine the directions in which the (very small) displacements of their nodes must occur. Since the finite element method is both linear and assumes no change in the angles of the elements, we can expect it to be equivalent to, indeed identical to, the approximate method. This is confirmed by Example 4.1.

In Section 2.1 we developed tests to distinguish between statically determinate and statically indeterminate structures, and also to identify structures which can move as mechanisms. Although the finite element method is equally applicable to statically determinate and statically indeterminate structures, it is still useful to be able to distinguish between them, and the same tests can be applied. It is also necessary to be able to identify mechanisms – which again give rise to singular overall stiffness matrices.

4.1.2 Computer program

Program SIPINJ (standing for Statically Indeterminate **PIN-J**ointed structure analysis) provides a FORTRAN implementation of the above finite element method. Although it is applicable to a wide range of both statically indeterminate and statically determinate plane structures, it is not as general as it could be. Features which have been deliberately omitted in the interests of clarity and simplicity include elastic supports, temperature changes of the elements, and any initial lack of fit between the elements in a statically indeterminate part of the structure: members are made exactly to their intended lengths. Consider, for example, the structure shown in Fig. 4.8 which initially has five members AB,

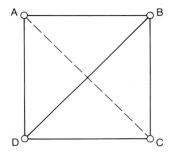

Fig. 4.8.

BC, CD, DA and BD and is statically determinate. If we now add a sixth member linking points A and C, this must have exactly the right distance between its pin holes to ensure a proper fit. If this is not the case, then we would have to first deform the structure to accommodate the lack of fit, as a result creating forces in the members before any external loading is applied. No allowance is made in the program for such initial loading of the members. Similarly, no attempt is made to investigate possible *buckling* of members of the structure which are in compression. We shall defer consideration of the phenomenon of buckling until Chapter 8.

Program SIPINJ follows very closely in terms of FORTRAN variable names and general layout the program SDPINJ for the analysis of statically determinate pin-jointed structures, and which was presented in Section 2.1.3. Indeed, program SIPINJ was developed from SDPINJ with only a minimum of changes to the coding. For example, wherever possible the same statement numbers have been used at equivalent points in the program.

```
      PROGRAM  SIPINJ
C
C  PROGRAM TO FIND THE DISPLACEMENTS OF THE NODES AND THE FORCES AND
C  STRESSES IN THE MEMBERS OF A STATICALLY INDETERMINATE
C  (OR STATICALLY DETERMINATE) PLANE PIN-JOINTED STRUCTURE.
C
      DIMENSION  X(50),Y(50),NPI(100),NPJ(100),AREA(100),E(100),
     1           OSTIFF(100,101),DELTA(100),FX(50),FY(50),
     2           NODE(50),ANG(50),CSVECT(4),T(100),STRESS(100)
      NNPMAX=50
      NELMAX=100
      PI=4.*ATAN(1.)
      OPEN(5,FILE='DATA')
      OPEN(6,FILE='RESULTS')
      WRITE(6,61)
  61  FORMAT('PLANE PIN-JOINTED STRUCTURE')
C
C  INPUT AND TEST THE NUMBER OF NODES (JOINTS).
      READ(5,*) NNP
      IF(NNP.GT.NNPMAX) THEN
        WRITE(6,62) NNP
  62    FORMAT(/ 'NUMBER OF NODES = ',I4,' TOO LARGE - STOP')
        STOP
      END IF
C
C  INPUT AND OUTPUT THE CARTESIAN COORDINATES OF THE NODES.
      READ(5,*)  (X(I),Y(I),I=1,NNP)
      WRITE(6,63)  (I,X(I),Y(I),I=1,NNP)
  63  FORMAT(/ 'COORDINATES OF THE NODES'
     1 / 2('NODE    X            Y         ') / (2(I3,2E12.4,4X)))
C
C  INPUT AND TEST THE NUMBER OF ELEMENTS (MEMBERS).
      READ(5,*) NEL
      IF(NEL.GT.NELMAX) THEN
        WRITE(6,64) NEL
  64    FORMAT(/ 'NUMBER OF ELEMENTS = ',I4,' TOO LARGE - STOP')
        STOP
      END IF
C
C  INPUT AND OUTPUT THE NODES CONNECTED BY THE ELEMENTS, ALSO THE
C  ELEMENT CROSS-SECTIONAL AREAS AND YOUNGS MODULI.
      READ(5,*)  (NPI(M),NPJ(M),AREA(M),E(M),M=1,NEL)
      WRITE(6,65)  (M,NPI(M),NPJ(M),AREA(M),E(M),M=1,NEL)
  65  FORMAT(/ 'ELEMENT NODES, AREAS AND YOUNGS MODULI'
     1      / ' M    I    J       AREA            E'
     2      / (I4,2I5,2E15.4))
```

```
C
C   INITIALIZE THE OVERALL STIFFNESS COEFFICIENTS AND EXTERNAL FORCES
C   ON THE NODES.
      NEQN=2*NNP
      DO 1 IROW=1,NEQN
      DO 1 ICOL=1,NEQN
  1   OSTIFF(IROW,ICOL)=0.
      DO 2 I=1,NNP
      FX(I)=0.
  2   FY(I)=0.
C
C   INPUT AND TEST THE NUMBER OF NODES AT WHICH EXTERNAL FORCES ARE APPLIED.
      READ(5,*) NNPF
      IF(NNPF.GT.NNPMAX) THEN
         WRITE(6,66) NNPF
 66      FORMAT(/ 'NUMBER OF LOADED NODES = ',I4,' TOO LARGE - STOP')
         STOP
      END IF
C
C   INPUT AND OUTPUT THE FORCE COMPONENTS AT THE LOADED NODES.
      IF(NNPF.GT.0) THEN
         READ(5,*) (NODE(K),FX(NODE(K)),FY(NODE(K)),K=1,NNPF)
         WRITE(6,67) (NODE(K),FX(NODE(K)),FY(NODE(K)),K=1,NNPF)
 67      FORMAT(/ 'FORCE COMPONENTS AT THE LOADED NODES'
     1          / '  NODE',12X,'FX',13X,'FY' / (I5,5X,2E15.4))
      END IF
C
C   INPUT AND TEST THE NUMBERS OF NODES AT WHICH THE STRUCTURE IS PINNED
C   TO ITS SURROUNDINGS (NNPP), AND SUPPORTED ON ROLLERS (NNPR).
      READ(5,*) NNPP,NNPR
      NNPT=NNPP+NNPR
      IF(NNPT.GT.NNPMAX) THEN
         WRITE(6,68) NNPT
 68      FORMAT(/ 'NUMBER OF NODES AT WHICH STRUCTURE SUPPORTED = ',I4,
     1           ' TOO LARGE - STOP')
         STOP
      END IF
C
C   INPUT AND OUTPUT THE NODES AT WHICH THE STRUCTURE IS PINNED.
      IF(NNPP.GT.0) THEN
         READ(5,*) (NODE(K),K=1,NNPP)
         WRITE(6,69) (NODE(K),K=1,NNPP)
 69      FORMAT(/ 'NODES FORMING PINNED SUPPORTS : ',10I4 / (18I4))
      END IF
C
C   INPUT AND OUTPUT THE NODES AT WHICH STRUCTURE RESTS ON ROLLERS,
C   TOGETHER WITH THE ANGLES (IN DEGREES MEASURED COUNTERCLOCKWISE FROM
C   THE POSITIVE X-AXIS) AT WHICH THESE ROLLERS ARE FREE TO MOVE.
      IF(NNPR.GT.0) THEN
         READ(5,*) (NODE(NNPP+K),ANG(NNPP+K),K=1,NNPR)
         WRITE(6,70) (NODE(NNPP+K),ANG(NNPP+K),K=1,NNPR)
 70      FORMAT(/ 'NODES AND ANGLES OF ROLLER SUPPORTS'
     1          / '  NODE         ANG (DEG) ' / (I5,F17.2))
      END IF
C
C   COMPARE THE NUMBER OF FORCE UNKNOWNS WITH THE NUMBER OF EQUATIONS.
      NUNK=NEL+2*NNPP+NNPR
      IF(NUNK.EQ.NEQN) THEN
         WRITE(6,76)
 76      FORMAT(/ 'THE STRUCTURE IS STATICALLY DETERMINATE')
      END IF
      IF(NUNK.GT.NEQN) THEN
         WRITE(6,71)
 71      FORMAT(/ 'THE STRUCTURE IS STATICALLY INDETERMINATE')
      END IF
      IF(NUNK.LT.NEQN) THEN
         WRITE(6,72)
 72      FORMAT(/ 'THE STRUCTURE IS A MECHANISM - STOP'
     1          / '(FEWER FORCE UNKNOWNS THAN EQUILIBRIUM EQUATIONS)')
         STOP
      END IF
```

```
C
C  SET UP THE LINEAR EQUATIONS, FIRST CONSIDERING EACH ELEMENT IN TURN.
      DO 3 M=1,NEL
      I=NPI(M)
      J=NPJ(M)
      ELENGT=SQRT((X(J)-X(I))**2+(Y(J)-Y(I))**2)
      COSINE=(X(J)-X(I))/ELENGT
      SINE=(Y(J)-Y(I))/ELENGT
C
C  COMPUTE ELEMENT STIFFNESS COEFFICIENTS.
      CSVECT(1)=-COSINE
      CSVECT(2)=-SINE
      CSVECT(3)=COSINE
      CSVECT(4)=SINE
      FACT=E(M)*AREA(M)/ELENGT
      DO 3 IR=1,4
      DO 3 IC=1,4
      ESTIFF=FACT*CSVECT(IR)*CSVECT(IC)
C
C  ADD EACH ELEMENT STIFFNESS TO THE OVERALL STIFFNESS MATRIX.
      IF(IR.LE.2)  IROW=2*(I-1)+IR
      IF(IR.GE.3)  IROW=2*(J-1)+IR-2
      IF(IC.LE.2)  ICOL=2*(I-1)+IC
      IF(IC.GE.3)  ICOL=2*(J-1)+IC-2
    3 OSTIFF(IROW,ICOL)=OSTIFF(IROW,ICOL)+ESTIFF
C
C  STORE THE EXTERNAL FORCES APPLIED TO THE NODES.
      DO 6 I=1,NNP
      OSTIFF(2*I-1,NEQN+1)=FX(I)
    6 OSTIFF(2*I,NEQN+1)=FY(I)
C
C  IMPOSE ZERO DISPLACEMENTS AT THE PINNED SUPPORTS.
      IF(NNPP.GT.0) THEN
         DO 4 K=1,NNPP
         I=NODE(K)
         DO 11 J=1,NEQN+1
         OSTIFF(2*I-1,J)=0.
   11    OSTIFF(2*I,J)=0.
         OSTIFF(2*I-1,2*I-1)=1.
    4    OSTIFF(2*I,2*I)=1.
      END IF
C
C  IMPOSE DISPLACEMENT CONSTRAINTS AT THE ROLLER SUPPORTS.
      IF(NNPR.GT.0) THEN
         DO 5 K=NNPP+1,NNPT
         I=NODE(K)
         ANGR=ANG(K)*PI/180.
         SINA=SIN(ANGR)
         COSA=COS(ANGR)
         DO 12 J=1,NEQN+1
         OSTIFF(2*I-1,J)=OSTIFF(2*I-1,J)*COSA+OSTIFF(2*I,J)*SINA
   12    OSTIFF(2*I,J)=0.
         OSTIFF(2*I,2*I-1)=SINA
    5    OSTIFF(2*I,2*I)=-COSA
      END IF
C
C  SOLVE THE LINEAR ALGEBRAIC EQUATIONS BY GAUSSIAN ELIMINATION.
      CALL   SOLVE(OSTIFF,DELTA,NEQN,2*NNPMAX,2*NNPMAX+1,IFLAG)
C
C  A UNIT VALUE OF THE ILL-CONDITIONING FLAG IMPLIES THAT THE STRUCTURE
C  IS A MECHANISM.
      IF(IFLAG.EQ.1) THEN
         WRITE(6,73)
   73    FORMAT(/ 'ILL-CONDITIONED EQUATIONS - STOP'
     1          / '(THE STRUCTURE IS A MECHANISM)')
         STOP
      END IF
C
C  OUTPUT THE COMPUTED NODAL POINT DISPLACEMENTS.
      WRITE(6,74) (I,DELTA(2*I-1),DELTA(2*I),I=1,NNP)
   74 FORMAT(/ 'COMPUTED DISPLACEMENTS OF THE NODES'
     1       / 2('    NODE   U (X DIRN)    V (Y DIRN)') /
     2         (2(I8,2E14.4)))
```

```
C
C   COMPUTE ELEMENT FORCES AND STRESSES.
      DO 7 M=1,NEL
      I=NPI(M)
      J=NPJ(M)
      ELENGT=SQRT((X(J)-X(I))**2+(Y(J)-Y(I))**2)
      COSINE=(X(J)-X(I))/ELENGT
      SINE=(Y(J)-Y(I))/ELENGT
      DISPI=DELTA(2*I-1)*COSINE+DELTA(2*I)*SINE
      DISPJ=DELTA(2*J-1)*COSINE+DELTA(2*J)*SINE
      STRAIN=(DISPJ-DISPI)/ELENGT
      STRESS(M)=STRAIN*E(M)
    7 T(M)=STRESS(M)*AREA(M)
C
C   OUTPUT ELEMENT FORCES AND STRESSES.
      WRITE(6,75) (M,T(M),STRESS(M),M=1,NEL)
   75 FORMAT(/ 'COMPUTED FORCES AND STRESSES IN THE ELEMENTS'
     1           ' (TENSILE POSITIVE)'
     2         / 2('   ELEMENT      FORCE        STRESS  ')
     3         / (2(I8,2E14.4)))
      STOP
      END
```

The following list provides the definitions of the FORTRAN variables used, arranged in alphabetical order.

Variable	Type	Definition
ANG	real	Array storing angles (in degrees) at which roller supports are free to move
ANGR	real	Angle in radians of a roller support
AREA	real	Array storing the cross-sectional areas of the elements
COSA	real	Cosine of the angle at which a roller support is free to move
COSINE	real	Cosine of element angle θ_m
CSVECT	real	Array storing cosine and sine terms used in forming element stiffness matrices (equation (4.6))
DELTA	real	Array storing the displacements of the nodal points
DISPI	real	Displacement of the first node of an element in the direction along the element from the first node to the second node
DISPJ	real	Displacement of the second node of an element in the direction along the element from the first node to the second node
E	real	Array storing the Young's moduli of the elements
ELENGT	real	Element length, L_m
ESTIFF	real	Element stiffness coefficient
FACT	real	Common factor (the axial stiffness) in the element stiffness coefficients
FX, FY	real	Arrays storing the external force components in the global coordinate directions at the nodes
I	integer	Nodal point number (sometimes first node of an element)
IC	integer	Column number in element stiffness matrix
ICOL	integer	Column number in array OSTIFF
IFLAG	integer	Flag for a singular or very ill-conditioned coefficient matrix (normally returned as zero by SOLVE, one if the ill-conditioning test is satisfied)
IR	integer	Row number in element stiffness matrix
IROW	integer	Row number in array OSTIFF
J	integer	Nodal point number (second node of an element)
K	integer	A counter (used in READ and WRITE statements)
M	integer	Element number
NEL	integer	Number of elements

Variable	Type	Definition
NELMAX	integer	Maximum number of elements permitted by the array dimensions
NEQN	integer	Number of equations to be solved
NNP	integer	Number of nodal points
NNPF	integer	Number of nodes at which external forces are applied
NNPMAX	integer	Maximum number of nodal points permitted by the array dimensions
NNPP	integer	Number of nodes at which structure is pinned to its surroundings
NNPR	integer	Number of nodes at which structure is supported on rollers
NNPT	integer	Total number of nodes at which structure is supported
NODE	integer	Array storing node numbers
NPI	integer	Array storing the numbers of the first nodes of the elements
NPJ	integer	Array storing the numbers of the second nodes of the elements
NUNK	integer	Total number of unknowns
OSTIFF	real	Array storing the coefficients of the overall stiffness matrix $[K]$, extended to include external force vector $[F]$ as its last column
PI	real	The mathematical constant π
SINA	real	Cosine of the angle at which a roller support is free to move
SINE	real	Sine of element angle θ_m
STRAIN	real	Strain in an element
STRESS	real	Array storing the stresses in the elements
T	real	Array storing the forces in the elements
X	real	Array storing X global coordinates of the nodes
Y	real	Array storing Y global coordinates of the nodes

The arrays used in SIPINJ are dimensioned in such a way as to allow a structure with up to 50 nodes and 100 elements to be analyzed, implying that up to 100 equations (twice the number of nodes) may have to be solved. Note that array OSTIFF, which stores the overall stiffness matrix $[K]$ extended to include the force vector $[F]$ is allowed to have up to 101 columns. The array dimensions can of course be changed, provided the values of NNPMAX and NELMAX defined in the next two statements are also adjusted.

The program reads information from a file named DATA, and writes onto a file named RESULTS. All READ statements call for free format input data. After writing out a heading, the program first reads the number of nodal points in the structure to be analyzed, and then tests that this does not exceed the maximum allowed by the array dimensions. We should note that, as in program SDPINJ, the checking of input data in the program is comparatively rudimentary, for the same reasons. However, some of the more obvious data errors are trapped, and in all cases input data are immediately written out to allow them to be checked.

After reading in the global coordinates of the nodes, X and Y for node 1, then node 2 and so on, the program reads in and tests the number of elements, followed by the numbers of the first and second nodes, the cross-sectional area and Young's modulus of each of the elements in turn. All coefficients of the overall stiffness matrix and the external force components at the nodes are then set to zero in preparation for the assembly process to come. The number of nodes at which external forces are applied is read in, followed by the numbers of the nodes concerned, together with the components in the X and Y directions of the corresponding forces. Next, the numbers of nodes at which there are pinned and roller supports are read in, followed by the numbers of the nodes forming pinned supports, then the numbers of the nodes resting on rollers together with the angles at which these rollers are free to move.

The total numbers of unknown forces can now be calculated, one force in each element, two reaction components at each pinned support, and one at each roller support. Equations (2.6) to (2.8) are then used to determine whether, based on the numbers of unknown forces and equilibrium equations, the structure is apparently statically determinate, statically indeterminate or a mechanism, and an appropriate message is written out. If the structure is a mechanism, no further calculations are possible and execution is halted.

The coefficients of the overall stiffness matrix in equations (4.8) are assembled by considering each of the elements in turn. Our typical element, shown in Fig. 4.1 and numbered m, has first and second nodes i and j, respectively, which in the programs are recovered from arrays NPI and NPJ and stored in variables I and J. If L_m(ELENGT) is the length of the element, which is the distance between nodes i and j, then the cosine and sine of the element angle θ_m may be defined as

$$\cos \theta_m = \frac{X_j - X_i}{L_m}, \quad \sin \theta_m = \frac{Y_j - Y_i}{L_m} \tag{4.13}$$

With these trigonometric quantities known, the coefficients of the element stiffness matrix are computed according to equation (4.6). Each of these coefficients is added to the appropriate term in the overall stiffness matrix, according to the numbers of the nodes involved.

The program then enters components of any externally applied forces as required by equations (4.7) in the right-hand side vector $[F]$ (equations (4.8) and (4.9)), which is now the last column of $[K]$. Note that, unlike in program SDPINJ, this application of external forces must be carried out before the boundary conditions at pinned and roller supports are applied, because these conditions now involve major changes to the relevant equations. At each of the pinned supports, the program implements the changes to the stiffness coefficients required by equations (4.10) to impose zero displacements there. Similarly, at roller supports, the changes defined by equations (4.11) and (4.12) are made.

The equations can now be solved, using subroutine SOLVE which is described in detail in Appendix B. If a unit value of IFLAG is returned to SIPINJ, this means that SOLVE has detected a singular or very ill-conditioned coefficient matrix. The most likely reason for this is that, although equations (2.6) and (2.7) are satisfied, the structure is effectively a mechanism, due to poor arrangement of the elements (such as in Example 2.3d). The program writes out a message to this effect and terminates execution. Otherwise, the computed displacements of the nodes, which are the primary unknowns in the analysis, are written out. Then the forces and stresses in the elements are computed and written out.

We note that, except for roller support angles which must be supplied in degrees, no units are specified for either the input data or the computed results. This is because we may use any consistent set of units: all forces should be supplied to the program in the same units, as should all lengths, and the units of the element Young's moduli must be compatible with these. The results will be presented in the same set of units.

4.1.3 Practical applications

Let us now examine how program SIPINJ can be used to analyze pin-jointed structures with the aid of some typical problems. The first of these, which is very straightforward, has already been subjected to an equilibrium analysis to find the forces in the elements and the reactions at the supports in Example 2.1. The nodal point displacements were found by a rather tedious manual method in Section 3.4.1. Because it is a simple problem, we can also generate the finite element equations by hand, which serves to illustrate the process by which element stiffnesses are assembled to form the overall stiffness matrix.

EXAMPLE 4.1

Consider again the structure shown in Figs 2.1, 2.9 and 3.12, which is redrawn in Fig. 4.9 with the nodes and elements shown numbered in convenient, but arbitrary orders. Element numbers are shown ringed, while node numbers are unringed. It is convenient to define the origin for global coordinates X and Y at node 3 as shown. The cross-sectional areas of the steel members of the structure are each 325 mm² (0.5 in²). Figure 4.10 shows the file of input data used to set up this problem for solution by program SIPINJ.

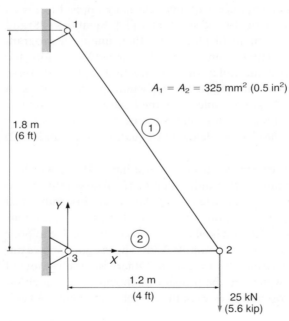

$A_1 = A_2 = 325$ mm² (0.5 in²)

1.8 m
(6 ft)

1.2 m
(4 ft)

25 kN
(5.6 kip)

Fig. 4.9.

The file has been annotated to indicate what each line of data represents. Meter and newton units are used for lengths and forces, respectively, and the value of Young's modulus for steel is obtained from Appendix A as 207×10^9 N/m². Note that, when entering data for the external load at node 2, the components are zero (in the X

```
3                                  (Number of nodes)
0. 1.8,1.2 0.,0. 0.                (Node coordinates)
2                                  (Number of elements)
1 2 325.0E-6 207.0E9      (Nodes, area and modulus of first element)
2 3 325.0E-6 207.0E9      (Nodes, area and modulus of second element)
1                 (Number of nodes subjected to external forces)
2 0. -25.0E3      (Number of loaded node, and force components)
2 0               (Numbers of pinned and roller supports)
1 3               (Node numbers of pinned supports)
```

Fig. 4.10.

direction) and $-25\,000$ N (in the Y direction). The file RESULTS produced by the program is shown in Fig. 4.11. The computed forces in the two elements are $30\,000$ N (tensile) and $-16\,670$ N (compressive). As we would expect, these values are in exact agreement with those obtained by a manual equilibrium analysis in Example 2.1. Similarly, the displacements of node 2 of -0.2973×10^{-3} m (-0.2973 mm or -0.012 in) and -0.1359×10^{-2} m (-1.359 mm or -0.054 in) in the global coordinate directions are identical to those obtained by the approximate manual method described in Section 3.4.1. As we have already anticipated, the present finite element method for finding displacements is equivalent to this linearized approximate method, which is perfectly adequate for the large majority of engineering structures.

```
PLANE PIN-JOINTED STRUCTURE

COORDINATES OF THE NODES
NODE     X           Y            NODE     X           Y
  1   0.0000E+00  0.1800E+01        2   0.1200E+01  0.0000E+00
  3   0.0000E+00  0.0000E+00

ELEMENT NODES, AREAS AND YOUNGS MODULI
  M    I    J        AREA            E
  1    1    2     0.3250E-03     0.2070E+12
  2    2    3     0.3250E-03     0.2070E+12

FORCE COMPONENTS AT THE LOADED NODES
  NODE          FX              FY
   2        0.0000E+00     -0.2500E+05

NODES FORMING PINNED SUPPORTS :     1   3

THE STRUCTURE IS STATICALLY DETERMINATE

COMPUTED DISPLACEMENTS OF THE NODES
     NODE   U (X DIRN)   V (Y DIRN)    NODE   U (X DIRN)    V (Y DIRN)
       1    0.0000E+00   0.0000E+00      2   -0.2973E-03   -0.1359E-02
       3    0.0000E+00   0.0000E+00

COMPUTED FORCES AND STRESSES IN THE ELEMENTS (TENSILE POSITIVE)
    ELEMENT     FORCE        STRESS    ELEMENT     FORCE        STRESS
       1     0.3005E+05   0.9245E+08      2    -0.1667E+05   -0.5128E+08
```

Fig. 4.11.

With such a simple structure, we can follow the assembly and solution of the linear finite element equations in detail. Figure 4.12 shows the two elements of the structure separately, also their dimensions and angles relative to the positive X direction (note that the first nodes of the elements have been chosen as nodes 1 and 2, respectively). For the first element

$$\frac{E_1 A_1}{L_1} = \frac{207 \times 10^9 \times 0.325 \times 10^{-3}}{2.163} = 31.10 \times 10^6 \text{ N/m}$$

$$c = \cos \theta_1 = 0.5547 \quad \text{and} \quad s = \sin \theta_1 = -0.8321$$

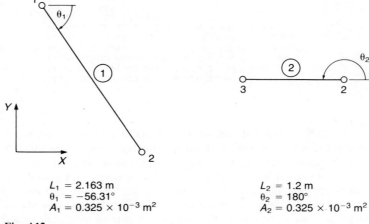

$L_1 = 2.163$ m
$\theta_1 = -56.31°$
$A_1 = 0.325 \times 10^{-3}$ m²

$L_2 = 1.2$ m
$\theta_2 = 180°$
$A_2 = 0.325 \times 10^{-3}$ m²

Fig. 4.12.

and, from equation (4.6), its stiffness matrix is

$$[k]_1 = 31.10 \times 10^6 \begin{bmatrix} 0.3077 & -0.4616 & -0.3077 & 0.4616 \\ -0.4616 & 0.6924 & 0.4616 & -0.6924 \\ -0.3077 & 0.4616 & 0.3077 & -0.4616 \\ 0.4616 & -0.6924 & -0.4616 & 0.6924 \end{bmatrix} \text{N/m}$$

which, according to equations (4.3) and (4.5), means that

$$\begin{bmatrix} U_1^{(1)} \\ V_1^{(1)} \\ U_2^{(1)} \\ V_2^{(1)} \end{bmatrix} = 10^6 \begin{bmatrix} 9.569 & -14.36 & -9.569 & 14.36 \\ -14.36 & 21.53 & 14.36 & -21.53 \\ -9.569 & 14.36 & 9.569 & -14.36 \\ 14.36 & -21.53 & -14.36 & 21.53 \end{bmatrix} \begin{bmatrix} u_1 \\ v_1 \\ u_2 \\ v_2 \end{bmatrix}$$

Similarly, for the second element

$$\frac{E_2 A_2}{L_2} = \frac{207 \times 10^9 \times 0.325 \times 10^{-3}}{1.2} = 56.06 \times 10^6 \text{ N/m}$$

$$c = \cos\theta_2 = -1.0 \quad \text{and} \quad s = \sin\theta_2 = 0$$

and its stiffness matrix is

$$[k]_2 = 56.06 \times 10^6 \begin{bmatrix} 1.0 & 0.0 & -1.0 & 0.0 \\ 0.0 & 0.0 & 0.0 & 0.0 \\ -1.0 & 0.0 & 1.0 & 0.0 \\ 0.0 & 0.0 & 0.0 & 0.0 \end{bmatrix} \text{N/m}$$

which means that

$$
\begin{bmatrix} U_2^{(2)} \\ V_2^{(2)} \\ U_3^{(2)} \\ V_3^{(2)} \end{bmatrix} = 10^6 \begin{bmatrix} 56.06 & 0.0 & -56.06 & 0.0 \\ 0.0 & 0.0 & 0.0 & 0.0 \\ -56.06 & 0.0 & 56.06 & 0.0 \\ 0.0 & 0.0 & 0.0 & 0.0 \end{bmatrix} \begin{bmatrix} u_2 \\ v_2 \\ u_3 \\ v_3 \end{bmatrix}
$$

The sums of the forces transmitted to the elements at the three nodes are therefore given by

$$
\begin{bmatrix} U_1^{(1)} \\ V_1^{(1)} \\ U_2^{(1)} + U_2^{(2)} \\ V_2^{(1)} + V_2^{(2)} \\ U_3^{(2)} \\ V_3^{(2)} \end{bmatrix}
$$

$$
= 10^6 \begin{bmatrix} 9.569 & -14.36 & -9.569 & 14.36 & 0 & 0 \\ -14.36 & 21.53 & 14.36 & -21.53 & 0 & 0 \\ -9.569 & 14.36 & 9.569+56.06 & -14.36+0 & -56.06 & 0 \\ 14.36 & -21.53 & -14.36+0 & 21.53+0 & 0 & 0 \\ 0 & 0 & -56.06 & 0 & 56.06 & 0 \\ 0 & 0 & 0 & 0 & 0 & 0 \end{bmatrix} \begin{bmatrix} u_1 \\ v_1 \\ u_2 \\ v_2 \\ u_3 \\ v_3 \end{bmatrix}
$$

Note that, for this simple structure, only in the case of the stiffness coefficients multiplying u_2 and v_2 (displacements of the only node which is common to both elements) are there contributions from both element stiffness matrices (some of which happen to be zero). Equating these summed forces to the external forces applied at the nodes, and modifying the first pair and the last pair of the resulting equations according to equations (4.10) to give the required displacement boundary conditions at nodes 1 and 3, we obtain

$$
10^6 \begin{bmatrix} 1 & 0 & 0 & 0 & 0 & 0 \\ 0 & 1 & 0 & 0 & 0 & 0 \\ -9.569 & 14.36 & 65.63 & -14.36 & -56.06 & 0 \\ 14.36 & -21.53 & -14.36 & 21.53 & 0 & 0 \\ 0 & 0 & 0 & 0 & 1 & 0 \\ 0 & 0 & 0 & 0 & 0 & 1 \end{bmatrix} \begin{bmatrix} u_1 \\ v_1 \\ u_2 \\ v_2 \\ u_3 \\ v_3 \end{bmatrix} = \begin{bmatrix} 0 \\ 0 \\ 0 \\ -25 \times 10^3 \\ 0 \\ 0 \end{bmatrix}
$$

In the program these equations are solved by the full Gaussian elimination method described in Appendix B. Since we know that $u_1 = v_1 = u_3 = v_3 = 0$, however, we can reduce the six equations to just two

$$
10^6 \begin{bmatrix} 65.63 & -14.36 \\ -14.36 & 21.53 \end{bmatrix} \begin{bmatrix} u_2 \\ v_2 \end{bmatrix} = \begin{bmatrix} 0 \\ -25 \times 10^3 \end{bmatrix}
$$

The first of these gives

$$v_2 = \frac{65.63}{14.36} u_2 = 4.570\, u_2$$

Substituting this expression for v_2 into the second equation

$$-14.36\, u_2 + (21.53 \times 4.570)u_2 = -25 \times 10^{-3}$$

from which

$$u_2 = \frac{-25 \times 10^{-3}}{21.53 \times 4.570 - 14.36} = \underline{-0.2975 \times 10^{-3}\, \text{m}}$$

and

$$v_2 = 4.570\, u_2 = 4.570 \times 0.2975 \times 10^{-3} = \underline{-1.360 \times 10^{-3}\, \text{m}}$$

These results differ slightly from the computed values of -0.2973×10^{-3} m and -1.359×10^{-3} m, respectively, due to the fact that here we have only retained four significant figures at each stage of the arithmetic operations, and have quoted the results to the same precision.

EXAMPLE 4.2

Consider again the structure shown in Figs 2.12 and 2.25, which is redrawn in Fig. 4.13. Assuming the cross-sectional area of each of the steel members is 1700 mm² (2.6 in²) find the displacements of the nodes, and in particular the downward displacement of node 7 at which the external load of 108 kN is applied. Find the effect on this displacement, and on the maximum stress in the members, of modifying the structure by adding two members of the same cross-sectional area linking nodes 1 to 3 and 3 to 5, respectively (shown by dotted lines in Fig. 4.13). Both additional members are exactly the right length when the structure is in its unloaded state.

    This example provides a further test for the program on a statically determinate problem where the forces in the members have been determined by another method, and

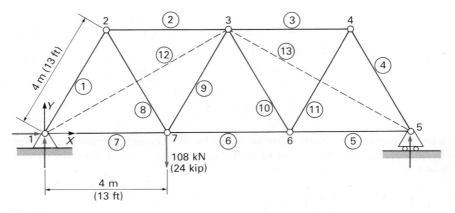

Fig. 4.13.

then demonstrates its application to a statically indeterminate problem. Taking the unmodified statically determinate form of the structure first, Fig. 4.14 (which is very similar to Fig. 2.26 for the previous computer solution to this problem) shows the file of input data supplied to program SIPINJ. Meter and newton units are used for lengths and forces. The file RESULTS produced by the program is shown in Fig. 4.15. The

```
7
0. 0.,2. 3.4641,6. 3.4641,10. 3.4641,12. 0.,8. 0.,4. 0.
11
1 2 1.7E-3 207.0E9    2 3 1.7E-3 207.0E9    3 4 1.7E-3 207.0E9
4 5 1.7E-3 207.0E9    5 6 1.7E-3 207.0E9    6 7 1.7E-3 207.0E9
7 1 1.7E-3 207.0E9    2 7 1.7E-3 207.0E9    7 3 1.7E-3 207.0E9
3 6 1.7E-3 207.0E9    6 4 1.7E-3 207.0E9
1
7 0. -108.0E3
1 1
1
5 0.
```

Fig. 4.14.

```
PLANE PIN-JOINTED STRUCTURE

COORDINATES OF THE NODES
NODE     X           Y          NODE      X           Y
  1   0.0000E+00  0.0000E+00      2   0.2000E+01  0.3464E+01
  3   0.6000E+01  0.3464E+01      4   0.1000E+02  0.3464E+01
  5   0.1200E+02  0.0000E+00      6   0.8000E+01  0.0000E+00
  7   0.4000E+01  0.0000E+00

ELEMENT NODES, AREAS AND YOUNGS MODULI
  M    I    J       AREA          E
  1    1    2    0.1700E-02    0.2070E+12
  2    2    3    0.1700E-02    0.2070E+12
  3    3    4    0.1700E-02    0.2070E+12
  4    4    5    0.1700E-02    0.2070E+12
  5    5    6    0.1700E-02    0.2070E+12
  6    6    7    0.1700E-02    0.2070E+12
  7    7    1    0.1700E-02    0.2070E+12
  8    2    7    0.1700E-02    0.2070E+12
  9    7    3    0.1700E-02    0.2070E+12
 10    3    6    0.1700E-02    0.2070E+12
 11    6    4    0.1700E-02    0.2070E+12

FORCE COMPONENTS AT THE LOADED NODES
 NODE           FX            FY
   7        0.0000E+00    -0.1080E+06

NODES FORMING PINNED SUPPORTS :    1

NODES AND ANGLES OF ROLLER SUPPORTS
 NODE        ANG (DEG)
   5           .00

THE STRUCTURE IS STATICALLY DETERMINATE

COMPUTED DISPLACEMENTS OF THE NODES
    NODE   U (X DIRN)   V (Y DIRN)    NODE   U (X DIRN)   V (Y DIRN)
      1   0.0000E+00   0.0000E+00       2   0.1575E-02  -0.2001E-02
      3   0.6300E-03  -0.3274E-02       4   0.1575E-03  -0.1273E-02
      5   0.1418E-02   0.0000E+00       6   0.1181E-02  -0.2410E-02
      7   0.4725E-03  -0.3728E-02

COMPUTED FORCES AND STRESSES IN THE ELEMENTS (TENSILE POSITIVE)
  ELEMENT    FORCE        STRESS    ELEMENT     FORCE        STRESS
      1   -0.8314E+05  -0.4890E+08      2   -0.8314E+05  -0.4890E+08
      3   -0.4157E+05  -0.2445E+08      4   -0.4157E+05  -0.2445E+08
      5    0.2078E+05   0.1223E+08      6    0.6235E+05   0.3668E+08
      7    0.4157E+05   0.2445E+08      8    0.8314E+05   0.4890E+08
      9    0.4157E+05   0.2445E+08     10   -0.4157E+05  -0.2445E+08
     11    0.4157E+05   0.2445E+08
```

Fig. 4.15.

calculated forces in the elements are identical to those obtained previously using program SDPINJ and displayed in Fig. 2.27. The maximum tensile stress of 48.90 MN/m^2 (7.1 ksi) occurs in element 8, and compressive stresses of the same maximum magnitude occur in elements 1 and 2. The required nodal displacements are also shown, the vertical displacement of node 7 (which is also the largest displacement) is 3.73 mm (0.15 in) downwards. Incidentally, it is worth noting that the horizontal displacement of node 5, which is supported on a roller, is 1.42 mm (0.056 in) in the positive X direction. When we first looked at this problem in Example 2.2, we said that it was necessary to treat this as a roller rather than a pinned support to avoid making the structure statically indeterminate, in the expectation that under the influence of the applied loading the support points would not remain at the same horizontal distance apart unless constrained to do so. The computed displacement of the roller support,

```
PLANE PIN-JOINTED STRUCTURE

COORDINATES OF THE NODES
NODE     X            Y          NODE     X            Y
  1   0.0000E+00   0.0000E+00      2   0.2000E+01   0.3464E+01
  3   0.6000E+01   0.3464E+01      4   0.1000E+02   0.3464E+01
  5   0.1200E+02   0.0000E+00      6   0.8000E+01   0.0000E+00
  7   0.4000E+01   0.0000E+00

ELEMENT NODES, AREAS AND YOUNGS MODULI
  M    I    J      AREA            E
  1    1    2    0.1700E-02     0.2070E+12
  2    2    3    0.1700E-02     0.2070E+12
  3    3    4    0.1700E-02     0.2070E+12
  4    4    5    0.1700E-02     0.2070E+12
  5    5    6    0.1700E-02     0.2070E+12
  6    6    7    0.1700E-02     0.2070E+12
  7    7    1    0.1700E-02     0.2070E+12
  8    2    7    0.1700E-02     0.2070E+12
  9    7    3    0.1700E-02     0.2070E+12
 10    3    6    0.1700E-02     0.2070E+12
 11    6    4    0.1700E-02     0.2070E+12
 12    1    3    0.1700E-02     0.2070E+12
 13    3    5    0.1700E-02     0.2070E+12

FORCE COMPONENTS AT THE LOADED NODES
  NODE         FX            FY
   7        0.0000E+00    -0.1080E+06

NODES FORMING PINNED SUPPORTS :    1

NODES AND ANGLES OF ROLLER SUPPORTS
  NODE     ANG (DEG)
   5         .00

THE STRUCTURE IS STATICALLY INDETERMINATE

COMPUTED DISPLACEMENTS OF THE NODES
   NODE   U (X DIRN)   V (Y DIRN)   NODE   U (X DIRN)   V (Y DIRN)
     1   0.0000E+00   0.0000E+00      2   0.1602E-02  -0.1802E-02
     3   0.8424E-03  -0.2571E-02      4   0.5321E-03  -0.1070E-02
     5   0.1765E-02   0.0000E+00      6   0.1367E-02  -0.1910E-02
     7   0.6579E-03  -0.3224E-02

COMPUTED FORCES AND STRESSES IN THE ELEMENTS (TENSILE POSITIVE)
  ELEMENT    FORCE        STRESS      ELEMENT    FORCE        STRESS
     1    -0.6683E+05  -0.3931E+08      2    -0.6683E+05  -0.3931E+08
     3    -0.2730E+05  -0.1606E+08      4    -0.2730E+05  -0.1606E+08
     5     0.3505E+05   0.2062E+08      6     0.6235E+05   0.3668E+08
     7     0.5788E+05   0.3405E+08      8     0.6683E+05   0.3931E+08
     9     0.5788E+05   0.3405E+08     10    -0.2730E+05  -0.1606E+08
    11     0.2730E+05   0.1606E+08     12    -0.2825E+05  -0.1662E+08
    13    -0.2472E+05  -0.1454E+08
```

Fig. 4.16.

which is nearly 40% of the maximum nodal displacement, merely confirms this conclusion.

Now, in order to analyze the modified structure, it is only necessary to add the two elements numbered 12 and 13 in Fig. 4.13, linking nodes 1 to 3 and 3 to 5 respectively, the remainder of the input data being unchanged. The results produced by the program for this modified structure are shown in Fig. 4.16. Adding the two extra members makes the structure statically indeterminate. The lengths of these members are such that the structure is initially stess-free. The maximum tensile stress still occurs in element 8, but is reduced (from 48.90 MN/m^2 (7.1 ksi) to 39.31 MN/m^2 (5.7 ksi)) and the maximum compressive stresses in elements 1 and 2 are also of magnitude 39.31 MN/m^2. Also, the vertical displacement of node 7 is reduced (from 3.73 mm (0.15 in) to 3.22 mm (0.13 in)) downwards.

EXAMPLE 4.3

The wheel shown in Fig. 4.17 consists of a small central hub connected to a rigid rim of internal diameter 51 in (1.3 m) by 8 radial spokes equally spaced around the circumference. The spokes, which are 0.16 in (4 mm) diameter and made from aluminum alloy, are initially adjusted so that each carries a tensile force of 270 lb (1.2 kN). The hub is then loaded vertically downwards by a force W as shown, with the rim resting on a rigid horizontal surface. Assuming the spokes may be treated as being pin-jointed to both the hub and the rim, find the maximum value of the load on the hub if the maximum allowable tensile stress in the spokes is 29 ksi (200 MN/m^2), and none of them must go into compression. The latter restriction is to avoid any possibility of buckling of the spokes. Find also the displacement of the hub when the maximum allowable load is applied.

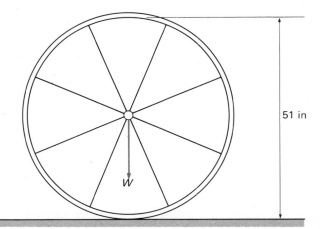

Fig. 4.17.

This is a very statically indeterminate problem, involving a large number of redundant members (six of the eight spokes could in principle be removed without the structure becoming a mechanism). Also, the structure is symmetrical in terms of both geometry and loading about its vertical center line, and we can take advantage of this to reduce the size of the problem actually analyzed. The assumption of a rigid rim represents an idealization of an actual rim. Now, although program SIPINJ is not capable of accommodating initial internal loading of the structure, in this case we can compute the forces and stresses in the spokes due to the external load, assuming the

51 in

spokes to be initially stress-free, and simply add these to the values imposed by pretensioning. The cross-sectional area of each of the spokes is

$$A = \frac{\pi}{4} \times 0.16^2 = 0.020\,11 \text{ in}^2$$

and the initial tensile stress is

$$\sigma = \frac{270}{0.020\,11} \text{ psi} = 13.43 \text{ ksi}$$

This means that, due to the external loading, the maximum compressive stress in the spokes must not exceed 13.43 ksi (to avoid the stress due to both forms of loading becoming compressive), while the maximum tensile stress must not be greater than 15.57 ksi (to avoid the total tensile stress exceeding 29 ksi).

Figure 4.18 shows a suitable model for one symmetrical half of the structure, with only four spokes. Since the rim is assumed to be rigid, we can take the spokes to be pin-jointed to fixed support points at the rim. Also, since the hub is small, we can take the spokes to be pin-jointed together at a point there. Due to the symmetry of the system, this point is constrained to move only in the vertical direction by what we can think of as a roller support. Now, we are required to find the magnitude of the load W applied to the hub, and it is therefore convenient to apply a unit load (1 lb). Considering only the half of the structure shown in Fig. 4.18, we must therefore apply a vertical load of 0.5 lb. There may also be a horizontal load, but this is the reaction force at the vertical roller support. Note, however, that there is no shear force between the two symmetrical halves, which would contribute to the vertical loading on each half. This is consistent with the principle expressed in Section 2.2.1 that there are no shear forces on a plane of mirror symmetry through a body.

Figure 4.18 shows a convenient numbering system for the 5 nodes and 4 elements, also a global coordinate system with its origin at the center of the wheel hub. The

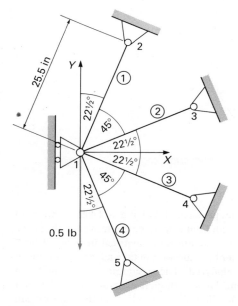

Fig. 4.18.

value of Young's modulus for aluminum alloy is given by Appendix A as 10×10^6 psi. Figure 4.19 shows the file of input data used to set up this problem for solution by program SIPINJ, in which lb and in units are used for forces and lengths, respectively. The file RESULTS produced by the program is shown in Fig. 4.20, computed displacements, forces and stresses being in inch, lb and psi units. For a vertical load of 1 lb on the hub, the maximum tensile stress (in element 1) is 11.49 psi and the maximum compressive stress (in element 4) is -11.49 psi. To avoid exceeding the maximum allowable tensile stress (in element 1), the maximum load on the hub is therefore

$$W = \frac{15.57 \times 10^3}{11.49} = 1360 \text{ lb}$$

```
5
0. 0., 9.758 23.56, 23.56 9.758, 23.56 -9.758, 9.758 -23.56
4
1 2 0.02011 10.0E6    1 3 0.02011 10.0E6
1 4 0.02011 10.0E6    1 5 0.02011 10.0E6
1
1 0. -0.5
4 1
2 3 4 5
1 90.
```

Fig. 4.19.

```
PLANE PIN-JOINTED STRUCTURE

COORDINATES OF THE NODES
NODE     X            Y          NODE     X            Y
   1  0.0000E+00   0.0000E+00      2  0.9758E+01   0.2356E+02
   3  0.2356E+02   0.9758E+01      4  0.2356E+02  -0.9758E+01
   5  0.9758E+01  -0.2356E+02

ELEMENT NODES, AREAS AND YOUNGS MODULI
   M    I    J       AREA            E
   1    1    2    0.2011E-01     0.1000E+08
   2    1    3    0.2011E-01     0.1000E+08
   3    1    4    0.2011E-01     0.1000E+08
   4    1    5    0.2011E-01     0.1000E+08

FORCE COMPONENTS AT THE LOADED NODES
   NODE          FX             FY
     1        0.0000E+00    -0.5000E+00

NODES FORMING PINNED SUPPORTS :     2    3    4    5

NODES AND ANGLES OF ROLLER SUPPORTS
   NODE        ANG (DEG)
     1           90.00

THE STRUCTURE IS STATICALLY INDETERMINATE

COMPUTED DISPLACEMENTS OF THE NODES
     NODE   U (X DIRN)    V (Y DIRN)   NODE   U (X DIRN)    V (Y DIRN)
        1  -0.6173E-11   -0.3170E-04      2   0.0000E+00    0.0000E+00
        3   0.0000E+00    0.0000E+00      4   0.0000E+00    0.0000E+00
        5   0.0000E+00    0.0000E+00

COMPUTED FORCES AND STRESSES IN THE ELEMENTS (TENSILE POSITIVE)
  ELEMENT    FORCE        STRESS     ELEMENT    FORCE        STRESS
     1     0.2310E+00   0.1149E+02      2     0.9566E-01   0.4757E+01
     3    -0.9566E-01  -0.4757E+01      4    -0.2310E+00  -0.1149E+02
```

Fig. 4.20.

while to avoid any of the spokes (particularly element 4) going into compression the maximum load is

$$W = \frac{13.43 \times 10^3}{11.49} = 1170 \text{ lb}$$

Since the latter condition is somewhat more restrictive, the maximum value of the load is 1170 lb (5.20 kN).

We also need to find the maximum displacement of the hub. For a downward vertical load on the hub of 1 lb, the computed downward displacement of the hub (node 1) is 0.3170×10^{-4} in. Therefore, for the maximum allowable load on the hub of 1170 lb, the displacement is $1170 \times 0.3170 \times 10^{-4} = 0.037$ in (0.94 mm) in the downward vertical direction.

4.1.4 Design of pin-jointed structures

Having now developed a method of analysis and computed program for finding the forces, stresses and deformations in plane pin-jointed structures, both statically determinate and statically indeterminate, we can begin to think seriously about using the program as an aid to designing such structures. In the examples we have considered so far, we have been given the geometry of the framework and have found the stresses and displacements resulting from the applied loading. In the context of design, however, we would be given only the overall geometrical requirements, such as the span of a bridge structure or the height of a tower or crane, together with the load to be carried. Details of the sizes and arrangements of the members would not be given, although previous experience of similar structures would provide some guidance as to the type of structure geometry likely to be suitable for the particular application.

The computer program can be used to explore a range of possible geometries with a view to finding an optimum solution to the design problem. Among the criteria for such an optimum is likely to be the need to minimize the weight of material used. While this suggests that all the members should carry stresses up to the maximum allowable levels, this can lead to a requirement for almost every member to have a different cross-sectional area. In the interests of minimizing cost, it is generally preferable to make members from readily available standard sections, manufactured in a relatively small number of sizes.

Although we are in a position to examine stresses and deformations in pin-jointed structures, we are not yet able to predict when members may fail in compression at relatively low stress levels due to the phenomenon of elastic instability known as *buckling*. This we will consider in Chapter 8, after which we will be fully equipped to tackle the design of pin-jointed structures.

4.2 Other statically indeterminate systems

Many practical solid mechanics problems are statically indeterminate. We have just considered the particular case of plane pin-jointed structures, and we will meet other more complicated examples in later chapters. In the meantime, let us examine some relatively straightforward statically indeterminate problems which are natural extensions of situations we have already met. Because the most difficult parts of these problems lie in using the physical principles of the mechanics of solids to set up the relevant equations, which are then easy to solve, they do not lend themselves to computer methods.

4.2.1 Some typical problems

A problem we met in Section 1.5, and which we used to distinguish between statically determinate and statically indeterminate situations, is that of an elevator cage supported by two cables which are not necessarily identical. We are now in a position to solve such a problem.

EXAMPLE 4.4

Find expressions for the tensions in the two cables supporting the rigid cage of weight W shown in Fig. 4.21a, given that the rotation of the cage is prevented by the guide rollers. The left-hand cable has a cross-sectional area of A_1, is made from a material of Young's modulus E_1 and is subject to a tension of T_1, while the area, modulus and tension associated with the right-hand cable are A_2, E_2 and T_2, respectively. The lengths of the cables, L_1 and L_2, are the same, and their weights may be neglected.

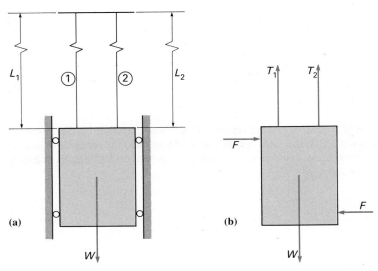

Fig. 4.21.

As with all mechanics of solids problems, we must apply the three principles of equilibrium of forces, compatibility of strains (geometry of deformation) and material stress–strain characteristics to obtain a full solution. Figure 4.21b shows the free body diagram for the cage. Equilibrium of forces acting on the cage in the vertical direction requires that

$$T_1 + T_2 = W \tag{4.14}$$

The geometry of deformation in this problem requires that, because there is no rotation of the rigid cage, both cables must extend by the same amount, say δ. Finally, the elastic stress–strain properties of the cables allow us to relate this extension to the tensions they carry. For example, the tensile stress in the left-hand cable is $\sigma_1 = T_1/A_1$, the strain is

$$e_1 = \frac{\sigma_1}{E_1} = \frac{T_1}{A_1 E_1}$$

and hence the extension is

$$\delta = e_1 L_1 = \frac{T_1 L_1}{A_1 E_1}$$

Similarly, for the right-hand cable

$$\delta = \frac{T_2 L_2}{A_2 E_2}$$

Consequently, we may express the tensions in the cables in terms of the common extension as

$$T_1 = \frac{A_1 E_1 \delta}{L_1} \quad \text{and} \quad T_2 = \frac{A_2 E_2 \delta}{L_2}$$

or

$$T_1 = k_1 \delta \quad \text{and} \quad T_2 = k_2 \delta \tag{4.15}$$

where

$$k_1 = \frac{A_1 E_1}{L_1} \quad \text{and} \quad k_2 = \frac{A_2 E_2}{L_2} \tag{4.16}$$

are the *stiffnesses* of the cables (the concept of stiffness was introduced in Section 4.1.1, and stiffness was defined as the ratio between applied force and the resulting change of length). Substituting these expressions for tensions in the cables into equilibrium equation (4.14), we have

$$(k_1 + k_2)\delta = W$$

from which

$$\delta = \frac{W}{k_1 + k_2}$$

and

$$T_1 = \frac{W k_1}{(k_1 + k_2)} \quad \text{and} \quad T_2 = \frac{W k_2}{(k_1 + k_2)} \tag{4.17}$$

These are the required expressions for the cable tensions.

These results can be generalized to systems where there are more than two elastic supports (in either tension or compression), effectively acting in parallel with each other. If there were, say, n cables supporting the cage in Fig. 4.21, then the tension in the ith cable with a stiffness of k_i would be

$$T_i = \frac{Wk_i}{(k_1 + k_2 + k_3 + \ldots + k_n)} \qquad (4.18)$$

where the summation of stiffnesses in the denominator is carried out over all n cables. A rather more sophisticated example of this type was provided by the spoked wheel in Example 4.3. Another type of statically indeterminate system can occur when elastic components are connected not in parallel but in series with each other, as in the following example.

EXAMPLE 4.5

Figure 4.22a shows two long solid cylinders just fitting between two rigid surfaces. The left-hand one has a constant cross-sectional area of A_1, and is made from a material of Young's modulus E_1, while the area and modulus of the right-hand one are A_2 and E_2, respectively. If the distance between the rigid surfaces is reduced by an amount δ, find expressions for the amounts by which the lengths of the cylinders are reduced.

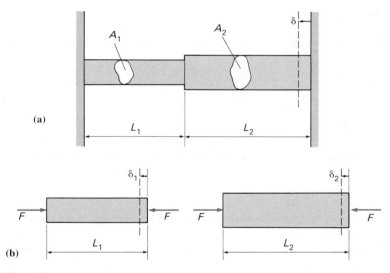

Fig. 4.22.

We apply first the principle of equilibrium of forces. Figure 4.22b shows the free body diagrams for the two cylinders, from which we can deduce that for them to be in equilibrium the same compressive force must act through each. Let us take the magnitude of this force to be F as shown. Ignoring local effects at the ends of the cylinders, the problem becomes one-dimensional and the uniaxial compressive stresses in the cylinders are of magnitude $\sigma_1 = F/A_1$ and $\sigma_2 = F/A_2$, respectively. Using stress–strain characteristics, the magnitude of the strain in the first cylinder is

$$e_1 = \frac{\sigma_1}{E_1} = \frac{F}{A_1 E_1}$$

and its reduction in length is

$$\delta_1 = e_1 L_1 = \frac{F L_1}{A_1 E_1} = \frac{F}{k_1}$$

where k_1 is stiffness defined as in equations (4.16). Similarly, the reduction in length of the second cylinder is

$$\delta_2 = \frac{F L_2}{A_2 E_2} = \frac{F}{k_2}$$

Finally, we must consider the geometry of deformation, which in this case requires that the sum of the individual reductions in length must equal the total reduction imposed on the system

$$\delta_1 + \delta_2 = \delta$$

Hence

$$F\left(\frac{1}{k_1} + \frac{1}{k_2}\right) = \delta$$

from which the force F may be found, giving the following expressions for the individual length reductions

$$\delta_1 = \frac{\dfrac{\delta}{k_1}}{\left(\dfrac{1}{k_1} + \dfrac{1}{k_2}\right)} \quad \text{and} \quad \delta_2 = \frac{\dfrac{\delta}{k_2}}{\left(\dfrac{1}{k_1} + \dfrac{1}{k_2}\right)} \tag{4.19}$$

These results can be generalized to systems where there are more than two elastic components (in either tension or compression), in series with each other. If there were, say, n cylinders between the rigid surfaces in Fig. 4.22, then the reduction in length of the ith one with a stiffness of k_i would be

$$\delta_i = \frac{\dfrac{\delta}{k_i}}{\left(\dfrac{1}{k_1} + \dfrac{1}{k_2} + \dfrac{1}{k_3} + \dots + \dfrac{1}{k_n}\right)} \tag{4.20}$$

where the summation of reciprocals of stiffnesses in the denominator is carried out over all n cylinders. We should note that the reciprocal of a stiffness is a

flexibility (ratio between displacement and the applied force which produces it).
So, if we define f_i as the flexibility of the ith cylinder

$$f_i = \frac{1}{k_i} = \frac{L_i}{A_i E_i} \tag{4.21}$$

equation (4.20) can be rewritten as

$$\delta_i = \frac{\delta f_i}{(f_1 + f_2 + f_3 + \ldots + f_n)} \tag{4.22}$$

In Examples 4.4 and 4.5, we adopted two somewhat different approaches to the solution of statically indeterminate problems. In the first case we chose a displacement variable (the common extension, δ, of the cables) as the primary unknown. We then established an equation for this unknown which ensured that equilibrium was satisfied. The tensions in the cables were then obtained from δ. Such an approach is referred to as an *equilibrium method* (sometimes *displacement method* since a displacement is the primary unknown). In Example 4.5, on the other hand, we chose a force variable (the common force, F, in the two cylinders) as the primary unknown. We then set up an equation for this unknown which ensured that the geometry of deformation (compatibility of strains) was satisfied. The changes in lengths of the cylinders were then obtained from F. Such an approach is referred to as a *compatibility method* (sometimes *force method* since a force is the primary unknown). The reasons we made the particular choices in the two cases should be understood. In Example 4.4 there was only one displacement to be found (δ) and two forces (T_1 and T_2), but in Example 4.5 the reverse was true (displacements δ_1 and δ_2 against force F). Either method can be used in a given problem, but one is usually more straightforward than the other, involving fewer primary unknowns. Looking back to the finite element method we developed for plane pin-jointed structures in Section 4.1.1, by choosing displacements of the nodal points as the primary unknowns, we were electing to use an equilibrium method.

EXAMPLE 4.6

The rigid bar AB shown in Fig. 4.23a is pinned at A and is initially horizontal, supported by two vertical solid circular rods CD and EB. Rod CD is made of brass, 1.0 m (40 in) long, 25 mm (1 in) diameter, and EB is made of steel, 1.5 m (60 in) long and 20 mm (0.8 in) diameter. Find the elongation of EB and the stress in each rod when the force of 150 kN (34 kip) is applied as shown. The weights of AB, CD and EB may be neglected.

With two elastic supporting members in parallel, the equilibrium method of analysis suggests itself. In this case, however, we cannot assume the same downward displacement for the attachment points D and B. What we can assume, however, is a single angle of rotation for the rigid bar AB about the pivot point A. Let this angle be θ in the clockwise direction. We may assume this angle to be small, such that the rods CD and EB remain effectively vertical. Also, the geometry of deformation defines the extensions of rods CD and EB as

(a)

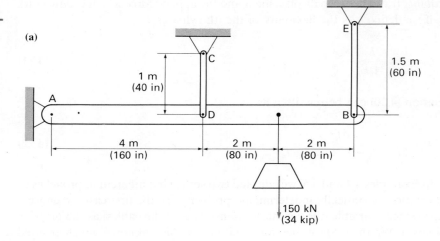

(b)

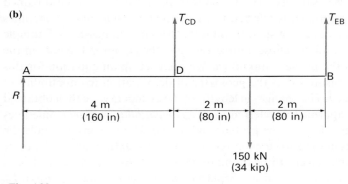

Fig. 4.23.

SI units

$$\delta_{CD} = 4\theta \quad \text{and} \quad \delta_{EB} = 8\theta$$

where θ is in radians. Therefore, the tensions in these rods are

$$T_{CD} = 4k_{CD}\theta \quad \text{and} \quad T_{EB} = 8k_{EB}\theta$$

where the stiffnesses are given by equation (4.16) as

$$k_{CD} = \frac{\pi \times 0.025^2}{4} \times \frac{103 \times 10^9}{1.0} = 50.56 \times 10^6 \text{ N/m}$$

and

$$k_{EB} = \frac{\pi \times 0.020^2}{4} \times \frac{207 \times 10^9}{1.5} = 43.35 \times 10^6 \text{ N/m}$$

The values of Young's modulus for brass and steel are obtained from Appendix A.

US customary units

$$\delta_{CD} = 160\,\theta \quad \text{and} \quad \delta_{EB} = 320\,\theta$$

where θ is in radians. Therefore, the tensions in these rods are

$$T_{CD} = 160\,k_{CD}\,\theta \quad \text{and} \quad T_{EB} = 320\,k_{EB}\,\theta$$

where the stiffnesses are given by equation (4.16) as

$$k_{CD} = \frac{\pi \times 1^2}{4} \times \frac{15 \times 10^6}{40} = 0.2945 \times 10^6 \text{ lb/in}$$

and

$$k_{EB} = \frac{\pi \times 0.8^2}{4} \times \frac{30 \times 10^6}{60} = 0.2513 \times 10^6 \text{ lb/in}$$

The values of Young's modulus for brass and steel are obtained from Appendix A.

Figure 4.23b shows the free body diagram for bar AB, which includes a reaction force at the pivot at A. Since we do not need to find this force, we may consider equilibrium of moments about A

$$150 \times 10^3 \times 6 - 4T_{CD} - 8T_{EB} = 0$$

or

$$(16\,k_{CD} + 64\,k_{EB})\,\theta = 9 \times 10^5$$

and

$$\theta = \frac{9 \times 10^5}{(16 \times 50.56 + 64 \times 43.35) \times 10^6}$$

$$= 2.512 \times 10^{-4} \text{ rad}$$

Having found the primary displacement unknown, we can now find the required extension of EB as

$$\delta_{EB} = 8\theta = 2.010 \times 10^{-3} \text{ m} = \underline{2.01 \text{ mm}}$$

Also, the tensions in the rods are

$$T_{CD} = 4k_{CD}\theta = 4 \times 50.56 \times 10^6 \times 2.512 \times 10^{-4}$$

$$= 50.80 \times 10^3 \text{ N}$$

$$T_{EB} = 8\,k_{EB}\theta = 8 \times 43.35 \times 10^6 \times 2.512 \times 10^{-4}$$

$$= 87.12 \times 10^3 \text{ N}$$

and the corresponding stresses are

$$\sigma_{CD} = \frac{50.80 \times 10^3}{\dfrac{\pi}{4} \times 0.025^2} = 103.5 \times 10^6 \text{ N/m}^2$$

$$= \underline{103 \text{ MN/m}^2}$$

and

$$\sigma_{EB} = \frac{87.12 \times 10^3}{\dfrac{\pi}{4} \times 0.020^2} = 277.3 \times 10^6 \text{ N/m}^2$$

$$= \underline{277 \text{ MN/m}^2}$$

Figure 4.23b shows the free body diagram for bar AB, which includes a reaction force at the pivot at A. Since we do not need to find this force, we may consider equilibrium of moments about A

$$34 \times 10^3 \times 240 - 160T_{CD} - 320T_{EB} = 0$$

or

$$(160^2 k_{CD} + 320^2\,k_{EB})\,\theta = 8.16 \times 10^6$$

and

$$\theta = \frac{8.16 \times 10^6}{(160^2 \times 0.2945 + 320^2 \times 0.2513) \times 10^6}$$

$$= 2.452 \times 10^{-4} \text{ rad}$$

Having found the primary displacement unknown, we can now find the required extension of EB as

$$\delta_{EB} = 320\,\theta = \underline{0.0785 \text{ in}}$$

Also, the tensions in the rods are

$$T_{CD} = 160\,k_{CD}\,\theta = 160 \times 0.2945 \times 10^6 \times 2.452$$

$$\times 10^{-4} \text{ lb} = 11.55 \text{ kip}$$

$$T_{EB} = 320\,k_{EB}\,\theta = 320 \times 0.2513 \times 10^6 \times 2.452$$

$$\times 10^{-4} \text{ lb} = 19.72 \text{ kip}$$

and the corresponding stresses are

$$\sigma_{CD} = \frac{11.55}{\dfrac{\pi}{4} \times 1^2} = \underline{14.7 \text{ ksi}}$$

and

$$\sigma_{EB} = \frac{19.72}{\dfrac{\pi}{4} \times 0.8^2} = \underline{39.2 \text{ ksi}}$$

These stresses are relatively high for the materials concerned – similar in magnitude to their yield stresses. The assumption that the bar AB is rigid in this problem is an idealization: in practice the bending deflections of the bar may not be negligible. Bending deflections are considered in Chapter 6.

EXAMPLE 4.7

Figure 4.24a shows a 12 mm (0.5 in) diameter mild steel bolt which is threaded with a pitch of 1 mm (0.040 in) and passes centrally through a copper tube of 15 mm (0.60 in) internal diameter, 25 mm (1 in) external diameter and 180 mm (7.2 in) long. The ends of the tube are closed by 10 mm (0.4 in) thick rigid washers which are secured by the nut on the threaded end of the bolt. After the nut has been adjusted to bring the components just into contact, it is then tightened by one eighth of a turn. Find the force induced in the bolt, and the stress in the copper tube.

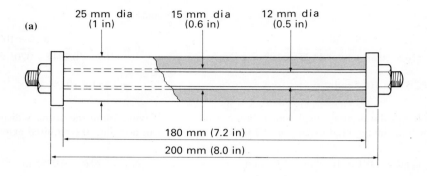

(a)

25 mm dia (1 in) 15 mm dia (0.6 in) 12 mm dia (0.5 in)

180 mm (7.2 in)

200 mm (8.0 in)

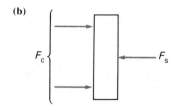

(b)

F_c F_s

Fig. 4.24.

This problem is similar to Example 3.1, except that here, instead of the bolt clamping rigid components together, one of them (the copper tube) is significantly deformable, making the problem statically indeterminate. With two elastic components, the bolt and the tube, in parallel, the equilibrium method of analysis is likely to be the more convenient. We can choose as an appropriate displacement unknown the amount, δ, by which the length of the copper tube is reduced by tightening the nut. The compressive force in the copper is therefore

$$F_c = k_c \delta$$

where, from equation (4.16)

$$k_c = \frac{A_c E_c}{L_c}$$

is the axial stiffness of the copper, A_c, E_c and L_c being its cross-sectional area, Young's modulus and length, respectively.

SI units

$$A_c = \frac{\pi}{4} \times (0.025^2 - 0.015^2) = 3.142 \times 10^{-4} \text{ m}^2$$

and, given the value of Young's modulus for copper as 120 GN/m^2 (Appendix A), this stiffness is

$$k_c = \frac{3.142 \times 10^{-4} \times 120 \times 10^9}{0.180} = 209.4 \times 10^6 \text{ N/m}$$

If the deformed length of the tube is $(0.180 - \delta)$ m, then because the washers are rigid the deformed length of the tube plus washers is $(0.200 - \delta)$ m. Consequently, the length $L_s = 0.200$ m of the steel bolt which was originally between the bolt head and nut (Fig. 4.24a) is increased to $(0.200 - \delta + 0.125 \times 10^{-3})$ m, 0.125×10^{-3} m being the length of the bolt drawn into the nut by its rotation (one eighth of its pitch of 1 mm). The extension of the bolt is therefore defined by the geometry of deformation of the system as $(0.125 \times 10^{-3} - \delta)$ m, and the tensile force in the steel is

$$F_s = k_s(0.125 \times 10^{-3} - \delta)$$

where

$$k_s = \frac{A_s E_s}{L_s}$$

is its axial stiffness. The cross-sectional area of the steel is

$$A_s = \frac{\pi}{4} \times 0.012^2 = 1.131 \times 10^{-4} \text{ m}^2$$

and the value of its Young's modulus is $E_s = 207$ GN/m^2 (Appendix A), so that

$$k_s = \frac{1.131 \times 10^{-4} \times 207 \times 10^9}{0.200} = 117.1 \times 10^6 \text{ N/m}$$

Figure 4.24b shows the free body diagram for the washer in contact with the nut in Fig. 4.24a, with the forces F_s and F_c acting on it. Now, for the washer to be in equilibrium

$$F_s = F_c$$

or

$$k_s(0.125 \times 10^{-3} - \delta) = k_c \delta$$

US customary units

$$A_c = \frac{\pi}{4} \times (1^2 - 0.6^2) = 0.5027 \text{ in}^2$$

and, given the value of Young's modulus for copper as 17×10^6 psi (Appendix A), this stiffness is

$$k_c = \frac{0.5027 \times 17 \times 10^6}{7.2} = 1.187 \times 10^6 \text{ lb/in}$$

If the deformed length of the tube is $(7.2 - \delta)$ in, then because the washers are rigid the deformed length of the tube plus washers is $(8.0 - \delta)$ in. Consequently, the length $L_s = 8.0$ in of the steel bolt which was originally between the bolt head and nut (Fig. 4.24a) is increased to $(8.0 - \delta + 0.005)$ in, 0.005 in being the length of the bolt drawn into the nut by its rotation (one eighth of its pitch of 0.04 in). The extension of the bolt is therefore defined by the geometry of deformation of the system as $(0.005 - \delta)$ in, and the tensile force in the steel is

$$F_s = k_s(0.005 - \delta)$$

where

$$k_s = \frac{A_s E_s}{L_s}$$

is its axial stiffness. The cross-sectional area of the steel is

$$A_s = \frac{\pi}{4} \times 0.5^2 = 0.1963 \text{ in}^2$$

and the value of its Young's modulus is $E_s = 30 \times 10^6$ psi (Appendix A), so that

$$k_s = \frac{0.1963 \times 30 \times 10^6}{8.0} = 0.7361 \times 10^6 \text{ lb/in}$$

Figure 4.24b shows the free body diagram for the washer in contact with the nut in Fig. 4.24a, with the forces F_s and F_c acting on it. Now, for the washer to be in equilibrium

$$F_s = F_c$$

or

$$k_s(0.005 - \delta) = k_c \delta$$

SI units

from which we may obtain the primary displacement unknown as

$$\delta = \frac{0.125 \times 10^{-3} \, k_s}{(k_s + k_c)} = \frac{0.125 \times 10^{-3} \times 117.1 \times 10^6}{(117.1 + 209.4) \times 10^6}$$

$$= 0.4483 \times 10^{-4} \, \text{m}$$

The tension in the bolt is therefore

$$F_s = 117.1 \times 10^6 \times (0.125 \times 10^{-3} - 0.4483$$

$$\times 10^{-4}) \, \text{N} = \underline{9.39 \, \text{kN}}$$

which compares with 14.6 kN (Example 3.1) when the same bolt is tightened onto rigid components. Finally, the stress in the copper tube is

$$\sigma_c = -\frac{F_c}{A_c} = -\frac{F_s}{A_c} = -\frac{9.39 \times 10^3}{3.142 \times 10^{-4}} \, \text{N/m}^2$$

$$= \underline{-29.9 \, \text{MN/m}^2}$$

US customary units

from which we may obtain the primary displacement unknown as

$$\delta = \frac{0.005 \, k_s}{(k_s + k_c)} = \frac{0.005 \times 0.7361 \times 10^6}{(1.187 + 0.7361) \times 10^6}$$

$$= 1.914 \times 10^{-3} \, \text{in}$$

The tension in the bolt is therefore

$$F_s = 0.7361 \times 10^6 \times (0.005 - 0.0019 \, 14) \, \text{lb}$$

$$= \underline{2.27 \, \text{kip}}$$

which compares with 3.68 kip (Example 3.1) when the same bolt is tightened onto rigid components. Finally, the stress in the copper tube is

$$\sigma_c = -\frac{F_c}{A_c} = -\frac{F_s}{A_c} = -\frac{2.27}{0.5027}$$

$$= \underline{-4.52 \, \text{ksi}}$$

4.2.2 Liquid-filled thin-walled vessels

In Sections 3.4.2 and 3.4.3, we considered the changes in dimensions and volumes of thin-walled cylindrical and spherical pressure vessels subject to internal pressure due to the fluids they contain. At that stage we confined our attention to relatively simple problems. For instance, in Example 3.12, which involved a specified volume of liquid being pumped into a vessel after it was just full at ambient pressure, we assumed the liquid to be incompressible. If it is compressible, then the increase in internal pressure not only causes the vessel to expand but also the liquid to contract, both of which effects help to accommodate the extra liquid.

Let us first consider the problem in general terms for any shape of vessel, although for the purposes of illustration it is convenient to show a spherical vessel as in Fig. 4.25. Figure 4.25a shows the vessel of internal volume V in its initial state, with atmospheric pressure acting on both its external and internal surfaces. As a result of overfilling the vessel, a hydrostatic pressure is created which acts throughout the liquid and over the internal surface of the vessel. Let this pressure be p above atmospheric pressure. Figure 4.25b illustrates the effect of this pressure on the vessel. While the solid line represents the vessel in its initial state, the dotted line shows the increased size (much exaggerated) associated with the pressure p. The new volume of the vessel is $V + \Delta V_V$. Figure 4.25c shows the liquid which is finally to be contained within the vessel. While the solid line represents the liquid at atmospheric pressure, the dotted line shows its reduced size due to the pressure p, the difference being ΔV_L. Now,

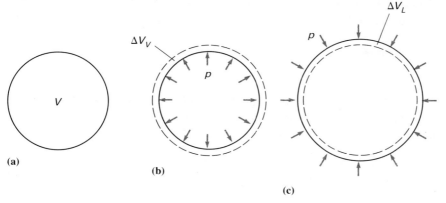

Fig. 4.25.

the initial volume of this liquid is $V + V_{in}$, where V_{in} is the volume of fluid pumped in, and the final volume is $V + V_{in} - \Delta V_L$. Note that, in making this statement, we are effectively measuring V_{in} at atmospheric pressure rather than at, say, pressure p. For liquids, which are only slightly compressible (compared with gases), $V_{in} \ll V$, and any variation of V_{in} with the pressure at which it is measured is very much smaller than V_{in} itself. The geometry of deformation of the system requires that the final volumes of the liquid and vessel are equal (the dotted circles in Figs 4.25a and b are of the same size), so that

$$V + V_{in} - \Delta V_L = V + \Delta V_V$$

$$V_{in} = \Delta V_V + \Delta V_L \tag{4.23}$$

Putting this equation into words, the volume of liquid pumped in is equal to the increase in the internal volume of the vessel due to the pressure generated, plus the reduction in volume of the total quantity of liquid due to the same pressure.

While there are two deformation variables, ΔV_V and ΔV_L, in the problem, there is only one force type variable, namely the pressure p. This suggests that we should use the compatibility method, with p as the primary variable. Provided there is no significant change in temperature during the pumping process, we can express the volumetric strain of the interior of the vessel as

$$e_{vol} = pC \tag{4.24}$$

where the constant C is given in terms of known dimensions and material properties by, for example, equation (3.26) for a cylindrical vessel or equation (3.31) for a spherical vessel (in either case with $\Delta T = 0$, since we are not yet considering the effects of temperature changes). Consequently

$$\Delta V_V = pCV \tag{4.25}$$

Now, in order to define the volumetric strain of the liquid, we may adapt equation (3.8) (which was introduced in the context of solid stress–strain behavior, but is equally applicable to liquids) to give

$$e_{\text{vol}} = -\frac{p}{K_L} \tag{4.26}$$

where K_L is the bulk modulus of the liquid, and $-p$ is the (compressive) hydrostatic stress acting in it. Consequently, the decrease in volume of the total quantity of liquid is

$$\Delta V_L = \frac{p(V + V_{\text{in}})}{K_L} \approx \frac{pV}{K_L} \tag{4.27}$$

Substituting the expressions for volume change given by equations (4.25) and (4.27) into equation (4.23), we obtain

$$pV\left(C + \frac{1}{K_L}\right) = V_{\text{in}}$$

and

$$p = \frac{V_{\text{in}}}{V\left(C + \dfrac{1}{K_L}\right)} \tag{4.28}$$

Having found the pressure, it is a straightforward matter to calculate changes in volume and dimensions of the vessel.

EXAMPLE 4.8

A spherical aluminum vessel of 350 mm (14 in) diameter and 4 mm (0.16 in) thick is just full of water under ambient conditions. If a further 5×10^4 mm^3 (50 cm^3 or 3 in^3) of water is pumped into the vessel, find the resulting internal pressure and change in diameter of the vessel. This is a repeat of Example 3.12, but with the liquid defined as water.

From equation (3.31), the constant C in equation (4.24) is

$$C = \frac{3D}{4Et}(1 - v) = \frac{3 \times 0.350 \times (1 - 0.34)}{4 \times 70 \times 10^9 \times 0.004} = 6.187 \times 10^{-10} \text{ m}^2/\text{N}$$

where the elastic properties of aluminum are obtained from Appendix A. From the same source, the bulk modulus of water is $K = 2.3$ GN/m^2. The volume of the vessel is

$$V = \frac{\pi D^3}{6} = \frac{\pi \times 350^3}{6} = 22.45 \times 10^6 \text{ mm}^3$$

Incidentally, we note that $V_{\text{in}}/V = 5 \times 10^4/22.45 \times 10^6 = 0.0022$, which confirms that

$V_{in} \ll V$ and justifies the approximation in equation (4.27). Using equation (4.28), the required pressure is

$$p = \frac{5 \times 10^4}{22.45 \times 10^6 \left(6.187 \times 10^{-10} + \dfrac{1}{2.3 \times 10^9}\right)} \ \text{N/m}^2 = \underline{2.11 \ \text{MN/m}^2} \ (310 \ \text{psi})$$

Had we assumed the water to be incompressible, we would have taken its bulk modulus to be infinite, the reciprocal of this modulus as zero, and

$$p = \frac{5 \times 10^4}{22.45 \times 10^6 \times 6.187 \times 10^{-10}} \ \text{N/m}^2 = 3.60 \ \text{MN/m}^2$$

which is identical to the result obtained in Example 3.12. Note what a significant effect the compressibility of the water has on the pressure generated. This is because the volumetric strains of the thin metal vessel and water are of the same order of magnitude.

In the present problem, the volumetric strain of the vessel is given by equation (4.24) as

$$e_{vol} = pC = 2.11 \times 10^6 \times 6.187 \times 10^{-10} = 1.31 \times 10^{-3}$$

from which the change in diameter may be obtained (using equation (3.30)) as

$$\Delta D = \frac{D e_{vol}}{3} = \frac{350 \times 1.31 \times 10^{-3}}{3} = \underline{0.153 \ \text{mm}} \ (0.0060 \ \text{in})$$

which compares with 0.260 mm (0.010 in) obtained in Example 3.12.

4.2.3 Resisted thermal expansion

Statically indeterminate problems can also arise in situations where there are components made from different materials, and which therefore tend to expand or contract by different amounts when the temperature of the system is changed. Let us consider some examples of this type, the first of which involves a liquid-filled vessel and therefore follows naturally from Example 4.8 above.

EXAMPLE 4.9

A thin-walled spherical steel vessel with a wall thickness of 2 mm (0.08 in) and a diameter of 250 mm (10 in) is closed and just full of water at 15°C (59°F). Calculate the internal pressure when the temperature of the vessel and water is raised to 75°C (167°F).

If the vessel and water are considered separately with no pressure acting on them, the rise in temperature would cause both the internal volume of the vessel and the volume of the water to increase, but by different amounts. Since the coefficient of volumetric thermal expansion of water ($\beta_w = 2.1 \times 10^{-4}\,°\text{C}^{-1}$, from Appendix A) exceeds that of steel ($\beta = 3\alpha = 0.33 \times 10^{-4}\,°\text{C}^{-1}$), the water tends to expand more than the interior of the vessel. This creates an internal pressure which increases the volume of the vessel over and above the effect of the temperature rise alone, and at the same time restricts the expansion of the water so that it remains within the vessel. In other words, the geometry of deformation of the system requires that the final volumes of the vessel and water must be identical. Since the initial volumes are also identical, this implies that each suffers the same volumetric strain due to the combined effects of the temperature rise and internal pressure.

Equation (3.31) defines the volumetric strain of the interior of the spherical vessel as

$$e_{vol} = \frac{3pD}{4Et}(1 - v) + 3\alpha\Delta T$$

where p is the internal pressure, D, t and E are the diameter, thickness and Young's modulus of the vessel, respectively, and ΔT is the temperature rise. Adding the thermal strain term to equation (3.8), the volumetric strain of the water is given by

$$e_{vol} = -\frac{p}{K_w} + \beta_w \Delta T$$

where K_w is its bulk modulus. Equating these two expressions for volumetric strain, and taking values for the material properties from Appendix A, we obtain an equation for the unknown pressure

$$\left(\frac{3 \times 250 \times (1 - 0.3)}{4 \times 207 \times 10^9 \times 2} + \frac{1}{2.3 \times 10^9}\right)p = (2.1 - 0.33) \times 10^{-4} \times (75 - 15)$$

$$7.518 \times 10^{-10}\, p = 0.01062$$

$$p = 14.13 \times 10^6 \text{ N/m}^2 = \underline{14.1 \text{ MN/m}^2} \text{ (2.0 ksi)}$$

EXAMPLE 4.10

In Example 4.7, after the nut has been tightened the temperature of the whole system is raised by 50°C (90°F). Assuming that the rigid washers have negligibly small coefficients of thermal expansion, calculate the final tension in the steel bolt, and the stress in the copper tube.

Assuming the steel and copper remain elastic, we may use the principle of superposition (Section 3.2.1) to consider the effects of tightening the nut and increasing the temperature separately. The first of these we dealt with in Example 4.7, so now let us examine the effect of the temperature rise alone. Figures 4.24a and b show the system and the free body diagram for one of the rigid washers, respectively. We again choose a displacement as the primary unknown, in this case the amount, δ', by which the length of the copper tube is increased due to the temperature rise. The geometry of deformation of the system requires that this is also the amount by which the length of the bolt increases. The states of stress in the tube and bolt are uniaxial, and the strain in the copper tube is therefore given by stress–strain equations (3.16) as

$$\frac{\delta'}{L_c} = -\frac{F_c}{E_c A_c} + \alpha_c \Delta T$$

$$F_c = -k_c \delta' + E_c A_c \alpha_c \Delta T$$

Similarly, the strain in the steel bolt is

$$\frac{\delta'}{L_s} = \frac{F_s}{E_s A_s} + \alpha_s \Delta T$$

$$F_s = k_s \delta' - E_s A_s \alpha_s \Delta T$$

For equilibrium of forces acting on the washer in Fig. 4.24b, $F_s = F_c$ and

$$(k_s + k_c)\delta' = (E_s A_s \alpha_s + E_c A_c \alpha_c)\Delta T$$

Substituting numerical values (from Example 4.7 and Appendix A).

SI units

$$(117.1 + 209.4) \times 10^6 \delta' = (207 \times 1.131 \times 11 + 120$$

$$\times 3.142 \times 16) \times 10^{-1} \times 50$$

$$\delta' = \frac{43\,040}{326.5 \times 10^6} \text{ m} = 0.1318 \text{ mm}.$$

The tension in the bolt is therefore

$$F_s = 117.1 \times 10^6 \times 0.1318 \times 10^{-3} - 207 \times 1.131$$

$$\times 11 \times 10^{-1} \times 50 \text{ N}$$

$$= 2.557 \text{ kN}$$

and the stress in the copper tube is

$$\sigma_c = -\frac{F_c}{A_c} = -\frac{F_s}{A_c} = -\frac{2.557 \times 10^3}{3.142 \times 10^{-4}} \text{ N/m}^2$$

$$= -8.138 \text{ MN/m}^2$$

Superimposing these values on the results obtained in Example 4.7, we obtain

$$F_s = 9.39 + 2.56 = \underline{11.9 \text{ kN}}$$

$$\sigma_c = -29.9 - 8.14 = \underline{-38.0 \text{ MN/m}^2}$$

due to the combined effect of tightening the nut and increasing the temperature.

US customary units

$$(0.7361 + 1.187) \times 10^6 \delta' = (30 \times 0.1963 \times 6.5 + 17$$

$$\times 0.5027 \times 8.9) \times 90$$

$$\delta' = \frac{10\,290}{1.923 \times 10^6} = 5.351 \times 10^{-3} \text{ in}$$

The tension in the bolt is therefore

$$F_s = 0.7361 \times 10^6 \times 5.351 \times 10^{-3} - 30 \times 0.1963$$

$$\times 6.5 \times 90 \text{ lb}$$

$$= 0.494 \text{ kip}$$

and the stress in the copper tube is

$$\sigma_c = -\frac{F_c}{A_c} = -\frac{F_s}{A_c} = -\frac{0.494}{0.5027}$$

$$= -0.982 \text{ ksi}$$

Superimposing these values on the results obtained in Example 4.7, we obtain

$$F_s = 2.27 + 0.494 = \underline{2.76 \text{ kip}}$$

$$\sigma_c = -4.52 - 0.982 = \underline{-5.50 \text{ ksi}}$$

due to the combined effect of tightening the nut and increasing the temperature.

EXAMPLE 4.11

A brass ring having an internal diameter of 150 mm (6.0 in) and an external diameter of 154 mm (6.16 in) is to be shrunk on to a steel ring of the same axial length and having internal and external diameters of 140 and 150.05 mm (5.6 and 6.002 in), respectively. Find the temperature to which the brass ring must be heated so that it will just slide over

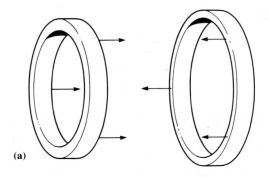

(a)

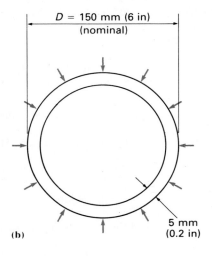

(b)

$D = 150$ mm (6 in)
(nominal)

5 mm
(0.2 in)

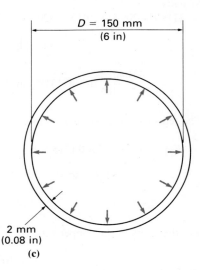

$D = 150$ mm
(6 in)

2 mm
(0.08 in)

(c)

Fig. 4.26.

the steel ring, as in Fig. 4.26a. When the assembly has returned to room temperature, find the interfacial pressure between the rings and the hoop stresses in them. Assume there are no axial stresses in the rings.

The first part of the problem involves allowing sufficient unrestricted thermal expansion of the brass ring to increase its internal diameter to match the outer diameter of the steel ring. From equations (3.21) and (3.24), when the ring is unstressed the hoop strain is given by

$$e_\theta = \frac{\Delta D}{D} = \alpha_b \Delta T$$

where ΔD is the change of (internal) ring diameter D, α_b is the coefficient of linear thermal expansion of brass and ΔT is the change in temperature of the ring. Now, the required change in diameter is $\Delta D = 0.05$ mm, $D = 150$ mm and $\alpha_b = 19 \times 10^{-6}\,°C^{-1}$ (Appendix A), from which

$$\Delta T = \frac{0.05}{150 \times 19 \times 10^{-6}} = \underline{17.5°C\ (31.5°F)}$$

We should note that from a practical point of view this is the minimum temperature rise necessary to allow the brass ring to slide over the steel ring. It allows no clearance between them, with a consequent risk of sticking if they are not perfectly aligned, leading to heat conduction from the brass to the steel and premature locking of the rings together. In this case a temperature rise of at least twice the calculated figure is to be preferred.

Cooling the combined rings back to room temperature introduces an interfacial pressure between them which tends to expand the brass ring and contract the steel ring (relative to their original dimensions at room temperature), resulting in a diameter of the interface of between 150 and 150.05 mm. Figures 4.26b and c show free body diagrams for the two rings. Since we wish to find the interfacial pressure, p, let us use it as our primary unknown and adopt the compatibility method of analysis. From equation (2.16), which is derived from equilibrium considerations, the hoop stress in the brass ring is

$$\sigma_{\theta b} = \frac{pD}{2t_b}$$

where t_b is its radial thickness (of 2 mm). With no axial stress and negligibly small radial stress, the hoop strain is given by the uniaxial stress–strain relationship

$$e_{\theta b} = \frac{\sigma_{\theta b}}{E_b} = \frac{pD}{2E_b t_b}$$

where E_b is the Young's modulus of brass. Consequently, the increase in the internal diameter of the ring relative to its original size at room temperature is

$$\Delta D_b = e_{\theta b} D = \frac{pD^2}{2E_b t_b}$$

Similarly, the *decrease* in the external diameter of the steel ring, subject to pressure p acting over its external surface of the same nominal diameter D is

$$\Delta D_s = \frac{pD^2}{2E_s t_s}$$

where E_s and t_s are the Young's modulus and radial thickness (nominally 5 mm) of the steel ring.

The geometry of deformation of the system requires that the sum of the increase in the internal diameter of the brass ring and the decrease in the diameter of the steel ring is equal to the initial difference of $\delta = 0.05$ mm between them. Hence

$$\frac{pD^2}{2}\left(\frac{1}{E_b t_b} + \frac{1}{E_s t_s}\right) = \delta$$

$$p\frac{150^2}{2}\left(\frac{1}{103 \times 10^9 \times 2} + \frac{1}{207 \times 10^9 \times 5}\right) = 0.05$$

$$p = \frac{0.05}{6.548 \times 10^{-8}} = 0.7636 \times 10^6 \text{ N/m}^2 = \underline{0.764 \text{ MN/m}^2} \text{ (110 psi)}$$

Also, the required hoop stresses in the rings are

$$\sigma_{\theta b} = \frac{0.7636 \times 10^6 \times 150}{2 \times 2} \text{ N/m}^2 = \underline{28.6 \text{ MN/m}^2} \text{ (4.1 ksi)}$$

and

$$\sigma_{\theta s} = -\frac{0.7636 \times 10^6 \times 150}{2 \times 5} \text{ N/m}^2 = \underline{-11.5 \text{ MN/m}^2} \text{ (1.7 ksi)}$$

Note that these results for interfacial pressure and stresses depend on the initial difference in diameter between the two rings, and not on the temperature increase applied to the brass ring in order to slide it over the steel ring.

EXAMPLE 4.12

The pressure vessel shown in Fig. 4.27 consists of an aluminum alloy cylinder of internal diameter 150 mm (6 in), wall thickness 3 mm (0.12 in) and length 250 mm (10 in). The ends of the cylinder are closed by flat plates which may be assumed to be rigid, and to have negligibly small coefficients of thermal expansion. These plates are held in position by four mild steel rods passing between them and secured by nuts on the outside. The diameter of each rod is 10 mm (0.4 in), and the distance between the nuts is 275 mm (11 in). With the cylinder open to the atmosphere at a temperature of 20°C (68°F), the nuts are adjusted to bring the end plates just into contact with the ends of the cylinder. Calculate the stresses in the cylinder and the rods when the entire system is then subjected to operating conditions of 1.2 MN/m² (170 psi) internal gage pressure and a temperature of 90°C (194°F).

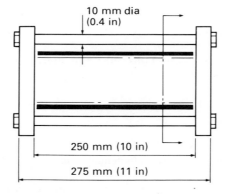
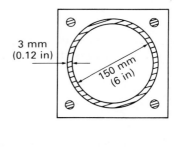

Fig. 4.27.

This is a more complicated example which illustrates not only the solution of statically indeterminate problems, but also the influence of end conditions on the axial stress in a thin-walled cylinder (a matter first discussed in Section 2.2.1). The hoop stress in the aluminum alloy cylinder, which depends on the internal pressure and not on the end conditions, may be found with the aid of equation (2.16) as

SI units

$$\sigma_{\theta a} = \frac{pR_i}{t} = \frac{1.2 \times 75}{3} = \underline{30.0 \text{ MN/m}^2}$$

US customary units

$$\sigma_{\theta a} = \frac{pR_i}{t} = \frac{170 \times 3.0}{0.12} \text{ psi} = \underline{4.25 \text{ ksi}}$$

Let us define δ as the amount by which the axial length of the cylinder increases due to the combined effect of internal pressure and change in temperature, and use this as the primary unknown of the problem. The geometry of deformation requires that δ is also the amount by which each of the steel rods increases in length. The axial strain in these rods is therefore

$$e_{zs} = \frac{\delta}{L_s} = \frac{\sigma_{zs}}{E_s} + \alpha_s \Delta T$$

where L_s is the length of the rods (275 mm or 11 in), σ_{zs} is the axial stress in them, E_s and α_s are the Young's modulus and coefficient of thermal expansion of steel and ΔT is the change in temperature (70°C or 126°F). The stress in the rods may therefore be expressed as

$$\sigma_{zs} = \frac{E_s \delta}{L_s} - E_s \alpha_s \Delta T$$

Similarly, with a biaxial state of stress in the wall of the aluminum alloy cylinder, the axial strain there is given by

$$e_{za} = \frac{\delta}{L_a} = \frac{1}{E_a}(\sigma_{za} - \nu_\alpha \sigma_{\theta a}) + \alpha_a \Delta T$$

where L_a is the length of the cylinder (250 mm), σ_{za} is the axial stress in it, E_a, ν_a and α_a are the Young's modulus, Poisson's ratio and coefficient of thermal expansion of aluminum alloy. The axial stress in the cylinder is therefore given by

$$\sigma_{za} = \nu_a \sigma_{\theta a} + \frac{E_a \delta}{L_a} - E_a \alpha_a \Delta T$$

Figure 4.28 shows the free body diagram for one of the rigid end plates together with short lengths of the cylinder and rods, with the pressure and stresses acting on them.

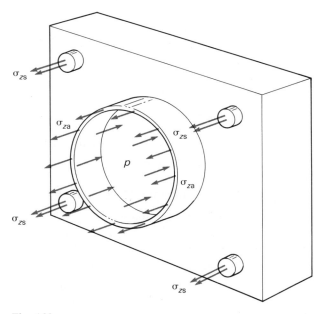

Fig. 4.28.

Note that the axial stress in the cylinder is shown as tensile although in practice it will be compressive if the cylinder is to be sealed. This conforms with the usual tensile positive sign convention. For this free body to be in equilibrium, there must be zero resultant force acting in the axial direction, and

$$\sigma_{za} A_a + \sigma_{zs} A_s - pA_i = 0$$

where A_a, A_s and A_i are the cross-sectional areas of the aluminium alloy in the cylinder, the steel in the four rods and the interior of the cylinder, respectively. Substituting the above expressions for the axial stresses into this equation, we obtain

$$(k_a + k_s)\delta = pA_i + (E_a A_a \alpha_a + E_s A_s \alpha_s)\Delta T - v_a \sigma_{\theta a} A_a$$

where k_a and k_s are the axial stiffnesses of the cylinder and rods, respectively. This equation can now be solved for δ. Using the given dimensions, and material properties from Appendix A, we have

SI units

$$A_a = \frac{\pi}{4} \times (156^2 - 150^2)\, \text{mm}^2 = 1442 \times 10^{-6}\, \text{m}^2$$

$$A_s = 4 \times \frac{\pi}{4} \times 10^2\, \text{mm}^2 = 314.2 \times 10^{-6}\, \text{m}^2$$

$$A_i = \frac{\pi}{4} \times 150^2\, \text{mm}^2 = 17\,670 \times 10^{-6}\, \text{m}^2$$

$$k_a = \frac{E_a A_a}{L_a} = \frac{68.9 \times 10^9 \times 1442 \times 10^{-6}}{250 \times 10^{-3}}\, \text{N/m}$$
$$= 397.4 \times 10^6\, \text{N/m}$$

$$k_s = \frac{E_s A_s}{L_s} = \frac{207 \times 10^9 \times 314.2 \times 10^{-6}}{275 \times 10^{-3}}\, \text{N/m}$$
$$= 236.5 \times 10^6\, \text{N/m}$$

$$E_a A_a \alpha_a \Delta T = 68.9 \times 10^9 \times 1442 \times 10^{-6} \times 23 \times 10^{-6}$$
$$\times 70 = 160.0 \times 10^3\, \text{N}$$

$$E_s A_s \alpha_s \Delta T = 207 \times 10^9 \times 314.2 \times 10^{-6} \times 11 \times 10^{-6}$$
$$\times 70 = 50.08 \times 10^3\, \text{N}$$

$$(397.4 + 236.5) \times 10^6 \delta = 1.2 \times 10^6 \times 17670 \times 10^{-6}$$
$$+ (160.0 + 50.08) \times 10^3 - 0.3$$
$$\times 30 \times 10^6 \times 1442 \times 10^{-6}$$

$$\delta = 0.3444 \times 10^{-3}\, \text{m}$$

US customary units

$$A_a = \frac{\pi}{4} \times (6.24^2 - 6.0^2) = 2.307\, \text{in}^2$$

$$A_s = 4 \times \frac{\pi}{4} \times 0.4^2 = 0.5027\, \text{in}^2$$

$$A_i = \frac{\pi}{4} \times 6.0^2 = 28.27\, \text{in}^2$$

$$k_a = \frac{E_a A_a}{L_a} = \frac{10 \times 10^6 \times 2.307}{10}$$
$$= 2.307 \times 10^6\, \text{lb/in}$$

$$k_s = \frac{E_s A_s}{L_s} = \frac{30 \times 10^6 \times 0.5027}{11}$$
$$= 1.371 \times 10^6\, \text{lb/in}$$

$$E_a A_a \alpha_a \Delta T = 10 \times 10^6 \times 2.307 \times 13 \times 10^{-6} \times 126$$
$$= 37.79 \times 10^3\, \text{lb}$$

$$E_s A_s \alpha_s \Delta T = 30 \times 10^6 \times 0.5027 \times 6.5 \times 10^{-6} \times 126$$
$$= 12.35 \times 10^3\, \text{lb}$$

$$(2.307 + 1.371) \times 10^6 \delta = 170 \times 28.27 + (37.79$$
$$+ 12.35) \times 10^3 - 0.3$$
$$\times 4.25 \times 10^3 \times 2.307$$

$$\delta = 0.014\,14\, \text{in}$$

Having found the primary unknown, we can now calculate the remaining required stresses (the hoop stress in the cylinder has already been found).

$$\sigma_{zs} = 207$$

$$\times 10^9 \left(\frac{0.3444 \times 10^{-3}}{275 \times 10^{-3}} - 11 \times 10^{-6} \times 70 \right) \text{N/m}^2$$

$$= 99.8 \text{ MN/m}^2$$

$$\sigma_{za} = 0.3 \times 30 \times 10^6 + 68.9$$

$$\times 10^9 \left(\frac{0.3444 \times 10^{-3}}{250 \times 10^{-3}} - 23 \times 10^{-6} \times 70 \right) \text{N/m}^2$$

$$= -7.0 \text{ MN/m}^2$$

Having found the primary unknown, we can now calculate the remaining required stresses (the hoop stress in the cylinder has already been found).

$$\sigma_{zs} = 30$$

$$\times 10^6 \left(\frac{0.014\,14}{11} - 6.5 \times 10^{-6} \times 126 \right) \text{psi}$$

$$= 14.0 \text{ ksi}$$

$$\sigma_{za} = 0.3 \times 4.25 \times 10^3 + 10$$

$$\times 10^6 \left(\frac{0.014\,14}{10} - 13 \times 10^{-6} \times 126 \right) \text{psi}$$

$$= -0.965 \text{ ksi}$$

As we anticipated, the axial stress in the cylinder is compressive, which makes it possible for the cylinder to be sealed against the rigid end plates.

Problems

SI METRIC UNITS – PIN-JOINTED STRUCTURES

Use program SIPINJ to solve the following plane framework problems. For all statically indeterminate structures it may be assumed that, in the absence of external loads, there are no internal forces in the members. Also, there are no changes in temperature.

4.1 Check the results of Problem 2.1 (*Hint*: to find only the forces in this statically determinate structure, any convenient values of overall dimensions, member cross-sectional areas and Young's moduli can be used).

4.2 In Problem 2.1, AB = 1.0 m, all the members are made of steel and have the same cross-sectional area of 500 mm². Find the maximum value of the load W if (i) the stress of greatest magnitude in the members is not to exceed 100 MN/m², and (ii) the vertical deflection of point C is not to exceed 5 mm.

4.3 Repeat Problem 4.2 with an additional member (with the same cross-sectional area) linking points A and D.

4.4 Check the results of Problem 2.3.

4.5 In Problem 2.3, AE = 1.5 m, the load P in 5 kN and all the members are made of steel and have the same cross-sectional area. Find the minimum

value of this area if (i) the stress of greatest magnitude in the members is not to exceed 150 MN/m², and (ii) the vertical displacement of point B is not to exceed 8 mm.

4.6 Check the results of Problem 2.6, taking the cross-sectional area of the steel members to be 1750 mm², and find the maximum vertical displacement of the structure.

4.7 Repeat Problem 4.6 with two additional members (with the same cross-sectional area) linking points B to G and G to D.

4.8 Determine the forces in all the members of the steel framework shown in Fig. P4.8. The cross-sectional area of each member is 1250 mm².

4.9 Check the results for forces in the members of Problem 2.7.

4.10 Repeat Problem 4.9, taking advantage of the symmetry of geometry and loading to analyze only half the structure.

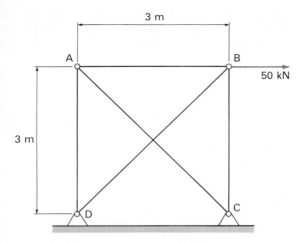

Fig. P4.8.

4.11 In Problem 2.9, the span AE is 4 m and each of the steel members has a cross-sectional area of 500 mm². Check the forces in the members, and find the vertical displacement of point G.

4.12 Repeat Problem 4.11, taking advantage of the symmetry of geometry and loading to analyze only half the structure.

4.13 Repeat Problem 4.12 with two additional members (with the same cross-sectional area) linking points C to H and C to F.

4.14 In Problem 1.4, all three members are made of steel, and the cross-sectional areas of AD, BD and CD are 100, 75 and 150 mm², respectively. Find the forces in the members and the displacements of point D.

4.15 In the structure shown in Fig. P4.15, all the members are made of aluminum and have cross-sectional areas of 250 mm². Find the stress of greatest magnitude in the members, and the vertical displacement of point F.

4.16 In Problem 2.11, points A and E are 500 mm apart, the members of the structure are made of brass and have cross-sectional areas of 700 mm². Check the previously calculated forces in the members (Problem 2.12), and find the change in the distance between points A and E.

4.17 Check the results for forces in the members of Problem 2.13 when the 100 kN load is applied at point H.

4.18 In Example 2.6, the cross-sectional areas of the steel members are all 400 mm². Check the forces in the members and find the resultant displacement of node 12.

4.19 In Problem 2.27, the cross-sectional areas of the steel members are all 500 mm². Check the forces in the members and find the vertical displacement of point Q.

4.20 Repeat Problem 4.19 with four additional members (with the same cross-sectional area) linking points J to B, B to H, H to D and D to F.

4.21 In Problem 2.29, the cross-sectional areas of the steel members are all 0.01 m². Check the forces in the members and find the greatest nodal point displacement in the vertical direction.

4.22 Repeat Problem 4.21 with six additional members (with the same cross-sectional area) linking points P to C, O to D, N to E, L to E, K to F and J to G.

SI METRIC UNITS – OTHER
STATICALLY INDETERMINATE SYSTEMS

4.23 The diameters of the two steel cables supporting the elevator cage in Fig. 4.21a are in the ratio 1 : 1.2. Find the corresponding ratio between the tensions in the cables.

4.24 In Fig. P4.24, a circular cylindrical copper specimen, C, is located between two cylindrical steel

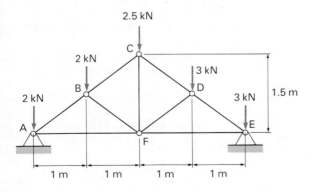

Fig. P4.15.

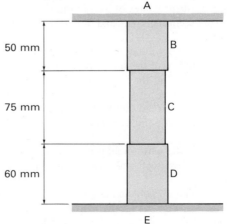

Fig. P4.24.

anvils, B and D, which themselves are located between the rigid crossheads A and E of a testing machine. The specimen and anvils have diameters of 25 mm and 30 mm, respectively. After the components of the system have been brought just into contact, the upper crosshead is lowered by a further 0.1 mm relative to the lower crosshead. If local effects at the ends of the components can be neglected, find the strain in the specimen. Compare this with the strain which would have been obtained if the anvils were rigid.

4.25 The cable used for an overhead electrical power transmission line consists of a core of sector-shaped aluminum conductors sheathed in strands of steel wire: the cable cross section is shown in Fig. P4.25. If there is no relative axial movement between the core and sheath when the cable is in tension, find the ratio between the stresses in the aluminum and steel.

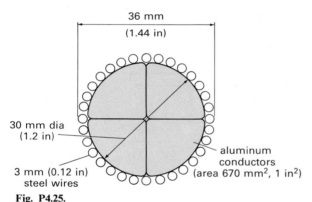

Fig. P4.25.

4.26 The bar AB, 2 m long, shown in Fig. P4.26 may be considered rigid, pinned at A and initially horizontal, supported by the vertical rod CD and a

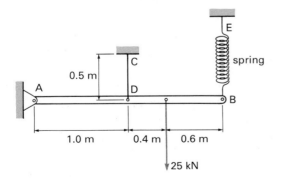

Fig. P4.26.

vertical spring (which is initially stress-free) linking B to fixed point E. Rod CD is steel, 0.5 m long and 8 mm diameter. Find the minimum stiffness of the spring EB if the stress in CD is not to exceed 150 MN/m² when the indicated force of 25 kN is applied.

4.27 In the sheathed cable whose cross section is shown in Fig. P4.25 the aluminum conductors have a total cross-sectional area of 670 mm², and the sheath consists of 32 steel wires each 3 mm in diameter. If the tension in the cable is 30 kN and there is no relative axial movement between the conductors and sheath, find the tensile forces and stresses in the aluminum and steel.

4.28 To the cable described in Problem 4.27, whose cross section is shown in Fig. P4.25, is added an outer layer of thermoplastic insulation with a nominal internal diameter of 36 mm and an outer diameter of 70 mm. If the Young's modulus of this material is 10 GN/m², find the tensile forces and stresses in the aluminum, steel and thermoplastic when the tension in the cable is 30 kN.

4.29 In the pressure vessel assembly of Example 4.12, shown in Fig. 4.27, the pitch of the threads on the steel rods is 0.4 mm. If, after the nuts have been adjusted to bring the end plates just into contact with the ends of the cylinder, each of the four nuts at one end of the vessel is tightened by a further quarter of a turn, calculate the stresses in the cylinder and in the rods both before and after the change in internal pressure and temperature.

4.30 The cylindrical tank with hemispherical ends described in Example 3.13 is to be filled not with air but with a liquid having a bulk modulus of K. Find an expression for the volume of liquid which must be pumped into the tank after it is just full at atmospheric pressure in order to give an internal gage pressure of magnitude p.

4.31 A steel tube of internal diameter 75 mm, wall thickness 1.5 mm and 1.25 m long is fitted with rigid end plugs and filled with oil at a pressure of 2 MN/m². If the pressure drops to 1.5 MN/m², determine the volume of oil leakage. Take the bulk modulus of the oil to be 2.8 GN/m², and assume that the plugs do not move relative to the tube.

4.32 In Problem 4.27 find the effect on the forces and stresses in the aluminum and steel of increasing the temperature of the cable by 20°C.

4.33 A sealed thin-walled spherical vessel of mean radius R and wall thickness t $(t \ll R)$ is initially unpressurized and just full of a liquid. The maximum allowable circumferential stress in the vessel wall is σ. Neglecting the effects of gravitational forces, show that the maximum temperature rise, ΔT, to which the vessel and its contents can be subjected is given by

$$\Delta T = \frac{\sigma}{(\beta - 3\alpha)}\left[\frac{2t}{KR} + \frac{3}{E}(1 - v)\right] \quad \text{for } \beta > 3\alpha$$

where: β = coefficient of volumetric thermal expansion of the liquid
K = bulk modulus of the liquid
E = Young's modulus of the vessel material
v = Poisson's ratio of the vessel material
α = coefficient of linear thermal expansion of the vessel material

4.34 In Problem 4.33, the vessel and liquid are maintained at the elevated temperature, and the pressure in the vessel is reduced to that of the surroundings by the release of some liquid. Find an expression for the fraction of the mass of liquid originally present which must be released.

4.35 In Example 4.12, the internal gage pressure is increased (with no further change in temperature) until the compressive axial stress in the cylinder vanishes and leakage is certain to occur. Calculate the pressure at which this happens.

4.36 In the device shown in Fig. P4.36, the rigid lever AB is connected to a rigid base by a steel rod at A, an aluminum rod at 10 mm from A and a spring 30 mm from A, as shown; both rods are 6 mm in diameter. At 20°C the lever is parallel to and 25 mm from the base and there are no forces in the rods or the spring. At 80°C the end B of the lever is required to have a displacement of 0.06 mm upwards from its position at 20°C. It can be assumed that the spring and the rods remain perpendicular to the base, and that the length of the spring is not affected by temperature. Find the stiffness of the spring to meet this specification, and the stresses induced in the steel and aluminum rods.

all dimensions in mm

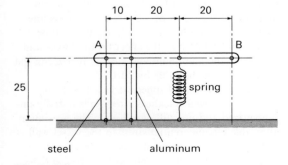

Fig. P4.36.

4.37 The rigid casting, marked by the letter A in Figure P4.37, is held by two 20 mm diameter steel bolts, B and C, against one end of an aluminum alloy tube D, which has internal and external diameters of 50 mm and 60 mm, respectively. The other end of the tube is in contact with a rigid surface, to which the bolts are also attached. The pitch of each bolt thread is 1.0 mm. After the casting has been brought just into contact with the tube, each of the two nuts E and F is tightened by a further quarter of a turn. If the temperature of the whole assembly is then raised by 30°C, calculate the stresses in the bolts and tube. The material of the casting may be assumed to have a negligibly small coefficient of thermal expansion.

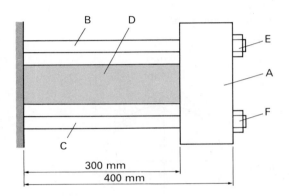

Fig. P4.37.

4.38 The materials testing equipment shown in Fig. P4.38 consists of two massive crossheads, labeled A, linked by two solid mild steel bars, labeled B, each of 100 mm diameter and effective length 1 m. Between the crossheads are located two brass anvil pieces, labeled C, between which is held an aluminum alloy specimen, D. The anvils are square in cross section, 100 mm by 100 mm, and each is 200 mm long, while the specimen is in the form of a circular cylinder of 100 mm diameter and 600 mm length. The equipment is adjusted until an axial compressive force of 500 kN is exerted on the specimen. A temperature rise of 80°C then occurs. Calculate the total axial force on the specimen following this temperature rise, and the total axial strain suffered by the specimen. The crossheads may be assumed to be rigid, and the other components of the system (bars, anvils and specimen) may be assumed to be in either uniaxial tension or uniaxial compression.

188

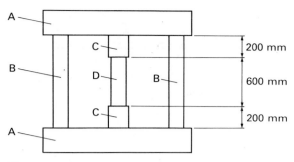

Fig. P4.38.

US CUSTOMARY UNITS – PIN-JOINTED STRUCTURES

Use program SIPINJ to solve the following plane framework problems. For all statically indeterminate structures it may be assumed that, in the absence of external loads, there are no internal forces in the members. Also, there are no changes in temperature.

4.39 Solve Problem 4.1.

4.40 In Problem 2.1, $AB = 40$ in, all the members are made of steel and have the same cross-sectional area of 0.78 in². Find the maximum value of the load W if (i) the stress of greatest magnitude in the members is not to exceed 15 ksi, and (ii) the vertical deflection of point C is not to exceed 0.2 in.

4.41 Repeat Problem 4.40 with an additional member (with the same cross-sectional area) linking points A and D.

4.42 Check the results of Problem 2.3.

4.43 In Problem 2.3, $AE = 60$ in, the load P is 1100 lb and all the members are made of steel and have the same cross-sectional area. Find the minimum value of this area if (i) the stress of greatest magnitude in the members is not to exceed 22 ksi, and (ii) the vertical displacement of point B is not to exceed 0.32 in.

4.44 Determine the forces in all the members of the framework shown in Fig. P4.44. The cross-sectional area of each member is 2 in².

4.45 Check the results for forces in the members of Problem 2.7.

4.46 Repeat Problem 4.45, taking advantage of the symmetry of geometry and loading to analyze only half the structure.

4.47 In Problem 1.25, all three members are made of steel, and the cross-sectional areas of AD, BD and CD are 0.16, 0.08 and 0.24 in², respectively. Find the forces in the members and the displacements of point D.

4.48 In the structure shown in Fig. P4.48, the steel members have cross-sectional areas of 2 in² and lengths of either 10 or $10\sqrt{2}$ ft. Find the stress of greatest magnitude in the members, and the vertical displacement of point D.

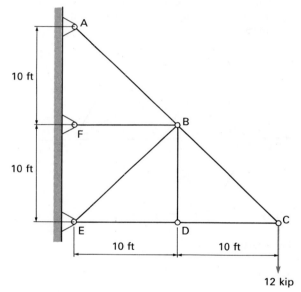

Fig. P4.44.

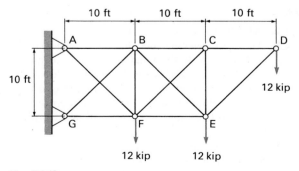

Fig. P4.48.

4.49 In Problem 2.46, points A and E are 20 in apart, the members of the structure are made of brass and have cross-sectional areas of 1 in². Check the previously calculated forces in the members (Problem 2.47), and find the change in the distance between points A and E.

4.50 Repeat Problem 4.49, taking advantage of the symmetry of geometry and loading to analyze only half the structure.

4.51 In Problem 2.62, the cross-sectional areas of the steel members are all 0.8 in². Check the forces in the members and find the vertical displacement of point E.

4.52 The structure shown in Fig. P4.52 has 12 aluminum alloy members AB, BC, etc., each of length 20 in and cross-sectional area 0.64 in², together with 4 steel members AC, BD, AE and FD each of cross-sectional area 0.16 in². Find the stresses of greatest magnitude in the aluminum alloy and steel members, and the relative displacement of points A and D.

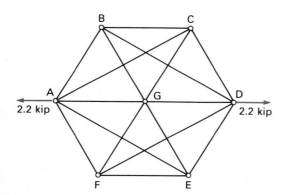

Fig. P4.52.

4.53 Repeat Problem 4.52 with the materials of the aluminum alloy and steel members interchanged.

US CUSTOMARY UNITS – OTHER
STATICALLY INDETERMINATE SYSTEMS

4.54 and 4.55 Solve Problems 4.23 and 4.25.

4.56 An object weighing 1000 lb is suspended from five fixed points A, B, C, D and E by 0.15 in diameter steel wires as shown in Fig. P4.56. If all the wires lie in a single vertical plane and are arranged to be just taut before the weight is applied, use program SIPINJ to find the tensions induced in them when it is applied.

4.57 In the sheathed cable whose cross section is shown in Fig. P4.25 the aluminum conductors have a total cross-sectional area of 1 in², and the

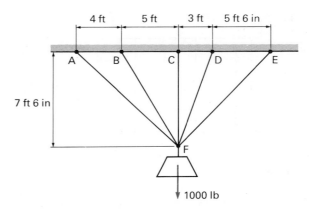

Fig. P4.56.

sheath consists of 32 steel wires each 0.12 in diameter. If the tension in the cable is 7 kip and there is no relative axial movement between the conductors and sheath, find the tensile forces and stresses in the aluminum and steel.

4.58 To the cable described in Problem 4.57, whose cross section is shown in Fig. P4.25, is added an outer layer of thermoplastic insulation with a nominal internal diameter of 1.44 in and an outer diameter of 2.8 in. If the Young's modulus of this material is 1500 ksi, find the tensile forces and stresses in the aluminum, steel and thermoplastic when the tension in the cable is 7 kip.

4.59 The stepped structural member shown in Fig. P4.59 joins two plates which may be considered rigid and held at a fixed distance apart. A total force of 15 kip is applied to a disk resting on the step as shown. Determine the magnitudes of the stresses in both portions of the member.

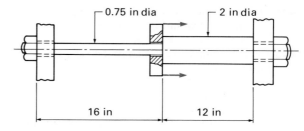

Fig. P4.59.

4.60 Solve Problem 4.29, taking the pitch of the threads on the steel rods as 0.016 in.

4.61 A spherical steel shell of diameter 10 in and thickness 0.1 in is just full of water at atmospheric pressure. Find the internal pressure when 1 in³ of water is pumped in, together with the change in diameter of the shell.

190

4.62 Solve Problem 4.30.

4.63 In Problem 4.57 find the effect of the forces and stresses in the aluminum and steel of increasing the temperature of the cable by 35°F.

4.64 to 4.66 Solve problems 4.33 to 4.35.

4.67 A steel tube with an outer diameter of 12 in and wall thickness 0.4 in is to have a brass tube of 0.25 in wall thickness shrunk onto it (by being first heated, then slid over the steel tube and finally cooled). The inner diameter of the brass tube is machined to be 0.008 in less than the outer diameter of the steel. Find the hoop stress in each of the two tubes after assembly, assuming that there are no axial stresses.

4.68 The temperature of the composite tube in Problem 4.67 is reduced by 40°F. Find the new hoop stresses in the brass and steel.

4.69 Repeat Problem 4.67 assuming the steel is not in the form of a tube but a solid shaft.

4.70 A steel rod has a diameter of 0.8 in. It is concentric with an aluminum alloy tube with an outer diameter of 2 in and an inner diameter of 1.6 in. The rod and tube are of the same length and both are bonded to rigid plates over the ends as shown in Fig. P4.70. At 70°F there is no stress in the steel or aluminum. Find the stresses in the steel and aluminum alloy when the temperature of the whole system is raised to 200°F.

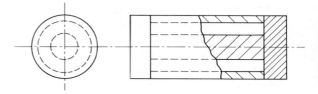

Fig. P4.70.

4.71 In Problem 4.70, an axial force is applied to the cover plates while the elevated temperature is maintained. Find the range of values of this force such that (a) the aluminum is never in tension, (b) the steel is never in compression.

5

Bending of Beams – Moments, Forces and Stresses

In all the types of problems we have considered in detail so far, stresses have been uniformly distributed over the components concerned. In most cases they have also been normal stresses, and frequently uniaxial. We now start to examine situations where the stresses are not uniform. An example is provided by the *bending* of a component. Although many components of engineering systems are subject to bending, the theory of bending is usually developed with reference to beams. Formally, we can define a *beam* as a *laterally loaded structural member, whose cross-sectional dimensions are small compared to its length*. Let us first review some engineering examples of beams.

5.1 Some practical examples of beams

One of the most familiar examples of beams is that of joists used to support the floors and ceilings of buildings. Figure 5.1 shows a typical arrangement of equally spaced horizontal joists with their ends resting on the walls of the building. On top of these is laid the floor, and from their undersides may be suspended the ceiling of the room below. In houses, joists are usually made from timber of rectangular cross section, but in larger buildings steel or reinforced concrete beams of more complex cross-sectional shape are used. In all cases, however, the length of the joist is much greater than the dimensions of its cross section, and the predominantly vertical loading is normal to this length. Similar examples are provided by the frames of buildings, the decks of bridges and other such structures.

Horizontal pipes carrying gases or liquids in, for example, chemical plants can be regarded as beams. They are loaded by their own weight together with

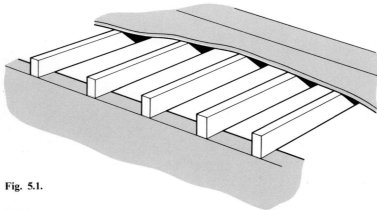

Fig. 5.1.

192

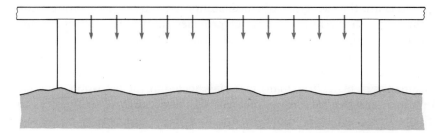

Fig. 5.2.

that of their contents, and are supported at intervals as shown in Fig. 5.2, each span being large compared with the pipe diameter. The pipe walls experience stresses due not only to bending but also to internal pressure.

Possibly less obvious examples of beams are masts, towers and tall chimneys. Figure 5.3 shows a vertical mast subjected to predominantly horizontal wind forces, although wire stays used to restrain the mast introduce some vertical loading. The cross section of the mast may well be circular, with a diameter which increases from tip to base, but remaining small compared to the height.

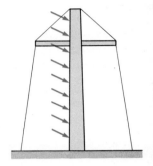

Fig. 5.3.

Another example is provided by the wings of an aircraft, Fig. 5.4. The length-to-depth ratio is large, and lateral loading is provided by the aerodynamic lift forces, together with the weight of engines and other attachments. Similarly, the hull of a ship, Fig. 5.5, can be regarded as a beam, subject to lateral weight and buoyancy forces.

There are two main questions which engineers ask about a beam. Firstly, at what load will it fail and, secondly, what deflections occur under moderate

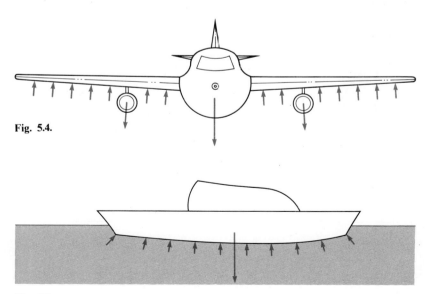

Fig. 5.4.

Fig. 5.5.

load? These are the two types of consideration we introduced in connection with the ladder shown in Fig. 1.1, which provides another example of a beam. As with other engineering structures, the relative importance as a form of external loading of the weight of a beam depends on its size. In the case of very large beams, such as bridges, most of the loading comes from the weight.

Beams can be made from many materials, including metals, wood, concrete and plastics, and can take many different geometric forms. Our object is to develop a theory which is applicable to all of these. Beams may be curved, although initially we will only consider straight beams.

5.1.1 Types of supports for beams

Beams are usually classified according to the way in which they are supported. Supports constrain the movement of beams, either in particular directions or in the rotational sense, or both. As a result, reaction forces and moments are induced, forces in the case of translational constraints and moments in the case of rotational constraints. Beams are said to be *simply supported* if their supports provide only translational constraints: reaction forces may be induced but not moments. The joists shown in Fig. 5.1 are supported by the walls in this way, as is the pipe in Fig. 5.2. Simple supports are of two main types, namely *pinned* where no translational movement of the beam is allowed, and *rolling* or *sliding*, where movement in one direction is permitted. These two types are equivalent to those illustrated in Fig. 2.20 for pin-jointed structures. Figures 5.6a and b show the ways in which we represent simple supports diagrammatically in our models of beams. While there may be a reaction force in any direction at a pinned support, at a roller support the reaction is in the direction normal to that in which it is free to move. Although the direction of movement is shown in Fig. 5.6b as being at an angle α to the x coordinate direction, in most practical cases $\alpha = 0$, and movement can only occur along the length of the beam. Note that it is customary to take the x axis as being along the beam. Figure 5.7 shows

(a)   (b)

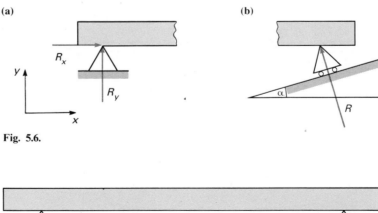

Fig. 5.6.

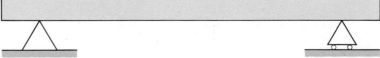

Fig. 5.7.

a typical model for a simply supported beam. In the absence of other information, we would normally assume that such a beam has one pinned support and one roller support, ensuring that in the absence of external loading there are no internal forces in the beam. We should also note that, while the diagrammatic representations of these simple supports might imply that they are only capable of providing reaction forces on the beam in the upward direction, we will normally assume that, when necessary, they can provide downward reactions to prevent the beam lifting off.

The second main type of beam support is usually referred to as *fixed* or *built-in*, providing both translational and rotational constraints: both reaction forces and moments may be induced. For example, the mast shown in Fig. 5.3 is fixed or built-in to the ground, typically by being cast in concrete. Also, the wings of the aircraft in Fig. 5.4 are fixed to the fuselage, both by continuity of structural members and by welding and riveting of the components together. Figure 5.8 shows how we represent this type of support diagrammatically, with the beam protruding from a rigid wall (the part of the beam within the wall

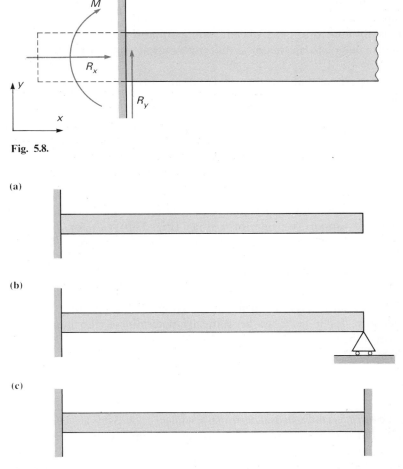

Fig. 5.8.

(a)

(b)

(c)

Fig. 5.9.

which is shown by a dotted outline in the figure is often omitted). Figure 5.9 shows the models for three types of beams with fixed supports. In Fig. 5.9a, the beam is built-in at one end and completely unsupported or *free* at the other. Such a beam is referred to as a *cantilever*. In Fig. 5.9b, the beam is built-in at one end and simply supported at the other, and may be referred to as a *propped cantilever*. In Fig. 5.9c, the beam is built-in at both ends and is described as a *fixed* or *encastré* beam.

In practice, deciding whether a beam support is of the simple or fixed type, or whether it must be treated as an intermediate case (allowing some rotation and providing some reaction moment) may require experience and engineering judgement. For example, although the pipe in Fig. 5.2 is likely to be clamped to its supports by means of steel straps, these would not be capable of providing significant reaction movements, and the pipe can be treated as simply supported. Also, the joists shown in Fig. 5.1 rest on walls at their ends. While these supports are more reasonably treated as simple rather than fixed, they are not strictly of either the pinned or roller types (Fig. 5.6), and both ends of each joist experience some friction which opposes horizontal movement. Since this longitudinal loading is not significant in relation to the lateral loads, however, it is permissible to treat such a beam as being simply supported as in Fig. 5.7.

The pipe example in Fig. 5.2 also demonstrates that the number of supports to a beam is not limited to two. A beam resting on more than two supports is referred to as *continuous*, in the sense that it is continuous across at least one support.

For the purposes of analysis, it is useful to be able to classify a beam as being either *statically determinate* or *statically indeterminate*. A beam is statically determinate if all the external forces and moments acting on it can be found from equilibrium considerations alone. Internal forces, moments and stresses can then be found by the methods described in this chapter, without having to consider the deformation of the beam. The analysis of statically indeterminate beams, which requires the determination of beam deflections, is postponed until Chapter 6. Usually, the only initially unknown forces and moments acting on a beam are the reactions at the supports. The numbers of unknowns for the various types of support are one for a roller (Fig. 5.6b), two for a pinned support (Fig. 5.6a) and three for a fixed support (Fig. 5.8). Suppose that the total number of unknowns for all the supports is N_r (adopting the same notation as for pin-jointed structures in Section 2.1.1). For a single body (the beam) subjected to any two-dimensional system of external forces and moments, there are only three independent equations of equilibrium (Section 1.2.4). Therefore, for a beam to be statically determinate, the number of unknowns must be equal to the number of equilibrium equations available for finding them, and

$$N_r = 3 \tag{5.1}$$

If $N_r > 3$ the beam is statically indeterminate, while if $N_r < 3$ it is a mechanism, in the sense that it is free to move. In many cases, beams experience no axial loading, and the equation of equilibrium in this direction provides only trivial information (that a support reaction component is zero). Nevertheless, in applying test equation (5.1) to a general problem, both the unknown force and the equation should be included.

Consider, for example, the simply supported beam shown in Fig. 5.7: with one pinned (two unknowns) and one roller (one unknown) support, $N_r = 3$, and it is statically determinate. Similarly, the cantilever shown in Fig. 5.8 is also statically determinate, because it has one fixed support with three unknowns, so that again $N_r = 3$. On the other hand, the propped cantilever and fixed beams shown in Figs 5.9b and c have $N_r = 4$ and $N_r = 6$, respectively, and are both statically indeterminate. If the beam shown in Fig. 5.7 had at its left-hand end not a pinned support but a roller, the number of unknowns would be reduced to two, creating a mechanism: the beam would be free to move in the horizontal direction. We may conclude that only two types of beams are statically determinate: beams simply supported at two points (one pinned and one roller), and cantilevers.

Equation (5.1), which provides a simple test for statical determinacy, is only applicable to single continuous beams. For example, an apparently continuous beam might be freely hinged (pin-jointed) at some position along its length. Such an arrangement must be regarded either as two linked beams (with two reaction force components between them) or as a single beam with four equations available, three equilibrium equations and one imposing the condition of zero moment at the hinge.

5.1.2 Types of loads on beams

By definition, a beam is a structural member subject to lateral loads. It may, however, also experience axial loading. The combination of lateral and axial loading is considered in Section 5.4, and in Chapter 8 we examine the potentially unstable behavior of bending members under compressive axial forces.

A commonly-occurring type of lateral load on a beam is a *distributed force*, acting over at least part of its length. For example, the beams shown in Figs 5.1 to 5.5 are all subject to distributed lateral forces: the weights of the joists and flooring in Fig. 5.1, the weight of pipe and its contents in Fig. 5.2, the wind force in Fig. 5.3, aerodynamic lift forces in Fig. 5.4, and weight and buoyancy forces in Fig. 5.5. We represent distributed lateral forces on a model of a beam as shown in Fig. 5.10. Figure 5.10a shows the general case of a force of intensity w per unit length acting over part of the beam, where w is a function, $w(x)$, of position x along the beam, measured from its left-hand end: we say that the force is *non-uniformly distributed*. Figure 5.10b, on the other hand, shows the special case of a *uniformly distributed force*, of intensity w per unit length, which is now independent of x.

Another type of lateral load is a *concentrated force*. By this we mean that the length of beam over which the force acts is so small in relation to its total length that we can model the force as though it is applied at a point in the two-dimensional view of the beam. For example, the weights of the aircraft engines in Fig. 5.4 can be regarded as concentrated forces acting downwards on the wings. Similarly, the reactions acting upwards on the pipe at its supports in Fig. 5.2 and on the joists at the walls in Fig. 5.1 can be regarded as

(a)

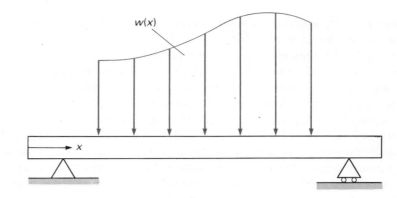

(b)

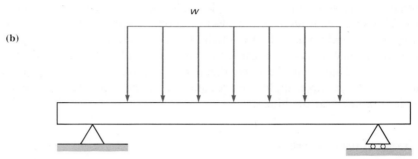

Fig. 5.10.

(a)

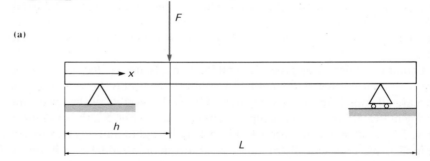

(b)

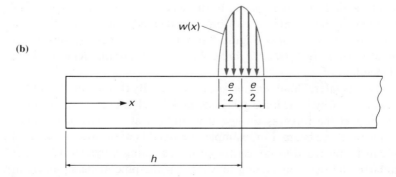

Fig. 5.11.

concentrated. Indeed, we have already used in Example 1.4 the concentrated force approximation for joist supports. We represent a concentrated lateral force on a model of a beam as shown in Fig. 5.11a. It is important to realize that such a representation is merely a convenient approximation for the purposes of analyzing overall behavior of the beam. It does not provide a satisfactory model of behavior in the local region of application of the force to the beam surface. A force acting at a point, which has zero area, implies a locally infinite stress, and real engineering materials are incapable of sustaining infinite stresses. All forces must act over finite areas of contact, even if these areas are very small. For example, even if a beam rests on an apparently knife-edged support, this support and the region of the beam surface to which it is applied must deform to bring finite areas of material into contact with each other.

The real distribution of force associated with the concentrated force in Fig. 5.11a is likely to be of the form shown in Fig. 5.11b, with $w(x)$ acting over a small length, e, of the beam. The total force, F, associated with this distribution is given by

$$F = \int_{h-e/2}^{h+e/2} w(x)\,dx \tag{5.2}$$

Provided $e \ll L$, where L is the length of the beam, we may use the concentrated force approximation of Fig. 5.11a.

Other forms of beam loads are moments, which in the absence of axial forces are applied as couples. For example, a fixed beam support (Fig. 5.8) will usually provide a reaction moment at an end of a beam. It is also possible to apply a couple at some position along a beam. In Fig. 5.3, for example, a difference in tension in the wire stays on either side of the mast would be transmitted to the mast as a couple (together with lateral and axial forces). Aligning the mast with the x axis, this physical situation can be represented as in Fig. 5.12a: equal and opposite forces of magnitude F acting a distance d apart give a couple of magnitude $C = Fd$. This can be represented as a concentrated couple applied to the beam as in Fig. 5.12b. The real distribution of lateral force creating this couple is likely to be of the form shown in Fig. 5.12c, with distributions $w(x)$ acting on opposite sides of the beam over small lengths, e. We can therefore express the couple, C, as

$$C = \int_{h-e}^{h+e} w(x)(h-x)\,dx \tag{5.3}$$

Provided $e \ll L$, where L is the length of the beam, we may use the concentrated couple approximation of Fig. 5.12b for the purposes of analyzing forces and stresses in the beam as a whole.

5.2 Shear forces and bending moments in beams

The bending of beams creates internal stress distributions which are more complicated than those we encountered in pin-jointed structures. In order to

(a)

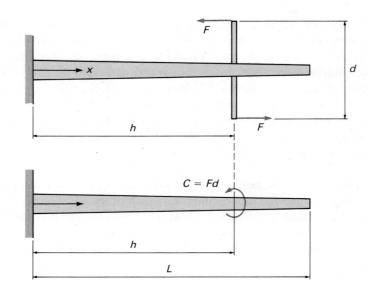

(b)

(c)

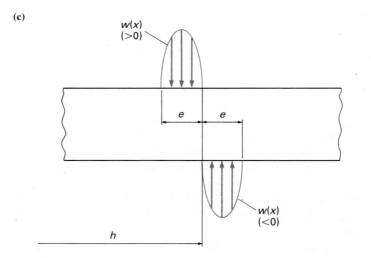

Fig. 5.12.

find these stresses, we first consider internal forces and moments in the beams, which are the resultants of the stresses.

5.2.1   Definition of shear force and bending moment

Let us consider the simply supported horizontal beam AB of negligible weight and length L shown in Fig. 5.13a, which is subjected to a concentrated vertical force of magnitude F at its center, C. Since the loading and the arrangement of the supports are symmetrical about the center of the beam, the vertical reaction forces at the supports are identical, and equal to $F/2$ in the upward direction. There are no external forces acting on the beam in the horizontal direction. Imagine now that we make a vertical cut through the beam at the section

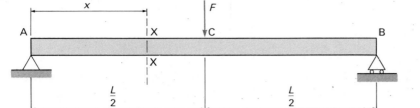

(b)

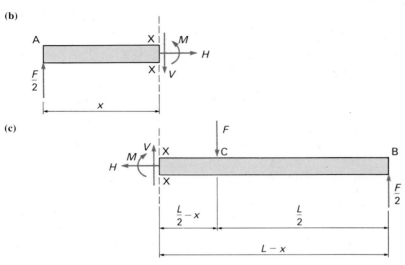

(c)

Fig. 5.13.

marked XX, at a distance x from the left-hand end of the beam, where $x < L/2$. The portion of the beam to the left of this section can be considered as a free body, as shown in Fig. 5.13b. Note that, because the support at A is a simple one, there is no external moment acting on the beam there. The distributions of stresses at section XX can be represented in terms of resultant forces in the horizontal and vertical directions, together with a moment, here shown as H, V, and M, respectively, which are external loads acting on the free body.

For equilibrium of forces acting on the free body in the horizontal and vertical directions

$$H = 0 \quad \text{and} \quad V = \frac{F}{2} \tag{5.4}$$

We note that, in the absence of external axial forces on the beam, no internal resultant forces in this direction are generated at any cross section. Vertical force V, which acts parallel to the surface of section XX, is the *shear force* in the beam at this section. For equilibrium of moments acting on the free body, taken about any point on section XX in Fig. 5.13b

$$M = \frac{Fx}{2} \tag{5.5}$$

This moment, M, is the *bending moment* in the beam at the section concerned.

Suppose that, instead of taking the portion of the beam to the left of section XX, we had considered that to the right, as shown in Fig. 5.13c. The resultant forces and moment at section XX are again of magnitudes H, V and M, respectively, but in the opposite directions (in accordance with Newton's third law). For equilibrium of forces acting on the free body in the horizontal and vertical directions, $H = 0$ and

$$V + \frac{F}{2} = F$$

or

$$V = \frac{F}{2}$$

For equilibrium of moments acting on the free body, taken about any point on section XX in Fig. 5.13c

$$M + F\left(\frac{L}{2} - x\right) = \frac{F}{2}(L - x)$$

or

$$M = \frac{Fx}{2}$$

As we should expect, the same expressions for forces and moment are obtained. It is important to recognize that we can find the shear force and bending moment at a particular beam cross section by considering the loads acting on the portion of the beam to either the left or the right of the section.

We can formally define the shear force and bending moment in a beam as follows

The shear force at any section is equal in magnitude to the sum of all the vertical components of the external forces acting to one side of that section.

The bending moment at any section is equal in magnitude to the sum of all the external moments acting to one side of that section.

The directions of the shear force and bending moment must be such as to ensure that the portion of the beam considered is in equilibrium.

At this stage we must adopt appropriate sign conventions for shear force and bending moment, also for coordinate directions, beam deflection and lateral forces, and always adhere to them when we analyze beam behavior. This is because the forms of the equations we will derive to link lateral force to shear force and shear force to bending moment, also to determine stresses and beam deflections, all depend on the conventions chosen. Unfortunately, there are no generally agreed conventions, mainly because there is no one set which is ideal for all aspects of beam analysis. The criteria we would like to satisfy might include the following.

1. Coordinates to be arranged in the conventional way (x horizontal and positive from left to right, and y vertical with upwards positive).
2. Lateral deflection to be positive in the positive y direction.
3. Normal stresses to follow the usual tensile positive convention.
4. Loads and deflections to be positive in the direction in which they usually occur in practice (vertically downward in the cases of forces and deflections).
5. Directions of shear forces to be consistent with the convention for shear stresses (which we have yet to consider).
6. Mathematical equations describing bending theory to contain a minimum number of negative signs (specifically, equations (5.10), (5.11), (5.33) and (6.4)).

It is impossible to satisfy all of these simultaneously, and a compromise must be found. The conventions we will use represent one such compromise, and have the merit of being quite widely used. In adopting them, we will satisfy criteria (1), (2) and (3), partially satisfy (4) (for forces but not for deflections), not satisfy (5), and retain some negative signs in conflict with criterion (6). We will note the latter as they occur.

Figure 5.14 illustrates our sign conventions. We choose to use coordinate x in the horizontal direction along the beam, measured from the left-hand end of the beam, and coordinate y in the upward vertical direction, as shown in Fig. 5.14a. The origin for y we will define when we come to consider stresses in beams in Section 5.3. We also choose lateral beam displacement or deflection, v, to be positive in the upward vertical direction. Bearing in mind the predominant direction of lateral loading of beams, we choose distributed and concentrated lateral forces to be positive in the downward direction as shown in Fig. 5.14b. As for shear force and bending moment, we adopt sign conventions which are consistent with the results derived from Fig. 5.13. In other words, the directions assigned to shear force V and bending moment M in Figs 5.13b and 5.13c are defined as positive. Let us try to generalize these definitions. Of the two, the shear force sign convention is perhaps the more difficult to appreciate and remember. Figures 5.14c and d show two possible ways of doing this – which to use is largely a matter of personal preference. If we isolate a very short element of the beam by taking two adjacent vertical sections through it as in Fig. 5.14c, then a positive shear force is one which, acting on each of the cut faces of the element, tends to rotate it in the clockwise direction. Alternatively, we can think of a single cut being made through the beam at the section of interest, as in Fig. 5.14d. If the external loading on the portion of the beam to the left of the chosen section would tend to raise this portion, then the shear force at the section is positive. Equilibrium of lateral loads on the beam as a whole then demands that the external loading on the portion of the beam to the right would tend to lower that portion. Finally, if we again isolate a short

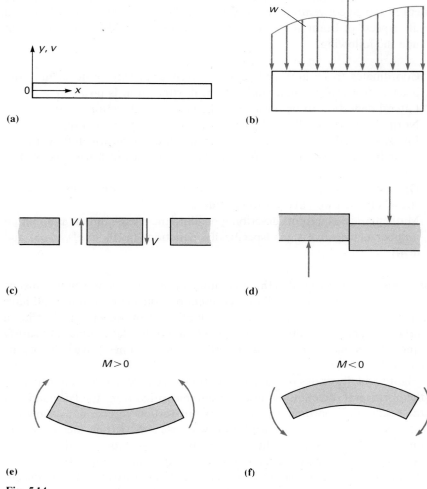

Fig. 5.14.

element of the beam as in Fig. 5.14e, we can define a positive bending moment as that which causes *sagging* of the element. In other words, the top surface of the element tends to become concave, and its bottom surface to become convex. The reverse tendency is referred to as *hogging*, which is illustrated in Fig. 5.14f, and is associated with a negative bending moment.

Having established definitions and sign conventions for shear forces and bending moments in a beam, we are now ready to determine their distributions along the length of the beam, provided the beam is statically determinate. When plotted as functions of axial position, these distributions are referred to as *shear force and bending moment diagrams*. Our motivation for studying them is twofold. Firstly, in order to find the stresses in a beam (as in Section 5.3), we need to know the shear forces and bending moments. In particular, in order to find the greatest stresses, we need to know the greatest values of shear force and bending moment. Secondly, when we come to determine the distribution of deflection along a beam in Chapter 6, we must first know the bending moment distribution.

Let us now consider as examples some simple cases involving the different types of beams and loadings which may be applied to them, including the one we started to analyze in Fig. 5.13.

EXAMPLE 5.1: Simply supported beam with a concentrated central force

Draw the shear force and bending moment diagrams for the beam shown in Fig. 5.13a (repeated in Fig. 5.16a), and find the greatest absolute values of shear force and bending moment.

We have already established in equations (5.4) and (5.5) expressions for the shear force and bending moment as functions of x over the left-hand half of the beam ($x < L/2$). We can do the same for the right-hand half: Fig. 5.15 shows the free body formed to the right of section XX when $x > L/2$, with the shear force, V, and bending moment, M, marked according to the sign conventions of Fig. 5.14. Although, according to the definitions of shear force and bending moment, we could consider external loads to either the left or right-hand sides of section XX, it is somewhat easier in this case to consider those to the right. The only vertical component of external force in this region is $F/2$ upwards (the support reaction at B). Since this tends to raise the right-hand portion of the beam, the sign convention illustrated in Fig. 5.14d requires that $V = -F/2$. We would of course have arrived at exactly the same conclusion had we considered equilibrium of vertical forces acting on the free body in Fig. 5.15. Also, the moment of the support reaction about section XX is $F(L - x)/2$, and is the only contribution to the bending moment there. Since it tends to cause sagging of the beam, the resulting bending moment is positive, and

$$M = \frac{F(L - x)}{2} \tag{5.6}$$

With general expressions for the shear force distributions over both halves of the beam, we can now plot the shear force diagram as in Fig. 5.16b. While the forces in each half are constant and of magnitude $F/2$, they are of opposite signs, and there is an abrupt change of magnitude F at the center point C. This change is due to the concentrated force which is applied at this point, which is also the reason for our having to consider the

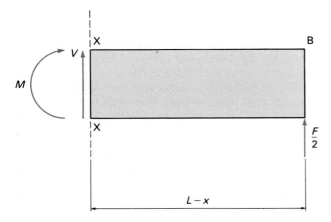

Fig. 5.15.

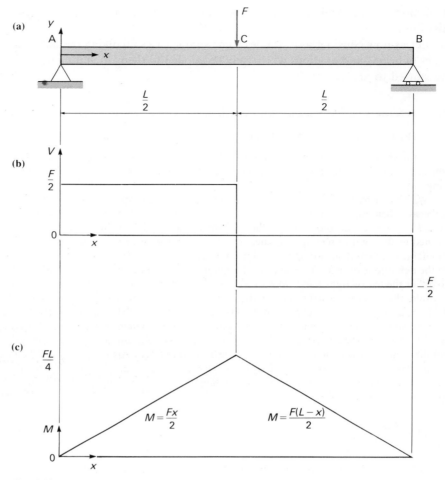

(a)

(b)

(c)

Fig. 5.16.

two halves of the beam separately. We can also plot the bending moment distribution as in Fig. 5.16c, the linear distributions over the left and right-hand halves of the beam being defined by equations (5.5) and (5.6), respectively. From zero values at the two ends of the beam which rests on simple supports, the bending moment rises to a maximum at the center of the beam where the concentrated force is applied. Although the bending moment distribution itself is continuous at the center of the beam, the slope of the distribution suffers an abrupt change there, again due to the concentrated force.

   In seeking the greatest absolute values of shear force and bending moment, we are concerned with the values of largest magnitude, regardless of their signs (which are consequences of the particular sign conventions chosen). However, for reasons which should become clearer when we consider stresses in beams in Section 5.3, in the case of bending moments it is sometimes important to know both the greatest sagging and the greatest hogging values. In this example, the greatest absolute shear force and bending moment are $F/2$ and $FL/4$, respectively. Note also that we deliberately use the word greatest rather than maximum in this context. This is because the latter tends to be associated with the idea of a *mathematical maximum* or local peak in a distribution. As we shall see in later examples, if a mathematical maximum exists in a shear force or bending moment distribution, it does not necessarily correspond to the greatest value.

There are two further features to note about the results of this example. Firstly, while the geometry and loading of the beam (Fig. 5.16a) are symmetrical about its center, as is the bending moment distribution (Fig. 5.16c), the plotted shear force distribution (Fig. 5.16b) is not. In fact, the plotted distribution is antisymmetrical about the center, which is a consequence of the way we have chosen to define shear force, and in particular its sign. Secondly, the step change in the shear force at the center is an idealization resulting from the representation of a concentrated force by a force acting at a point. If the real distribution of force is as shown in Fig. 5.11, repeated in Fig. 5.17a, then the shear force

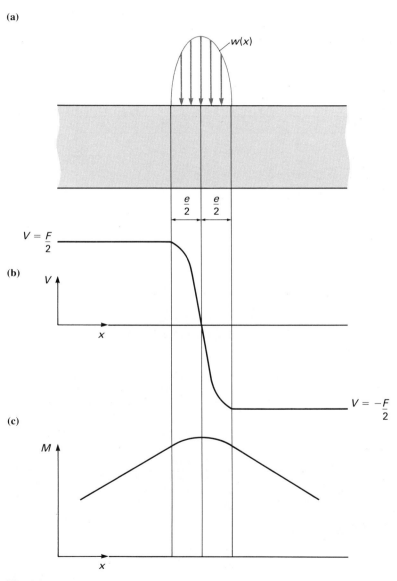

(a)

$w(x)$

$\dfrac{e}{2}$ $\dfrac{e}{2}$

$V = \dfrac{F}{2}$

(b)

V

x

$V = -\dfrac{F}{2}$

(c)

M

x

Fig. 5.17.

varies as in Fig. 5.17b and the bending moment as in Fig. 5.17c. Provided the
length e is small, however, a step change is a reasonable approximation for
most practical purposes.

EXAMPLE 5.2: Simply supported beam with a uniformly distributed force

Draw the shear force and bending moment diagrams for the simply supported beam **AB**
of negligible weight and length L shown in Fig. 5.18a, which is subjected to a uniformly
distributed vertical force of intensity w per unit length over the whole of its length. Find
the greatest absolute values of shear force and bending moment.

This is another simple case of a commonly occurring type of beam and loading. Due
to the symmetry of loading and arrangement of the supports, the upward vertical
reactions on the beam at the supports are the same, and equal to $wL/2$. There are no
forces acting on the beam in the horizontal direction. Because the applied force is

(a)

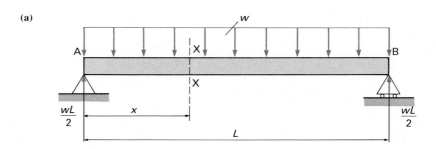

(b)

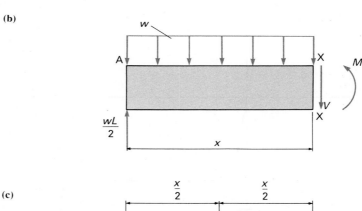

(c)

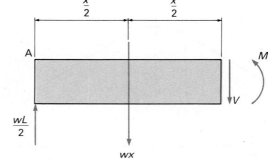

Fig. 5.18.

continuous rather than concentrated, we can establish expressions for the shear force and bending moment which are valid over the whole length of the beam. Imagine that we cut the beam at the section marked XX in Fig. 5.18a, at a distance x from the left-hand end of the beam, and consider the portion of the beam to the left of this section as a free body as in Fig. 5.18b. The external forces acting on this portion are the upward reaction of $wL/2$ at the support, and the downward distributed force of intensity w per unit length. The resultant of this distributed force is wx, acting downward at the center of the portion concerned, namely at a distance of $x/2$ from the end of the beam and from section XX, as shown in Fig. 5.18c. Equating the shear force, V, to the sum of the external forces in the upward vertical direction

$$V = \frac{wL}{2} - wx = w\left(\frac{L}{2} - x\right) \tag{5.7}$$

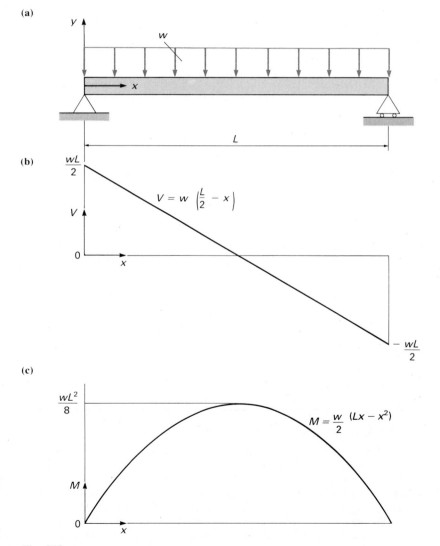

(a)

(b)

(c)

Fig. 5.19.

and equating the bending moment, M, to the sum of the external moments in the sagging sense about the cut face

$$M = \frac{wLx}{2} - \frac{wx^2}{2} = \frac{w}{2}(Lx - x^2) \tag{5.8}$$

The distributions represented by these equations are drawn as shear force and bending moment diagrams in Fig. 5.19. The greatest absolute values of shear force and bending moment are $wL/2$ and $wL^2/8$, respectively.

There are several features to notice in Fig. 5.19. Firstly, whereas a concentrated force (Fig. 5.16) gave rise to mainly constant shear force and linear bending moment distributions, a uniformly distributed load gives rise to linear shear force and quadratic (parabolic) bending moment distributions. Once again we note the symmetry of the bending moment and antisymmetry of the plotted shear force distributions for a beam and loading which are symmetrical about the center. In this case there is a true mathematical maximum to the bending moment distribution, which is also the greatest bending moment. In other words, if we differentiate with respect to x the expression for moment given by equation (5.8) we obtain

$$\frac{\mathrm{d}M}{\mathrm{d}x} = w\left(\frac{L}{2} - x\right) \tag{5.9}$$

which takes the value zero at $x = L/2$. This confirms that the maximum bending moment occurs at the center of the beam. The center is also the position at which the shear force passes through zero. Indeed, we can observe that the expression for bending moment gradient in equation (5.9) is identical to that for shear force in equation (5.7). The fact that the shear force is zero at the center of the beam, which is also a plane of symmetry, provides further confirmation of the general principle first expressed in Section 2.2.1 that there is no shear force on a plane of mirror symmetry through a body. The same was also true in Example 5.1, although there the presence of the concentrated force at the plane of symmetry introduced a complication in the form of a step change in the shear force which passed through zero. The shear force distribution shown in Fig. 5.17b for a real concentrated force provides a more satisfactory confirmation.

EXAMPLE 5.3: Pure bending of a simply supported beam

Draw the shear force and bending moment diagrams for the simply supported beam of negligible weight shown in Fig. 5.20a, which is subjected to externally applied moments of magnitude M_0 in the sagging sense at each of its ends.

This is a particularly straightforward example. With only the two moments applied, which are in equilibrium with each other, there are no reactions at the supports. Consequently, the shear force is zero everywhere along the beam. Similarly, the bending moment is constant and equal to M_0 everywhere, as shown in Fig. 5.20c. This situation, with constant bending moment and zero shear force, is referred to as *pure bending*.

(a)

(b)

(c)

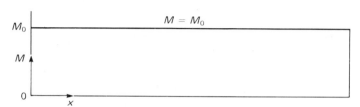

Fig. 5.20.

Examples 5.1, 5.2 and 5.3 represent three cases of important types of loading, namely concentrated force, uniformly distributed force and end moment, applied to a simply supported beam. Indeed, we can regard them as standard cases which can be applied to particular problems either individually or superimposed on each other (as, for instance, in Example 5.7). Similar standard cases exist for the only other type of statically determinate beam, the cantilever, and shear force and bending moment diagrams for these are shown in Figs 5.21 to 5.23. Figure 5.21 shows a cantilever beam of length L with a concentrated lateral force of F applied at its free end, which gives rise to a constant shear force (of magnitude F), and a bending moment distribution which varies linearly from zero at the free end to a hogging value of FL at the fixed end. Figure 5.22 shows a cantilever, also of length L, with a uniformly distributed lateral force of intensity w. This results in a shear force distribution which varies linearly from zero at the free end to wL (the total external load on the beam) at the fixed end, also a bending moment distribution which varies quadratically from zero at the free end to a hogging value of $wL^2/2$ (the total load multiplied by the distance from its line of action) at the fixed end. Note that the slope of the bending moment distribution is zero at the free end. Figure 5.23 shows a cantilever subjected to a moment at its free end, which is another example of pure bending identical in terms of shear force and bending moment distributions to Fig. 5.20.

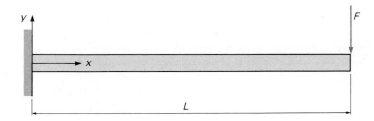

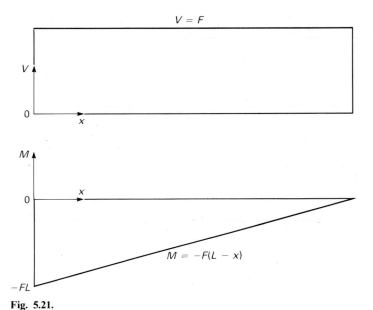

Fig. 5.21.

5.2.2 Relationships between distributed lateral force, shear force and bending moment

From the ways we define shear force and bending moments, and from the examples of shear force and bending moment distribution we have considered so far, it is becoming clear that lateral force, shear force and bending moment are closely linked to each other. For instance, we have already noted in connection with the simply supported beam carrying a uniformly distributed lateral force in Example 5.2, that the derivative with respect to axial position along the beam of the bending moment distribution, equation (5.9), is identical to the shear force defined by equation (5.7). Let us now try to establish general relationships between lateral force, shear force and bending moment.

To do this, we need to consider a short element of a beam subject to any form of lateral loading. Such an element, of infinitesimal length δx, is shown in Fig. 5.24. We represent the loading as a distributed force of intensity w per unit length, a quantity which in general is a function of axial distance, x, along the beam. Provided δx is small, however, the amount by which w varies over the element is also small, and may be neglected. The variations of shear force and bending moment cannot in general be neglected in this way, and we assume

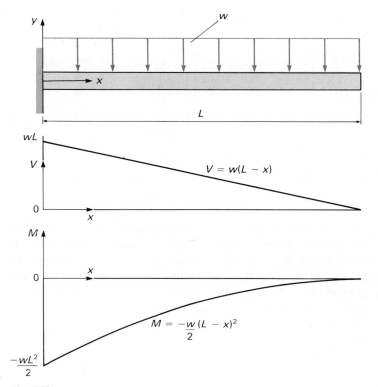

Fig. 5.22.

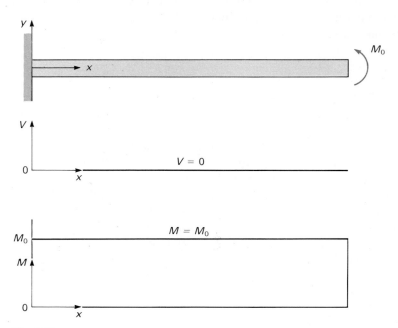

Fig. 5.23.

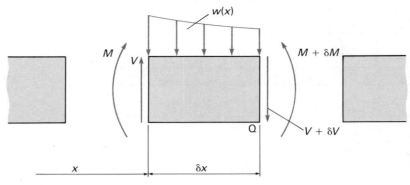

Fig. 5.24.

that values of V and M, respectively, at the left-hand end of the element increase to $V + \delta V$ and $M + \delta M$ at the right-hand end, as shown. Note that the directions of forces and moments in Fig. 5.24 are all positive according to the established sign conventions, also that the infinitesimal changes of shear force and bending moment are such as to imply increases in the positive x direction.

We now consider equilibrium of forces and moments acting on the element. For equilibrium of forces in the vertical direction

$$V - (V + \delta V) - w\,\delta x = 0$$

$$\frac{\delta V}{\delta x} = -w$$

and we note that it is the change in shear force, rather than V itself, which must balance the applied lateral force. This is why it is necessary to treat V, but not w, as a varying quantity in deriving the equation. Taking the limit as $\delta x \rightarrow 0$, we obtain the differential relationship

$$\frac{\mathrm{d}V}{\mathrm{d}x} = -w(x) \tag{5.10}$$

which implies that the slope of the shear force distribution is equal in magnitude to, but opposite in sign, from the intensity of the distributed lateral force at any position along the beam. The negative sign is a consequence of our particular choice of sign conventions for shear force and lateral force. We have in fact already met an illustration of this result in Example 5.2, which concerned a simply supported beam subject to a uniformly distributed lateral load. Equation (5.7) for the shear force distribution can be differentiated with respect to x to give equation (5.10).

For equilibrium of moments applied to the element shown in Fig. 5.24, about the point Q

$$M - (M + \delta M) + V\delta x - \frac{w(\delta x)^2}{2} = 0$$

Note that the resultant lateral force is of magnitude $w\,\delta x$, and acts at the center of the element, a horizontal distance of $\delta x/2$ from point Q. Rearranging this equation

$$\frac{\delta M}{\delta x} = V - \frac{w\,\delta x}{2}$$

and taking the limit as $\delta x \to 0$, which eliminates the term involving the lateral force, we obtain the differential relationship

$$\frac{dM}{dx} = V(x) \tag{5.11}$$

This implies that the slope of the bending moment distribution is equal to the shear force at any position along the beam, and confirms the relationship we illustrated in Example 5.2. It also means that mathematical maximum and minimum values of bending moment, which are located where the slope of the bending moment distribution is zero, occur at points where the shear force is zero.

Equations (5.10) and (5.11) can also be used in other ways. Equation (5.10), for example, can be expressed as the integral relationship

$$V_2 - V_1 = \int_{x_1}^{x_2} -w(x)\,dx \tag{5.12}$$

where V_1 and V_2 are the shear forces at the beam cross sections at $x = x_1$ and $x = x_2$, respectively. This means that the change in the shear force between two positions along the beam is equal in magnitude to, but opposite in sign from, the area under the curve representing the distributed lateral force between these positions, as illustrated in Fig. 5.25. If A_w is the shaded area under the curve in Fig. 5.25a, then in the shear force distribution in Fig. 5.25b, $V_2 - V_1 = -A_w$. Similarly, Equation (5.11) can be expressed as the integral relationship

$$M_2 - M_1 = \int_{x_1}^{x_2} V(x)\,dx \tag{5.13}$$

where M_1 and M_2 are the bending moments at $x = x_1$ and $x = x_2$, respectively. This means that the change in the bending moment between two positions along the beam is equal to the area under the shear force distribution between these positions. If A_V is the shaded area under the curve in Fig. 5.25b, then in the bending moment distribution in Fig. 5.25c, $M_2 - M_1 = A_V$.

In deriving equations (5.10) to (5.13), the only form of external loading we have considered is distributed lateral force, and not concentrated forces or couples. We have already examined the effect of a concentrated lateral force in Example 5.1, where we saw that this led to an abrupt step change in both the

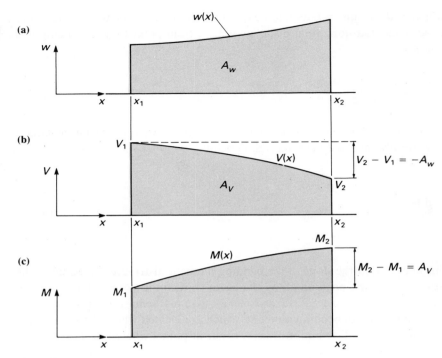

Fig. 5.25.

shear force distribution and the slope of the bending moment distribution (Fig. 5.16). These are entirely consistent with equations (5.10) and (5.11): a concentrated force is equivalent to a distributed force of locally infinite intensity, leading to an infinite slope of the shear force distribution (which is a vertical line), a nonunique value of shear force at the point concerned, and hence a nonunique value of the slope of the bending moment distribution there. A more realistic representation of a typical concentrated lateral force as being distributed over a small but finite length of beam, together with the corresponding shear force distribution which obeys equation (5.10), is shown in Fig. 5.17. The case of a concentrated applied couple we will consider in Example 5.6. Indeed, we are now ready to examine some further examples of shear force and bending moment diagrams for statically determinate beams. Before doing so, we can identify the steps which are necessary to tackle such problems manually: a computer method is described in Section 5.2.3. These steps are as follows.

1 Find the reaction forces and moments at the supports (a step which requires only equilibrium equations for a statically determinate beam).
2 Find the shear force at suitable positions along the beam. These include positions at either side of each support and each point of application of concentrated force, and at points of abrupt change of applied distributed force. Values at other intermediate points may also be helpful.
3 Draw the shear force diagram with the aid of equations (5.10) and (5.12).

4   Locate points of zero shear force, which will be points of maximum or minimum bending moment.

5   Find the bending moment at suitable positions along the beam. These include the support points, points of application of concentrated force, points of abrupt change of applied distributed force, points of maximum and minimum bending moment, and positions at either side of any concentrated couple. Values at other intermediate points may also be helpful.

6   Draw the bending moment diagram with the aid of equations (5.11) and (5.13).

EXAMPLE 5.4

Draw the shear force and bending moment diagrams for the beam of negligible weight shown in Fig. 5.26, and find the greatest absolute value of shear force and the greatest sagging and hogging bending moments. The beam, which is 8 m (24 ft) long, is simply supported at its left-hand end, point A, and at 2 m (6 ft) from its right-hand end, point B. It carries a concentrated downward force of 18 kN (4 kip) at point C, which is 2 m (6 ft) from A, together with uniformly distributed downward forces of intensity 24 kN/m (1.6 kip/ft) between points C and B, and 6 kN/m (0.4 kip/ft) between B and D, at the right-hand end of the beam.

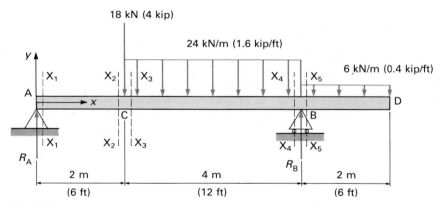

Fig. 5.26.

This is a typical example of the type met in practice, involving a mixture of concentrated and distributed external forces and concentrated reactions, also a beam which overhangs its support at one end. We start by finding the reactions at the supports. Although the support at A is capable of providing a reaction component in the horizontal direction, since there are no other external forces in the horizontal direction this component is zero, and the equation of equilibrium of forces in this direction is trivial. We are then left with just two independent equilibrium equations for the beam from which to find the vertical reactions R_A and R_B at the support points A and B, respectively. To find R_A, we can use the equation of equilibrium of moments about point B (which does not involve the reaction R_B).

SI units

$$R_A \times 6 - 18 \times 4 - (24 \times 4) \times 2 + (6 \times 2) \times 1 = 0$$

The uniformly distributed force of 24 kN/m acting over the 4 m length of beam between points C and B has a resultant of 24×4 kN which acts 2 m to the left of B, and the uniformly distributed force of 6 kN/m acting over the 2 m length of beam between points B and D has a resultant of 6×2 kN which acts 1 m to the right of B. From this equation, we obtain R_A as

$$R_A = \tfrac{1}{6}(72 + 192 - 12) = 42 \text{ kN}$$

Similarly, to find R_B, we can use the equation of equilibrium of moments about point A

$$18 \times 2 + (24 \times 4) \times 4 - R_B \times 6 + (6 \times 2) \times 7 = 0$$

from which

$$R_B = \tfrac{1}{6}(36 + 384 + 84) = 84 \text{ kN}$$

It is useful to check these results by considering equilibrium of forces in the vertical direction

$$R_A + R_B = 18 + (24 \times 4) + (6 \times 2) = 126 \text{ kN}$$

which is satisfied by $R_A = 42$ kN and $R_B = 84$ kN.

We now find values of the shear force at suitable positions along the beam. Starting from the left-hand end of the beam, we first consider a section just to the right of the support at A, which is labeled $X_1 X_1$ in Fig. 5.26. The only external force acting on the beam to the left of this section is the support reaction of 42 kN in the upward vertical direction, which according to our sign convention implies a shear force of $+ 42$ kN. Moving along the beam to section $X_2 X_2$, which is just to the left of the concentrated force at point C, the total external force acting on the beam to the left of this section remains at 42 kN. Since there is no distributed lateral load on the beam between A and C, equation (5.10) tells us that the gradient of the shear force distribution is zero, and equation (5.12) gives us the same information that there is no change in the shear force. We can therefore start to draw the shear force diagram as in Fig. 5.27(SI)b (Fig. 5.27(SI)a merely repeats the geometry and loading of the beam) by means of a horizontal line at $V = + 42$ kN between A and C. We now consider the section marked $X_3 X_3$ in Fig. 5.26, which is just to the right of the concentrated force at C. The total external force acting on the beam to the left of this section now includes the concentrated

US customary units

$$R_A \times 18 - 4 \times 12 - (1.6 \times 12) \times 6 + (0.4 \times 6) \times 3 = 0$$

The uniformly distributed force of 1.6 kip/ft acting over the 12 ft length of beam between points C and B has a resultant of 1.6×12 kip which acts 6 ft to the left of B, and the uniformly distributed force of 0.4 kip/ft acting over the 6 ft length of beam between points B and D has a resultant of 0.4×6 kip which acts 3 ft to the right of B. From this equation, we obtain R_A as

$$R_A = \tfrac{1}{18}(48 + 115.2 - 7.2) = 8.667 \text{ kip}$$

Similarly, to find R_B, we can use the equation of equilibrium of moments about point A

$$4 \times 6 + (1.6 \times 12) \times 12 - R_B \times 18 + (0.4 \times 6) \times 21 = 0$$

from which

$$R_B = \tfrac{1}{18}(24 + 230.4 + 50.4) = 16.933 \text{ kip}$$

It is useful to check these results by considering equilibrium of forces in the vertical direction

$$R_A + R_B = 4 + (1.6 \times 12) + (0.4 \times 6) = 25.6 \text{ kip}$$

which is satisfied by $R_A = 8.667$ kip and $R_B = 16.933$ kip.

We now find values of the shear force at suitable positions along the beam. Starting from the left-hand end of the beam, we first consider a section just to the right of the support at A, which is labeled $X_1 X_1$ in Fig. 5.26. The only external force acting on the beam to the left of this section is the support reaction of 8.667 kip in the upward vertical direction, which according to our sign convention implies a shear force of $+ 8.667$ kip. Moving along the beam to section $X_2 X_2$, which is just to the left of the concentrated force at point C, the total external force acting on the beam to the left of this section remains at 8.667 kip. Since there is no distributed lateral load on the beam between A and C, equation (5.10) tells us that the gradient of the shear force distribution is zero, and equation (5.12) gives us the same information that there is no change in the shear force. We can therefore start to draw the shear force diagram as in Fig. 5.27(US)b; (Fig. 5.27(US)a merely repeats the geometry and loading of the beam) by means of a horizontal line at $V = + 8.667$ kip between A and C. We now consider the section marked $X_3 X_3$ in Fig. 5.26, which is just to the right of the concentrated force at C. The total external force acting on the beam to the left of this section now

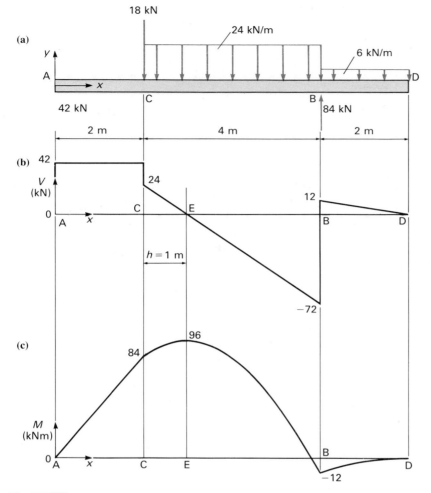

Fig. 5.27(SI)

force of 18 kN, and is $(42 - 18)$ kN in the upward vertical direction, implying a shear force of $+24$ kN, together with a step change from $+42$ kN to $+24$ kN at point C.

As we move along the beam towards the second support at point B, we progressively include more of the 24 kN/m distributed force, until at section X_4X_4 in Fig. 5.26, which is just to the left of B, the total upward force is $(42 - 18 - 24 \times 4)$ kN, a shear force of -72 kN. Since the distributed force is of constant intensity, the shear force varies linearly from $+24$ kN just to the right of C to -72 kN just to the left of B, as shown in Fig. 5.27(SI)b. Equation (5.10) tells us that the slope of the shear force distribution in this region is constant at -24 kN/m, and equation (5.12) gives the change in shear force as $-24 \times 4 = -96$ kN. As we

includes the concentrated force of 4 kip, and is $(8.667 - 4)$ kip in the upward vertical direction, implying a shear force of $+4.667$ kip, together with a step change from $+8.667$ kip to $+4.667$ kip at point C.

As we move along the beam towards the second support at point B, we progressively include more of the 1.6 kip/ft distributed force, until at section X_4X_4 in Fig. 5.26, which is just to the left of B, the total upward force is $(8.667 - 4 - 1.6 \times 12)$ kip, a shear force of -14.533 kip. Since the distributed force is of constant intensity, the shear force varies linearly from $+4.667$ kip just to the right of C to -14.533 kip just to the left of B, as shown in Fig. 5.27(US)b. Equation (5.10) tells us that the slope of the shear force distribution in this region is constant at -1.6 kip/ft, and equation (5.12) gives the change in shear force as

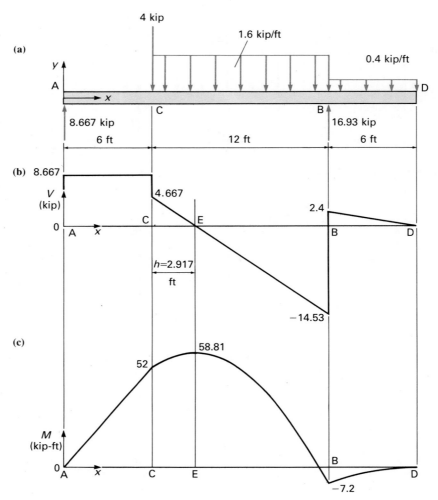

Fig. 5.27 (US)

SI units

pass to section X_5X_5 in Fig. 5.26, which is just to the right of the support at B, we include the concentrated upward support reaction of 84 kN acting to the left of the section, indicating a step change in shear force from -72 kN to $+12$ kN. Finally, as we move to the free end of the beam at D, where there is no concentrated lateral force, the shear force must fall to zero. With a uniformly distributed force over this part of the beam, the fall is linear. Equation (5.10) gives the constant gradient of the shear force distribution as -6 kN/m, which results in a fall from 12 kN to exactly zero over the 2 m length. If we failed to get this agreement at the end of the beam it would mean that we had made an error, either in the calculation of the reactions at the

US customary units

$-1.6 \times 12 = -19.2$ kip. As we pass to section X_5X_5 in Fig. 5.26, which is just to the right of the support at B, we include the concentrated upward support reaction of 16.933 kip acting to the left of the section, indicating a step change in shear force from -14.533 kip to $+2.4$ kip. Finally, as we move to the free end of the beam at D, where there is no concentrated lateral force, the shear force must fall to zero. With a uniformly distributed force over this part of the beam, the fall is linear. Equation (5.10) gives the constant gradient of the shear force distribution as -0.4 kip/ft, which results in a fall from 2.4 kip to exactly zero over the 6 ft length. If we failed to get this agreement at the end of the beam it would mean that we had made an error, either in the calculation of the reactions at the supports, or in drawing earlier parts of the shear force diagram.

Before considering bending moments, we locate points where the shear force is zero, which will be points of maximum or minimum bending moment. In Fig. 5.27(SI)b there are three such points: the support point B, the end of the beam D, and point E which lies between C and B. Suppose that E is located a distance h to the right of C as shown. Between C and E the shear force changes from $+24$ kN to zero in a region where equation (5.10) gives the gradient of the shear force distribution as -24 kN/m. Therefore $h = 1$ m, and point E is 3 m from the left-hand end of the beam.

Now we can find the bending moment at suitable points along the beam. At the ends of the beam, A and D, no external moments are applied and the bending moments are zero. At support point B, we can find the bending moment as the total external moment acting on the beam to the right of the section concerned, which is that due to the 6 kN/m distributed force acting over a 2 m length of beam having a resultant with a line of action 1 m from B

$$M_B = -(6 \times 2) \times 1 = -12 \text{ kNm}$$

This moment is negative because it causes a hogging effect at B. Similarly, the bending moment at point C is equal to the total external moment acting on the beam to the left of point C, which is that due to the reaction of 42 kN at A

$$M_C = 42 \times 2 = 84 \text{ kNm}$$

Finally, the bending moment at E can be found in a similar way, taking into account the reaction at A, the concentrated force of 18 kN at C and the uniformly distributed force of 24 kN/m acting on the 1 m length of beam between C and E

$$M_E = 42 \times 3 - 18 \times 1 - (24 \times 1) \times 0.5$$
$$= 96 \text{ kNm}$$

Note that, while the support reaction creates a sagging moment at E, the other two forces create hogging moments.

We can now construct the bending moment diagram shown in Fig. 5.27(SI)c. Between A and C, the shear force is constant at $+42$ kN, implying a bending moment distribution with a constant slope of $+42$ kNm/m, taking the bending moment from zero at A to 84 kN at C. Between C and B, the shear force varies linearly from $+24$ kN to -72 kN, which means

supports, or in drawing earlier parts of the shear force diagram.

Before considering bending moments, we locate points where the shear force is zero, which will be points of maximum or minimum bending moment. In Fig. 5.27(US)b there are three such points: the support point B, the end of the beam D, and point E which lies between C and B. Suppose that E is located a distance h to the right of C as shown. Between C and E the shear force changes from $+4.667$ kip to zero in a region where equation (5.10) gives the gradient of the shear force distribution as -1.6 kip/ft. Therefore $h = 4.667/1.6 = 2.917$ ft, and point E is 8.917 ft from the left-hand end of the beam.

Now we can find the bending moment at suitable points along the beam. At the ends of the beam, A and D, no external moments are applied and the bending moments are zero. At support point B, we can find the bending moment as the total external moment acting on the beam to the right of the section concerned, which is that due to the 0.4 kip/ft distributed force acting over a 6 ft length of beam having a resultant with a line of action 3 ft from B

$$M_B = -(0.4 \times 6) \times 3 = -7.2 \text{ kip-ft}$$

This moment is negative because it causes a hogging effect at B. Similarly, the bending moment at point C is equal to the total external moment acting on the beam to the left of point C, which is that due to the reaction of 8.667 kip at A

$$M_C = 8.667 \times 6 = 52 \text{ kip-ft}$$

Finally, the bending moment at E can be found in a similar way, taking into account the reaction at A, the concentrated force of 4 kip at C and the uniformly distributed force of 1.6 kip/ft acting on the 2.917 ft length of beam between C and E

$$M_E = 8.667 \times 8.917 - 4 \times 2.917 - (1.6 \times 2.917)$$
$$\times 1.4585 = 58.81 \text{ kip-ft}$$

Note that, while the support reaction creates a sagging moment at E, the other two forces create hogging moments.

We can now construct the bending moment diagram shown in Fig. 5.27(US)c. Between A and C, the shear force is constant at $+8.667$ kip, implying a bending moment distribution with a constant slope of $+8.667$ kip-ft/ft, taking the bending moment from zero at A to 52 kip-ft at C. Between C and B, the shear force varies linearly from $+4.667$ kip to -14.53 kip,

SI units

that the bending moment distribution along the beam is a quadratic function of x, with a slope which varies from $+24$ kNm/m at C to -72 kNm/m at B. Note the abrupt change in slope at C (from $+42$ kNm/m to $+24$ kNm/m). We have already located the maximum bending moment of 96 kNm between C and B at point E. Between B and D, the shear force varies linearly from $+12$ kN to zero, giving a quadratic bending moment distribution with a slope which varies from $+12$ kNm/m at B to zero at the end of the beam. Point D is therefore a point of mathematical minimum bending moment.

From Fig. 5.27(SI)b, the greatest absolute value of shear force is 72 kN, just to the left of support point B, while the greatest sagging and hogging bending moments are $+96$ kNm at E and -12 kNm at B. In this case the mathematical maximum bending moment is also the greatest in magnitude.

US customary units

which means that the bending moment distribution along the beam is a quadratic function of x, with a slope which varies from $+4.667$ kip-ft/ft at C to -14.53 kip-ft/ft at B. Note the abrupt change in slope at C (from $+8.667$ to $+4.667$ kip-ft/ft). We have already located the maximum bending moment of 58.81 kip-ft between C and B at point E. Between B and D, the shear force varies linearly from $+2.4$ kip to zero, giving a quadratic bending moment distribution with a slope which varies from $+2.4$ kip-ft/ft at B to zero at the end of the beam. Point D is therefore a point of mathematical minimum bending moment.

From Fig. 5.27(US)b, the greatest absolute value of shear force is 14.53 kip, just to the left of support point B, while the greatest sagging and hogging bending moments are $+58.81$ kip-ft at E and -7.2 kip-ft at B. In this case the mathematical maximum bending moment is also the greatest in magnitude.

EXAMPLE 5.5

Draw the shear force and bending moment diagrams for the cantilever beam of negligible weight shown in Fig. 5.28, and find the greatest absolute value of shear force and the greatest sagging and hogging bending moments. The beam, which is 3 m (9.8 ft) long, is built into a vertical wall at its left-hand end, point A. It carries a distributed downward force which varies linearly in intensity from 15 kN/m (1 kip/ft) at A to zero at its right-hand end, point B. There is a concentrated upward force of magnitude 10 kN (2.2 kip) at C, which is 2 m (6.6 ft) from A.

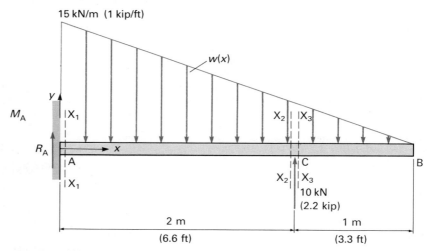

Fig. 5.28.

A practical example of a distributed lateral force with an intensity which varies linearly is that due to hydrostatic pressure in a liquid. The intensity of force acting on, say, the vertical wall of a tank containing water is directly proportional to the depth below the surface of the water.

First we must find the vertical reaction R_A and the moment M_A at the support at A. Note that we assume the moment to be in the sagging sense to be consistent with our sign convention. The linear distribution of lateral force, which must be such that the values of the force intensity at the left and right-hand ends of the beam are 15 kN/m and zero, respectively, is defined by the relationship

$$w(x) = 15 - 5x \text{ kN/m}$$

where x is the distance measured along the beam from its left-hand end. For equilibrium of forces acting on the beam in the vertical direction

$$R_A + 10 = \int_0^3 w(x)\,dx = \int_0^3 (15 - 5x)\,dx$$

$$= \left[15x - \frac{5x^2}{2} \right]_0^3 = 15 \times 3 - \frac{5 \times 3^2}{2} = 22.5 \text{ kN}$$

$$R_A = 12.5 \text{ kN}$$

Note that, since the distributed force varies linearly, we could have found its resultant of 22.5 kN as the product of the average intensity of 7.5 kN/m and the length of the beam of 3 m. The above method is more general, and is applicable to any form of distribution. Similarly, for equilibrium of moments acting on the beam about the point A

$$M_A + \int_0^3 w(x)x\,dx - 10 \times 2 = 0$$

$$M_A = 20 - \int_0^3 (15x - 5x^2)\,dx$$

$$= 20 - \left[\frac{15x^2}{2} - \frac{5x^3}{3} \right]_0^3 = 20 - \frac{15 \times 3^2}{2} + \frac{5 \times 3^3}{3}$$

$$= -2.5 \text{ kNm}$$

Once again, since the distributed force varies linearly, we could have found its moment about A of -22.5 kNm as the product of the resultant of 22.5 kN and one third of the length of the beam.

Perhaps the above integral expressions for the total force and moment due to an arbitrarily distributed lateral force require more explanation. Suppose we have such an arbitrary distribution acting over a beam of length L, as shown in Fig. 5.29. Consider an infinitesimal length, δx, of beam, which is at a distance x from the left-hand end of the beam. The force intensity there is $w(x)$ per unit length, and the force on length δx is therefore $w(x)\delta x$. The total downward vertical force on the beam can be found by summing this expression over the entire length of the beam

$$\text{total force} = \sum w(x)\delta x = \int_0^L w(x)\,dx \tag{5.14}$$

Similarly, the moment in the clockwise direction about the left-hand end of the beam of the force of $w(x)\delta x$ acting on length δx is $w(x)x\delta x$. The total clockwise moment is therefore

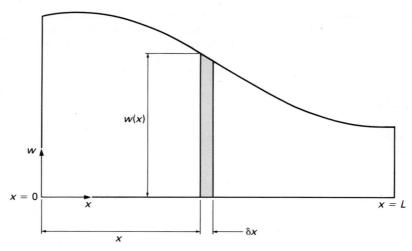

Fig. 5.29.

$$\text{total moment} = \sum w(x)\,x\,\delta x = \int_0^L w(x)\,x\,\mathrm{d}x \tag{5.15}$$

Returning to the particular example under consideration, our next step is to find the shear force at suitable positions along the beam. If we consider section $X_1 X_1$ in Fig. 5.28, which is just to the right of the support at A, the only external force acting on the beam to the left of this section is the support reaction of 12.5 kN in the upward vertical direction, which implies a shear force of $+ 12.5$ kN. Moving along the beam to section $X_2 X_2$, which is just to the left of the concentrated force at point C, the total external force acting on the beam to the left of this section now includes part of the distributed force. Using equation (5.12)

$$V_2 - V_1 = \int_0^2 (5x - 15)\,\mathrm{d}x = \left[\frac{5x^2}{2} - 15x\right]_0^2 = \frac{5 \times 2^2}{2} - 15 \times 2$$

$$V_2 = 12.5 + 10 - 30 = -7.5 \text{ kN}$$

which is the shear force at section $X_2 X_2$. Moving to section $X_3 X_3$, which is just to the right of point C, a step change in the shear force has occurred due to the concentrated force of 10 kN, which increases the shear force to $+ 2.5$ kN. Finally, at the end B of the beam, the shear force is zero. Also, since the intensity of the distributed force is a linear function of x, the shear force distributions between A and C, and between C and B, are quadratic functions. Figure 5.30b shows the shear force diagram.

Before drawing the bending moment diagram, it is useful to locate the points of zero shear force. One of these is at B, and the shear force distribution also passes through zero at C. The point which is of particular interest to us, however, is D, which lies between A and C. To find this point, we first obtain a general expression for the shear force distribution between A and C as a function of x, with the aid of equation (5.10)

$$\frac{\mathrm{d}V}{\mathrm{d}x} = -w(x) = 5x - 15$$

(a)

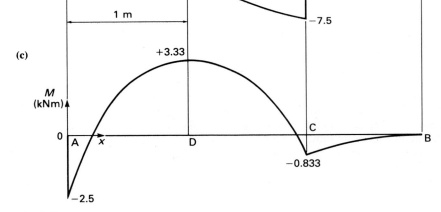

Fig. 5.30.

Since $V = +12.5$ kN at $x = 0$, this expression can be integrated with respect to x to give

$$V = 12.5 + \frac{5x^2}{2} - 15x$$

To find where $V = 0$, we must solve the quadratic equation

$$2.5x^2 - 15x + 12.5 = 0$$

or

$$x^2 - 6x + 5 = 0$$

which can be factorized

$$(x - 5)(x - 1) = 0$$

The two possible roots are $x = 1$ m and $x = 5$ m, only the first of which lies between A and C (the $x = 5$ root represents the point where the quadratic shear force distribution between A and C in Fig. 5.30b would, if continued, recross the x axis 2 m beyond the end of the beam). We conclude that point D lies midway between A and C, and corresponds to a mathematical maximum or minimum in the bending moment distribution.

We can find this bending moment distribution, again between points A and C, by integrating the shear force distribution with the aid of equation (5.11)

$$\frac{\mathrm{d}M}{\mathrm{d}x} = V(x) = 12.5 + 2.5x^2 - 15x$$

Since $M = M_A = -2.5$ kNm at $x = 0$,

$$M = -2.5 + 12.5x + \frac{2.5x^3}{3} - \frac{15x^2}{2}$$

Therefore, the bending moment at D is

$$M_D = -2.5 + 12.5 \times 1 + \frac{2.5 \times 1^3}{3} - \frac{15 \times 1^2}{2} = 3.333 \text{ kNm}$$

and at C is

$$M_C = -2.5 + 12.5 \times 2 + \frac{2.5 \times 2^3}{3} - \frac{15 \times 2^2}{2} = -0.8333 \text{ kNm}$$

The bending moment at B is zero, and we may therefore draw the bending moment diagram as in Fig. 5.30c. Since the shear force distributions between A and C, and between C and B, are quadratic functions of x, the bending moment distributions over the same ranges are cubic functions. Note that the slope of the bending moment distribution is zero not only at D but also at B.

From Fig. 5.30b, the greatest absolute value of shear force is 12.5 kN (2.8 kip), at support point A, while the greatest sagging and hogging bending moments are $+3.33$ kNm (2.5 kip-ft) at D and -2.5 kNm (-1.8 kip-ft) at A. In this case the mathematical maximum bending moment is also the greatest in magnitude.

EXAMPLE 5.6: Concentrated couple on a simply supported beam

Draw the shear force and bending moment diagrams for the beam of negligible weight shown in Fig. 5.31, which is of length L and is simply supported at its ends, A and B. It is subjected to a concentrated couple of magnitude C in the counterclockwise direction at point D, which is a distance a from A and b ($= L - a$) from B.

We have already considered in Section 5.1.2 concentrated couples as a possible form of loading on beams. These can arise in various ways, including torsional loading applied to the side of a beam (as in Example 1.3, Fig. 1.11), and in situations such as that shown in Fig. 5.12a, with an actual lateral force distribution of the form shown in Fig. 5.12c. For some purposes at least, it is convenient to represent a concentrated couple as acting at a single beam cross section, as in Fig. 5.12b, and again here in Fig. 5.31.

We start by finding the vertical reactions, R_A and R_B, at the supports. For equilibrium of forces acting on the beam in the vertical direction in Fig. 5.31

$$R_A + R_B = 0$$

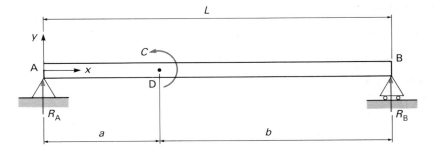

Fig. 5.31.

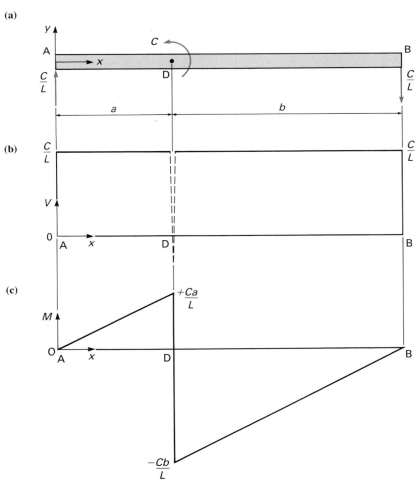

Fig. 5.32.

and for equilibrium of moments about **B**

$$R_A L - C = 0$$

Hence

$$R_A = -R_B = \frac{C}{L}$$

(a)

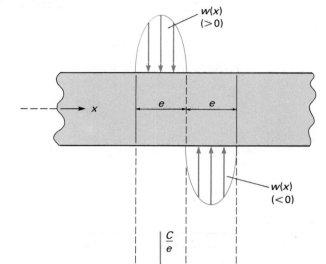

(b)

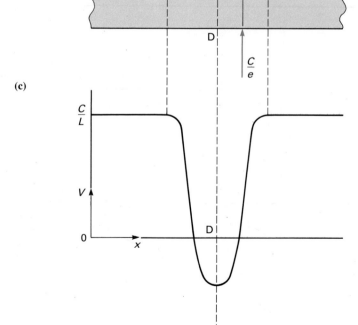

(c)

Fig. 5.33.

which, since there are no distributed lateral forces on the beam, implies a constant shear force distribution of magnitude $+C/L$ along the beam, except perhaps at the point D where the couple is applied, as shown in Fig. 5.32b. Note that the support reaction at B is actually in the downward direction: we assume that the simple support there is capable of providing a reaction in this direction (otherwise equilibrium would be impossible). Let us leave what happens to the shear force at D until we have determined the bending moment distribution. The bending moment is zero at the ends of the beam which are simply supported, and according to equation (5.11) is linearly distributed with a slope of $+C/L$ between A and D and between D and B. Figure 5.32c shows the bending moment distribution: at D there is a step change from $+Ca/L$ to $-Cb/L$, a change of magnitude $C(a+b)/L$ or C, the applied couple. Now, such a change implies an infinite negative slope to the bending moment distribution, and therefore an infinite negative shear force. This local effect on the shear force distribution is indicated by the dotted lines in Fig. 5.32b. It is, however, an artificial consequence of the concentrated couple approximation, and would not occur in practice.

If the concentrated couple is produced by large opposing lateral forces on the beam, the actual loading is likely to be of the form shown in Fig. 5.12c, which is repeated in Fig. 5.33a. The resultants of the upward and downward lateral force distributions would act at a distance apart of the order of e, and would then be of magnitude C/e as shown in Fig. 5.33b, in order to give a couple of magnitude C. In the limit as $e \to 0$, we would model a concentrated couple of finite magnitude C as the product of infinitely large force C/e and an infinitely small distance e. At point D in Fig. 5.32b the shear force would then be infinite (and negative). Returning to the loading shown in Fig. 5.33a, the corresponding shear force distribution would be of the form indicated in Fig. 5.33c. What Figs 5.32b and 5.33c have in common is that the area enclosed between the locally disturbed part of the shear force distribution and the line $V = C/L$ is finite and equal to the moment of the applied concentrated couple, C (in accordance with equation (5.13)). A more detailed treatment of the local region of application of a concentrated couple, which would be necessary to find the stresses there, is outside the scope of the present form of beam analysis.

EXAMPLE 5.7: Use of superposition

Draw the shear force and bending moment diagrams for the cantilever beam of negligible weight shown in Fig. 5.34, and find the greatest absolute value of shear force and the greatest sagging and hogging bending moments. The beam, which is 2 m (6.6 ft) long, is built into a vertical wall at its left-hand end, point A. It carries a uniformly distributed downward force of intensity 4 kN/m (270 lb/ft), and an upward vertical concentrated force of 2 kN (450 1b) at its right-hand end, point B.

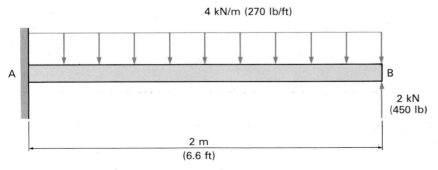

Fig. 5.34.

Although we could proceed to solve this problem in the usual way, it is worth noting that it is really a simple combination of two cases we have already considered, namely (i) a cantilever with a uniformly distributed lateral force (Fig. 5.22 with $L = 2$ m and $w = 4$ kN/m) and (ii) a cantilever with a concentrated force at its free end (Fig. 5.21 with $L = 2$ m and $F = -2$ kN). Consequently, we can use the principle of superposition to combine not only the loadings but also the shear force and bending moment distributions. This is conveniently done diagrammatically, as shown in Fig. 5.35. The three columns of loading, shear force and bending moment diagrams correspond to the actual problem, case (i) and case (ii), respectively, the first being obtained as the sum of the second and third in each case.

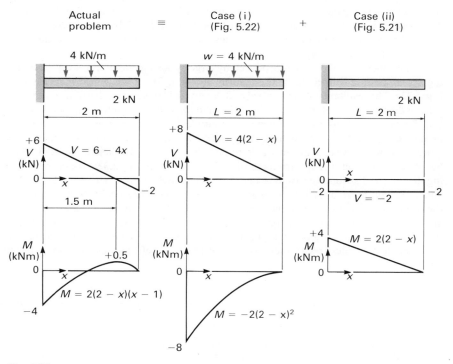

Fig. 5.35.

The greatest absolute value of shear force is 6 kN (1.3 kip), which occurs at the built-in end of the beam. The greatest hogging bending moment is -4 kNm (-2.9 kip-ft), also at the built-in end. The greatest sagging bending moment is 0.5 kNm (0.37 kip-ft), a mathematical maximum occurring at 1.5 m from the built-in end. This is a good example of a problem where the mathematical maximum value of bending moment is *not* the greatest in magnitude.

EXAMPLE 5.8

Two slings are to be used in lifting a large graphite rod of length L (of the order of several meters) and uniform cross section, the rod remaining horizontal during the lift. Calculate the most suitable positions of the slings assuming that failure would be by bending of the rod under its own weight.

This is a very practical example of the application of bending moment diagrams. It is also a straightforward example of situations where we wish to minimize the greatest bending moment in a beam in order to minimize the greatest bending stresses and hence the likelihood of failure. Since the rod is of uniform cross section, its weight is uniformly distributed along its length: let this force be of intensity w per unit length. In view of this uniform loading, we will assume that the slings should be symmetrically located relative to the center of the rod. Figure 5.36a shows the arrangement, with the slings at the same distance h from the ends. We are seeking the optimum value of this distance as a proportion of the total length L.

Figures 5.36b and c show the shear force and bending moment diagrams. Note that we have a maximum sagging bending moment, M_E, at the center of the rod, at point E,

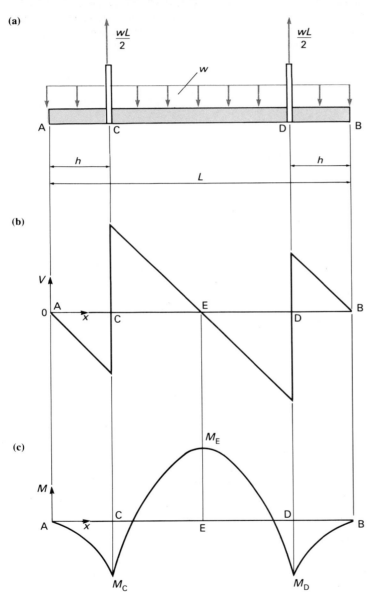

Fig. 5.36.

and that the greatest hogging bending moments, M_C and M_D, occur at the slings, points C and D (these moments are equal in magnitude). Now, consider what happens as we move the slings towards the ends of the rod ($h \to 0$): we finally reach the situation shown in Fig. 5.19 (a uniformly loaded beam simply supported at its ends), with a maximum sagging bending moment of $wL^2/8$ at the center and zero moments at the slings. When we go to the other extreme of locating both of the slings at the center of the rod ($h \to L/2$), we have effectively two cantilevers, each of length $L/2$ protruding horizontally from the slings. Figure 5.22 shows the relevant situation, the greatest bending moment at the slings again being of magnitude $wL^2/8$ (the maximum shown in Fig. 5.22 of $wL^2/2$ is for a cantilever of length L), but this time in the hogging sense. For an intermediate position of the slings (Fig. 5.36), if we move the slings closer together we reduce M_E but increase the magnitude of M_C (and M_D), while if we move them further apart we achieve the reverse effect. Since we wish to minimize the greatest bending moment, regardless of whether this greatest moment is sagging or hogging, the optimum position of the slings is achieved when the greatest sagging and greatest hogging moments are equal in magnitude

$$|M_E| = |M_C| \quad \text{or} \quad M_E = -M_C$$

Now

$$M_E = \frac{wL}{2}\left(\frac{L}{2} - h\right) - \frac{w}{2}\left(\frac{L}{2}\right)^2 = \frac{wL^2}{8} - \frac{wLh}{2}$$

and

$$M_C = -\frac{wh^2}{2}$$

so that when these moments are equal in magnitude

$$\frac{wL^2}{8} - \frac{wLh}{2} = \frac{wh^2}{2}$$

or

$$4h^2 + 4Lh - L^2 = 0$$

which is a quadratic equation for h in terms of L, with roots

$$h = [-2L \pm \sqrt{(4L^2 + 4L^2)}]/4 = L(-1 \pm \sqrt{2})/2$$

$$= 0.207L \quad \text{or} \quad -1.207L$$

Only the first of these is a physically possible solution, and we conclude that the slings should be located at distances of $0.207L$ from the ends of the rod. If this is done, the magnitude of the greatest bending moment, such as M_C, is

$$|M_C| = \frac{wh^2}{2} = \frac{0.207^2 wL^2}{2} = \frac{wL^2}{46.7}$$

which is only about 17% of the greatest bending moment (of $wL^2/8$) obtained with the slings in either of their extreme positions.

5.2.3 Computer method for shear force and bending moment distributions

Although shear force and bending moment distributions in statically determinate beams can be determined by the manual methods we have been considering, it is often more convenient to use a computer method. Once again, we can adopt a finite element approach, much as we did for pin-jointed structures. We divide the beam into a number of elements arranged along its length. Figure 5.37 shows a typical element, numbered m and of length L_m. We can think of this element as being joined to its immediate neighbors via its nodal points i and j. Our approach, like those used for pin-jointed structures, will involve considering equilibrium of forces, and in this case equilibrium of moments also, at the nodes.

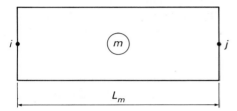

Fig. 5.37.

In one respect, the finite element treatment of beam problems is more straightforward than that of pin-jointed structures, in that the elements are arranged to lie one after the other along the length of the beam. On the other hand, the way in which we divide the system into elements is less clearcut, because we are now dealing with a continuous beam. Since we intend to consider equilibrium conditions at the nodes, however, it is necessary to arrange for nodes to be at points where there are concentrated forces, couples and other abrupt changes in loading. A minimum requirement is therefore for element interfaces to occur at the following:

(a) the supports,
(b) points of application of concentrated lateral forces,
(c) points of application of concentrated couples,
(d) points at which there are abrupt changes in the intensity of distributed lateral forces.

Whether it is necessary for any further subdivision depends on how distributed lateral forces vary along the beam, and on the form of such variation which can be accurately modeled by the type of elements we decide to use.

In fact, we will use elements over each of which the distributed lateral force is of constant intensity. This does not prevent us analyzing beams with continuously varying distributed forces, but, as we shall find, the solutions obtained are only approximate, and a larger number of elements is required to obtain results accurate enough for engineering purposes. This idea of element refinement to improve accuracy is a feature of almost all finite element methods (and other numerical techniques), which are applied to problems where exact solution is not possible. The reasons for choosing such simple 'constant force' elements here are, firstly, to keep the analysis and computer program as straightforward as possible and, secondly, to be consistent with the method developed for statically indeterminate beams in Chapter 6.

Following our usual sign conventions, we can show the distributed lateral force, of constant intensity w, the shear forces, V, and bending moments, M, acting on the typical element as in Fig. 5.38. In the cases of shear forces and bending moments, we assign subscripts i and j to indicate which nodes, and therefore which ends, of the element they refer to: we assume that nodes i and j are at the left and right-hand ends, respectively. Note the use of a superscript on the nodal point shear forces and bending moments to indicate the element number. This is because these quantities are not necessarily unique at a nodal point, due to a concentrated force or couple applied there. We use *global coordinate* X to define axial position along the beam, with some convenient origin relative to the beam as a whole (typically, but not necessarily, its left-hand end). We also introduce a *local coordinate*, x, which is parallel to X, but with its origin at the left hand node, i, of the element. Now, using differential equation (5.10) which relates shear force to distributed lateral force along a beam, together with the fact that $V = V_i^{(m)}$ at $x = 0$, we can obtain an expression for the shear force variation over the element as

$$V(x) = V_i^{(m)} - wx \tag{5.16}$$

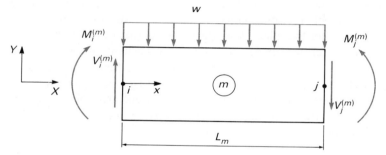

Fig. 5.38.

and the shear force at node j as

$$V_j^{(m)} = V_i^{(m)} - wL_m \tag{5.17}$$

Similarly, since $M = M_i^{(m)}$ at $x = 0$, differential equation (5.11) which links shear force and bending moment along a beam gives the variation of bending moment over the element as

$$M(x) = M_i^{(m)} + V_i^{(m)}x - \frac{wx^2}{2} \tag{5.18}$$

and the bending moment at node j as

$$M_j^{(m)} = M_i^{(m)} + V_i^{(m)}L_m - \frac{wL_m^2}{2} \tag{5.19}$$

When we come to consider equilibrium at the nodes, it will be convenient to work with forces and moments of couples associated with the elements connected there, which are all in the same direction and sense, respectively. For forces we choose as positive the upward vertical direction defined by global coordinate Y, and for couples we choose the counterclockwise sense as positive. Figure 5.39 shows the forces, Q, and couples, N, *acting on the element* in these directions at the nodes. Comparing Figs 5.38 and 5.39, we can see that there are some very simple relationships between these new variables and the shear forces and bending moments, as follows

$$Q_i^{(m)} = V_i^{(m)}, \quad Q_j^{(m)} = -V_j^{(m)}, \quad N_i^{(m)} = -M_i^{(m)}, \quad N_j^{(m)} = M_j^{(m)} \quad (5.20)$$

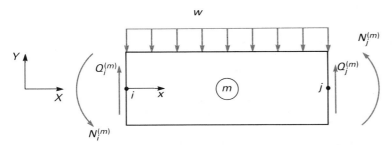

Fig. 5.39.

Since we are intending to find shear force and bending moment distributions along beams, it is appropriate to have these quantities, and in particular their values at the nodal points, as the primary unknowns to be determined directly by the analysis. As we have seen in the earlier Examples, however, shear forces and bending moments are not necessarily unique at every point along a beam. If there are concentrated forces or couples applied, then there are step changes in the shear force or bending moment distributions at the points of application. Consequently, we must allow for there to be two values each of shear force and bending moment at every node, except those at the ends of the beam (where there is only one of each). These are the limiting values as the node is approached from the left and right-hand sides, and are the values at the node in the two elements which are connected together there. In other words, there are four unknowns associated with every element (two shear forces and two bending moments). In addition, for a statically determinate beam subject only to couples and lateral forces (with no axial loading), it follows from equation (5.1) that there are exactly two unknown reactions (force or moment) at the supports. If there are N_e elements along the beam, the total number of unknowns is therefore $(4N_e + 2)$.

In every element, we have relationships between the nodal values of shear forces and bending moments in the form of equations (5.17) and (5.19), which can be rearranged in a more convenient form for computing purposes, with unknown quantities on the left-hand side, and knowns on the right

$$V_i^{(m)} - V_j^{(m)} = w L_m \quad (5.21)$$

$$M_i^{(m)} - M_j^{(m)} + V_i^{(m)} L_m = \frac{w L_m^2}{2} \tag{5.22}$$

With N_e elements, there is a total of $2 N_e$ such equations. Also, at every node there are two equilibrium equations, one equating the sum of the external lateral forces to the sum of the lateral forces transmitted to the elements at that node

$$\Sigma Q_i^{(m)} = -F_i - R_i \tag{5.23}$$

and the other defining a similar balance for the external couples and moments of couples transmitted to the elements

$$\Sigma N_i^{(m)} = C_i + R_i' \tag{5.24}$$

In these equations, F_i and R_i are the concentrated external force and concentrated reaction force at a support, respectively, either or both of which may be zero at a particular node, while C_i and R_i' are the concentrated external couple and reaction moment at a built-in support, respectively. The negative signs in the force equilibrium equation are due to the external forces following the beam sign convention of positive downwards, opposite to that of the Q forces. In the moment equilibrium equation, all couples are assumed to be positive in the counterclockwise direction. Now, it is again convenient to have all unknowns on the left-hand sides of the equations, and all the knowns on the right

$$\Sigma Q_i^{(m)} + R_i = -F_i \tag{5.25}$$

$$\Sigma N_i^{(m)} - R_i' = C_i \tag{5.26}$$

Note that the element force and moment summations are carried out over only the two elements connected together at the particular node, and over only one element in the case of the nodes at the ends of the beam. Equations (5.25) and (5.26) are the equilibrium conditions for typical node i. If there are N_n nodes along the beam, there are $2 N_n$ such equations, and a total of $(2 N_e + 2 N_n)$ equations. Now, for an arrangement of elements lying one after the other along a beam

$$N_n = N_e + 1 \tag{5.27}$$

and

$$2 N_e + 2 N_n = 4 N_e + 2$$

which indicates that the number of equations available is equal to the number of unknowns to be found.

The equations we have generated, in the form of equations (5.21), (5.22), (5.25) and (5.26), are linear algebraic equations, which we can express in matrix form as

$$[A][V] = [F] \qquad (5.28)$$

where $[V]$ is a vector containing the unknown shear forces, bending moments, and reaction forces and moments at the supports. Similarly, $[F]$ is a vector containing the known external forces and couples, while $[A]$ is a square matrix of known coefficients. Let us choose to arrange the unknowns in the following order: for each element in turn, taken in the order in which they are numbered, the shear force and bending moment (in that order) for the left-hand node, followed by the shear force and bending moment for the right-hand node. These are followed finally by the two support reactions (in the order in which the supports are defined, taking force before moment at a built-in support). In symbol form

$$[V]^T = [V_1^{(1)} M_1^{(1)} V_2^{(1)} M_2^{(1)} V_2^{(1)} M_2^{(1)} \ldots V_{Nn}^{(Ne)} M_{Nn}^{(Ne)} R_1 R_2]$$

where the superscript T indicates transposition from a column to a row vector to permit more compact presentation. Let us also choose to arrange the equations in the following order: equilibrium equations (forces followed by moments) for node 1, equations linking shear forces and bending moments (in that order) at the nodes of element 1, equilibrium equations for node 2, and so on to the end of the beam.

Program SDBEAM (standing for Statically Determinate BEAM analysis) provides a FORTRAN implementation of the above analysis. As far as possible it follows in terms of FORTRAN variable names and general layout the programs SDPINJ and SIPINJ for the analysis of pin-jointed structures (Sections 2.1.3 and 4.1.2).

```
      PROGRAM  SDBEAM
C
C  PROGRAM TO FIND THE SHEAR FORCE AND BENDING MOMENT DISTRIBUTIONS
C  ALONG A STATICALLY DETERMINATE BEAM, SUBJECT TO LATERAL FORCES
C  AND CONCENTRATED COUPLES.
C
      DIMENSION  X(51),NPI(50),NPJ(50),A(202,203),VMR(202),F(51),
     1           C(51),W(50),NODE(51),NELEM(50),SF(50,2),BM(50,2),
     2           R(2),XBMMAX(50),BMMAX(50)
      NNPMAX=51
      NELMAX=50
      OPEN(5,FILE='DATA')
      OPEN(6,FILE='RESULTS')
      WRITE(6,61)
   61 FORMAT('STATICALLY DETERMINATE BEAM')
C
C  INPUT AND TEST THE NUMBER OF NODAL POINTS.
      READ(5,*) NNP
      IF(NNP.GT.NNPMAX) THEN
        WRITE(6,62) NNP
   62   FORMAT(/ 'NUMBER OF NODES = ',I4,' TOO LARGE - STOP')
        STOP
      END IF
C
C  INPUT AND OUTPUT THE AXIAL COORDINATES OF THE NODES, WHICH MUST
C  BE ARRANGED IN ORDER FROM LEFT TO RIGHT ALONG THE BEAM.
      READ(5,*) (X(I),I=1,NNP)
      WRITE(6,63) (I,X(I),I=1,NNP)
   63 FORMAT(/ 'COORDINATES OF THE NODES'
     1 / 4('NODE     X       ') / (4(I3,E12.4,2X)))
```

```
C
C   DEFINE THE NUMBER OF ELEMENTS.
        NEL=NNP-1
C
C   DEFINE THE NODES CONNECTED BY THE ELEMENTS, BOTH NODES AND ELEMENTS
C   BEING NUMBERED FROM LEFT TO RIGHT ALONG THE BEAM.
C   ALSO INITIALIZE THE DISTRIBUTED FORCES ON THE ELEMENTS.
        DO 1 M=1,NEL
        NPI(M)=M
        NPJ(M)=M+1
    1   W(M)=0.
C
C   INITIALIZE THE EQUATION COEFFICIENTS AND EXTERNAL FORCES AND COUPLES
C   CONCENTRATED AT THE NODES.
        NEQN=4*NEL+2
        DO 2 IROW=1,NEQN
        DO 2 ICOL=1,NEQN
    2   A(IROW,ICOL)=0.
        DO 3 I=1,NNP
        F(I)=0.
    3   C(I)=0.
C
C   INPUT AND TEST THE NUMBER OF NODES AT WHICH EXTERNAL CONCENTRATED
C   LATERAL FORCES ARE APPLIED.
        READ(5,*) NNPF
        IF(NNPF.GT.NNPMAX) THEN
        WRITE(6,64) NNPF
   64   FORMAT(/ 'NUMBER OF NODES WITH CONCENTRATED FORCES = ',I4,
    1             ' TOO LARGE - STOP')
        STOP
        END IF
C
C   INPUT AND OUTPUT THE FORCES AT THESE NODES (DOWNWARDS POSITIVE).
        IF(NNPF.GT.0) THEN
        READ(5,*) (NODE(K),F(NODE(K)),K=1,NNPF)
        WRITE(6,65) (NODE(K),F(NODE(K)),K=1,NNPF)
   65   FORMAT(/ 'CONCENTRATED (DOWNWARD) LATERAL FORCES'
    1         / '  NODE',12X,'F' / (I5,5X,E15.4))
        END IF
C
C   INPUT AND TEST THE NUMBER OF NODES AT WHICH EXTERNAL CONCENTRATED
C   COUPLES ARE APPLIED.
        READ(5,*) NNPC
        IF(NNPC.GT.NNPMAX) THEN
        WRITE(6,66) NNPC
   66   FORMAT(/ 'NUMBER OF NODES WITH CONCENTRATED COUPLES = ',I4,
    1             ' TOO LARGE - STOP')
        STOP
        END IF
C
C   INPUT AND OUTPUT THE COUPLES AT THESE NODES (COUNTERCLOCKWISE
C   POSITIVE).
        IF(NNPC.GT.0) THEN
        READ(5,*) (NODE(K),C(NODE(K)),K=1,NNPC)
        WRITE(6,67) (NODE(K),C(NODE(K)),K=1,NNPC)
   67   FORMAT(/ 'CONCENTRATED (COUNTERCLOCKWISE) COUPLES'
    1         / '  NODE',12X,'C' / (I5,5X,E15.4))
        END IF
C
C   INPUT AND TEST THE NUMBER OF ELEMENTS OVER WHICH EXTERNAL
C   DISTRIBUTED LATERAL FORCES ARE APPLIED.
        READ(5,*) NELW
        IF(NELW.GT.NELMAX) THEN
        WRITE(6,68) NELW
   68   FORMAT(/ 'NUMBER OF ELEMENTS WITH DISTRIBUTED FORCES = ',I4,
    1             ' TOO LARGE - STOP')
        STOP
        END IF
C
C   INPUT AND OUTPUT THE FORCE INTENSITIES ON THESE ELEMENTS (DOWNWARDS
C   POSITIVE).
        IF(NELW.GT.0) THEN
        READ(5,*) (NELEM(K),W(NELEM(K)),K=1,NELW)
        WRITE(6,69) (NELEM(K),W(NELEM(K)),K=1,NELW)
```

```
  69     FORMAT(/ 'DISTRIBUTED (DOWNWARD) LATERAL FORCES'
       1         / ' ELEMENT',12X,'W' / (I5,5X,E15.4))
         END IF
C
C  INPUT AND TEST THE NUMBERS OF NODES AT WHICH THE BEAM IS SIMPLY
C  SUPPORTED (NNPSS), AND BUILT IN (NNPBI).
         READ(5,*) NNPSS,NNPBI
         NNPT=NNPSS+NNPBI
         IF(NNPT.GT.NNPMAX) THEN
            WRITE(6,70) NNPT
  70     FORMAT(/ 'NUMBER OF NODES AT WHICH BEAM IS SUPPORTED = ',I4,
       1             ' TOO LARGE - STOP')
            STOP
         END IF
C
C  INPUT AND OUTPUT THE NODES AT WHICH THE BEAM IS SIMPLY SUPPORTED.
         IF(NNPSS.GT.0) THEN
            READ(5,*) (NODE(K),K=1,NNPSS)
            WRITE(6,71) (NODE(K),K=1,NNPSS)
  71     FORMAT(/ 'NODES AT WHICH BEAM IS SIMPLY SUPPORTED : ',7I4 /
       1            (18I4))
         END IF
C
C  INPUT AND OUTPUT THE NODES AT WHICH THE BEAM IS BUILT IN.
         IF(NNPBI.GT.0) THEN
            READ(5,*) (NODE(NNPSS+K),K=1,NNPBI)
            WRITE(6,72) (NODE(NNPSS+K),K=1,NNPBI)
  72     FORMAT(/ 'NODES AT WHICH BEAM IS BUILT IN : ',8I4 / (18I4))
         END IF
C
C  TEST THE NUMBER OF UNKNOWN SUPPORT REACTION FORCES AND MOMENTS.
         NUNK=NNPSS+2*NNPBI
         IF(NUNK.GT.2) THEN
            WRITE(6,73)
  73     FORMAT(/ 'THE BEAM IS STATICALLY INDETERMINATE - STOP')
            STOP
         END IF
         IF(NUNK.LT.2) THEN
            WRITE(6,74)
  74     FORMAT(/ 'THE BEAM IS A MECHANISM - STOP')
            STOP
         END IF
C
C  SET UP THE LINEAR EQUATIONS, FIRST CONSIDERING EACH ELEMENT IN TURN.
         DO 4 M=1,NEL
         I=NPI(M)
         J=NPJ(M)
         ELENGT=X(J)-X(I)
C
C  TEST FOR NEGATIVE ELEMENT LENGTH (WHICH RESULTS FROM NODAL
C  COORDINATES NOT BEING ENTERED IN ORDER FROM LEFT TO RIGHT ALONG THE
C  BEAM.
         IF(ELENGT.LT.0.) THEN
            WRITE(6,75) M,I,J
  75     FORMAT(/ 'ELEMENT ',I2,' WITH NODES ',I2,' AND ',I2,
       1             ' HAS NEGATIVE LENGTH - STOP' /
       2             '(NODAL COORDINATES NOT ENTERED IN ORDER FROM LEFT TO'
       3             ' RIGHT ALONG THE BEAM)')
            STOP
         END IF
C
C  CONTRIBUTION OF ELEMENT TO FORCE EQUILIBRIUM EQUATION AT ITS FIRST
C  NODE.
         A(4*(M-1)+1,4*(M-1)+1)=1.
C
C  CONTRIBUTION OF ELEMENT TO MOMENT EQUILIBRIUM EQUATION AT ITS FIRST
C  NODE.
         A(4*(M-1)+2,4*(M-1)+2)=-1.
C
C  COEFFICIENTS OF EQUATION LINKING SHEAR FORCES AT THE TWO NODES.
         A(4*(M-1)+3,4*(M-1)+1)=1.
         A(4*(M-1)+3,4*(M-1)+3)=-1.
         A(4*(M-1)+3,NEQN+1)=W(M)*ELENGT
```

```
C
C   COEFFICIENTS OF EQUATION LINKING BENDING MOMENTS AT THE TWO NODES.
        A(4*(M-1)+4,4*(M-1)+1)=ELENGT
        A(4*(M-1)+4,4*(M-1)+2)=1.
        A(4*(M-1)+4,4*(M-1)+4)=-1.
        A(4*(M-1)+4,NEQN+1)=0.5*W(M)*ELENGT**2
C
C   CONTRIBUTION OF ELEMENT TO FORCE EQUILIBRIUM EQUATION AT ITS SECOND
C   NODE.
        A(4*M+1,4*(M-1)+3)=-1.
C
C   CONTRIBUTION OF ELEMENT TO MOMENT EQUILIBRIUM EQUATION AT ITS SECOND
C   NODE.
    4   A(4*M+2,4*(M-1)+4)=1.
C
C   ADD APPLIED CONCENTRATED FORCES AND COUPLES TO THE LINEAR EQUATIONS.
        DO 5 I=1,NNP
        A(4*(I-1)+1,NEQN+1)=-F(I)
    5   A(4*(I-1)+2,NEQN+1)=C(I)
C
C   ADD COEFFICIENTS OF UNKNOWN SIMPLE SUPPORT REACTION FORCES TO THE
C   LINEAR EQUATIONS.
        IF(NNPSS.GT.0) THEN
          DO 6 K=1,NNPSS
          I=NODE(K)
    6     A(4*(I-1)+1,4*NEL+K)=1.
        END IF
C
C   ADD COEFFICIENTS OF UNKNOWN BUILT-IN SUPPORT REACTION FORCES AND
C   MOMENTS TO THE LINEAR EQUATIONS.
        IF(NNPBI.GT.0) THEN
          DO 7 K=1,NNPBI
          I=NODE(K)
          A(4*(I-1)+1,4*NEL+NNPSS+2*(K-1)+1)=1.
    7     A(4*(I-1)+2,4*NEL+NNPSS+2*(K-1)+2)=-1.
        END IF
C
C   SOLVE THE LINEAR ALGEBRAIC EQUATIONS BY GAUSSIAN ELIMINATION.
        CALL   SOLVE(A,VMR,NEQN,4*NELMAX+2,4*NELMAX+3,IFLAG)
C
C   STOP IF A UNIT VALUE OF THE ILL-CONDITIONING FLAG IS DETECTED.
        IF(IFLAG.EQ.1) THEN
          WRITE(6,76)
   76     FORMAT(/ 'EQUATIONS ARE ILL-CONDITIONED - STOP')
          STOP
        END IF
C
C   STORE SHEAR FORCES AND BENDING MOMENTS IN CONVENIENT ARRAYS.
        DO 8 M=1,NEL
        SF(M,1)=VMR(4*(M-1)+1)
        BM(M,1)=VMR(4*(M-1)+2)
        SF(M,2)=VMR(4*(M-1)+3)
    8   BM(M,2)=VMR(4*(M-1)+4)
C
C   ALSO THE REACTION FORCES AND MOMENTS AT THE SUPPORTS.
        R(1)=VMR(4*NEL+1)
        R(2)=VMR(4*NEL+2)
C
C   OUTPUT THE COMPUTED SHEAR FORCES AND BENDING MOMENTS AT THE NODES
C   OF EACH ELEMENT.
        WRITE(6,77) (M,NPI(M),NPJ(M),SF(M,1),SF(M,2),BM(M,1),BM(M,2),
    1               M=1,NEL)
   77   FORMAT(/ 'COMPUTED SHEAR FORCES AND BENDING MOMENTS AT THE'
    1           ' NODES OF EACH ELEMENT' /
    2           ' ELEM NODES      SF AT I        SF AT J  '
    3           '     BM AT I        BM AT J  ' / (3I4,4E15.4))
C
C   OUTPUT THE COMPUTED REACTION FORCES AT THE SIMPLE SUPPORTS.
        IF(NNPSS.GT.0) THEN
          WRITE(6,79) (NODE(K),R(K),K=1,NNPSS)
   79     FORMAT(/ 'COMPUTED (DOWNWARD) REACTION FORCES AT THE'
    1             ' SIMPLE SUPPORTS' / ' NODE      FORCE ' / (I5,E15.4))
        END IF
```

```
C
C   OUTPUT THE COMPUTED REACTION FORCES AT THE BUILT-IN SUPPORTS.
        IF(NNPBI.GT.0) THEN
            WRITE(6,80) (NODE(NNPSS+K),R(NNPSS+2*(K-1)+1),K=1,NNPBI)
   80       FORMAT(/ 'COMPUTED (DOWNWARD) REACTION FORCES AT THE'
        1   ' BUILT-IN SUPPORTS' / ' NODE      FORCE ' / (I5,E15.4))
C
C   OUTPUT THE COMPUTED REACTION MOMENTS AT THE BUILT-IN SUPPORTS.
            WRITE(6,81) (NODE(NNPSS+K),R(NNPSS+2*(K-1)+2),K=1,NNPBI)
   81       FORMAT(/ 'COMPUTED (COUNTERCLOCKWISE) MOMENTS AT THE'
        1   ' BUILT-IN SUPPORTS' / ' NODE      MOMENT ' / (I5,E15.4))
        END IF
C
C   SEARCH FOR MATHEMATICAL MAXIMUM OR MINIMUM VALUES OF BENDING
C   MOMENT WITHIN ELEMENTS (WHERE SHEAR FORCE ZERO).
        NBMMAX=0
        DO 9 M=1,NEL
        IF(SF(M,1)*SF(M,2).LE.0.) THEN
            NBMMAX=NBMMAX+1
            I=NPI(M)
            J=NPJ(M)
            IF(SF(M,1)*SF(M,2).LT.0.) THEN
                XLOCAL=SF(M,1)/W(M)
            ELSE
                IF(SF(M,1).EQ.0.) XLOCAL=0.
                IF(SF(M,2).EQ.0.) XLOCAL=X(J)-X(I)
            END IF
            XBMMAX(NBMMAX)=X(I)+XLOCAL
            BMMAX(NBMMAX)=BM(M,1)+SF(M,1)*XLOCAL-0.5*W(M)*XLOCAL**2
        END IF
    9   CONTINUE
C
C   OUTPUT MATHEMATICAL MAXIMUM OR MINIMUM VALUES OF BENDING MOMENT.
        IF(NBMMAX.GT.0) THEN
            WRITE(6,82) (XBMMAX(IBMMAX),BMMAX(IBMMAX),IBMMAX=1,NBMMAX)
   82       FORMAT(/ 'MATHEMATICAL MAXIMUM OR MINIMUM VALUES OF BENDING'
        1           ' MOMENT' / ' AXIAL POSITION      MOMENT ' /
        2           (2E15.4))
        END IF
C
C   FIND GREATEST ABSOLUTE VALUE OF SHEAR FORCE AND GREATEST SAGGING
C   AND HOGGING BENDING MOMENTS.
        GRSF=0.
        XGRSF=0.
        GRBMS=0.
        XGRBMS=0.
        GRBMH=0.
        XGRBMH=0.
        IF(NBMMAX.GT.0) THEN
            DO 10 IBMMAX=1,NBMMAX
            IF(BMMAX(IBMMAX).GT.GRBMS) THEN
                GRBMS=BMMAX(IBMMAX)
                XGRBMS=XBMMAX(IBMMAX)
            END IF
            IF(BMMAX(IBMMAX).LT.GRBMH) THEN
                GRBMH=BMMAX(IBMMAX)
                XGRBMH=XBMMAX(IBMMAX)
            END IF
   10       CONTINUE
        END IF
        DO 11 M=1,NEL
        DO 11 IEND=1,2
        IF(IEND.EQ.1) INODE=NPI(M)
        IF(IEND.EQ.2) INODE=NPJ(M)
        IF(ABS(SF(M,IEND)).GT.GRSF) THEN
            GRSF=ABS(SF(M,IEND))
            XGRSF=X(INODE)
        END IF
        IF(BM(M,IEND).GT.GRBMS) THEN
            GRBMS=BM(M,IEND)
            XGRBMS=X(INODE)
        END IF
        IF(BM(M,IEND).LT.GRBMH) THEN
            GRBMH=BM(M,IEND)
            XGRBMH=X(INODE)
        END IF
```

```
   11   CONTINUE
C
C   OUTPUT GREATEST ABSOLUTE VALUE OF SHEAR FORCE AND GREATEST SAGGING
C   AND HOGGING BENDING MOMENTS, TOGETHER WITH THEIR POSITIONS.
      WRITE(6,83) GRSF,XGRSF,GRBMS,XGRBMS,GRBMH,XGRBMH
   83   FORMAT(/ 'GREATEST SHEAR FORCE =',E12.4,'   AT X =',E12.4
      1 /'GREATEST SAGGING BENDING MOMENT =',E12.4,'   AT X =',E12.4
      2 /'GREATEST HOGGING BENDING MOMENT =',E12.4,'   AT X =',E12.4)
      STOP
      END
```

The following list provides the definitions of the FORTRAN variables used, arranged is alphabetical order.

Variable	Type	Definition
A	real	Coefficient matrix $[A]$ extended to include vector $[F]$ as its last column
BM	real	Array storing the bending moments at the nodes of the elements (first subscript defines the element number, second subscript the node number within the element)
BMMAX	real	Array storing mathematical maximum and minimum values of bending moment within elements
C	real	Array storing the external concentrated couples (counterclockwise positive) acting on the beam at the nodes
ELENGT	real	Element length
F	real	Array storing the external concentrated forces (downwards positive) acting on the beam at the nodes
GRBMH	real	Greatest value of bending moment (hogging)
GRBMS	real	Greatest value of bending moment (sagging)
GRSF	real	Greatest value of shear force
I	integer	Nodal point number (sometimes first node of an element)
IBMMAX	integer	Counter for mathematical maximum or minimum values of bending moments
ICOL	integer	Column number in array A
IEND	integer	Counter for the (two) ends of an element
IFLAG	integer	Flag for a singular or very ill-conditioned coefficient matrix (normally returned as zero by SOLVE, one if the ill-conditioning test is satisfied)
INODE	integer	Node number
IROW	integer	Row number in array A
J	integer	Nodal point number (second node of an element)
K	integer	A counter (used in READ and WRITE statements)
M	integer	Element number
NBMMAX	integer	number of mathematical maximum and minimum values of bending moment within elements
NEL	integer	Number of elements
NELEM	integer	Array storing element numbers
NELMAX	integer	Maximum number of elements permitted by the array dimensions
NELW	integer	Number of elements over which external distributed forces act
NEQN	integer	Number of equations to be solved
NNP	integer	Number of nodal points
NNPBI	integer	Number of nodes at which the beam is built-in
NNPC	integer	Number of nodes at which external concentrated couples are applied
NNPF	integer	Number of nodes at which external concentrated forces are applied
NNPMAX	integer	Maximum number of nodal points permitted by the array dimensions
NNPSS	integer	Number of nodes at which the beam is simply supported
NNPT	integer	Total number of nodes at which the beam is supported
NODE	integer	Array storing node numbers
NPI	integer	Array storing the numbers of the first nodes of the elements
NPJ	integer	Array storing the numbers of the second nodes of the elements
NUNK	integer	Total number of unknown support reaction forces and moments

Variable	Type	Definition
R	real	Array storing the reactions (forces or moments) at the supports
SF	real	Array storing the shear forces at the nodes of the elements (first subscript defines the element number, second subscript the node number within the element)
VMR	real	Array storing the vector of unknown shear forces, bending moments and support reactions, $[V]$
W	real	Array storing the intensities of the external distributed forces (downwards positive) acting on the elements
X	real	Array storing X global coordinates of the nodes
XBMMAX	real	Array storing global coordinates of mathematical maximum and minimum values of bending moment within elements
XGRBMH	real	Position (X coordinate) of the point of greatest hogging bending moment
XGRBMS	real	Position (X coordinate) of the point of greatest sagging bending moment
XGRSF	real	Position (X coordinate) of the point of greatest shear force
XLOCAL	real	Local coordinate, x, within an element

The arrays used in SDBEAM are dimensioned in such a way as to allow a beam with up to 50 elements and 51 nodes to be analyzed, implying that up to 202 equations (four times the number of elements, plus two) may have to be solved. Note that array A, which stores the coefficient matrix $[A]$ extended to include vector $[F]$ is allowed to have up to 203 columns. The array dimensions can of course be changed, provided the values of NNPMAX and NELMAX defined in the next two statements are also adjusted.

The program reads information from a file named DATA, and writes onto a file named RESULTS. All READ statements call for free format input data. After writing out a heading, the program first reads the number of nodal points along the beam to be analyzed, and then tests that this does not exceed the maximum allowed by the array dimensions. We should note that, as in programs SDPINJ and SIPINJ, the checking of input data in the program is comparatively rudimentary, for the same reasons. Some of the more obvious data errors are trapped, and in all cases input data is immediately written out to allow it to be checked.

The program reads in and tests the number of nodes, followed by the global axial coordinates of the nodes, X, for node 1, then node 2 and so on. It is assumed that the nodes are numbered consecutively from left to right along the beam, and the axial coordinates must reflect this ordering (if they do not, execution will fail at a later point in the program). Although we do not have to make this assumption, and the program could be coded to accept an arbitrary numbering system, there is little to be gained with a geometrically simple system such as a beam. By assuming we know the node numbering system, it is then not necessary to enter as data the nodes of the elements, which is a significant saving of labor. The program also numbers the elements from left to right along the beam, which means that the element numbered m has as its nodal points the nodes m and $m + 1$, as shown by Fig. 5.40. At the same time, the distributed forces on the elements are set to zero.

All coefficients of matrix $[A]$ and the external concentrated forces and couples at the nodes are then set to zero in preparation for the assembly process to come. The number of nodes at which external concentrated forces are

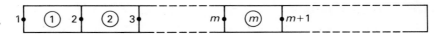

Fig. 5.40.

applied is read in, followed by the numbers of the nodes concerned together with the corresponding forces, taking the downward direction as positive. Then, the number of nodes at which external concentrated couples are applied is read in, followed by the numbers of the nodes concerned together with the corresponding couples, taking the counterclockwise direction as positive. The final possible form of loading is that due to external distributed forces, and the number of elements over which such forces are applied is read in, followed by the numbers of the elements concerned together with the corresponding force intensities (values of force per unit length), again taking the downward direction as positive.

Next, the numbers of nodes at which the beam is simply supported and built-in are read in, followed by the numbers of the nodes forming simple supports, then the numbers of the nodes where the beam is built-in. Note that no distinction is made between pinned and rolling simple supports (Fig. 5.6). This is because the program does not accept axial loading of the beam so that the distinction is unnecessary. It also means that the formal test for whether a beam is statically determinate, equation (5.1), must be modified, since both a reaction force component and the equilibrium equation in the axial direction have been eliminated. The program calculates the total number of support reaction forces and moments (one for a simple support, and two for a built-in support). The beam is statically indeterminate, statically determinate or a mechanism according to whether this number is greater than, equal to or less than two. Only if the beam is statically determinate is the analysis allowed to proceed.

A convenient way to assemble the coefficients of the linear equations is by considering each of the elements in turn. Our typical element, shown in Fig. 5.37 and numbered m, has first and second nodes i and j, respectively, which in the program are recovered from arrays NPI and NPJ and stored in variables I and J. The length of the element is found as the distance between nodes i and j. If this length is negative, the global coordinates of the nodes must have been entered in an incorrect order, and execution is halted.

Bearing in mind the order we have already defined for the unknowns and for the equations in the set (5.28), we can expect the typical element m to make the following contributions, which are implemented in the program.

1 To the equation numbered $4(m-1)+1$, which is the force equilibrium equation (5.25) for its first node, it contributes $+Q_i^{(m)}$ or, according to equations (5.20), $+V_i^{(m)}$. This is the unknown numbered $4(m-1)+1$, and the corresponding coefficient in $[A]$ is $+1$.

2 To the equation numbered $4(m-1)+2$, which is the moment equilibrium equation (5.26) for its first node, it contributes $+N_i^{(m)}$ or, according to equations (5.20), $-M_i^{(m)}$. This is the unknown numbered $4(m-1)+1$, and the corresponding coefficient in $[A]$ is -1.

3   It contributes the whole of the equation numbered $4(m-1)+3$, which is the equation (5.21) linking the shear forces at its nodes. The coefficients of $V_i^{(m)}$ and $V_j^{(m)}$, which are the unknowns numbered $4(m-1)+1$ and $4(m-1)+3$, are $+1$ and -1, respectively, and the contribution to $[F]$, which is now the last column of $[A]$, is $+wL_m$.

4   It contributes the whole of the equation numbered $4(m-1)+4$, which is the equation (5.22) linking the shear forces at its nodes. The coefficients of $M_i^{(m)}$, $M_j^{(m)}$ and $V_i^{(m)}$, which are the unknowns numbered $4(m-1)+2$, $4(m-1)+4$ and $4(m-1)+1$, are $+1$, -1, and L_m, respectively, and the contribution to $[F]$, which is now the last column of $[A]$, is $+wL_m^2/2$.

5   To the equation numbered $4m+1$, which is the force equilibrium equation (5.25) for its second node, it contributes $+Q_j^{(m)}$ or, according to equations (5.20), $-V_j^{(m)}$. This is the unknown numbered $4(m-1)+3$, and the corresponding coefficient in $[A]$ is -1.

6   To the equation numbered $4m+2$, which is the moment equilibrium equation (5.26) for its second node, it contributes $+N_j^{(m)}$ or, according to equations (5.20), $+M_j^{(m)}$. This is the unknown numbered $4(m-1)+4$, and the corresponding coefficient in $[A]$ is $+1$.

The program then adds known concentrated forces and moments at the nodes to the equilibrium equations: values at node i contribute to the vector $[F]$ in the equations numbered $4(i-1)+1$ and $4(i-1)+2$, which now forms the last column of matrix $[A]$. Reaction forces and moments at the supports, which are unknowns, are dealt with differently. Acting at node i, they contribute to the equations numbered $4(i-1)+1$ and $4(i-1)+2$, and from equations (5.25) and (5.26) they result in coefficients of $+1$ and -1, respectively, being added to the relevant columns of $[A]$. The numbering of these columns depends on the order in which the support points were defined in the input data: simple supports first, followed by built-in ones.

The equations can now be solved, using subroutine SOLVE which is described in detail in Appendix B. If a unit value of IFLAG is returned to SDBEAM, this means that SOLVE has detected a singular or very ill-conditioned coefficient matrix. This is unlikely to happen in the present program, but it writes out a warning message and terminates execution. Otherwise, the computed shear forces and bending moments at both nodes of each of the elements are transferred to arrays SF and BM, also support reaction forces and moments to array R, and are then written out. The program then searches for mathematical maximum or minimum values of bending moment, examining each element in turn. Such a value is only obtained within an element if the shear force passes through zero between the two nodes concerned, or is zero at one of them. These conditions can be detected by the product of the shear forces at the nodes being either less than or equal to zero, and the local coordinate of the required point can be found with the aid of equation (5.16) as

$$x = \frac{V_i^{(m)}}{w}$$

and the global coordinate by adding that of node *i*. The greatest absolute value of shear force and the greatest sagging and hogging bending moments are then found, first by searching the mathematical maximum and minimum values (in the case of bending moment), and then the values at the nodal points. These greatest values are then written out. Note that the searching process involves first setting the greatest values to zero and the corresponding positions (*X* coordinates) to zero. This can mean that if, for example, there are no hogging bending moments along the beam the initial settings would be unchanged and the result for the greatest hogging moment would be shown as zero at *X* = 0.

We note that no units are specified for either the input data or the computed results. This is because we may use any consistent set of force and length units.

We will now examine how program SDBEAM can be used to analyze statically determinate beams with the aid of two problems, which we have already solved manually in Examples 5.4 and 5.5.

EXAMPLE 5.9

Consider again the beam shown in Fig. 5.26, using program SDBEAM to find the distributions of shear force and bending moment in US customary units.

The beam is redrawn in Fig. 5.41 with the nodes and elements shown numbered in the orders assumed by the program. The choice of nodes and elements is determined by the positions of the supports, the concentrated applied force and the abrupt change in the distributed lateral force. Element numbers are shown ringed, while node numbers are unringed. It is convenient to define the origin for global coordinate *X* at node 1 as shown. Figure 5.42 shows the file of input data used to set up this problem for solution by program SDBEAM, ft and kip units being used for length and force, respectively.

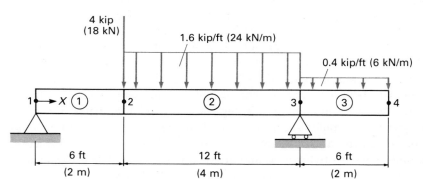

Fig. 5.41.

```
4                              (Number of nodes)
0.  6.  18.  24.               (Node coordinates)
1                              (Number of concentrated lateral forces)
2  4.                          (Node number and concentrated force)
0                              (Number of concentrated couples)
2                              (Number of elements with distributed lateral forces)
2  1.6 ,  3  0.4               (Element numbers and force intensities)
2  0                           (Numbers of simple and built-in supports)
1  3                           (Node numbers of the simple supports)
```

Fig. 5.42.

The file has been annotated to indicate what each line of data represents. The file RESULTS produced by the program is shown in Fig. 5.43. As we would expect, the computed shear forces and bending moments at the ends of the elements, the reactions at the supports and the maximum and greatest shear forces and bending moments are all identical to those obtained by manual methods in Example 5.4, and plotted in Fig. 5.27(US).

```
STATICALLY DETERMINATE BEAM

COORDINATES OF THE NODES
NODE    X        NODE    X        NODE    X        NODE    X
  1  0.0000E+00    2  0.6000E+01    3  0.1800E+02    4  0.2400E+02

CONCENTRATED (DOWNWARD) LATERAL FORCES
   NODE           F
     2         0.4000E+01

DISTRIBUTED (DOWNWARD) LATERAL FORCES
  ELEMENT          W
     2         0.1600E+01
     3         0.4000E+00

NODES AT WHICH BEAM IS SIMPLY SUPPORTED :    1    3

COMPUTED SHEAR FORCES AND BENDING MOMENTS AT THE NODES OF EACH ELEMENT
 ELEM  NODES      SF AT I        SF AT J        BM AT I        BM AT J
   1   1   2    0.8667E+01     0.8667E+01     0.0000E+00     0.5200E+02
   2   2   3    0.4667E+01    -0.1453E+02     0.5200E+02    -0.7200E+01
   3   3   4    0.2400E+01     0.0000E+00    -0.7200E+01     0.0000E+00

COMPUTED (DOWNWARD) REACTION FORCES AT THE SIMPLE SUPPORTS
   NODE      FORCE
     1    -0.8667E+01
     3    -0.1693E+02

MATHEMATICAL MAXIMUM OR MINIMUM VALUES OF BENDING MOMENT
  AXIAL POSITION        MOMENT
     0.8917E+01       0.5881E+02
     0.2400E+02       0.4768E-05

GREATEST SHEAR FORCE =  0.1453E+02   AT X =  0.1800E+02
GREATEST SAGGING BENDING MOMENT =  0.5881E+02   AT X =  0.8917E+01
GREATEST HOGGING BENDING MOMENT = -0.7200E+01   AT X =  0.1800E+02
```

Fig. 5.43.

EXAMPLE 5.10

Consider again the beam shown in Fig. 5.28, using program SDBEAM to find the distributions of shear force and bending moment.

The beam is redrawn in Fig. 5.44 with the simplest possible arrangement of three nodes and two elements shown numbered in the orders assumed by the program. Node 1 is again chosen as the origin for global coordinate X.

This is an example with a distributed lateral force of varying intensity acting on the beam, which we know cannot be modeled exactly using the present finite element approach. We must assume a distributed lateral force of constant intensity over each element, and we therefore take the actual force intensity at the center of each element as the value applicable to the entire element. The centers of elements 1 and 2 are at $X = 1$ m and 2.5 m, at which $w = 10$ kN/m and 2.5 kN/m, respectively. Figure 5.45 shows the file of input data used to set up this problem for solution by program SDBEAM, meter and kN units being used for length and force. Note that the concentrated force of 10 kN at node 2 is in the upward direction, and therefore negative. The file RESULTS produced by the program is shown in Fig. 5.46.

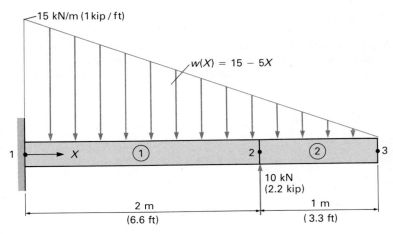

Fig. 5.44.

```
3
0.  2.  3.
1
2  -10.
0
2
1  10. ,  2  2.5
0  1
1
```

Fig. 5.45.

```
STATICALLY DETERMINATE BEAM

COORDINATES OF THE NODES
NODE     X          NODE     X         NODE     X        NODE     X
 1  0.0000E+00    2  0.2000E+01     3  0.3000E+01

CONCENTRATED (DOWNWARD) LATERAL FORCES
   NODE          F
    2          -0.1000E+02

DISTRIBUTED (DOWNWARD) LATERAL FORCES
   ELEMENT        W
     1          0.1000E+02
     2          0.2500E+01

NODES AT WHICH BEAM IS BUILT IN :    1

COMPUTED SHEAR FORCES AND BENDING MOMENTS AT THE NODES OF EACH ELEMENT
 ELEM  NODES       SF AT I         SF AT J         BM AT I         BM AT J
   1    1    2    0.1250E+02    -0.7500E+01    -0.6250E+01    -0.1250E+01
   2    2    3    0.2500E+01     0.0000E+00    -0.1250E+01     0.0000E+00

COMPUTED (DOWNWARD) REACTION FORCES AT THE BUILT-IN SUPPORTS
   NODE      FORCE
    1      -0.1250E+02

COMPUTED (COUNTERCLOCKWISE) MOMENTS AT THE BUILT-IN SUPPORTS
   NODE     MOMENT
    1      0.6250E+01

MATHEMATICAL MAXIMUM OR MINIMUM VALUES OF BENDING MOMENT
   AXIAL POSITION        MOMENT
      0.1250E+01      0.1563E+01
      0.3000E+01      0.0000E+00

GREATEST SHEAR FORCE = 0.1250E+02   AT X =  0.0000E+00
GREATEST SAGGING BENDING MOMENT = 0.1563E+01  AT X = 0.1250E+01
GREATEST HOGGING BENDING MOMENT = -0.6250E+01  AT X = 0.0000E+00
```

Fig. 5.46.

Comparing the results with those shown in Fig. 5.30, it is interesting to note that while the shear forces at the nodal points have been computed exactly, the bending moments certainly have not. That the shear forces are exact is due to the fact that, by taking the average intensity of distributed force over each element as the actual value at its center, for this linearly varying force distribution the correct total force on each element is obtained. The same exactness would not be obtained for a simply supported beam, however, because the reaction forces at the supports cannot be obtained without using moment equilibrium. The computed greatest sagging and hogging bending moments of $+1.563$ kNm and -6.250 kNm compare very unfavorably with the exact values of $+3.333$ kNm and -2.5 kNm. We conclude that we must use more elements to better represent the distributed lateral force.

Figures 5.48 and 5.49 show the input data and results obtained when using six elements, of constant length 0.5 m, along the beam as in Fig. 5.47. Once again, the constant intensity of distributed force over each element is taken as the actual value at its center. While the computed shear forces at the nodes remain exact, the values of the greatest sagging and hogging bending moments have improved to $+3.125$ kNm and -2.813 kNm, respectively. Note also that the position of the greatest sagging moment, which is a mathematical maximum, is now determined exactly. This is only because we happened to have located a node at this point, which also explains why two mathematical maxima are show there – one at the end of each of the elements which meet at the node concerned.

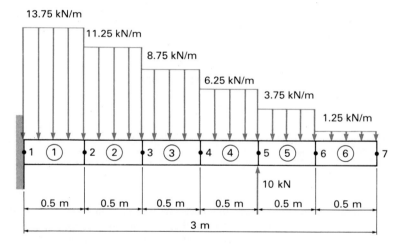

Fig. 5.47.

```
7
0.  0.5  1.  1.5   2.   2.5   3.
1
5  -10.
0
6
1  13.75,  2  11.25,  3  8.75,  4  6.25,  5  3.75,  6  1.25
0  1
1
```

Fig. 5.48.

STATICALLY DETERMINATE BEAM

```
COORDINATES OF THE NODES
NODE     X          NODE     X          NODE     X          NODE     X
  1  0.0000E+00       2  0.5000E+00       3  0.1000E+01       4  0.1500E+01
  5  0.2000E+01       6  0.2500E+01       7  0.3000E+01

CONCENTRATED (DOWNWARD) LATERAL FORCES
   NODE          F
     5        -0.1000E+02

DISTRIBUTED (DOWNWARD) LATERAL FORCES
   ELEMENT        W
      1        0.1375E+02
      2        0.1125E+02
      3        0.8750E+01
      4        0.6250E+01
      5        0.3750E+01
      6        0.1250E+01

NODES AT WHICH BEAM IS BUILT IN :    1

COMPUTED SHEAR FORCES AND BENDING MOMENTS AT THE NODES OF EACH ELEMENT
   ELEM  NODES      SF AT I        SF AT J        BM AT I        BM AT J
    1   1   2    0.1250E+02     0.5625E+01    -0.2813E+01     0.1719E+01
    2   2   3    0.5625E+01     0.0000E+00     0.1719E+01     0.3125E+01
    3   3   4    0.0000E+00    -0.4375E+01     0.3125E+01     0.2031E+01
    4   4   5   -0.4375E+01    -0.7500E+00     0.2031E+01    -0.9375E+00
    5   5   6    0.2500E+01     0.6250E+00    -0.9375E+00    -0.1563E+00
    6   6   7    0.6250E+00     0.0000E+00    -0.1563E+00     0.0000E+00

COMPUTED (DOWNWARD) REACTION FORCES AT THE BUILT-IN SUPPORTS
   NODE        FORCE
     1       -0.1250E+02

COMPUTED (COUNTERCLOCKWISE) MOMENTS AT THE BUILT-IN SUPPORTS
   NODE        MOMENT
     1        0.2813E+01

MATHEMATICAL MAXIMUM OR MINIMUM VALUES OF BENDING MOMENT
   AXIAL POSITION          MOMENT
       0.1000E+01        0.3125E+01
       0.1000E+01        0.3125E+01
       0.3000E+01        0.0000E+00

GREATEST SHEAR FORCE =  0.1250E+02    AT X =  0.0000E+00
GREATEST SAGGING BENDING MOMENT =  0.3125E+01    AT X =  0.1000E+01
GREATEST HOGGING BENDING MOMENT = -0.2813E+01    AT X =  0.0000E+00
```

Fig. 5.49.

It appears that further element refinement is necessary if bending moments are to be determined accurately. Table 5.1 shows not only the greatest values obtained from Figs 5.46 and 5.49, but also those from further similar computations using 15 and 30 elements along the beam, in each case using elements of constant length. With 15 elements, the accuracy (relative to the exact solution) is of the order of 1–2%, and with 30 elements this improves to 0.2–0.5%. This progressive improvement of the results with

Table 5.1 Effect of element refinement on computed bending moments

No. elements	Greatest sagging moment (kNm)	% error	Greatest hogging moment (kNm)	% error
2	1.563	53	−6.250	150
6	3.125	6.2	−2.813	12.5
15	3.300	1.0	−2.550	2.0
30	3.325	0.2	−2.512	0.5
exact	3.333	—	−2.500	—

element refinement is typical of approximate finite element methods (where the elements are not capable of representing internal variations exactly). If we did not have the exact solution for comparison, we would have to refine the elements until we were satisfied that there was no further significant change in the results.

251

Stresses due to bending

We see from these two examples that the main use for a computer program such as SDBEAM is for problems where all the details of the beam geometry and loading are defined numerically. It offers particular advantages for more complicated problems (with large numbers of applied forces and couples), where manual methods become increasingly laborious. It is not suitable, however, for problems such as those in Examples 5.1 to 5.3, where results are required in the form of general algebraic expressions for shear force and bending moment. Nor is it very suitable for problems such as Example 5.8, where the positions of the supports are to be varied.

5.3 Stresses due to bending

So far in this chapter, we have been concerned with finding the distributions of shear force and bending moment in statically determinate beams. In Chapter 6 we will establish methods for doing the same for statically indeterminate beams. Irrespective of the type of beam involved, however, shear forces and bending moments represent only the resultants of the internal stress distributions in the beam. In general, both normal and shear stresses are involved, and are associated with the bending moment and shear force, respectively. We must determine these stresses if we are to assess the ability of the beam to support the loads applied to it.

5.3.1 Normal stresses due to bending

In order to find the normal stresses in a beam caused by bending, we start by considering a state of pure bending. We recall from Example 5.3 that such a state exists if the bending moment is constant over at least part of the beam, and the shear force is zero. We must take into account the way in which the beam deforms due to the application of this moment. Figure 5.50 shows the beam, with an initially rectangular grid of lines drawn on it, both before and after deformation has taken place. The amount of deformation involved is exaggerated for clarity. We assume, and for pure bending it is perfectly true, that the lines which were originally horizontal change to being arcs of circles. Similarly, the lines which were originally vertical lie along radii of these circles, and meet at a common point which is the *center of curvature* for the beam.

Figure 5.51a shows a short length of beam in its undeformed state, with two adjacent cross sections marked as A_1A_2 and B_1B_2. After the bending moment M is applied, these become the sections $A_1'A_2'$ and $B_1'B_2'$ shown in Fig. 5.51b. In this deformed state, the piece of beam concerned may well have been bodily displaced and rotated from its original position. The effect of the deformation is to reduce the length of the line A_1B_1 at the top surface of the beam, and to

(a)

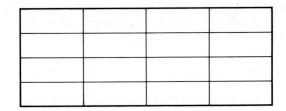

(b)

Fig. 5.50.

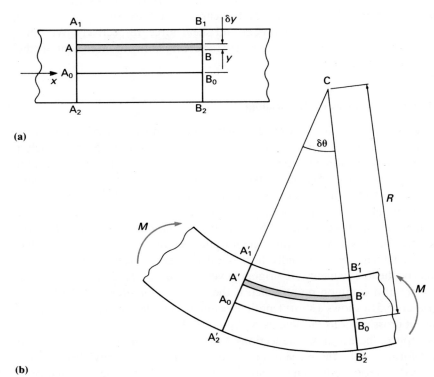

(a)

(b)

Fig. 5.51.

increase the length of the line A_2B_2 at its bottom surface. We can therefore expect that there is a line, A_0B_0, at some intermediate position whose length does not change. This line lies in the *neutral surface* of the beam, which extends not only along the length of the beam but also across the width of the beam, at right angles to the view shown. The neutral surface is the surface at which there is no axial strain due to bending. Although we introduced the vertical coordinate for a beam as y in Section 5.2.1, we deferred defining a precise origin for it. This we can now do, by choosing to define the origin at the neutral surface of the beam.

Let us now formally list our assumptions, including those we have already made concerning the deformation of the beam.

1 Plane cross sections remain plane: lines $A_1'A_2'$ and $B_1'B_2'$ remain straight after bending.
2 Longitudinal lines, such as A_0B_0, A_1B_1 and A_2B_2, all bend to circular arcs with the same center of curvature, the point C.
3 Lateral normal stresses σ_y and σ_z are negligibly small compared with the axial stress σ_x, implying a state of uniaxial normal stress everywhere in the beam.
4 The curvature of the beam is small: the distance of the center of curvature, C, from the beam is large compared to the depth of the beam (in Fig. 5.51b the curvature is exaggerated for clarity).
5 The beam material remains linearly elastic.
6 The Young's modulus of elasticity, E, of the beam material is the same throughout the beam, and is also the same in tension and compression. In Section 5.3.6, we will see how to deal with composite beams made from more than one material.

In order to proceed, we first define the radius R shown in Fig. 5.51b as the *radius of curvature of the neutral surface*, and the angle subtended by the two cross sections at the center of curvature as $\delta\theta$. Now, let us consider the typical longitudinal element of material marked as AB in Fig. 5.51a, which is at a height of y above the neutral surface, and which deforms to the arc $A'B'$ in Fig. 5.51b. From the geometry of deformation, we can deduce that the longitudinal strain in this element is

$$e_x = \frac{A'B' - A_0B_0}{A_0B_0}$$

where

$$A_0B_0 = R\delta\theta \quad \text{and} \quad A'B' = (R - y)\delta\theta$$

and

$$e_x = \frac{(R - y)\delta\theta - R\delta\theta}{R\delta\theta} = \frac{-y}{R} \tag{5.29}$$

Since we have assumed that a state of uniaxial normal stress exists everywhere, we can obtain an expression for the longitudinal stress from the stress–strain relationship, equation (3.5), as

$$\sigma_x = e_x E = \frac{-yE}{R} \qquad (5.30)$$

This is the normal stress due to bending, often referred to as the *bending stress*. We note from equations (5.29) and (5.30) that both the bending stress and corresponding strain vary linearly with position y through the depth of the beam, both being zero at the neutral surface. The presence of the negative signs in these equations is a consequence of our choice of sign conventions for coordinates and bending moments: a positive (sagging) bending moment which causes the type of deformation shown in Fig. 5.51b results in the upper part of the beam, where y is positive, being put into compression. Figure 5.52 shows the form of bending stress distribution obtained in a beam of rectangular cross section.

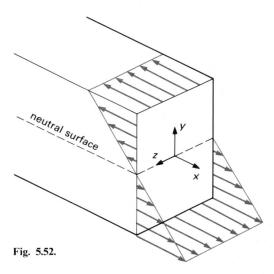

Fig. 5.52.

Using the assumed geometry of deformation and the stress–strain relationship, we have determined the form of the bending stress distribution. However, we have not as yet related the magnitudes of these stresses to the applied bending moment, nor have we defined the position of the neutral surface. We must now consider equilibrium of both forces and moments applied to the beam cross section. In the absence of any external axial (longitudinal) force applied to the beam, the resultant of the bending stress distribution at any beam cross section, such as that shown in Fig. 5.52, must be zero. In the case of pure bending there is certainly no external axial force acting on the beam, and the same is true of many other bending problems of practical interest. The combination of bending and axial loading is considered in Section 5.4. For equilibrium of moments, the stress distribution at the beam cross section, which is equivalent to a couple, must have a resultant moment which is equal in magnitude to the bending moment at that section.

Let us examine first a beam whose cross section need not be simple in shape, such as that shown in Fig. 5.52, but which is nevertheless symmetrical about the plane of bending (the plane of Fig. 5.51). Figure 5.53b shows such a cross section (which is symmetrical about the plane $z = 0$), while Fig. 5.53a shows a side view of the same beam equivalent to Fig. 5.51a, and Fig. 5.53c shows the linear bending stress distribution. The line of intersection of the neutral surface with the beam cross section is known as the *neutral axis*. Consider the small shaded strip in the y, z plane of the cross section, which is at a distance y above the neutral axis, is of thickness δy in this direction, and is of width b in the z direction. In general, this width is a function of y. The longitudinal force (normal to the cross section) acting on this small strip is equal to the bending stress acting on it multiplied by its area, or $\sigma_x b \, \delta y$. The total longitudinal force acting on the cross section can be found by summing this expression over all such strips which make up the cross-sectional area

$$\text{Total force} = \sum \sigma_x b \delta y = \int_{-c_2}^{+c_1} \sigma_x b(y) \, dy$$

(a)   (b)   (c)

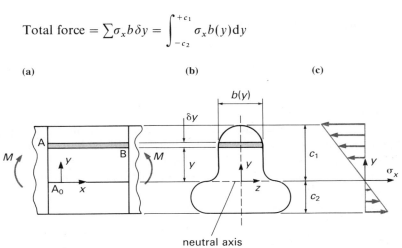

Fig. 5.53.

where c_1 and c_2 are the vertical distances from the neutral axis to the highest and lowest points of the cross section, respectively. Given the distribution of bending stress from equation (5.30), this becomes

$$\text{Total force} = \int_{-c_2}^{+c_1} \frac{-yE}{R} b(y) \, dy$$

from which we deduce that, since both E and R are independent of y

$$\int_{-c_2}^{+c_1} b(y) \, y \, dy = 0 \qquad (5.31)$$

The integral on the left-hand side of this equation is the *first moment of area* of the cross section about the neutral axis. Moments of area are discussed in

Appendix C, where it is shown that for this first moment to be zero the neutral axis must pass through the center of area or *centroid* of the cross section. Also, because the cross section has a vertical axis of symmetry, the centroid must lie on this axis.

We have therefore defined the position of the neutral axis. For the rectangular beam cross section shown in Fig. 5.52, it passes through the geometric center, and $c_1 = c_2$. The same is not true, however, for the cross section shown in Fig. 5.53b. The moment, according to the sagging positive sign convention, of the longitudinal force acting on the shaded strip in this view about the neutral axis is $-(\sigma_x b\delta y)y$, the negative sign being due to the fact that a tensile bending stress at a positive value of y would give a moment in the hogging sense. The total moment can be found by summing this expression over all such strips which make up the cross-sectional area

$$\text{Total moment} = \sum - \sigma_x by\delta y = \int_{-c_2}^{+c_1} - \sigma_x b(y)y \, dy = \int_{-c_2}^{+c_1} \frac{yE}{R} b(y)y \, dy$$

and, since this must be equal to the bending moment, M, at the cross section, we have

$$M = \frac{E}{R} \int_{-c_2}^{+c_1} b(y)y^2 \, dy \qquad (5.32)$$

Now, the integral in this equation is the *second moment of area* of the cross section *about the neutral axis*, which is given the symbol I. It is quite commonly referred to as the *moment of inertia* of the cross section, although the units are $(\text{length})^4$ rather than the mass $\times$ $(\text{length})^2$ required for a true moment of inertia. The reason for the confusion is that the moment of inertia of a thin lamina of uniform thickness whose shape is that of the cross section is mI, where m is the mass per unit area of the lamina.

Although we derived equation (5.32) by taking moments about the neutral axis, we would have obtained exactly the same result if we had taken moments about any other axis parallel to the neutral axis. For example, taking moments about the bottom edge of the cross section in Fig. 5.53b

$$M = \int_{-c_2}^{+c_1} \frac{yE}{R} b(y)(y + c_2) \, dy$$

$$= \frac{E}{R} \int_{-c_2}^{+c_1} b(y)y^2 \, dy + \frac{Ec_2}{R} \int_{-c_2}^{+c_1} b(y)y \, dy$$

According to equation (5.31), the second of these two integrals is zero, leaving equation (5.32) for the bending moment. What this means physically is that, because the bending stress distribution has a zero resultant force (in the longitudinal direction), it must be equivalent to a couple, which has the same moment about any axis. Figure 5.54 shows a view of the bending stress distribution over the beam cross section shown in Fig. 5.53b. If we consider the separate parts of this distribution above and below the neutral axis, each has a

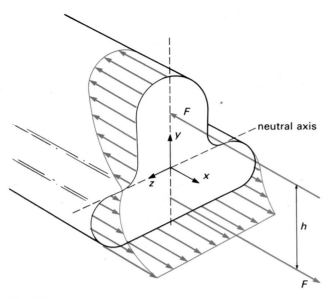

Fig. 5.54.

resultant of the same magnitude, say F, but in opposite directions. If the vertical distance between the lines of action of these forces is h, then a couple of moment Fh is created, which must be equal to the bending moment, M.

Equation (5.32) can be written as

$$M = \frac{E}{R} I$$

and, combining this result with equation (5.30), we have

$$\frac{M}{I} = \frac{E}{R} = \frac{\sigma_x}{-y} \qquad (5.33)$$

These equations are the main results from the simple theory of bending we have developed. Although they have been derived for pure bending of a beam, and provide a very good model for pure bending of beams and other components, they can also be applied without serious loss of accuracy to many situations where there is lateral loading. A significant effect of lateral loading is, as we shall see in Section 5.3.2, to introduce shear stresses in the beam. So, by using equations (5.33) even when there is such loading, we are *assuming that the distribution of normal bending stresses is not affected by the shear stresses.* Provided the component concerned is a true beam, with cross-sectional dimensions much smaller than its length, the assumption is a good one. More exact treatments for beam bending, in which this assumption is not made, are discussed in Chapter 10, Section 10.3.

The distribution of bending stress in a beam, which was first obtained in equation (5.30), can now be expressed in terms of the bending moment with the aid of equations (5.33) as

$$\sigma_x = \frac{-yM}{I} \tag{5.34}$$

In order to assess the strength of a component in bending, we need to know the maximum bending stresses. These occur at the points in the cross section furthest from the neutral axis, which are at its upper and lower surfaces. Although many beams used in practice are symmetrical about the neutral axis (with $c_1 = c_2$ in Fig. 5.53b), so that the maximum tensile and maximum compressive bending stresses are equal in magnitude, this is not always the case. If c is the larger of the two dimensions c_1 and c_2, the magnitude of the maximum stress is given by

$$\sigma_{max} = \left| \frac{cM}{I} \right| \tag{5.35}$$

Whether this is tensile or compressive depends on both the form of the cross section and the sign of the bending moment. The distinction is only likely to be important for materials which have a different maximum allowable stress in tension and compression, such as brittle materials like cast iron and concrete (typical properties are given in Appendix A: concrete is usually assumed to have no tensile strength at all). Under these circumstances, we must examine both the magnitude and the sign of the maximum stresses at the upper and lower surfaces. The maximum bending stress given by equation (5.35) is the maximum at the particular beam cross section. The greatest value anywhere along a beam of constant cross section is found by considering the section at which the greatest bending moment occurs. If we need to distinguish between the greatest tensile and compressive stresses, we may need to consider not just the bending moment of greatest magnitude, but also the greatest sagging and hogging moments separately. It is for this reason that in Section 5.2, and in program SDBEAM, we found both of these values. If the cross section varies, then I, c and M all vary with position along the beam in equation (5.35), and the greatest bending stress is more difficult to determine.

We note that in equation (5.35) the ratio I/c depends only upon the geometry of the beam cross section. It is referred to as the *elastic section modulus, S*

$$S = \frac{I}{c} \tag{5.36}$$

and the magnitude of the maximum bending stress is then given by

$$\sigma_{max} = \left| \frac{M}{S} \right| \tag{5.37}$$

Before we can start to solve problems, we need to know how to calculate second moments of area of beam cross sections. While methods for doing this for relatively complex shapes are described in Section 5.3.3, let us consider here the straightforward case of a beam of rectangular cross section shown in Fig. 5.55a. The constant breadth and depth of the beam are b and d, respectively, and the neutral axis passes through the geometric center of the section. The second moment of area about the neutral axis is therefore

$$I = \int_{-d/2}^{+d/2} b y^2 \, \mathrm{d}y = \left[b \frac{y^3}{3} \right]_{-d/2}^{+d/2} = \frac{bd^3}{12} \tag{5.38}$$

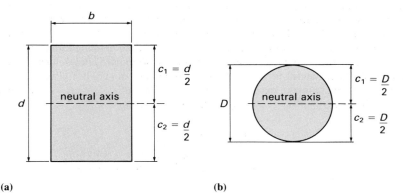

Fig. 5.55.

Also, the elastic section modulus is

$$S = \frac{I}{c} = \frac{2I}{d} = \frac{bd^2}{6} \tag{5.39}$$

Since the maximum bending stress is inversely proportional to S, it is also inversely proportional to the breadth, b, of the beam, but inversely proportional to the *square* of its depth. Increasing the depth of a rectangular beam is a more effective way of increasing its resistance to failure than increasing the breadth (the same is true to even greater extent for the beam stiffness, which we will find in Chapter 6 is proportional to I). It is for this reason that, for example, the floor joists shown in Fig. 5.1 are arranged with the larger of their two cross-sectional dimensions vertical.

Another commonly-occurring case is that of a beam of circular cross section, shown in Fig. 5.55b with a diameter of D. Once again, the neutral axis passes through the geometric center of the cross section, and its second moment of area about this axis is derived in Appendix C as

$$I = \frac{\pi D^4}{64} \tag{5.40}$$

The elastic section modulus is therefore

$$S = \frac{I}{c} = \frac{2I}{D} = \frac{\pi D^3}{32}$$

(5.41)

Results for other relatively simple shapes are given in Appendix C, Table C.1.

EXAMPLE 5.11

A timber floor joist is 6 m (20 ft) long and is simply supported at its ends. It has a rectangular cross section with a depth of 300 mm (12 in) and breadth of 100 mm (4 in), and is subjected to a uniformly distributed lateral force of intensity 2.5 kN/m (170 lb/ft), due to floor loading and the weight of the joist itself. Find the maximum bending stress.

This is a straightforward example of the use of equation (5.35) to find maximum bending stresses. For a beam of uniform cross section, the maximum bending stress occurs at the cross section which carries the greatest absolute bending moment. Since the joist, which is of length $L = 6$ m (20 ft), is simply supported and subjected to a uniformly distributed lateral force of intensity $w = 2.5$ kN/m (170 lb/ft), we may use Fig. 5.19 to find this moment, occurring at the mid point, as

SI units

$$M = \frac{wL^2}{8} = \frac{2.5 \times 6^2}{8} = 11.25 \text{ kNm}$$

From equation (5.38), the second moment of area of the joist cross section about its neutral axis is

$$I = \frac{bd^3}{12} = \frac{0.1 \times 0.3^3}{12} = 2.25 \times 10^{-4} \text{ m}^4$$

and the maximum distance from the neutral axis is $c = 0.15$ m. Equation (5.35) then gives the maximum bending stress as

$$\sigma_{max} = \frac{cM}{I} = \frac{0.15 \times 11.25 \times 10^3}{2.25 \times 10^{-4}} \text{ N/m}^2$$

$$= 7.5 \text{ MN/m}^2$$

With a sagging maximum bending moment, a compressive longitudinal stress of this magnitude occurs at the top surface of the joist at its mid point, and a tensile stress of the same magnitude at the bottom surface.

US customary units

$$M = \frac{wL^2}{8} = \frac{170 \times 20^2}{8} \text{ lb ft} = 8.5 \text{ kip-ft}$$

From equation (5.38), the second moment of area of the joist cross section about its neutral axis is

$$I = \frac{bd^3}{12} = \frac{4 \times 12^3}{12} = 576 \text{ in}^4$$

and the maximum distance from the neutral axis is $c = 6$ in. Equation (5.35) then gives the maximum bending stress as

$$\sigma_{max} = \frac{cM}{I} = \frac{6 \times 8.5 \times 12}{576}$$

$$= 1.06 \text{ ksi}$$

With a sagging maximum bending moment, a compressive longitudinal stress of this magnitude occurs at the top surface of the joist at its mid point, and a tensile stress of the same magnitude at the bottom surface.

5.3.2 Shear stresses due to bending

We have seen that the bending moment at a particular cross section of a beam is the resultant of the bending stress distribution there. Similarly, the shear force is the resultant of the shear stress distribution, as indicated in Fig. 5.56. If we

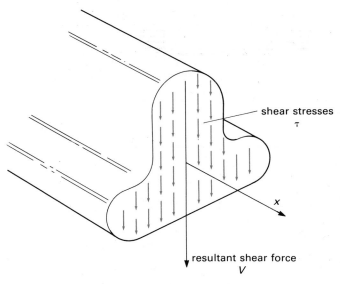

shear stresses
τ

x

resultant shear force
V

Fig. 5.56.

assume that the shear stress is uniformly distributed over the cross section, which is of area A, we can express this stress as $\tau = V/A$. However, it turns out that this assumption is not a very realistic one, and we must investigate the problem in rather more detail.

Shear stresses exist, not only over the plane of the cross section normal to the axis of the beam, but also on horizontal planes parallel to the axis. To be convinced of the truth of this statement, we need to consider a small element of material removed from, say, the top surface of the beam, as shown in Fig. 5.57. The length of the element is δx, and its lower surface is parallel to the neutral surface and at a height h above it. The element, which is shown shaded in Fig. 5.57a, and as a free body in Fig. 5.57b, is subjected to bending stress distributions over each of its end faces, and a shear stress distribution over its lower surface. Now, the bending stress, σ_x, is a function of y, the distance from the neutral surface, and in general is also a function of axial distance, x, along the beam. If the bending stress does vary with x, then in order for the element to be in equilibrium in the axial direction, a shear stress, say τ', must be applied to the lower surface of this element as shown. Indeed, we can express this equilibrium condition as

$$\int_h^{c_1} \sigma_x(x+\delta x, y)b(y)\,\mathrm{d}y - \int_h^{c_1} \sigma_x(x, y)b(y)\,\mathrm{d}y + \tau'b(h)\delta x = 0$$

where $b(y)$ is the breadth of the beam at a distance y from the neutral axis. Using equation (5.34), the bending stresses at the ends of the element are given by

$$\sigma_x(x, y) = -\frac{yM(x)}{I} \quad \text{and} \quad \sigma_x(x+\delta x, y) = -\frac{yM(x+\delta x)}{I}$$

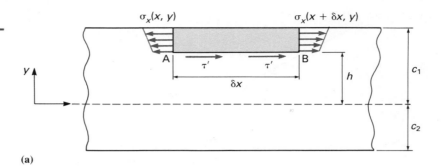

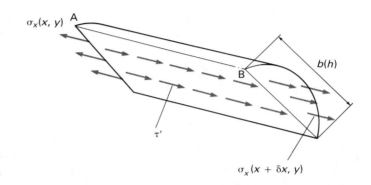

(b)

Fig. 5.57.

and

$$\tau'b(h) = \frac{[M(x + \delta x) - M(x)]}{\delta x} \frac{1}{I} \int_h^{c_1} yb(y)\,dy$$

Now, the integral in this result is the first moment of area about the neutral axis of the region of the cross section above the line $y = h$, and may be represented by the symbol Q

$$Q(h) = \int_h^{c_1} yb(y)\,dy \tag{5.42}$$

In view of equation (5.31), it could equally well be defined as the first moment of area about the neutral axis of the region *below* the line $y = h$. As the length of the element, δx, tends to zero, the expression involving the bending moments at each of its ends, $[M(x + \delta x) - M(x)]/\delta x$, becomes the derivative dM/dx, which according to equation (5.11) is equal to the shear force, V. Hence

$$\tau' = \frac{V(x)Q(h)}{Ib(h)}$$

$$\tau'(x, y) = \frac{V(x)Q(y)}{Ib(y)} \tag{5.43}$$

We note that, at $y = c_1$ and $y = -c_2$, $Q(y) = 0$, and therefore the shear stress is also zero. This is to be expected because, at the upper and lower surfaces of the beam, there can be no shear stress on the exposed surfaces. We also note that $Q(y)$ has its maximum value at the neutral axis, which is therefore likely to be the position of maximum shear stress. So, the region of the cross section close to the neutral axis, although it does not carry large bending stresses, does have an important role to play in carrying shear stresses. If the beam is subjected to pure bending, with no lateral loading, then the shear force is zero (Figs 5.20 and 5.23), and there are no shear stresses.

The importance of shear stresses in beams can be demonstrated in practical terms by the two beam arrangements shown in Fig. 5.58. In Fig. 5.58a the beam is a continuous solid, while in Fig. 5.58b it has the same dimensions but is made up of flat strips which are not bonded together. When the latter is subjected to lateral loading, no shear stresses are transmitted across the interfaces between the strips, which consequently slide relative to one another. Evidence for this sliding is provided by the fact that the ends of the strips take up the stepped formation shown: the plane ends do not remain plane as in the case of the solid beam. The result is that, for a given load, the compound beam deflects much more than the solid one. This effect may be desirable, and is utilized in, for example, the design of leaf springs.

(a)

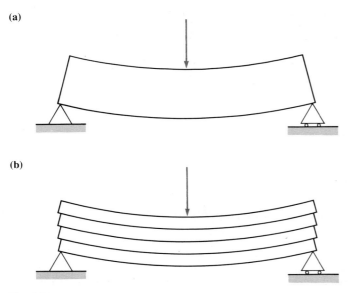

(b)

Fig. 5.58.

So far we have treated shear stresses on vertical cross-sectional planes of a beam and those on longitudinal planes parallel to the neutral surface (τ and τ' in Figs 5.56 and 5.57) as though they are independent quantities. We can demonstrate a direct link between them with the aid of Fig. 5.59, which shows a small element of a beam with dimensions δx, δy and b in the x, y and z directions, respectively. The stresses acting on the faces of this element are shown in Fig. 5.59b. For equilibrium of moments acting about the center, O, of the element, we require that

$$[\tau(x, y) + \tau(x + \delta x, y)]\,\delta y\, b\, \frac{\delta x}{2} = [\tau'(x, y) + \tau'(x, y + \delta y)]\,\delta x\, b\, \frac{\delta y}{2}$$

(a)

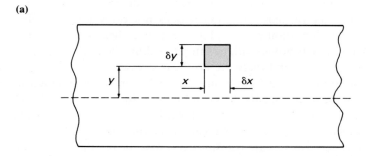

(b)

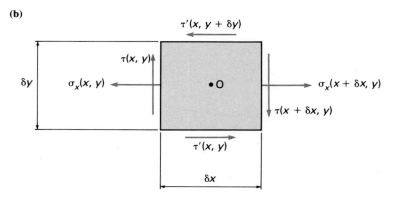

Fig. 5.59.

and as δx and δy tend to zero this simplifies to

$$\tau(x, y) = \tau'(x, y) \tag{5.44}$$

This means that equation (5.43) also defines the shear stress distribution over the cross section of a beam.

Let us consider the particular case of a beam with a rectangular cross section of depth d and breadth b. Using equation (5.42)

$$Q(h) = b \int_h^{d/2} y \, dy = \frac{b}{2}\left[\left(\frac{d}{2}\right)^2 - h^2\right]$$

and the shear stress distributions at a particular beam cross section are therefore given by equation (5.43) as

$$\tau(y) = \tau'(y) = \frac{V}{2I}\left[\left(\frac{d}{2}\right)^2 - y^2\right] = \frac{6V}{bd^3}\left[\left(\frac{d}{2}\right)^2 - y^2\right] \qquad (5.45)$$

We note that the distributions are parabolic, as shown in Fig. 5.60, going to zero at the upper and lower surfaces of the beam, and with a maximum at the neutral surface, where $y = 0$, of

$$\tau_{max} = \tau'_{max} = \frac{3}{2}\frac{V}{bd} \qquad (5.46)$$

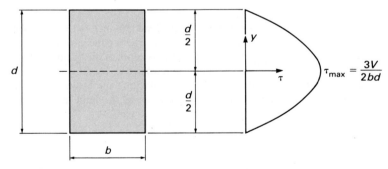

Fig. 5.60.

In other words, the maximum value of shear stress is 50% greater than the average value obtained by dividing the shear force by the cross-sectional area of the beam.

Shear stresses in typical beams, whose lengths are large compared to their cross-sectional dimensions, are generally much less likely to cause failure than the bending stresses. This is because they are substantially smaller in magnitude than the bending stresses, with maximum values which occur in regions of low bending stress. We can confirm this assertion with the aid of the following two examples.

EXAMPLE 5.12

A timber floor joist is 6 m (20 ft) long and is simply supported at its ends. It has a cross section with a depth of 300 mm (12 in) and breadth of 100 mm (4 in), and is subjected to a uniformly distributed lateral force of intensity 2.5 kN/m (170 lb/ft), due to floor loading and the weight of the joist itself. Find the maximum shear stress.

The beam and its loading are the same as in Example 5.11, and we now wish to find the maximum shear stress rather than the maximum bending stress. For a beam of uniform cross section, the maximum shear stress occurs at the cross section which carries the greatest absolute shear force. Since the joist, which is of length $L = 6$ m (20 ft), is simply supported and subjected to a uniformly distributed lateral force of intensity $w = 2.5$ kN/m (170 lb/ft), we may use Fig. 5.19 to find this force, occurring at either end of the beam, as

SI units

$$V = \frac{wL}{2} = \frac{2.5 \times 6}{2} = 7.5 \text{ kN}$$

Equation (5.46) then gives the maximum shear stress in the beam as

$$\tau_{max} = \frac{3}{2} \frac{7.5 \times 10^3}{0.1 \times 0.3} \text{ N/m}^2 = \underline{0.375 \text{ MN/m}^2}$$

This compares with a maximum bending stress of 7.5 MN/m², which is twenty times greater.

US customary units

$$V = \frac{wL}{2} = \frac{170 \times 20}{2} = 1700 \text{ lb}$$

Equation (5.46) then gives the maximum shear stress in the beam as

$$\tau_{max} = \frac{3}{2} \frac{1700}{4 \times 12} = \underline{53.1 \text{ psi}}$$

This compares with a maximum bending stress of 1060 psi, which is twenty times greater.

EXAMPLE 5.13

Figure 5.61 shows a cantilever beam of negligible weight and length L with a uniform rectangular cross section which is of depth d. If the beam carries a concentrated lateral force of magnitude F at its free end, find the ratio of length to depth of the beam at which the maximum bending and shear stresses are equal in magnitude.

From Fig. 5.21 we see that the shear force is constant along the beam, while the bending moment varies linearly and takes its greatest value at the built-in end. The magnitudes of the greatest shear force and bending moment are

$$V = F \quad \text{and} \quad M = FL$$

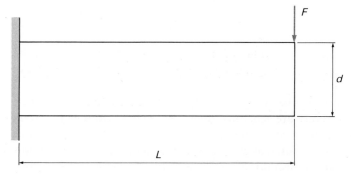

Fig. 5.61.

Since the beam is of uniform cross section, the maximum bending and shear stresses occur at cross sections which carry the greatest bending moment and shear force, respectively. Equation (5.35) gives the maximum bending stress as

$$\sigma_{max} = \frac{dM}{2I} = \frac{FLd}{2I}$$

and equation (5.46) gives the maximum shear stress as

$$\tau_{max} = \frac{V}{2I}\left(\frac{d}{2}\right)^2 = \frac{Fd^2}{8I}$$

Equating these two expressions for maximum stresses, we find that

$$\frac{L}{d} = \frac{1}{4}$$

which demonstrates that shear stresses are generally only significant in magnitude in very short beams. We should note, however, that with such a short beam the simple theory of bending, on which the expressions for maximum stresses are based, no longer provides an accurate description of beam behavior.

5.3.3 Beams of various cross sections

So far we have considered only beams of very simple cross section, mainly solid rectangular. Although such sections are used in practice, particularly for timber beams which have to be cut from solid, they do not represent very economical use of materials such as steel and concrete which can be fabricated, and whose densities are high enough to make the weight of large beams a significant source of lateral loading.

One way to improve the performance of a rectangular section is to make it hollow, as shown in Fig. 5.62. Beams with this form of cross section are often referred to as *box girders*, and are widely used in engineering structures. Provided both the inner and outer rectangles forming this section are symmetrical about its neutral axis, we can find the second moment of area about this axis by subtracting the second moment of the inner rectangle from that of the outer one. Using equation (5.38)

$$I = \frac{1}{12}(b_2 d_2^3 - b_1 d_1^3) \tag{5.47}$$

and the elastic section modulus is

$$S = \frac{2I}{d_2} = \frac{1}{6d_2}(b_2 d_2^3 - b_1 d_1^3) \tag{5.48}$$

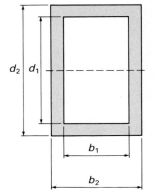

Fig. 5.62.

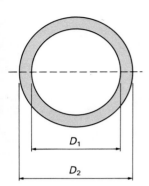

Fig. 5.63.

As an example, if $b_1 = 0.8b_2$ and $d_1 = 0.8d_2$, the reduction in the cross-sectional area of material in the beam relative to a solid beam with the same external dimensions is 64%. On the other hand, the reduction in both the second moment of area and the elastic section modulus is only 41%.

Another commonly-occurring example of a hollow sectioned beam is a pipe of uniform wall thickness, as in Fig. 5.63. The second moment of area about the neutral axis, and the elastic section modulus, are given by

$$I = \frac{\pi}{64}(D_2^4 - D_1^4) \tag{5.49}$$

and

$$S = \frac{2I}{D_2} = \frac{\pi}{32 D_2}(D_2^4 - D_1^4) \tag{5.50}$$

Returning to the hollow rectangular section shown in Fig. 5.62, the reason why it represents a relatively efficient use of material is that a comparatively large proportion of the material is concentrated at the maximum distance from the neutral axis, in the two horizontal walls. Another essentially similar form of cross section which offers the same benefit is that shown in Fig. 5.64a. Sections of this type are usually manufactured in steel by rolling, and are often referred to as *rolled steel joists* or *I sections*. They are manufactured in a wide range of standard sizes, with geometric properties (including second moments of area and elastic section moduli) being available as design data (from, for example, American or British Standards). Two horizontal *flanges* are separated by a relatively thin vertical *web*. The arrangement is usually symmetrical, and hence

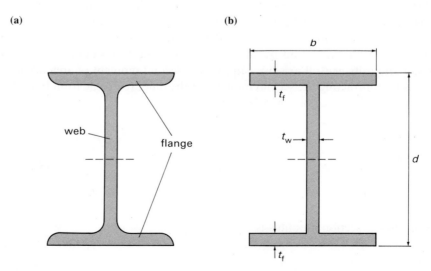

(a)   **(b)**

Fig. 5.64.

with the neutral axis halfway between the upper and lower surfaces of the beam. The main function of the flanges, which concentrate material as far from the neutral axis as possible, is to carry the bending stresses. The web serves to connect the flanges and therefore to carry the shear stresses, which are greatest at the neutral axis.

In order to assess the geometric properties of an I section, it is convenient to simplify its actual shape to make the thicknesses of the flanges and web uniform at t_f and t_w, respectively, as shown in Fig. 5.64b. The overall depth and breadth are d and b. There are at least two ways in which we can find the second moment of area about the neutral axis. Firstly, we can treat the section as a rectangle of depth d and breadth b, from which are subtracted two rectangles of depth $(d - 2t_f)$ and breadth $(b - t_w)/2$ (which are equivalent to a single rectangle of breadth $(b - t_w)$), all rectangles being symmetrical about the neutral axis. Hence

$$I = \frac{bd^3}{12} - \tfrac{1}{12}(b - t_w)(d - 2t_f)^3 \tag{5.51}$$

Alternatively, the section can be divided into three rectangular subregions as shown in Fig. 5.65a. The first of these, which is the web shown in isolation in

(a)

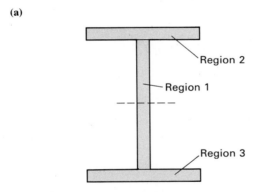

(b)   **(c)**

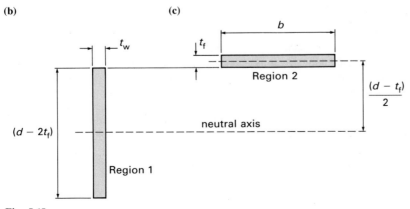

Fig. 5.65.

Fig. 5.65b, is symmetrical about the neutral axis, with a second moment of area of

$$I_1 = \frac{t_w(d - 2t_f)^3}{12}$$

The second and third subregions are the flanges, and are effectively identical, a typical one being shown in isolation in Fig. 5.65c. Although rectangular, this subregion is not symmetrical about the neutral axis, and we must use the parallel axis theorem (Appendix C) to find its second moment of area about the neutral axis of the beam cross section as

$$I_2 = \frac{bt_f^3}{12} + bt_f \frac{(d - t_f)^2}{4}$$

The total second moment of area is found by summing the contributions from the three subregions

$$I = I_1 + 2I_2$$

$$= \frac{t_w(d - 2t_f)^3}{12} + \frac{b}{12}[2t_f^3 + 6t_f(d - t_f)^2]$$

$$= \frac{bd^3}{12} + \frac{t_w(d - 2t_f)^3}{12} + \frac{b}{12}[8t_f^3 - 12t_f^2 d + 6t_f d^2 - d^3]$$

$$= \frac{bd^3}{12} + \frac{t_w(d - 2t_f)}{12} + \frac{b}{12}(2t_f - d)^3$$

$$= \frac{bd^3}{12} - \tfrac{1}{12}(b - t_w)(d - 2t_f)^3$$

which, as expected, is identical to equation (5.51).

EXAMPLE 5.14

The beam in Example 5.4 is of I section, of depth $d = 252$ mm (10 in), breadth $b = 203$ mm (8 in), flange thickness $t_f = 13.5$ mm (0.53 in) and web thickness $t_w = 8.0$ mm (0.32 in). Find the maximum bending and shear stresses in the beam, together with the corresponding stress distributions over its cross section.

All we need from the earlier example in order to find the maximum stresses are the greatest absolute values of shear force and bending moment in the beam, which are 72 kN (16 kip) and 96 kNm (71 kip-ft), respectively. The beam section is a standard one, with relatively wide flanges. Equation (5.51) gives the second moment of area about the neutral axis as

$$I = \frac{203 \times 252^3}{12} - \tfrac{1}{12}(203 - 8)(252 - 2 \times 13.5)^3$$

$$= 85.6 \times 10^6 \text{ mm}^4$$

$$S = \frac{2I}{d} = \frac{2 \times 85.6 \times 10^6}{252} = 680 \times 10^3 \text{ mm}^3$$

Equation (5.37) then gives the maximum bending stress in terms of the greatest bending moment as

$$\sigma_{\max} = \frac{M}{S} = \frac{96 \times 10^3}{680 \times 10^{-6}} \text{N/m}^2 = \underline{141 \text{ MN/m}^2} \text{ (20 ksi)}$$

Bearing in mind that the greatest bending moment is in the sagging sense, Fig. 5.66 shows the linear distribution of bending stresses over the corresponding beam cross section.

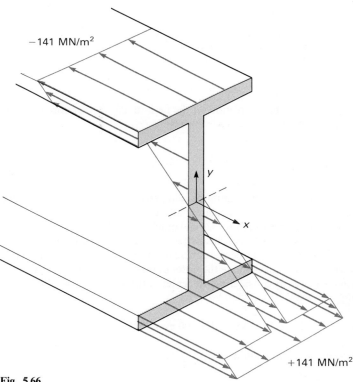

Fig. 5.66.

Figure 5.67 shows the beam cross section with dimensions indicated. The line AB through its center is the neutral axis. The first moment of area about this line, $y = 0$, of the region above it, which can be divided into two rectangles, **ABDC** and the flange region, is

$$Q(0) = (8 \times 112.5) \times 112.5 \times 0.5 + (203 \times 13.5)(112.5 + 13.5 \times 0.5)$$

$$= 50\,635 + 326\,800 = 377\,400 \text{ mm}^3$$

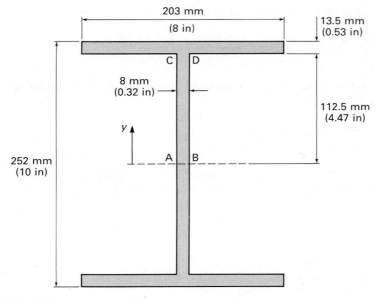

Fig. 5.67.

The maximum shear stress in the y direction, at the neutral axis, is therefore given by equation (5.43) in terms of the greatest shear force as

$$\tau_{max} = \frac{72 \times 10^3 \times 377\,400 \times 10^{-9}}{85.6 \times 10^{-6} \times 0.008}\,\text{N/m}^2 = \underline{39.7\ \text{MN/m}^2}\ (5.8\ \text{ksi})$$

which is small compared with the maximum bending stress. In order to see how the shear stress decreases with y, we can recalculate it at the top of the web, just below the line CD. Here the value of Q is reduced to $326\,800$ mm^3, and hence the shear stress to 34.4 MN/m^2. The analysis, which assumes shear stress to be constant across the width of the section, is not valid at the line CD, because there is no shear stress on the underside of the flange. Nevertheless, we can estimate the order of magnitude of the shear stresses in the y direction in the flange by considering the center of the flange, at $y = 119.25$ mm. Here the value of Q is

$$Q = (203 \times 6.75)(119.25 + 6.75 \times 0.5) = 168\,000\ \text{mm}^3$$

and the shear stress in the y direction is

$$\tau = \frac{72 \times 10^3 \times 168\,000 \times 10^{-9}}{85.6 \times 10^{-6} \times 0.203}\,\text{N/m}^2 = 0.70\ \text{MN/m}^2$$

which is very small.

In the flanges, the main directions of shear stresses acting on the plane of the beam cross section are in fact along the flanges, as shown in Fig. 5.68a. The forms of shear stress variation along the web and flanges are as illustrated in Fig. 5.68b: at the section where the shear force is greatest, the values of shear stress in the y direction in the web at points P and Q have been calculated as 39.7 MN/m^2 and 34.4 MN/m^2, respectively. The shear stresses acting in the z direction along the flanges make no contribution to the resultant shear force on the cross section in the y direction. Indeed, for most purposes we

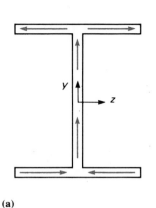

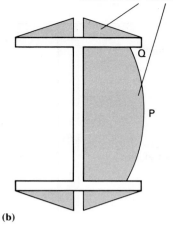

shear stress variations

(a) (b)

Fig. 5.68.

may assume that this entire shear force is supported by the web, and that for such an I section the distribution of shear stress in the web is uniform. In the present example, the average shear stress so calculated is

$$\tau_{ave} = \frac{72 \times 10^3}{0.225 \times 0.008} \, \text{N/m}^2 = 40 \, \text{MN/m}^2 \, (5.8 \, \text{ksi})$$

which is very close to the previously calculated maximum value.

The beam cross sections we have considered so far have been symmetrical, not only about the plane of bending but also about the neutral axis. Figure 5.69 shows an example of a section where this is no longer true, although there is still symmetry about the plane of bending. Before we can find the second moment of area of such a section, we must first locate the neutral axis. We may divide the cross section into three rectangular subregions with areas of $A_1 = b_1 d_1$, $A_2 = b_2 d_2$ and $A_3 = b_3 d_3$, as shown, and (as discussed in Appendix C) take first moments of area about, say, the lower edge EF of the section

$$(A_1 + A_2 + A_3)c_2 = A_1 \frac{d_1}{2} + A_2 \left(d_1 + \frac{d_2}{2} \right) + A_3 \left(d_1 + d_2 + \frac{d_3}{2} \right) \quad (5.52)$$

From this equation c_2, the height of the neutral axis above the lower edge, can be found. We then find the second moment of area of the whole cross section about the neutral axis by summing the second moments of area of the three subregions about this axis, which may be found with the aid of the parallel axis theorem (Appendix C)

$$I = I_1 + I_2 + I_3$$

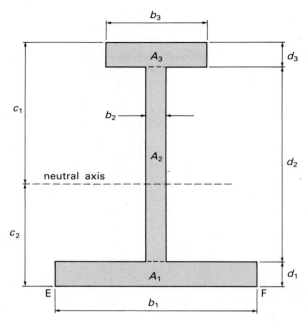

Fig. 5.69.

where

$$I_1 = \frac{b_1 d_1^3}{12} + A_1\left(c_2 - \frac{d_1}{2}\right)^2$$

$$I_2 = \frac{b_2 d_2^3}{12} + A_2\left(d_1 + \frac{d_2}{2} - c_2\right)^2$$

$$I_3 = \frac{b_3 d_3^3}{12} + A_3\left(d_1 + d_2 + \frac{d_3}{2} - c_2\right)^2$$

EXAMPLE 5.15

A water trough 6 m (20 ft) long is simply supported at the ends. It is made of steel plates 600 mm (24 in) wide and 15 mm (0.6 in) thick formed into a V and welded at the bottom as shown in Fig. 5.70. If the trough is completely full of water, determine the maximum tensile and compressive bending stresses at the middle cross section.

The trough may be treated as a simply supported beam subject to a uniformly distributed lateral force (the combined weight of water and steel), but whose cross section is not symmetrical about its neutral axis. From Fig. 5.19, the bending moment at the middle of the beam, which is also the greatest bending moment, is $+ wL^2/8$ (in the sagging sense), where $L = 6$ m (20 ft) is the length of the beam, and w is the intensity of the distributed lateral force.

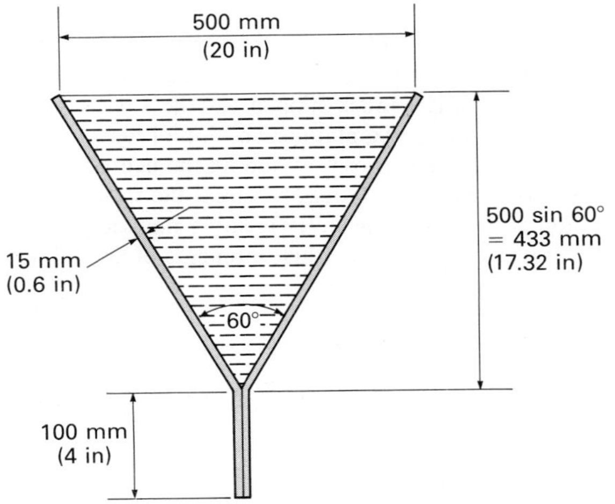

Fig. 5.70.

SI units

The cross-sectional area of the two steel plates is $2 \times 600 \times 15 = 18\,000$ mm². The volume of steel per meter length of the trough is therefore $18\,000 \times 10^{-6}$ m³/m, and, taking the density of steel as 7850 kg/m³ (Appendix A), the weight of steel per unit length is

$$w_{steel} = 18\,000 \times 10^{-6} \times 7850 \times 9.81 = 1386 \text{ N/m}$$

Similarly, the cross-sectional area of water is that of a triangle with a base of 500 mm and a height of 433 mm, which is $0.5 \times 500 \times 433 = 0.10825 \times 10^6$ mm², and the weight of water per unit length of trough is

$$w_{water} = 0.108\,25 \times 1000 \times 9.81 = 1062 \text{ N/m}$$

Note that the weight of the steel contributes more to the beam loading than does the weight of the water. The total distributed lateral force is

$$w = 1386 + 1062 = 2448 \text{ N/m}$$

and the bending moment at the middle section is

$$M = \frac{2448 \times 6^2}{8} = 11\,020 \text{ Nm}$$

Now, in order to find the bending stresses, we must first find the position of the neutral axis of the beam cross section, and then the second moment of area about this axis. In these calculations, we consider only the steel walls of the trough and not the water, which

US customary units

The cross-sectional area of the two steel plates is $2 \times 24 \times 0.6 = 28.8$ in². The volume of steel per foot length of the trough is therefore $28.8 \times 12 = 345.6$ in³/ft, and, taking the density of steel as 0.284 lb/in³ (Appendix A), the weight of steel per unit length is

$$w_{steel} = 345.6 \times 0.284 = 98.15 \text{ lb/ft}$$

Similarly, the cross-sectional area of water is that of a triangle with a base of 20 in and a height of 17.3 in, which is $0.5 \times 20 \times 17.3 = 173$ in², and the weight of water per unit length of trough is

$$w_{water} = 173 \times 12 \times 0.036 = 74.74 \text{ lb/ft}$$

Note that the weight of the steel contributes more to the beam loading than does the weight of the water. The total distributed lateral force is

$$w = 98.15 + 74.74 = 172.9 \text{ lb/ft}$$

and the bending moment at the middle section is

$$M = \frac{172.9 \times 20^2}{8} \text{ lb ft} = 8.645 \text{ kip-ft}$$

Now, in order to find the bending stresses, we must first find the position of the neutral axis of the beam cross section, and then the second moment of area about this axis. In these calculations, we consider only the steel walls of the trough and not the water, which contributes nothing to the bending strength. While we can consider the sloping walls as rectangles inclined to the horizontal, it is much simpler to take advantage of the fact that the horizontal distribution of material is irrelevant for the purposes of finding the neutral axis and second moment of area, and treat the trough cross section as shown in Fig. 5.71. Treating this section as two rectangles, and taking first moments of area about the lower edge

$$(100 \times 30 + 433 \times 34.64)c_2$$
$$= 100 \times 30 \times 50 + 433 \times 34.64 \times (100 + 0.5 \times 433)$$

contributes nothing to the bending strength. While we can consider the sloping walls as rectangles inclined to the horizontal, it is much simpler to take advantage of the fact that the horizontal distribution of material is irrelevant for the purposes of finding the neutral axis and second moment of area, and treat the trough cross section as shown in Fig. 5.71. Treating this section as two rectangles, and taking first moments of area about the lower edge

$$(4 \times 1.2 + 17.32 \times 1.386)c_2$$
$$= 4 \times 1.2 \times 2 + 17.32 \times 1.386 \times (4 + 0.5 \times 17.32)$$

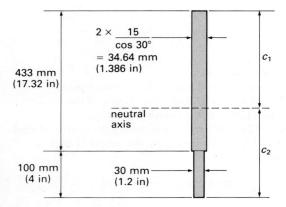

Fig. 5.71.

which gives the height of the neutral axis above the lower edge as

$$c_2 = 272.1 \text{ mm}$$

The distance from the neutral axis to the top edge of the trough is therefore

$$c_1 = 100 + 433 - c_2 = 260.9 \text{ mm}$$

The second moment of area about the neutral axis is

$$I = \frac{30 \times 100^3}{12} + 100 \times 30 \times (c_2 - 50)^2 + \frac{34.64 \times 433^3}{12}$$
$$+ 34.64 \times 433 \times \left(100 + \frac{433}{2} - c_2\right)^2$$
$$= 414.4 \times 10^6 \text{ mm}^4$$

which gives the height of the neutral axis above the lower edge as

$$c_2 = 10.88 \text{ in}$$

The distance from the neutral axis to the top edge of the trough is therefore

$$c_1 = 4 + 17.32 - c_2 = 10.44 \text{ in}$$

The second moment of area about the neutral axis is

$$I = \frac{1.2 \times 4^3}{12} + 4 \times 1.2 \times (c_2 - 2)^2 + \frac{1.386 \times 17.32^3}{12}$$
$$+ 1.386 \times 17.32 \times \left(4 + \frac{17.32}{2} - c_2\right)^2$$
$$= 1061 \text{ in}^4$$

276

Using equation (5.34), the maximum tensile bending stress at the middle cross section of the trough, which occurs at the lower edge of the trough plates, is

$$\sigma_x = \frac{c_2 M}{I} = \frac{272.1 \times 10^{-3} \times 11\,020}{414.4 \times 10^{-6}}\,\text{N/m}^2$$

$$= 7.24\ \text{MN/m}^2$$

and the maximum compressive stress, at the top edge, is

$$\sigma_x = \frac{-c_1 M}{I} = \frac{-260.9 \times 10^{-3} \times 11\,020}{414.4 \times 10^{-6}}\,\text{N/m}^2$$

$$= -6.94\ \text{MN/m}^2$$

Using equation (5.34), the maximum tensile bending stress at the middle cross section of the trough, which occurs at the lower edge of the trough plates, is

$$\sigma_x = \frac{c_2 M}{I} = \frac{10.88 \times 8.645 \times 12}{1061}$$

$$= 1.06\ \text{ksi}$$

and the maximum compressive stress, at the top edge, is

$$\sigma_x = \frac{-c_1 M}{I} = \frac{-10.44 \times 8.645 \times 12}{1061}$$

$$= -1.02\ \text{ksi}$$

5.3.4 Unsymmetrical bending

When we were developing the simple theory of bending in Section 5.3.1, we made the assumption that the cross section of the beam concerned is symmetrical about the plane of bending (as in Fig. 5.53). Let us now consider what happens if this is not the case. Figure 5.72 shows a beam of arbitrary cross section, with the usual coordinates x, y and z arranged with their origin at the centroid, C, of a typical cross section. If we assume, as we did for beams of symmetrical cross section, that bending is confined to the x, y plane, then the z axis is the neutral axis, and the distribution of bending stress is given by equation (5.34) as

$$\sigma_x = \frac{-yM}{I} \tag{5.34}$$

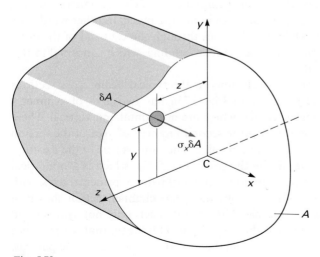

Fig. 5.72.

277

This result was derived by considering equilibrium of moments about the neutral axis. We should also consider equilibrium of moments about the y axis, the externally applied moment about this axis being zero. The force in the x direction acting on a small element, δA, of the total cross-sectional area A is $\sigma_x \delta A$ (Fig. 5.72), and the total moment of such forces about the y axis is

$$\int_A z\sigma_x \, \mathrm{d}A = \int_A z\left(\frac{-yM}{I}\right)\mathrm{d}A = 0$$

which implies that

$$P_{yz} = \int_A yz \, \mathrm{d}A = 0 \tag{5.53}$$

where P_{yz} is a second moment of area involving the product of distances y and z from the axes. In the context of moments of inertia it is referred to as a *product of inertia*. Although we shall not attempt to prove the results, such a second moment has some important properties. In particular, for any shape of cross section there is always (at least) one pair of mutually perpendicular axes, y and z, for which equation (5.53) holds. These are referred to as the *principal axes* of the cross section. Also, if the section has an axis of symmetry, this axis is a principal axis. Further, the second moments of area about the two principal axes are the maximum and minimum values for any possible axes through the centroid of the section.

So, we must now consider the practical implications of equation (5.53) for a beam of any cross section. If the cross section is symmetrical about the y axis, which lies in the assumed plane of bending, the equation is satisfied, and the beam is in moment equilibrium about this axis, and the previously derived theory of bending is valid. If it is not symmetrical, then the equation may not be satisfied, in which case equilibrium is not possible without some externally applied moment about the y axis, for bending to be confined to the x, y plane. What happens in practice is that, in the absence of such a moment, bending occurs not only in the x, y plane, but also in the x, z plane. In other words, the neutral axis of bending, while it still passes through the centroid of the cross section, is no longer horizontal. The plane of bending, which is normal to this axis, is no longer vertical.

Let us try to relate these conclusions to the typical beam cross sections shown in Fig. 5.73. The rectangle in Fig. 5.73a has two axes of symmetry passing through its geometric center, which are horizontal and vertical. These are the principal axes, about which the second moment of area of the section takes either a maximum or a minimum value. The same is true of the I section shown in Fig. 5.73b. In the case of the circle, Fig. 5.73c, which is symmetrical about any diameter, there is an infinite number of pairs of principal axes, with the same second moment of area about each. The channel section shown in Fig. 5.73d is an interesting example. Although it is certainly not symmetrical about the y axis, it is symmetrical about the z axis. This means that the z axis is a principal axis, from which it follows that the y axis must also be a principal axis, and beam behavior can be analyzed using the simple bending theory. The angle

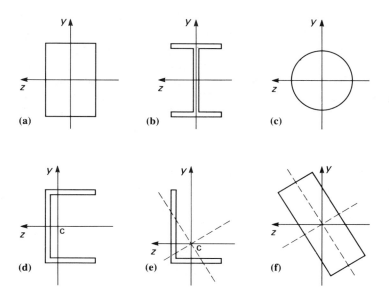

Fig. 5.73.

section shown in Fig. 5.73e has no planes of symmetry, and the principal axes, which are shown as dotted lines, are inclined to the y and z axes. In this case, lateral loading in the y direction would cause the beam to deflect in both the y and z directions. The same is true of the rectangle shown in Fig. 5.73f, with its principal axes rotated from the vertical and horizontal. In other words, unsymmetrical bending can result from either unsymmetrical beam section geometry or from loading which is unsymmetrical with respect to the principal axes of this section.

Although a full treatment of the stresses due to unsymmetrical bending is outside the scope of this book, in principle it is straightforward. We take the bending moment at a particular beam cross section and resolve it into its components about the principal axes. To each of these components we can apply simple bending theory to find the corresponding bending stress distributions. Provided the beam material remains linearly elastic, these distributions can be superimposed to find the total stresses.

5.3.5 Reinforcement of beams

In practice we may be faced with the situation where a beam is not strong enough to support the loading we wish to apply to it. That is, the maximum bending stress at the cross section required to take the greatest bending moment exceeds the maximum allowable stress. Much more rarely, the maximum shear stress in the beam might be excessive. The beam must be strengthened by increasing its elastic section modulus. The most obvious way to do this is by using a beam of larger cross section. This is not always practical, however, and is not likely to be an ideal solution in terms of weight and cost, mainly because the whole beam is thereby strengthened, even in regions of low bending moment.

One approach which is sometimes used to strengthen an I section beam is to cut it along its web as shown in Fig. 5.74a, and to weld the two halves together again as in Fig. 5.74b leaving a row of hexagonally shaped holes through the web. By increasing the distance between the flanges, the second moment of area about the neutral axis and the elastic section modulus are both increased without increasing the weight of the beam. The weakening of the web, which is only required to support relatively low shear stresses, is not significant.

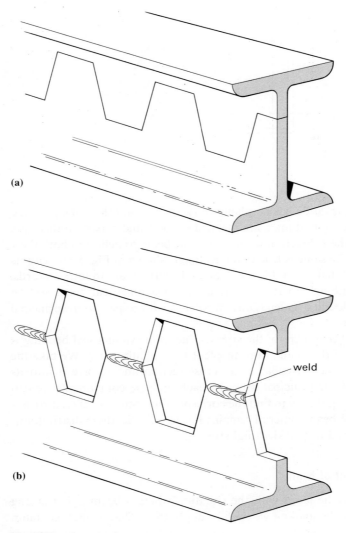

(a)

(b)

weld

Fig. 5.74.

A commonly used method of beam reinforcement is to weld plates of the same material to the flanges as shown in Fig. 5.75. This has the effect of adding material to the beam cross section where it can be most useful in supporting bending stresses – as far from the neutral axis as possible. The welds are necessary to prevent relative displacement in the axial direction between the

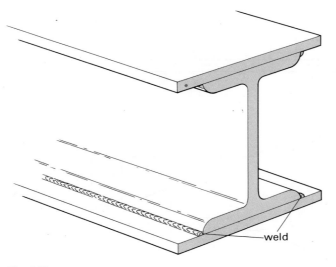

Fig. 5.75.

plates and the beam, thereby ensuring that the arrangement behaves as though it were a continuous solid beam. The plates need only be added in regions of the beam where the bending moment is high.

EXAMPLE 5.16

A simply supported steel I section beam, whose depth is 530 mm (21 in), and second moment of area is 404×10^{-6} m^4 (970 in^4) is to carry a uniformly distributed lateral force of intensity 120 kN/m (8.2 kip/ft) over a span of 5 m (16 ft). If the maximum allowable bending stress is 160 MN/m^2 (23 ksi), show that some reinforcement of the beam will be necessary, and demonstrate how this can be done using steel plates which are 8 mm (0.32 in) thick.

In this Example, we are not told whether the distributed lateral force includes the weight of the beam itself. Since the weight would be only about 1 kN/m, however, we will ignore it. Similarly, we will ignore the weight of the reinforcing plates.

First we can find the greatest bending moment the beam can carry without reinforcement, with the aid of equation (5.35), as

$$M = \frac{I\sigma_{max}}{c} = \frac{404 \times 10^{-6} \times 160 \times 10^6}{265 \times 10^{-3}} \text{ Nm} = 244 \text{ kNm (180 kip-ft)}$$

The distribution of bending moment along the beam is as shown in Fig. 5.19c, with a maximum value at the center of the beam, point C, of $wL^2/8$, where $w = 120$ kN/m and $L = 5$ m, which is 375 kNm. Figure 5.76 shows the shear force and bending moment distributions for the present problem. The bending moment exceeds the value of 244 kNm over a substantial part of the central region of the beam, specifically between points D and E. If these are each at a distance of h from the center of the beam, the shear force at each is of magnitude $120h$. Therefore, since the change in the bending moment between, say, points D and C is equal to the area under the shear force distribution between these points (which is shown shaded)

$$375 - 244 = \tfrac{1}{2} \times h \times 120h$$

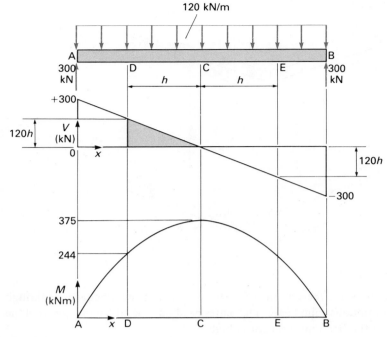

Fig. 5.76.

and

$$h = 1.48 \text{ m} \ (4.9 \text{ ft})$$

In other words, the distance between points D and E is 2.96 m, and this is the minimum length of beam over which reinforcing plates must be added to the flanges.

The arrangement of the flange plates is shown in Fig. 5.77. Since we are assuming that the maximum allowable bending stress applies equally to tensile and compressive stresses, it is appropriate to maintain the symmetry of the beam by adding plates of the same width, b, to each flange. We can find the modified second moment of area about the

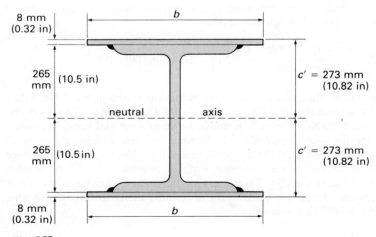

Fig. 5.77.

neutral axis by adding the contributions of the plates using the parallel axis theorem (Appendix C)

$$I' = 404 \times 10^{-6} + 2\left[\frac{b \times 0.008^3}{12} + (b \times 0.008)\left(0.265 + \frac{0.008}{2}\right)^2\right]$$

$$= 404 \times 10^{-6} + 1.16 \times 10^{-3}b \text{ m}^4$$

where b is in meters. Again using equation (5.35), the minimum value of this second moment of area required to carry the maximum bending moment at the center of the beam is

$$I' = \frac{Mc'}{\sigma_{max}} = \frac{375 \times 10^3 \times 0.273}{160 \times 10^6} \text{ m}^4 = 640 \times 10^{-6} \text{ m}^4$$

Comparing these two expressions for I', we can deduce the required plate width b to be

$$b = \frac{(640 - 404) \times 10^{-6}}{1.16 \times 10^{-3}} = 0.203 \text{ m} = 203 \text{ mm (8 in)}$$

In this Example, we adopted the most straightforward arrangement of plates by adding a single plate to each flange over the relevant part of the beam, and making them wide enough to allow the beam to support the greatest bending moment at its center. This level of reinforcement could have been reduced away from the center of the beam. One way to do this is as shown in Fig. 5.78, with a single pair of narrower plates attached between points D and E, and a second pair added on top of the first over a shorter length of beam at its center.

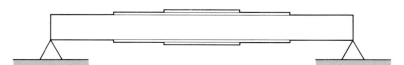

Fig. 5.78.

The form of the reinforced beam shown in Fig. 5.78 is an attempt to make the bending strength of the beam follow the bending moment distribution imposed on it. The logical extension of this trend is to design a beam with a continuously varying elastic section modulus which follows exactly the bending moment distribution. The maximum bending stress at every beam cross section is then the same, implying efficient use of material. Such a constant strength beam could have a cross section in the form of an I, with varying depth, as shown in Fig. 5.79.

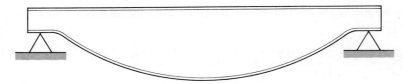

Fig. 5.79.

5.3.6 Composite beams

With the reinforced beams we have just been considering, the reinforcing plates were of the same material as the original beam, and we could treat them as continuous homogeneous solids. In some circumstances, however, composite beams are made from two or more materials with quite different physical properties. For example, in aerospace applications where high strength and low weight are of greater importance than cost, beams may be made with thin outer skins of high strength materials, such as carbon fibers, separated by relatively thick cores of low density foamed polymers. While the skins take the bending stresses, the cores have only to support the much lower shear stresses. Another important example, and one which we will consider in some detail, is that of reinforced concrete beams, which are widely used in buildings and other structures.

In order to analyze the behavior of composite beams, we first make the assumption that the materials are bonded rigidly together so that there can be no relative axial movement between them. This means that the first five of the six assumptions we made for homogeneous beams in Section 5.3.1 are still valid; in particular, that plane beam cross sections remain plane after bending. The one assumption that is no longer valid is that the Young's modulus is the same throughout the beam.

Let us consider, for example, the composite beam shown in Fig. 5.80, which is rectangular in cross section and of breadth b. The core of depth d_1 of one

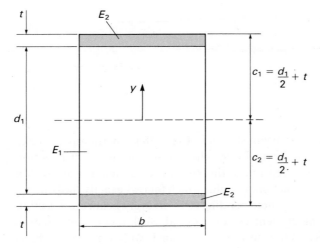

Fig. 5.80.

material having a Young's modulus of E_1 is bonded to outer layers each of thickness t of a second material having a Young's modulus of E_2. Provided plane beam cross sections remain plane after bending, equation (5.29) defining the distribution of normal *strain* over the beam cross section, still applies

$$e_x = \frac{-y}{R} \tag{5.29}$$

where R is the radius of curvature of the neutral surface. The distribution of bending *stress* is therefore given by

$$\sigma_x = e_x E(y) = \frac{-yE(y)}{R} \tag{5.54}$$

where $E(y) = E_1$ for $y < d_1/2$ and $E(y) = E_2$ for $y \geq d_1/2$.

The bending stress is directly proportional not only to the distance from the neutral axis but also to the Young's modulus of the material concerned. If $E_2 > E_1$ in Fig. 5.80, the outer layers will carry much greater bending stresses than the core material.

Equations (5.31) and (5.32) defining the position of the neutral axis and the bending moment, respectively, become

$$\int_{-c_2}^{+c_1} b(y) E(y) y \, dy = 0 \tag{5.55}$$

and

$$M = \frac{1}{R} \int_{-c_2}^{+c_1} b(y) E(y) y^2 \, dy \tag{5.56}$$

Now, we can return these equations to forms effectively the same as those for a homogeneous beam by considering the equivalent beam of a single material. Suppose we choose to work in terms of the core material of modulus E_1. We then define an equivalent breadth, $b_e(y)$, of beam as the actual breadth if the material at that vertical position is of modulus E_1, and as the actual breadth multiplied by the modulus ratio E_2/E_1 if the material is of modulus E_2. In the present case

$$b_e(y) = b \text{ for } y < d_1/2 \quad \text{and} \quad b_e(y) = bE_2/E_1 \text{ for } y \geq d_1/2$$

and Fig. 5.81 shows the equivalent beam of core material. Equations (5.55) and (5.56) become

$$\int_{-c_2}^{+c_1} b_e(y) y \, dy = 0 \tag{5.57}$$

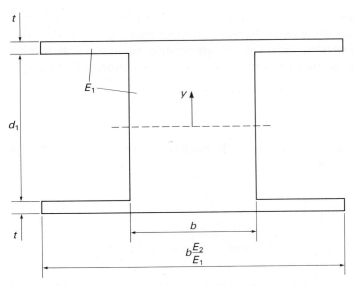

Fig. 5.81.

and

$$M = \frac{E_1}{R} \int_{-c_2}^{+c_1} b_e(y)y^2 \, \mathrm{d}y \tag{5.58}$$

which means that we can work in terms of the equivalent beam cross section to locate the actual neutral axis and determine the second moment of area about this axis. Other bending equations such as (5.33) (with $E = E_1$ and $I = I_e$, the second moment of area about the neutral axis of the equivalent cross section), (5.34) and (5.35) are also applicable, with the exception that bending stresses in the second material of modulus E_2 must be obtained from those for the equivalent beam by multiplying by the modulus ratio E_2/E_1. This means that stresses in material furthest from the neutral axis are not necessarily the greatest (but only in the present case if $E_2 < E_1$).

We have made the assumption that the two materials are rigidly bonded together. In practice, we would attempt to achieve this by welding, or by the use of an adhesive or fasteners such as nails, screws, bolts or rivets. Although shear stresses at the interface are likely to be small in relation to the shear strength of the beam materials concerned, the strength of the weld, adhesive or fasteners may well be substantially less, and we need to investigate the possibility of shear failure and debonding of the materials. We can adapt equation (5.43) to give the shear force over the interface per unit length of beam as

$$\begin{array}{l} \text{Interface shear force} \\ \text{(per unit length of beam)} \end{array} = \frac{V(x)\,Q_e(y)}{I_e} \tag{5.59}$$

where y is the distance of the interface from the neutral axis, $Q_e(y)$ is the first moment of area about the neutral axis of the region *of the equivalent beam cross section* above the interface, and $V(x)$ is the shear force at the particular beam cross section. From the interface shear force can be obtained the average shear stress in an adhesive by dividing by the breadth of the actual beam, or the average shear stress in fasteners by dividing by the total cross-sectional area of the fasteners crossing the interface per unit length of beam.

EXAMPLE 5.17

A timber beam of rectangular cross section of breadth $b = 50$ mm (2 in) and depth $d_1 = 100$ mm (4 in) has steel plates of thickness $t = 5$ mm (0.2 in) bonded to its upper and lower surfaces, as shown in Fig. 5.80. At a particular cross section, it is subjected to a bending moment of 1.5 kNm (1.1 kip-ft) and a shear force of 3 kN (670 lb). Find the magnitudes of the maximum bending stresses in the timber and steel at this section, and the average shear stress over either of the interfaces between the two materials. If, instead of adhesive bonding, each of the steel plates is attached to the beam by means of a single row of screws of 6 mm (0.24 in) diameter, for which the maximum allowable shear stress is 100 MN/m² (15 ksi) find the maximum distance apart along the beam the screws must be placed.

Given that the values of Young's modulus for timber and steel (Appendix A) are $E_1 = 12$ GN/m² and $E_2 = 207$ GN/m², implying a modulus ratio of $E_2/E_1 = 17.2$, Fig. 5.82 shows the cross section of the equivalent timber beam. The second moment of area of this section about its neutral axis is

$$I_e = \frac{50 \times 100^3}{12} + 2\left[\frac{862.5 \times 5^3}{12} + (862.5 \times 5) \times 52.5^2\right] \text{mm}^4$$

$$= 27.96 \times 10^{-6} \text{ m}^4$$

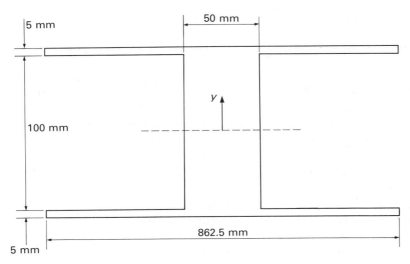

Fig. 5.82.

Therefore, the magnitude of the maximum bending stress in the timber, at $y = 50$ mm, is

$$\sigma_{\text{max, t}} = \frac{yM}{I_e} = \frac{0.050 \times 1.5 \times 10^3}{27.96 \times 10^{-6}} \text{ N/m}^2 = \underline{2.68 \text{ MN/m}^2} \text{ (390 psi)}$$

and in the steel, at $y = 55$ mm, is

$$\sigma_{\text{max, s}} = \frac{yM}{I_e} \frac{E_2}{E_1} = \frac{0.055 \times 1.5 \times 10^3 \times 17.2}{27.96 \times 10^{-6}} \text{ N/m}^2 = \underline{50.8 \text{ MN/m}^2} \text{ (7.4 ksi)}$$

The first moment of area of the upper steel region of the equivalent beam cross section about the neutral axis is

$$Q_e = (862.5 \times 5) \times 52.5 \text{ mm}^3 = 0.2264 \times 10^{-3} \text{ m}^3$$

From equation (5.59), the interface shear force per unit length of beam is

$$\text{shear force} = \frac{VQ_e}{I_e} = \frac{3 \times 10^3 \times 0.2264 \times 10^{-3}}{27.96 \times 10^{-6}} = 24\,290 \text{ N/m}$$

and the average interface shear stress is found by dividing this figure by the actual beam width at the interface as

$$\tau = \frac{24\,290}{0.050} \text{ N/m}^2 = \underline{0.49 \text{ MN/m}^2} \text{ (70 psi)}$$

If screws of 6 mm diameter are used to attach the steel plates, then the maximum shear force in each screw is the product of its cross-sectional area and the maximum allowable shear stress

$$\text{maximum screw shear force} = \pi \times 0.003^2 \times 100 \times 10^6 = 2827 \text{ N}$$

and the maximum spacing of the screws along the beam is

$$\text{maximum screw spacing} = \frac{2827}{24\,290} \text{ m} = \underline{116 \text{ mm}} \text{ (4.6 in)}$$

Composite beams need not be made up of horizontal layers of material as in the above example. For example, a beam might have stiffening plates attached to its sides as shown in Fig. 5.83a. Once again, the equivalent beam of the main beam material can be formed by scaling the breadth of the plate material in proportion to the modulus ratio, as in Fig. 5.83b. Bearing in mind that the strain at any level is the same in both materials, the bending stresses in them are in proportion to their moduli.

Concrete, which is widely used in buildings and other structures, is brittle and is much weaker in tension than in compression. Indeed, it is often assumed for design purposes that it has zero tensile strength. Since tensile loading is unavoidable in beams, it is necessary to reinforce concrete beams in regions where they are expected to be in tension. Steel is the most commonly used

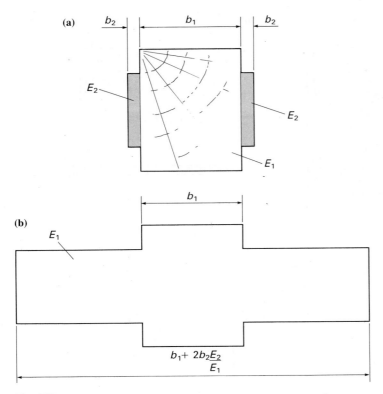

b_2 b_1 b_2

E_2

E_2

E_1

(b)

E_1

b_1

$$b_1 + 2b_2\frac{E_2}{E_1}$$

Fig. 5.83.

material for this purpose, both because it is structural material in its own right, and because it has essentially the same coefficient of thermal expansion as concrete (Appendix A). This is important to minimize the internal stresses generated by changes in temperature.

For a concrete beam subject to sagging bending moments, a typical form of steel reinforcement is as shown in cross section in Fig. 5.84a. Two steel rods are embedded in the concrete towards the bottom of its rectangular cross section. It is customary to ignore the tensile strength of the concrete, and so construct an equivalent concrete beam cross section as shown in Fig. 5.84b, with two separate regions: the concrete in compression above the neutral axis, and the concrete equivalent of the steel in tension at some distance below the neutral axis. It is assumed that the concrete is sufficiently strong in shear to ensure that relative axial movement of these two regions is prevented, so that the assumption of plane cross sections remaining plane is applicable. It is also usual to assume that all the steel cross-sectional area is concentrated at the same distance below the neutral axis, at the level of the centers of the steel rods, and that the tensile stress in the rods is constant. The stress distribution is therefore as shown in Fig. 5.84c, with the usual linear variation of compressive stress in the concrete above the neutral axis. We do not know in advance, however, where the neutral axis will be, but we can locate it by taking first moments of equivalent area.

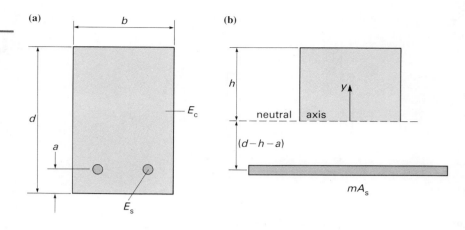

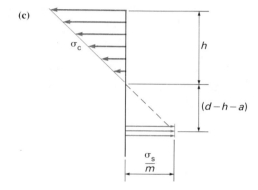

Fig. 5.84.

The overall breadth and depth of the beam are b and d, respectively, and the centers of the steel rods, which have a combined area of A_s, are at a distance of a above the lower edge of the cross section. The ratio between the Young's moduli of steel and concrete is $m = E_s/E_c$. If the neutral axis is at a distance h below the upper edge of the cross section, then taking first moments of equivalent area about this axis gives

$$hb\frac{h}{2} = mA_s(d - h - a) \tag{5.60}$$

from which h can be found. Also, the second moment of area about the neutral axis of the equivalent beam is

$$I_e = \frac{bh^3}{12} + hb\left(\frac{h}{2}\right)^2 + mA_s(d - h - a)^2 \tag{5.61}$$

which can be used to find the bending stresses in the concrete and steel. The uniform tensile stress in the steel, found by multiplying the stress in the equivalent concrete beam at a distance $(d - h - a)$ below the neutral axis by the modulus ratio, is

$$\sigma_s = m(d - h - a)\frac{M}{I_e} \qquad (5.62)$$

where M is the bending moment, and the maximum compressive stress in the concrete, at the upper surface of the beam is

$$\sigma_{max, c} = -h\frac{M}{I_e} \qquad (5.63)$$

EXAMPLE 5.18

A reinforced concrete beam of rectangular cross section of breadth $b = 150$ mm (6 in) and depth $d = 350$ mm (14 in) has two steel reinforcing rods of 25 mm (1 in) diameter with their centers a distance $a = 50$ mm (2 in) above the lower surface of the concrete. If the maximum allowable tensile stress in the steel is 150 MN/m² (22 ksi) and the maximum allowable compressive stress in the concrete is 10 MN/m² (1.5 ksi) find the greatest sagging bending moment the beam can support, and the corresponding radius of curvature.

SI units
The total cross-sectional area of the steel rods is

$$A_s = 2 \times \pi \times 12.5^2 = 981.7 \text{ mm}^2$$

and, from the data given in Appendix A, the modulus ratio is

$$m = \frac{E_s}{E_c} = \frac{207}{13.8} = 15$$

Therefore, using equation (5.60) to find the position of the neutral axis

$$\frac{150\,h^2}{2} = 15 \times 981.7 \times (350 - h - 50)$$

which is a quadratic equation for h, the only relevant solution of which is $h = 163.6$ mm. Substituting this value into equation (5.61)

$$I_e = \frac{150 \times 163.6^3}{12} + \frac{150 \times 163.6^3}{4} + 15 \times 981.7$$

$$\times (300 - 163.6)^2$$

$$= 492.9 \times 10^6 \text{ mm}^4 = 492.9 \times 10^{-6} \text{ m}^4$$

US customary units
The total cross-sectional area of the steel rods is

$$A_s = 2 \times \pi \times 0.5^2 = 1.571 \text{ in}^2$$

and, from the data given in Appendix A, the modulus ratio is

$$m = \frac{E_s}{E_c} = \frac{30}{2.0} = 15$$

Therefore, using equation (5.60) to find the position of the neutral axis

$$\frac{6h^2}{2} = 15 \times 1.571 \times (14 - h - 2)$$

which is a quadratic equation for h, the only relevant solution of which is $h = 6.546$ in. Substituting this value into equation (5.61)

$$I_e = \frac{6 \times 6.546^3}{12} + \frac{6 \times 6.546^3}{4} + 15$$

$$\times 1.571 \times (12 - 6.546)^2$$

$$= 1262 \text{ in}^4$$

SI units

Using equation (5.62), the bending moment required to generate the maximum allowable tensile stress in the steel is

$$M_1 = \frac{I_e \sigma_s}{m(d-h-a)} = \frac{492.9 \times 10^{-6} \times 150 \times 10^6}{15 \times 136.4 \times 10^{-3}} \text{ Nm}$$

$$= 36.1 \text{ kNm}$$

while, from equation (5.63), that required to generate the maximum allowable compressive stress in the concrete is

$$M_2 = \frac{I_e \sigma_{max, c}}{h} = \frac{492.9 \times 10^{-6} \times 10 \times 10^6}{163.6 \times 10^{-3}} \text{ Nm}$$

$$= 30.1 \text{ kNm}$$

The second value is somewhat more restrictive than the first, and the required greatest bending moment is therefore 30.1 kNm. The radius of curvature with this bending moment is given by equation (5.33) as

$$R = \frac{E_c I_e}{M_2} = \frac{13.8 \times 10^9 \times 492.9 \times 10^{-6}}{30.1 \times 10^3} = \underline{226 \text{ m}}$$

US customary units

Using equation (5.62), the bending moment required to generate the maximum allowable tensile stress in the steel is

$$M_1 = \frac{I_e \sigma_s}{m(d-h-a)} = \frac{1262 \times 22}{15 \times 5.454 \times 12}$$

$$= 28.3 \text{ kip-ft}$$

while, from equation (5.63), that required to generate the maximum allowable compressive stress in the concrete is

$$M_2 = \frac{I_e \sigma_{max, c}}{h} = \frac{1262 \times 1.5}{6.546 \times 12}$$

$$= 24.1 \text{ kip-ft}$$

The second value is somewhat more restrictive than the first, and the required greatest bending moment is therefore 24.1 kip-ft. The radius of curvature with this bending moment is given by equation (5.33) as

$$R = \frac{E_c I_e}{M_2} = \frac{2 \times 10^3 \times 1262}{24.1 \times 12^2} = \underline{727 \text{ ft}}$$

We note that the design of the beam is such that the greatest bending moments determined from the maximum allowable stresses in the steel and concrete are nearly the same. This means that, as an increasing bending moment is applied to the beam, both materials reach their maximum allowable stresses at nearly the same load. The ideal design is one in which the maximum values are reached at exactly the same load, as this represents the most economical use of material.

Also of interest in a steel reinforced concrete beam is the shear stress generated at the interface between the steel and concrete when the beam is loaded. We can derive an expression for this in much the same way we did for more general shear stresses due to bending (Section 5.3.2). Consider a small element of the steel reinforcement of length δx at an average distance of g below the neutral surface of the beam, as shown in Fig. 5.85. For convenience, we can treat all the steel from several reinforcing rods as being lumped together in a single unit having a total cross-sectional area of A_s, and a total perimeter (in the plane of the beam cross section) of p_s (note that the rod or rods need not be circular). For equilibrium of forces acting on the steel element

$$[\sigma_s(x+\delta x) - \sigma_s(x)]A_s = \tau p_s \delta x$$

where σ_s is the tensile stress in the steel and τ is the shear stress at the interface between the steel and concrete, both of which are assumed to be uniform. In

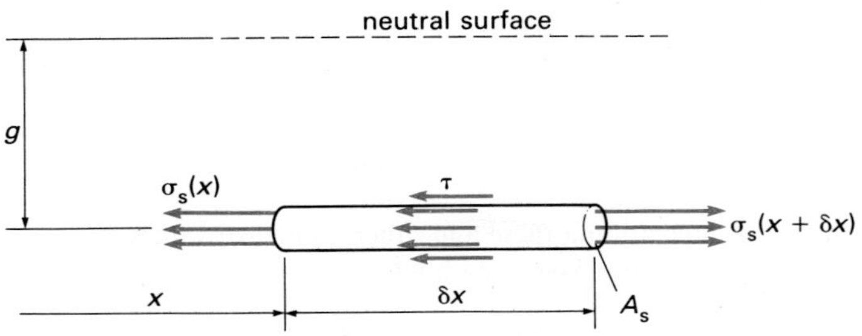

Fig. 5.85.

effect, we are using the mean values of these stresses, the amounts by which they vary at a particular cross section being small. The tensile stresses at the ends of the element are given by equation (5.62) as

$$\sigma_s(x) = mg \frac{M(x)}{I_e} \quad \text{and} \quad \sigma_s(x + \delta x) = mg \frac{M(x + \delta x)}{I_e}$$

and

$$\tau = \frac{[M(x + \delta x) - M(x)]}{\delta x} \frac{mg A_s}{p_s I_e}$$

As the length of the element, δx, tends to zero, this expression for the shear stress becomes

$$\tau = \frac{dM}{dx} \frac{mg A_s}{p_s I_e} = \frac{mg A_s V(x)}{p_s I_e} \tag{5.64}$$

where $V(x)$ is the shear force at the particular beam cross section.

EXAMPLE 5.19

The reinforced concrete beam considered in Example 5.18 is 3 m (9.8 ft) long and simply supported at its ends. It is subjected to a uniformly distributed lateral force, including its own weight, of intensity 17 kN/m (1.2 kip/ft). Find the maximum shear stress at the interface between the steel and concrete.

SI units
The maximum shear force, which occurs at either end of the beam (Fig. 5.19), is half the total lateral force, or 25.5 kN. The total perimeter of the two steel rods is

$$p_s = 2 \times \pi \times 25 = 157.1 \text{ mm}$$

US customary units
The maximum shear force, which occurs at either end of the beam (Fig. 5.19), is half the total lateral force, or 5.88 kip. The total perimeter of the two steel rods is

$$p_s = 2 \times \pi \times 1 = 6.283 \text{ in}$$

SI units

and the distance between them and the neutral surface of the beam is

$$g = (d - h - a) = 136.4 \text{ mm}$$

Substituting these and other numerical values into equation (5.64), we obtain

$$\tau = \frac{15 \times 0.1364 \times 981.7 \times 10^{-6} \times 25.5 \times 10^3}{0.1571 \times 492.9 \times 10^{-6}} \text{ N/m}^2$$

$$= 0.66 \text{ MN/m}^2$$

US customary units

and the distance between them and the neutral surface of the beam is

$$g = (d - h - a) = 5.454 \text{ in}$$

Substituting these and other numerical values into equation (5.64), we obtain

$$\tau = \frac{15 \times 5.454 \times 1.571 \times 5.88}{6.283 \times 1262} \text{ ksi}$$

$$= 95 \text{ psi}$$

5.3.7 Bending of initially curved beams

So far, we have assumed that beams undergoing bending are initially straight. In particular, in order to obtain equation (5.29) for the longitudinal bending strain in the beam, we effectively assumed that the radius of curvature of the neutral surface was infinite before bending and equal to R after bending. We can relax this assumption in order to treat initially curved beams, although we still assume that radii of curvature are large compared to the depth of the beam. Beams with radii of curvature comparable in magnitude to the beam depth require a more sophisticated form of analysis, which is outside the scope of this book.

Figure 5.86a shows a short length of an initially curved beam in its undeformed state, with two adjacent cross sections marked $A_1 A_2$ and $B_1 B_2$. After the bending moment M is applied, these become the sections $A'_1 A'_2$ and $B'_1 B'_2$ as shown in Fig. 5.86b. The corresponding situation for an initially straight beam is shown in Fig. 5.51. The line $A_0 B_0$ represents the neutral

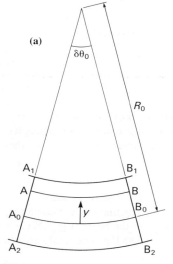
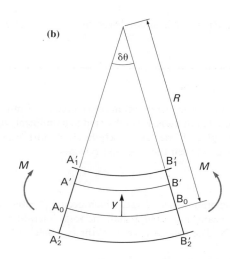

Fig. 5.86.

surface of the beam, while AB is a typical longitudinal element at a distance y above this surface. The strain in this element is

$$e_x = \frac{A'B' - AB}{AB}$$

where

$$AB = (R_0 - y)\delta\theta_0 \quad \text{and} \quad A'B' = (R - y)\delta\theta$$

In these expressions, R_0 and R are the radii of curvature of the neutral surface before and after bending, respectively, while $\delta\theta_0$ and $\delta\theta$ are the (small) angles subtended at the centers of curvature before and after bending. Since the length $A_0 B_0$ of the neutral surface is unchanged

$$R_0\delta\theta_0 = R\delta\theta$$

and the strain in the typical element is

$$e_x = \frac{-y(\delta\theta - \delta\theta_0)}{(R_0 - y)\delta\theta_0} \approx \frac{-y(\delta\theta - \delta\theta_0)}{R_0\delta\theta_0} = -y\left(\frac{1}{R} - \frac{1}{R_0}\right)$$

Note that the approximation is only valid if $y \ll R_0$, which is consistent with the simple beam theory assumptions. If it is not, then the strain is no longer a linear function of y. This result for the longitudinal strain reduces to the expected form for an initially straight beam when R_0 is infinite. In all cases the strain is proportional to the *change* of curvature, which is characterized by the reciprocal of the radius of curvature. The generalized form of simple bending equations (5.33) is

$$\frac{M}{I} = E\left(\frac{1}{R} - \frac{1}{R_0}\right) = \frac{\sigma_x}{-y} \tag{5.65}$$

EXAMPLE 5.20

The solid steel bar shown in Fig. 5.87 is of circular cross section with a diameter of 12 mm (0.48 in), and in its stress-free state takes the form of a semicircle of radius 500 mm (19.7 in). If this radius is reduced to 490 mm (19.3 in) by the application of bending moments to the ends of the bar, find the magnitude of these moments and the maximum bending stresses in the steel.

The bar is subjected to pure bending, with no shear force and the same bending moment at every cross section. From equation (5.40), the second moment of area of the cross section about its neutral axis is

$$I = \frac{\pi \times 12^4}{64} = 1018 \text{ mm}^4$$

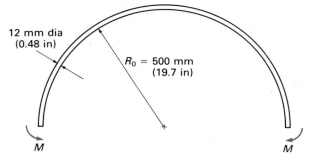

Fig. 5.87.

and the maximum distance from the neutral axis is $y = \pm 6$ mm. Since the change in radius of curvature is from $R_0 = 0.5$ m to $R = 0.49$ m, equation (5.65) gives the maximum bending stress as

$$\sigma_x = 0.006 \times 207 \times 10^9 \left(\frac{1}{0.49} - \frac{1}{0.5} \right) \text{N/m}^2 = \underline{50.7 \text{ MN/m}^2} \ (7.4 \text{ ksi})$$

and the bending moment as

$$M = 1018 \times 10^{-12} \times 207 \times 10^9 \left(\frac{1}{0.49} - \frac{1}{0.5} \right) = \underline{8.60 \text{ Nm}} \ (6.3 \text{ lb ft})$$

5.4 Combined bending and axial loads

Our treatment of bending has so far been confined to situations where the members concerned are subjected to externally applied couples and lateral forces, but not to axial forces. The normal stresses due to the latter are in the same direction as the bending stresses. Provided the material remains linearly elastic, however, it is possible to superimpose the two stress distributions.

Combined bending and axial loads can arise in a number of different ways. Figure 5.88 provides the straightforward example of a screw clamp. When such a device is tightened onto an object, equal and opposite forces, say F, are

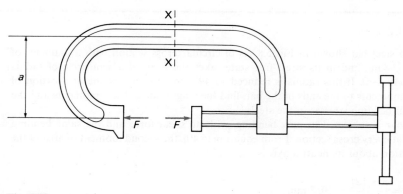

Fig. 5.88.

applied to the clamp as shown. A typical cross section of the clamp frame such as XX therefore experiences not only a bending moment, and hence a bending stress distribution, due to this force, but also a normal tensile force of magnitude F. In the presence of an axial force, we now have to be more careful how we define the bending moment at such a section. In Section 5.3.1 we were able to demonstrate that the bending moment is the same about any axis normal to the plane of bending, *provided the distribution of axial stress has a zero resultant force*. This is now no longer the case, and we must therefore define the bending moment to be *about the neutral axis of the cross section*. In the present example, if a is the distance between the line of action of F and the neutral surface of cross section XX, the bending moment there is $M = Fa$.

Figure 5.89a shows the bending stress distribution due to moment M, namely $\sigma_x = -yM/I$, where as usual y is the distance from the neutral axis, and I is the second moment of area of the cross section about this axis. Figure 5.89b shows the uniform distribution of tensile stress due to the axial force F. This is of magnitude F/A, where A is the area of the cross section. If these two distributions are superimposed, we obtain the one shown in Fig. 5.89c, with

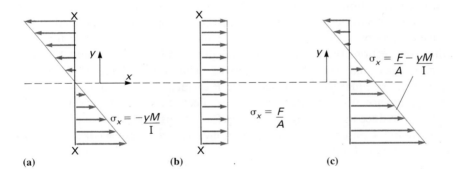

$$\sigma_x = \frac{F}{A} - y\frac{M}{I} \qquad (5.66)$$

Fig. 5.89.

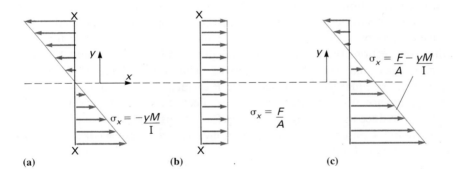

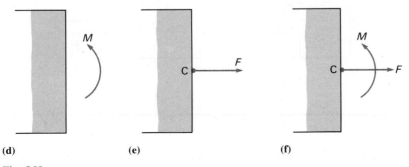

Note that the usual sign conventions are followed: F is positive if it is tensile, and M is positive if it is in the sagging sense. We also note in Fig. 5.89c that, although there is a position at which the combined axial stress is zero, this no longer occurs at $y = 0$. In fact, this position depends on the relative magnitudes of the two stress distributions, or more precisely on the magnitude of distance a relative to the depth of the cross section. Indeed, it is quite possible for the zero stress position to occur outside the cross section. This position is sometimes still referred to as the neutral axis, although it now depends not only on the cross section geometry but also on the applied loading. Here we shall reserve the term *neutral axis* for the *neutral axis of the cross section in the absence of a resultant axial force.*

Figures 5.89d to f show the resultants of the bending, axial and combined stress distributions. Note particularly that in Fig. 5.89e, because the distribution of stress is uniform, the resultant force F acts through the centroid, C, of the cross section, which is a point on the neutral axis. This serves to confirm that under the combined loading, in Fig. 5.89f, the bending moment applied to the section must be taken as the moment about the neutral axis. Another way of saying the same thing is that the combined stress distribution in Fig. 5.89c has a resultant force F acting through the centroid, together with a resultant couple M.

Combined bending and axial loading can also arise in other ways. For example, an object under nominally axial loading also suffers bending when the line of action of the axial force does not pass through the centroid of a particular cross section. Consider the bar shown in Fig. 5.90, which is rectangular in cross section and of depth d. Over part of its length, material is removed to a depth of h from the lower surface. The ends of the bar are subjected to axial tensile forces of magnitude F, which we assume to act through the centroids of the cross sections there. Now, if we consider a typical cross section YY in the region where material has been removed, while there is certainly a uniform tensile stress equal to the applied force F divided by the local cross-sectional area, there are also bending stresses. The effect of material removal at section YY is to reduce its depth by h, and therefore to raise the position of its neutral axis by $h/2$. Consequently, the bending moment *about the neutral axis of the cross section is $Fh/2$.*

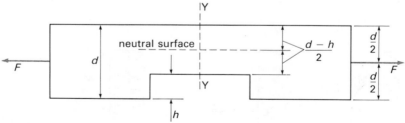

Fig. 5.90.

EXAMPLE 5.21

A tension member of rectangular cross section 100 mm (4 in) by 10 mm (0.4 in) is subjected to an axial tensile force of 65 kN (15 kip). At a section YY a slot 30 mm (1.2 in)

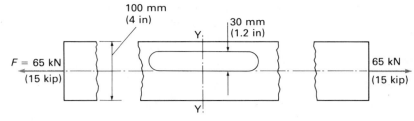

Fig. 5.91.

wide is cut in the member as shown in Fig. 5.91. Determine the distribution of axial stress across the section YY.

The fact that the slot is offset from the center of the member means that the neutral axis of section YY is also offset, thereby introducing some bending. Although shear stresses cannot be transmitted across the slot at this section, the assumption of plane sections remaining plane is still a reasonable one.

SI units

The cross-sectional area of the member at section YY is $A = 700$ mm². If the neutral axis of the section is at a height of c_2 above the lower edge of the section, taking first moments of area about this edge gives

$$700 \, c_2 = (50 \times 10) \times 25 + (20 \times 10) \times 90$$

$$c_2 = 43.57 \text{ mm}$$

which means that the distance between the neutral axis and the upper edge of the cross section is $c_1 = (100 - 43.57) = 56.43$ mm. Also, the second moment of area about the neutral axis is

$$I = \frac{10 \times 50^3}{12} + (10 \times 50) \times (43.57 - 25)^2 + \frac{10 \times 20^3}{12}$$

$$+ (10 \times 20) \times (43.57 - 90)^2$$

$$= 714.4 \times 10^3 \text{ mm}^4$$

and the bending moment about the neutral axis of the section is the product of the applied force, F, of 65 kN and the distance between its line of action through the center of the member and the neutral axis

$$M = -65 \times 10^3 \times (50 - 43.57) \times 10^{-3}$$

$$= -417.9 \text{ Nm (hogging)}$$

Using equation (5.66), the total axial stress at the

US customary units

The cross-sectional area of the member at section YY is $A = 1.12$ in². If the neutral axis of the section is at a height of c_2 above the lower edge of the section, taking first moments of area about this edge gives

$$1.12 \, c_2 = (2 \times 0.4) \times 1 + (0.8 \times 0.4) \times 3.6$$

$$c_2 = 1.743 \text{ in}$$

which means that the distance between the neutral axis and the upper edge of the cross section is $c_1 = (4 - 1.743) = 2.257$ in. Also, the second moment of area about the neutral axis is

$$I = \frac{0.4 \times 2^3}{12} + (0.4 \times 2) \times (1.743 - 1)^2 + \frac{0.4 \times 0.8^3}{12}$$

$$+ (0.4 \times 0.8) \times (1.743 - 3.6)^2$$

$$= 1.829 \text{ in}^4$$

and the bending moment about the neutral axis of the section is the product of the applied force, F, of 15 kip and the distance between its line of action through the center of the member and the neutral axis

$$M = -15 \times 10^3 \times (2 - 1.743)$$

$$= -3855 \text{ lb in (hogging)}$$

Using equation (5.66), the total axial stress at the

<table>
<tr><td>

SI units
upper edge of section YY is

$$\sigma_x = \frac{F}{A} - c_1\frac{M}{I} = \frac{65 \times 10^3}{700 \times 10^{-6}} - 0.056\,43$$

$$\times \frac{(-417.9)}{714.4 \times 10^{-9}} \text{ N/m}^2$$

$$= \underline{126 \text{ MN/m}^2}$$

and at the lower edge is

$$\sigma_x = \frac{F}{A} + c_2\frac{M}{I} = \frac{65 \times 10^3}{700 \times 10^{-6}} + 0.043\,57$$

$$\times \frac{(-417.9)}{714.4 \times 10^{-9}} \text{ N/m}^2$$

$$= \underline{67.4 \text{ MN/m}^2}$$

</td><td>

US customary units
upper edge of section YY is

$$\sigma_x = \frac{F}{A} - c_1\frac{M}{I} = \frac{15 \times 10^3}{1.12}$$

$$- 2.257 \times \frac{(-3855)}{1.829} \text{ psi}$$

$$= \underline{18.1 \text{ ksi}}$$

and at the lower edge is

$$\sigma_x = \frac{F}{A} + c_2\frac{M}{I} = \frac{15 \times 10^3}{1.12}$$

$$+ 1.743 \times \frac{(-3855)}{1.829} \text{ psi}$$

$$= \underline{9.72 \text{ ksi}}$$

</td></tr>
</table>

The variation of axial stress across the section is linear, as shown in Fig. 5.92. We note that in this case the bending moment is not large enough to give rise to compressive stresses. Nevertheless, the small displacement of the neutral axis away from the centerline of the member creates a nearly two to one ratio between the maximum and minimum axial stress.

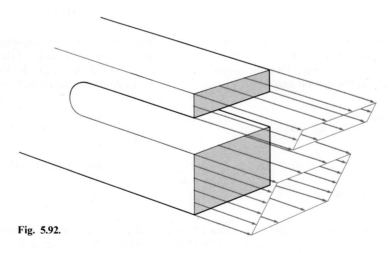

Fig. 5.92.

EXAMPLE 5.22

Figure 5.93 shows the main structural members of a testing machine which is designed to apply a vertical force between the points marked X and Y. The horizontal distance between the line XY and the nearest point, B, of the frame of the machine is 500 mm (20 in). The cross section of the frame at the plane marked CB is in the form of a T as

300

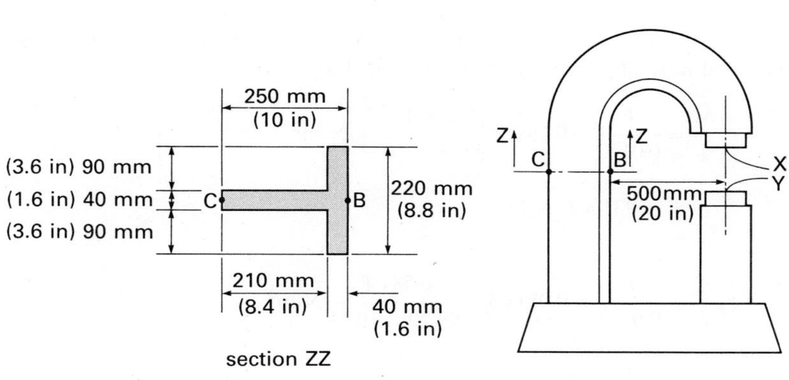

Fig. 5.93.

shown, with an overall depth between C and B of 250 mm (10 in), and a material thickness of 40 mm (1.6 in). Given that the maximum allowable stresses in the cast iron frame at the section through C and B are 50 MN/m² (7.5 ksi) in tension and 100 MN/m² (15 ksi) in compression, and assuming that the weights of the machine components may be neglected, calculate both the maximum compressive and maximum tensile forces which may be applied to objects placed between X and Y.

This example involves a number of features of bending problems: combined bending and axial loading, a cross section which is not symmetrical about its neutral axis, and the need to distinguish between tensile and compressive maximum stresses. The area of the frame cross section through C and B is

$$A = 220 \times 40 + 210 \times 40 = 17\,200 \text{ mm}^2 = 0.0172 \text{ m}^2$$

The neutral axis of this section is parallel to the edge of the cross section which passes through the point B, at a distance of, say, c_2 from it. Taking first moments of area about this edge, we obtain

$$17\,200\,c_2 = 220 \times 40 \times 20 + 210 \times 40 \times (40 + 105)$$

from which

$$c_2 = 81.05 \text{ mm}$$

Also, the distance of point C from the neutral axis is

$$c_1 = 250 - 81.05 = 169.0 \text{ mm}$$

and the second moment of area of the cross section about its neutral axis is

$$I = \frac{220 \times 40^3}{12} + 40 \times 220 \times (81.05 - 20)^2 + \frac{40 \times 210^3}{12}$$

$$+ 40 \times 210 \times (81.05 - 145)^2 \text{ mm}^4 = 99.19 \times 10^{-6} \text{ m}^4$$

The distance of the line XY from the neutral axis is 581.0 mm. Therefore, if a *compressive* force of F newtons is applied to an object between points X and Y, a bending moment of magnitude $M = 0.581\,F$ Nm is applied in the counterclockwise sense at the frame cross section we are considering, tending to cause compressive stresses at C and tensile stresses at B. The force F is also applied as a uniform tensile stress to the cross section. The total normal stresses at points C and B are

$$\sigma_C = \frac{F}{A} - c_1 \frac{M}{I} = \frac{F}{0.0172} - 0.169 \times \frac{0.581F}{99.19 \times 10^{-6}} \ \text{N/m}^2$$

$$= -931.8 \ F \ \text{N/m}^2$$

and

$$\sigma_B = \frac{F}{A} + c_2 \frac{M}{I} = \frac{F}{0.0172} + 0.081\,05 \times \frac{0.581\,F}{99.19 \times 10^{-6}} \ \text{N/m}^2$$

$$= +532.9 \ F \ \text{N/m}^2$$

Now, since the tensile stress at B is more than half the compressive stress at C, a maximum allowable stress is reached first at B, and the corresponding value of F is

$$\text{maximum } F \text{ (compressive)} = \frac{50 \times 10^6}{532.9} \ \text{N} = \underline{93.8 \text{ kN}} \ (21 \text{ kip})$$

On the other hand, if the force applied to the object between X and Y is tensile, the signs of all the stresses are reversed, and a maximum allowable (tensile) stress is reached first at point C, and the corresponding value of F is

$$\text{maximum } F \text{ (tensile)} = \frac{50 \times 10^6}{931.8} \ \text{N} = \underline{53.7 \text{ kN}} \ (12 \text{ kip})$$

Problems

SI METRIC UNITS – SHEAR FORCES AND BENDING MOMENTS

5.1 A light beam of length 3 m which is simply supported at its ends carries a concentrated lateral force of 7 kN at its center. Find the greatest absolute values of shear force and bending moment.

5.2 A light beam of length 3 m which is simply supported at its ends carries a total lateral force of 7 kN uniformly distributed along its length. Find the greatest absolute values of shear force and bending moment.

5.3 A light cantilever beam of length 3 m carries a concentrated lateral force of 7 kN at its free end. Find the greatest absolute values of shear force and bending moment.

5.4 A light cantilever beam of length 3 m carries a total lateral force of 7 kN uniformly distributed along its length. Find the greatest absolute values of shear force and bending moment.

5.5 Draw the shear force and bending moment diagrams for the light beam shown in Fig. P5.5, find expressions for the greatest absolute values of shear

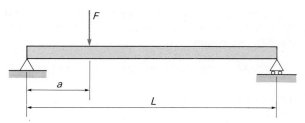

Fig. P5.5.

force and bending moment, and show that these are consistent with Fig. 5.16 when $a = L/2$.

5.6 Draw the shear force and bending moment diagrams for the light beam shown in Fig. P5.6, find expressions for the greatest absolute values of shear

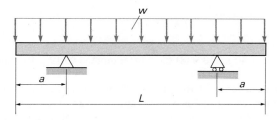

Fig. P5.6.

force and bending moment, and show that these are consistent with Fig. 5.19 when $a = 0$.

5.7 Draw the shear force and bending moment diagrams for the light beam shown in Fig. P5.7, find expressions for the greatest absolute values of shear force and bending moment, and show that these are consistent with Fig. 5.16 when $b = 0$, and consistent with Fig. 5.19 when $b = L$.

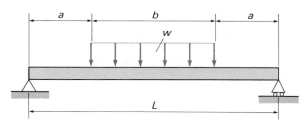

Fig. P5.7.

5.8 Draw the shear force and bending moment diagrams for the light beam shown in Fig. P5.8 by superimposing the distributions shown in Figs 5.16 and 5.19. Find expressions for the greatest absolute values of shear force and bending moment.

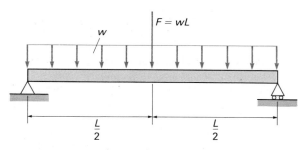

Fig. P5.8.

5.9 Draw the shear force and bending moment diagrams for the light cantilever beam shown in Fig. P5.9, find expressions for the greatest absolute values of shear force and bending moment, and show that these are consistent with Fig. 5.21 when $a = 0$.

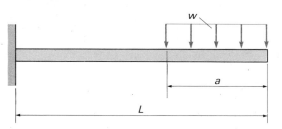

Fig. P5.9.

5.10 Draw the shear force and bending moment diagrams for the light cantilever beam shown in Fig. P5.10 by superimposing the distributions shown in Figs 5.21 and 5.22. Find expressions for the greatest absolute values of shear force and bending moment.

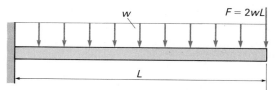

Fig. P5.10.

5.11 to 5.16 Draw the shear force and bending moments diagrams for the light beams shown in Figs P5.11 to P5.16, in each case finding the greatest absolute value of shear force and the greatest sagging and hogging bending moments.

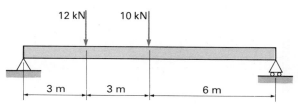

Fig. P5.11.

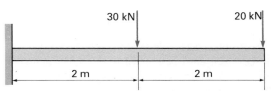

Fig. P5.12.

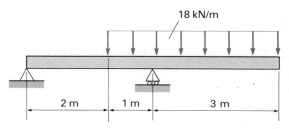

Fig. P5.13.

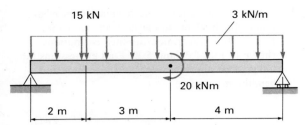

Fig. P5.14.

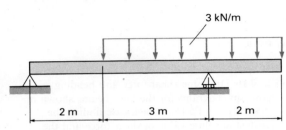

Fig. P5.15.

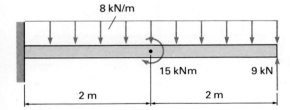

Fig. P5.16.

5.17 In Problem 5.15, find the lateral concentrated force which must be applied to the right-hand end of the beam to minimize the greatest absolute bending moment anywhere along it.

5.18 Use program SDBEAM to confirm the results of Example 5.7.

5.19 Use program SDBEAM to confirm the results of Example 5.6 for the particular case of $b = 3a$.

5.20 to 5.25 Use program SDBEAM to solve Problems 5.11 to 5.16.

5.26 Timber floor joists 6 m long are 300 mm deep and 100 mm broad, and are simply supported at the ends. The timber weighs 6 kN/m³, and the maximum allowable bending moment in each joist is 9 kNm. Determine how far apart, center to center, such joists should be placed when supporting a floor loaded with 2 kN/m².

5.27 Draw the shear force and bending moments diagrams for the light beam shown in Fig. P5.27, which carries a linearly varying distributed lateral force in addition to two concentrated forces. Find the greatest absolute value of shear force and the greatest sagging and hogging bending moments.

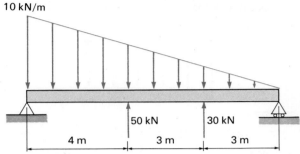

Fig. P5.27.

5.28 Use program SDBEAM to solve Problem 5.27.

5.29 Two vehicles approach each other across a bridge. The bridge may be treated as a simply supported light beam of length L, and the weights of the vehicles as concentrated forces on the beam. Vehicle 1 is of weight W and vehicle 2 of weight $2W$, and vehicle 1 travels at twice the speed of vehicle 2. The vehicles begin to cross simultaneously. Figure P5.29 shows the situation a short time after the vehicles begin to cross, when they have travelled distances of $2x$ and x, respectively, where $x < L/3$. Draw the shear force and bending moment diagrams for the bridge at this time. If the bridge will fail when subject to a bending moment of $WL/2$, show that failure will occur, and find the distance between the vehicles when it does. Find the bending moment the bridge must withstand if the vehicles are to cross without failure occurring.

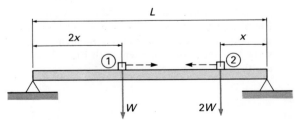

Fig. P5.29.

5.30 Use program SDBEAM to find the greatest absolute value of shear force and the greatest sagging and hogging bending moments in the light beam shown in Fig. P5.30.

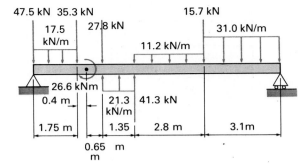

Fig. P5.30.

SI METRIC UNITS – STRESSES DUE TO BENDING

5.31 A long strip of steel 2 mm thick is to be bent into the form of a circle for ease of transportation. Determine the smallest radius of the circle if the normal stress in the steel is not to exceed 200 MN/m².

5.32 A 50 mm by 225 mm timber beam is used as a floor joist, and is simply supported at its ends. If the span is 6 m and the maximum allowable bending stress in the timber is 10 MN/m², find the maximum uniformly distributed lateral load the beam can support.

5.33 For the beam defined in Problem 5.32, calculate the greatest concentrated lateral load that can be supported at its center.

5.34 A box girder 7 m in length is simply supported at two positions, one at 1 m from its left-hand end

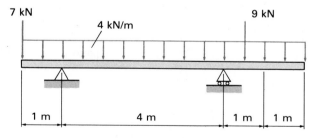

(a)

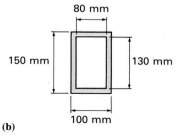

(b)

Fig. P5.34.

and the other at 2 m from its right-hand end, as shown in Fig. P5.34a. The cross section of the girder is a hollow rectangle, with an external width of 100 mm and height 150 mm, and internal width 80 mm and height 130 mm, giving a uniform wall thickness of 10 mm, as shown in Fig. P5.34b. The girder is subjected to a uniformly distributed downward vertical force, including its own weight, of intensity 4 kN/m over its entire length, together with concentrated vertical forces of 7 kN at the left hand end and 9 kN at 1 m from the right hand end. Find: (i) the magnitude and position of the greatest shear force, (ii) the magnitude and position of the greatest bending moment, (iii) the greatest bending stress, and (iv) the greatest shear stress in the beam.

5.35 A beam of rectangular cross section 100 mm deep and 30 mm wide is made of a material which is very weak in shear. If the maximum allowable shear stress is 1.0 MN/m², find the greatest shear force that can be carried by the beam.

5.36 For the beam defined in Problem 5.32, find the greatest shear stress when the beam carries its maximum permissible uniformly distributed lateral force.

5.37 The laminated timber beam shown in Fig. P5.37 is formed by nailing together three layers of timber each 50 mm thick and 150 mm wide. The nails are 4 mm diameter and are arranged in two rows along each of the upper and lower surfaces of the beam with axial spacings of 120 mm. If the maximum allowable shear stress in the nails is 50 MN/m², calculate the maximum shear force the beam can support.

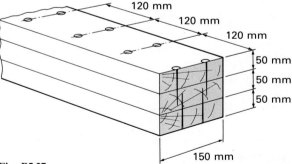

Fig. P5.37.

5.38 The I section steel beam shown in Fig. P5.38 is 933 mm deep and has flanges which are 423 mm wide and 42.7 mm thick. The thickness of the web is 24 mm. If the maximum allowable normal and shear stresses in the steel are 150 and 75 MN/m², respectively, find the greatest bending moment and shear force the beam can support.

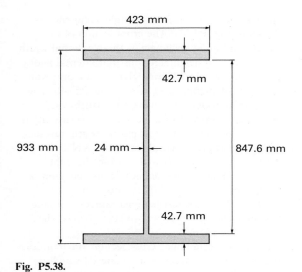

Fig. P5.38.

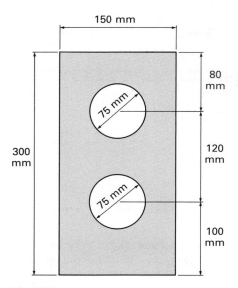

Fig. P5.41.

5.39 If the beam in Problem 5.38 is rotated through 90° about its axis, find the greatest bending moment it can support.

5.40 to 5.42 For each of the beam cross sections shown in Figs P5.40 to P5.42, find the position of the neutral axis for bending in the vertical plane, the second moment of area about this axis, and the greatest bending moment each can support if the maximum allowable bending stress is 100 MN/m² in each case.

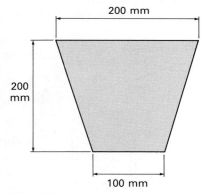

Fig. P5.40.

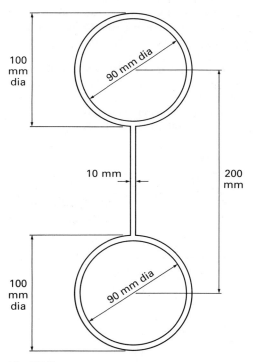

Fig. P5.42.

5.43 Figure P5.43 shows the cross section of a cast iron beam. The maximum allowable bending stresses in the beam are 100 MN/m² in compression and 30 MN/m² in tension. Find the greatest sagging bending moment that can be safely applied, and the corresponding radius of curvature of the beam.

5.44 A horizontal cantilever in the form of an inverted T section as shown in Fig. P5.44 is used to support a hoist for removing catalyst from a reactor. One end is built in, and the vertical load is applied 1.0 m along the cantilever. If the maximum

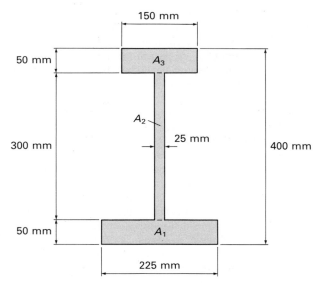

Fig. P5.43.

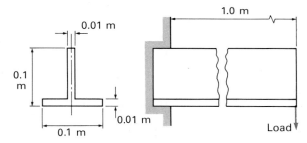

Fig. P5.44.

allowable normal stress in the material of which the cantilever is made is 330 MN/m², determine the greatest load that can be lifted from the reactor. Neglect the weight of the cantilever itself.

5.45 For the cantilever of Problem 5.44, find the maximum shear stress when the greatest load is being lifted.

5.46 A horizontal girder 10 m long is simply supported at its ends and carries two vertical concentrated forces of 60 kN each acting at a point 2 m either side of the center. The girder consists of a rolled steel joist 250 mm deep and 150 mm broad (with a second moment of area about its neutral axis of 80×10^6 mm⁴), with flanges reinforced by plates 225 mm wide and 10 mm thick. Determine how many plates are required at the center of the beam if the maximum bending stress is not to exceed 120 MN/m². Calculate also the minimum length of each plate. Neglect the weight of the girder, and

the contributions of the welds to the area of the cross section.

5.47 A 150 mm by 75 mm timber joist is strengthened by two 100 mm by 12.5 mm steel plates as shown in Fig. P5.47. If the maximum allowable bending stress in the timber is 7.0 MN/m², find the corresponding maximum bending stress in the steel and the maximum simply supported span over which a uniformly distributed force of intensity 8 kN/m can be safely carried.

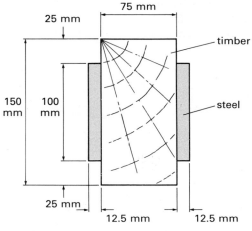

Fig. P5.47.

5.48 In Problem 5.47, if the steel plates are removed, find how broad the timber beam would have to be to give the same strength.

5.49 The simply supported reinforced concrete beam shown in Fig. P5.49 is 300 mm broad and has six 20 mm diameter steel rods at a depth of 750 mm from the top of its rectangular cross section. Assuming that concrete has no strength in tension, calculate the maximum bending stresses in the steel and concrete.

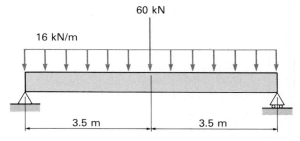

Fig. P5.49.

5.50 In Problem 5.49, find the maximum shear stress at the interface between the steel and concrete.

5.51 In Problem 5.49, if the maximum allowable tensile stress in the steel is 180 MN/m² and the maximum allowable compressive stress in the concrete is 10 MN/m², where along the span are two of the reinforcing rods no longer required?

5.52 The simply supported steel beam shown in Fig. P5.52 is of rectangular cross section with a depth of 75 mm and breadth of 25 mm. It is initially curved in the form of an arc of a circle such that at its center it is 50 mm lower than the supports which are 2 m apart. If hogging bending moments of 2 kNm are applied to the ends of the beam as shown, find the amount by which its center is displaced in the vertical direction.

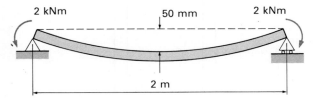

Fig. P5.52.

SI METRIC UNITS – COMBINED BENDING AND AXIAL LOADS

5.53 An open-link chain is formed by bending 20 mm diameter steel rods as shown in Fig. P5.53. If the maximum allowable stress in the steel is 120 MN/m², calculate the greatest tensile force *F* the chain can safely support.

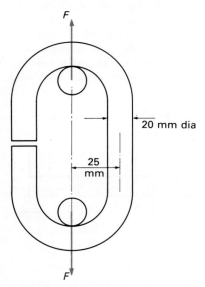

Fig. P5.53.

5.54 A tie bar of rectangular cross section, 20 mm by 60 mm, is subjected to an axial tensile force of 55 kN. It is necessary to remove material from one of the 20 mm faces for part of the length of the bar. Determine the greatest thickness which may be removed if the maximum longitudinal stress is not to exceed 120 MN/m².

5.55 A crane is made by supporting a length of channel section beam vertically, as shown in Fig. P5.55. It is simply supported in bearings at its top and bottom ends, which are 4 m apart, and all vertical force is transmitted through the bottom bearing. The channel section is 150 mm wide, 100 mm deep, and has a constant thickness of 20 mm. A bracket attached to the 150 mm face 3 m from the bottom end supports a vertical load whose line of action is 250 mm away from the face and in the plane of symmetry of the cross section. Find the maximum allowable value of this load if the tensile stress in the channel section is nowhere to exceed 50 MN/m².

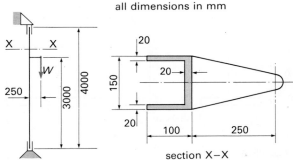

Fig. P5.55.

5.56 The vertical 10 m high mast shown in Fig. P5.56 consists of a hollow steel tube of circular cross section having a constant wall thickness of 12 mm, but with an external diameter which varies linearly with vertical position, from 75 mm at the tip to 150 mm at the base. The mast is subject to both wind loading and the tension in a wire attached to its tip. The wire is inclined at 10° to the vertical, and in the presence of the wind loading carries a tension of 5 kN. The horizontal component of this force is in the opposite direction to the distributed horizontal loading due to wind pressure. This pressure is of intensity 1.0 kN per unit projected area of the mast, meaning that if the external diameter of the mast is *d* meters, the local load per unit length applied to the mast is *d* kN/m. Show that, due to the wind pressure alone, the bending

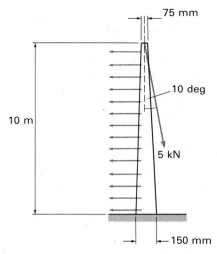

Fig. P5.56.

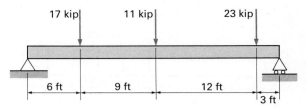

Fig. P5.67.

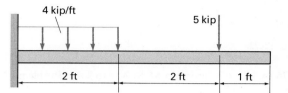

Fig. P5.68.

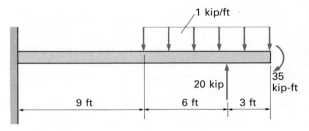

Fig. P5.69.

moment at the base of the mast would be 5.0 kNm. Taking this and the other forms of loading into account, calculate the maximum vertical tensile and compressive stresses at its base.

US CUSTOMARY UNITS – SHEAR FORCES AND BENDING MOMENTS

5.57 A light beam of length 12 ft which is simply supported at its ends carries a concentrated lateral force of 2 kip at its center. Find the greatest absolute values of shear force and bending moment.

5.58 A light beam of length 12 ft which is simply supported at its ends carries a total lateral force of 2 kip uniformly distributed along its length. Find the greatest absolute values of shear force and bending moment.

5.59 A light cantilever beam of length 12 ft carries a concentrated lateral force of 2 kip at its free end. Find the greatest absolute values of shear force and bending moment.

5.60 A light cantilever beam of length 12 ft carries a total lateral force of 2 kip uniformly distributed along its length. Find the greatest absolute values of shear force and bending moment.

5.61 to 5.66 Solve Problems 5.5 to 5.10.

5.67 to 5.72 Draw the shear force and bending moments diagrams for the light beams shown in Figs P5.67 to P5.72, in each case finding the greatest absolute value of shear force and the greatest sagging and hogging bending moments.

5.73 In Problem 5.70, find the lateral concentrated force which must be applied to the right-hand end

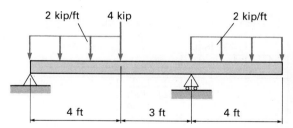

Fig. P5.70.

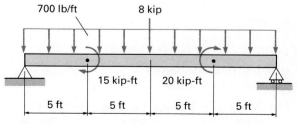

Fig. P5.71.

of the beam to minimize the greatest absolute bending moment anywhere along it.

5.74 and 5.75 Solve Problems 5.18 and 5.19.

309

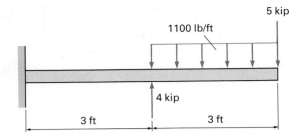

Fig. P5.72.

5.76 to 5.81 Use program SDBEAM to solve Problems 5.67 to 5.72.

5.82 A light beam 6 ft long is simply supported at each end and carries a distributed lateral force over the whole span. The distribution is a quadratic function of position along the beam, with a maximum intensity of 400 lb/ft at the center of the beam and falling to zero at its ends. Find the greatest absolute values of shear force and bending moment.

5.83 Use program SDBEAM to solve Problem 5.82.

5.84 Solve Problem 5.29.

5.85 Use program SDBEAM to find the greatest absolute value of shear force and the greatest sagging and hogging bending moments in the light beam shown in Fig. P5.85.

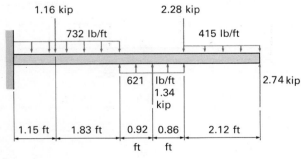

Fig. P5.85.

US CUSTOMARY UNITS –
STRESSES DUE TO BENDING

5.86 A long strip of steel 1/8 in thick is to be bent into the form of a circle for ease of transportation. Determine the smallest radius of the circle if the normal stress in the steel is not to exceed 30 ksi.

5.87 A 2 in by 9 in timber beam is used as a floor joist, and is simply supported at its ends. If the span is 15 ft and the maximum allowable bending

stress in the timber is 1.5 ksi, find the maximum uniformly distributed lateral load the beam can support.

5.88 For the beam defined in Problem 5.87, calculate the greatest concentrated lateral load that can be supported at its center.

5.89 It is proposed to make the discharge pipe from a vessel as shown in Fig. P5.89. The pipe has an internal diameter of 3 in, a thickness of 0.2 in, and is subjected to a total uniformly distributed lateral force of intensity 55 lb/ft over its 15 ft length. If the maximum normal stress the material will withstand is 26 ksi, comment on the proposed design.

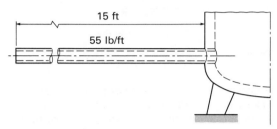

Fig. P5.89.

5.90 For the beam defined in Problem 5.87, find the greatest shear stress when the beam carries its maximum permissible uniformly distributed lateral force.

5.91 Find the greatest shear stress in the discharge pipe in Problem 5.89.

5.92 The box section beam shown in Fig. P5.92 is made from four planks of timber 1 in by 8 in glued together. If the beam is subjected to a shear force of 2.5 kip, find the mean shear stress in the adhesive layers.

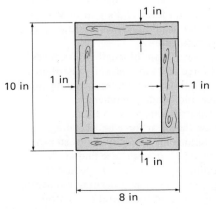

Fig. P5.92.

5.93 to 5.95 For each of the beam cross sections shown in Figs P5.93 to P5.95, find the position of the neutral axis for bending in the vertical plane, the second moment of area about this axis, and the greatest bending moment each can support if the maximum allowable bending stress is 22 ksi in each case.

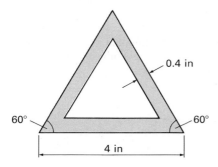

Fig. P5.93.

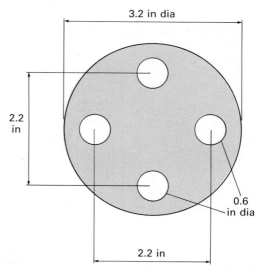

Fig. P5.95.

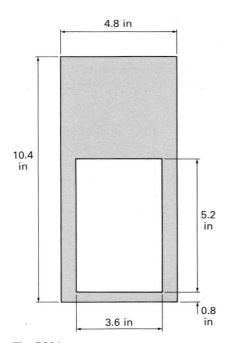

Fig. P5.94.

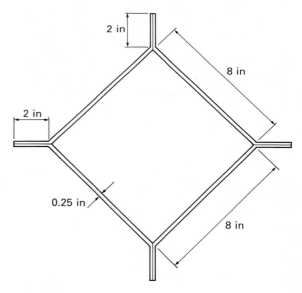

Fig. P5.96.

5.96 The box section beam shown in Fig. P5.96 is made from four steel plates 12 in wide and 1/4 in thick welded together. If the maximum allowable bending stress in the steel is 27 ksi, calculate the greatest bending moment which can be applied to the beam to cause bending in the vertical plane.

5.97 A timber beam having an 8 in by 12 in rectangular cross section is reinforced top and bottom by steel plates 8 in wide and 0.4 in thick as shown in Fig. P5.97. For bending in the vertical plane, find the greatest bending moment which the section can carry if the maximum allowable normal stresses in the steel and timber are 22 ksi and 1.5 ksi, respectively.

5.98 A simply supported reinforced concrete beam is to carry a uniformly distributed load, including its own weight, of 820 lb/ft over a 15 ft span. Design·

311

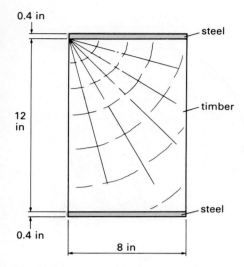

0.4 in

steel

12 in

timber

steel

0.4 in

8 in

Fig. P5.97.

the center section, with appropriate tensile reinforcement, given that the breadth of the beam is to be 6 in, the maximum allowable tensile stress in the steel is 22 ksi and the maximum allowable compressive stress in the concrete is 1.5 ksi. (Note that in an economical design, both the steel and concrete develop their maximum allowable stresses).

5.99 In Problem 5.98, find the maximum shear stress at the interface between the steel and concrete.

US CUSTOMARY UNITS – COMBINED BENDING AND AXIAL LOADS

5.100 The central portion of the bar of circular cross section shown in Fig. P5.100 is offset by an amount equal to its diameter. If the ends of the bar are subjected to uniform tensile stresses of 2 ksi, find the maximum and minimum normal stresses in the offset portion. Changes in geometry due to the loading may be ignored.

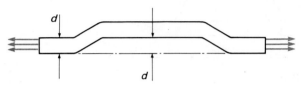

Fig. P5.100.

5.101 A tie bar of rectangular cross section, 1 in by 2.5 in, is subjected to an axial tensile force of 16 kip. It is necessary to remove material from one of the 1 in wide faces for part of the length of the

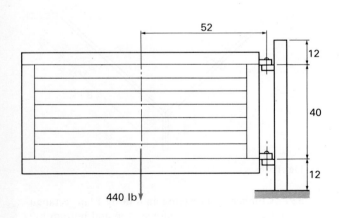

52

12

40

12

440 lb

dimensions in inches

(a)

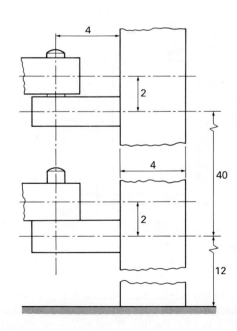

4

2

4

40

2

12

(b)

Fig. P5.103.

312

bar. Determine the greatest thickness which may be removed if the maximum longitudinal stress is not to exceed 18 ksi.

5.102 A cylindrical steel pressure vessel has an external diameter of 20 in and a wall thickness of 0.16 in. It is subjected to an internal gage pressure of 220 psi, and a bending moment of 52 kip-ft is applied at a particular section remote from the ends or other attachments. Calculate the greatest axial and hoop stresses at that section.

5.103 The gate shown in Fig. P5.103a weighs 440 lb. It is supported by two hinges and its center of gravity is at 52 in from the axis of rotation. The hinges are 40 in apart and attached to a hollow square section post with external dimensions of 4 in by 4 in and a uniform wall thickness of 0.12 in. The lower hinge is 12 in above ground level and the overall length of the post is 64 in above the same datum. Details of the hinges are shown in Fig. P5.103b. The hinges are adjusted so that the weight of the gate is supported entirely on the lower hinge. Draw the bending moment diagram for the post, and determine the greatest and least values of the vertical normal stress at the base of the post.

6

Bending of Beams – Deflections

In Chapter 5 we developed methods for finding the distributions of shear force and bending moment in statically determinate beams, based on equilibrium conditions for forces and moments. We also saw how to determine the distributions of bending and shear stresses in any beam, once the bending moments and shear forces are known. In many practical applications, the amount by which a beam deforms is as important as its ability to support the applied loads without failing. We must therefore develop methods for evaluating this deformation. Here it is not so much the internal strains within the beam which are important, but rather the lateral displacements or *deflections* of the whole beam cross section. If the beam is statically indeterminate, we cannot even find the forces, moments and stresses without also considering deflections.

There are two modes of deformation of a beam which contribute to its deflection. Consider, for example, the cantilever shown in Fig. 6.1a, which is subjected to a concentrated force at its free end. Under the action of the resulting bending moment distribution in the beam, it bends into the form of a curve, which is shown much exaggerated in Fig. 6.1b. It is this deformation that we considered when analyzing bending stresses in beams in Chapter 5: if the bending moment in the beam were constant, the curve would be an arc of a circle. The lateral deflection involved, such as v_b at the free end, is the *bending deflection* of the beam. On the other hand, the application of a concentrated force at the free end creates a shear force at each and every cross section of the beam which is equal in magnitude to the applied force. This is essentially the

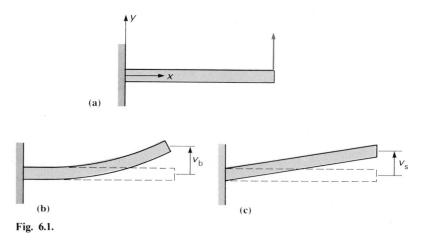

Fig. 6.1.

shearing situation we encountered in Fig. 1.23, and we can expect the beam to suffer shear deformation as shown in Fig. 6.1c, again much exaggerated. The lateral deflection involved, such as v_s at the free end, is the *shear deflection* of the beam. The total deflection is the sum of the contributions from bending and shear. What we will assume, however, is that the *shear deflection is negligibly small compared with the bending deflection.* In terms of the end deflections, we can express this symbolically as

$$v_s \ll v_b \tag{6.1}$$

Some justification for this assumption is provided in Example 6.2.

6.1 Relationship between curvature and bending moment

In order to determine the deflections of beams, we must establish a link between beam deformation and the applied loading. Because we have assumed shear deflections associated with shear forces to be negligible, we can anticipate that it is only bending moments which make a significant contribution. We have in fact already established a link between bending moment and a deformation variable, the radius of curvature, R, of the neutral surface of the beam. This was in equations (5.33), the relevant part of which may now be expressed as

$$\frac{1}{R} = \frac{M}{EI} \tag{6.2}$$

where M is the bending moment at a particular beam cross section, I is the second moment of area of this cross section about its neutral axis, and E is the Young's modulus of the beam material. The product EI is often referred to as the *flexural rigidity* (or bending stiffness) of a beam, and the reciprocal of the radius of curvature is often referred to as the curvature.

Now, before we can use equation (6.1) to determine beam deflections, we must establish a relationship between deflection and curvature. For lateral deflection, which is in the y direction, we use the symbol v. We have already established the sign convention for v, in Section 5.2.1, under which the deflection is regarded as positive when it is in the positive y direction, which is vertically upwards. Although we will usually refer simply to beam deflection, which by implication is the lateral displacement of the entire beam cross section, we should really be more precise and define it as the *lateral displacement of the neutral surface of the beam*. In practice, the internal lateral strains in a beam are so small as to have a negligible effect on lateral displacements of different parts of the cross section.

Figure 6.2 shows a small element of a beam, of length δx, located between the points A and B which lie on the neutral surface. The greatly exaggerated deflected shape of this neutral surface is also shown, with the displaced positions of points A and B being marked as A′ and B′. From points A′ and B′ are drawn both the tangents to the neutral surface, which meet at the point D, and the normals to the surface. Provided δx is small, the radius of curvature of the neutral surface at B′ is very nearly equal to the radius at A′, namely $R(x)$,

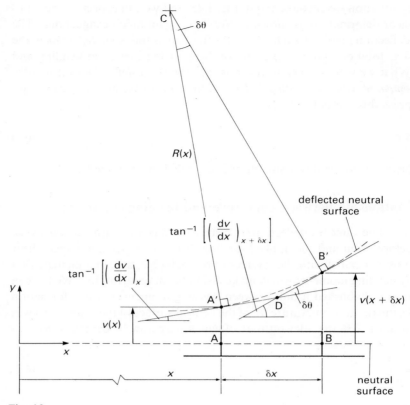

Fig. 6.2.

and the normals from A′ and B′ meet at point C, the center of curvature at point A′. If the very small angle between the tangents at point D is $\delta\theta$, then this is also the angle between the normals at point C, and

$$\delta\theta = \frac{\text{length of arc A′B′}}{R(x)} \approx \frac{\delta x}{R(x)}$$

Also, this angle between the tangents to the curve is the difference between the slopes of the neutral surface at point A′ and point B′ (more precisely, it is the difference between the angles whose trigonometric tangents are equal to the slopes, but both angles are very small)

$$\delta\theta \approx \left(\frac{dv}{dx}\right)_{x+\delta x} - \left(\frac{dv}{dx}\right)_{x}$$

Equating these two expressions for $\delta\theta$, we obtain

$$\frac{1}{R(x)} \approx \frac{\left(\dfrac{dv}{dx}\right)_{x+\delta x} - \left(\dfrac{dv}{dx}\right)_{x}}{\delta x}$$

and, as $\delta x \to 0$, the expression on the right-hand side becomes the derivative with respect to x of the slope, which is the second derivative with respect to x of the deflection, v, and

$$\frac{1}{R(x)} = \frac{d^2v}{dx^2} \tag{6.3}$$

Therefore, equation (6.2) becomes

$$\frac{d^2v}{dx^2} = \frac{M(x)}{EI(x)} \tag{6.4}$$

which is the general form of the *moment–curvature relationship*. Note that the second moment of area may vary with position x along the beam, which will be the case if the cross section of the beam varies. Such problems are discussed in Section 6.2.4. If I is constant, however, the equation is more conveniently expressed as

$$EI\frac{d^2v}{dx^2} = M(x) \tag{6.5}$$

Bearing in mind the differential relationships we established in equation (5.11) between shear force and bending moment, and in equation (5.10) between lateral force and shear force, we can differentiate equation (6.5) with respect to x to obtain

$$EI\frac{d^3v}{dx^3} = \frac{dM}{dx} = V(x) \tag{6.6}$$

and again to get

$$EI\frac{d^4v}{dx^4} = \frac{dV}{dx} = -w(x) \tag{6.7}$$

provided I does not vary along the beam. A potential advantage of equation (6.7) as a starting point for analyzing beam deflections is that it only requires a knowledge of the applied external loads, and not the internal shear forces or bending moments. On the other hand, it is a fourth-order differential equation which must be integrated four times to obtain a solution for deflections.

For manual calculations, it is generally more convenient to start from moment–curvature equations (6.4) or (6.5) – finding an expression for the bending moment distribution along the beam is usually not a difficult task.

Each of these equations is a second-order linear ordinary differential equation. Their forms are, however, particularly straightforward, with the second derivative of deflection with respect to distance x along the beam being equated to a known function of x. This function can be integrated, in many cases analytically, once to obtain the slope of the beam, and a second time to obtain the deflection. Each time we integrate, however, we will introduce an unknown *constant of integration*, in arriving at what in mathematical terms is called the general solution of the differential equation. The values of these constants must be found from the geometry of deformation imposed on the particular beam. Again in mathematical terms, the geometry of deformation defines the *boundary conditions* of the problem. For example, if the beam is simply supported, the deflections at the fixed supports are known, often as zero. Similarly, at the built-in end of a cantilever, both the deflection and slope are known. Other types of boundary conditions are also possible, but these are only relevant to statically indeterminate beams, and are considered in Section 6.3.

We are now ready to start solving problems involving beam deflections, and we will use two different approaches. Firstly, we will develop a manual method which involves integrating the appropriate form of differential moment–curvature equation, and apply this to both statically determinate and statically indeterminate problems. Then we will develop a numerical computer method which is applicable to any type of beam problem. There are a number of other methods, including graphical techniques, available, but these are mostly used when relatively complex problems have to be solved without the aid of a computer.

6.2 Deflection of statically determinate beams

If we initially restrict attention to statically determinate beams, we can define the distribution of bending moment along a given beam from equilibrium considerations alone, and use this distribution to set up the moment–curvature equation.

6.2.1 Solutions for some simple cases

We start with an example of pure bending.

EXAMPLE 6.1 Pure bending of a simply supported beam

Figure 6.3 shows a simply supported uniform beam of length L and negligible weight, with moments of magnitude M_0 applied in the sagging sense at its ends. Find the maximum absolute deflection of the beam, and the slopes at its ends.

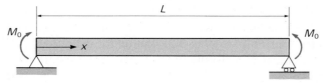

Fig. 6.3.

We found the shear force and bending moment distributions for this pure bending problem in Example 5.3. The bending moment distribution takes the constant form

$$M(x) = M_0$$

and moment–curvature equation (6.5) is therefore

$$EI\frac{d^2v}{dx^2} = M_0$$

where E is the Young's modulus of the beam material, and I is the constant second moment of area of its cross section about the neutral axis. This may be integrated, first to find the slope

$$EI\frac{dv}{dx} = M_0 x + A \tag{6.8}$$

where A is a constant of integration, and then again to find the deflection

$$EIv = M_0\frac{x^2}{2} + Ax + B \tag{6.9}$$

where B is a second constant of integration.

Now, the boundary conditions for this problem require that there is no deflection of the beam at either of the simple supports. At the left-hand support, $x=0$ and $v=0$, numerical values may be substituted into equation (6.9) to give

$$EI \times 0 = M_0 \times 0 + A \times 0 + B$$

from which $B = 0$. Similarly, at the right-hand support, $x=L$ and $v = 0$, giving

$$0 = M_0\frac{L^2}{2} + AL + 0$$

from which

$$A = -M_0\frac{L}{2}$$

Substituting these expressions for the constants A and B into equations (6.8) and (6.9), we obtain the distributions of the slope and deflection of the beam as

$$\frac{dv}{dx} = \frac{M_0}{EI}\left(x - \frac{L}{2}\right) \tag{6.10}$$

and

$$v = \frac{M_0}{2EI}(x^2 - xL) \tag{6.11}$$

These are plotted in Fig. 6.4: the slope distribution is linear and the deflection distribution is parabolic (quadratic). The maximum absolute deflection occurs at the

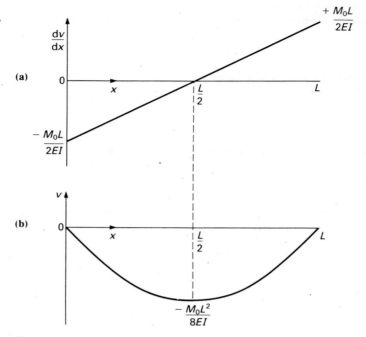

Fig. 6.4.

center of the beam (where the slope is zero), and is found by substituting $x = L/2$ into equation (6.11) as

$$v_{max} = -\frac{M_0 L^2}{8EI} \qquad (6.12)$$

where the negative sign implies that the deflection is in the downward direction. Also, the magnitude of the slope at either end of the beam is found by substituting either $x = 0$ or $x = L$ into equation (6.10) as

$$\text{end slope} = \frac{M_0 L}{2EI} \qquad (6.13)$$

As indicated in Fig. 6.4a, the sign of this slope is negative at the left-hand support and positive at the right-hand support.

It is worth noting that, *for this simple case of pure bending* (but not for other situations involving lateral loading) we can determine the deflected shape of the beam more directly. Under pure bending, the radius of curvature of the beam is constant along its length, and is given by equation (6.2) as

$$R = \frac{EI}{M_0}$$

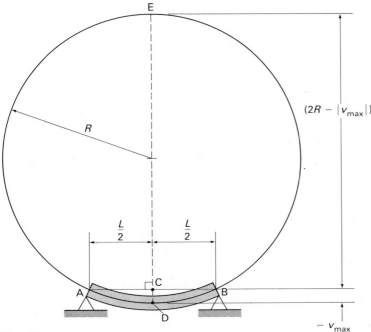

Fig. 6.5.

The beam takes the form of an arc of a circle, of radius R, as shown in Fig. 6.5. If ends of the beam are the points A and B, the center point of its neutral surface is C, which deflects to D, and the line DCE is a diameter of the circle which is normal to AB, then from the geometric properties of a circle we have

$$AC \times CB = DC \times CE$$

or

$$\frac{L}{2} \times \frac{L}{2} = (-v_{max}) \times (2R - |v_{max}|)$$

Now, since in practice $|v_{max}| \ll 2R$ (the maximum deflection is greatly exaggerated in Fig. 6.5), we may approximate $(2R - |v_{max}|)$ by $2R$, and therefore

$$v_{max} = -\frac{L^2}{8R} = -\frac{M_0 L^2}{8EI}$$

This is identical to the result obtained by integrating the moment–curvature equation.

Among statically determinate beam problems, those involving cantilevers are somewhat easier to solve than ones concerning simply supported beams, if only because both boundary conditions are applied at the same end of the beam, which may be chosen as the origin for x. Let us consider now two cases of laterally loaded cantilevers.

EXAMPLE 6.2 Cantilever with a concentrated end force

Figure 6.6 shows a uniform cantilever beam of length L and negligible weight, with a concentrated force of magnitude F applied at its free end. Find the deflection and slope of the beam at this end.

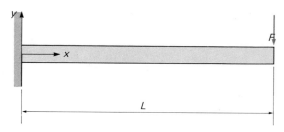

Fig. 6.6.

The shear force and bending moment distributions for this problem are illustrated in Fig. 5.21, and the bending moment distribution takes the form

$$M(x) = -F(L-x)$$

Therefore, using equation (6.5), the moment–curvature relationship for this problem is

$$EI\frac{d^2v}{dx^2} = Fx - FL$$

Repeated integration gives the general solutions for slope and deflection as

$$EI\frac{dv}{dx} = F\frac{x^2}{2} - FLx + A$$

and

$$EIv = F\frac{x^3}{6} - FL\frac{x^2}{2} + Ax + B$$

where A and B are constants of integration. We note that, while in the case of a constant bending moment distribution (Example 6.1) the slope varied linearly and the deflection quadratically with position along the beam, in the present problem with a linearly varying bending moment the slope and deflection are quadratic and cubic functions of x, respectively. The boundary conditions for this problem are that both the deflection and slope of the beam are zero at the built-in end of the cantilever, at $x=0$. Substituting these conditions into the above general solutions for the slope and curvature, we conclude that $A = B = 0$. The required slope and deflection at the free end of the cantilever, found by substituting $x = L$ into the general expressions, are therefore·

$$\left(\frac{dv}{dx}\right)_{x=L} = -\frac{FL^2}{2EI} \quad \text{and} \quad v(L) = -\frac{FL^3}{3EI} \tag{6.14}$$

At the beginning of this chapter, we made the assumption that shear deflections of beams are negligibly small compared with the bending deflections. Indeed, this assumption was made using the example of a cantilever with a concentrated force at its free end (Fig. 6.1 and equation (6.1)). Now that we have found the bending deflection of such a beam, we are in a position to test the validity of the assumption. To do this, we must also have an expression for the shear deflection at the end of the beam. Let us assume that the beam is of rectangular cross section of breadth b and depth d. With a constant shear force in the beam of magnitude F (Fig. 5.21), the mean shear stress is F/bd, and the mean shear strain is $\gamma = F/Gbd$, where G is the shear modulus of the beam material. Therefore, using equation (1.12), the shear deflection, v_s, of the free end of the cantilever is of the order of

$$v_s = \gamma L = \frac{FL}{Gbd}$$

Note that this is only an approximation, because the actual shear stress distribution is parabolic rather than uniform (equation (5.45)). From equation (6.14), the magnitude of the bending deflection, v_b, of the free end is

$$v_b = \frac{FL^3}{3EI} = \frac{4FL^3}{Ebd^3}$$

and the ratio between the shear and bending deflections is

$$\frac{v_s}{v_b} = \frac{Ed^2}{4GL^2} \approx \frac{1}{2}\left(\frac{d}{L}\right)^2 \tag{6.15}$$

The last approximation is based on the assumption that $E = 2G$, which actually corresponds to a zero value of Poisson's ratio, but is adequate for present purposes. For a typical beam whose length is at least ten times its depth this result shows that the shear deflection is less than 1% of the bending deflection, and may therefore be neglected. A more accurate analysis of this problem is presented in Chapter 10, Section 10.3.2, where a similar conclusion is reached concerning the relative magnitude of the shear and bending deflections.

EXAMPLE 6.3 Cantilever with a uniformly distributed force

Figure 6.7 shows a uniform cantilever beam of length L with a downward vertical uniformly distributed force (including its own weight) of intensity w per unit length. Find the deflection and slope of the free end of the beam.

The shear force and bending moment distributions for this problem are illustrated in Fig. 5.22, and the bending moment distribution takes the form

$$M(x) = -\frac{w}{2}(L - x)^2$$

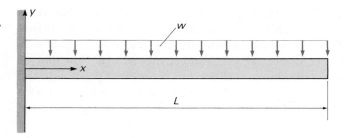

Fig. 6.7.

Therefore, using equation (6.5), the moment–curvature relationship is

$$EI\frac{d^2v}{dx^2} = -\frac{wx^2}{2} + wLx - \frac{wL^2}{2}$$

Repeated integration gives the general solutions for slope and deflection as

$$EI\frac{dv}{dx} = -\frac{wx^3}{6} + \frac{wLx^2}{2} - \frac{wL^2x}{2} + A$$

and

$$EIv = -\frac{wx^4}{24} + \frac{wLx^3}{6} - \frac{wL^2x^2}{4} + Ax + B$$

where A and B are constants of integration. In this case, a quadratic variation of bending moment results in a cubic variation of slope and a quartic variation of deflection. The boundary conditions for this problem are again that deflection and slope of the beam are zero at the built-in end of the cantilever, at $x = 0$, from which $A = B = 0$. The required slope and deflection at the free end of the cantilever, found by substituting $x = L$ into the general expressions, are therefore

$$\left(\frac{dv}{dx}\right)_{x=L} = -\frac{wL^3}{6EI} \quad \text{and} \quad v(L) = -\frac{wL^4}{8EI} \tag{6.16}$$

The same method can be applied to cantilevers with nonuniformly distributed loading. Provided we can integrate (twice) analytically the expression for the bending moment, we can find the slopes and deflections. Results for a number of commonly occurring types of cantilever loading are summarized in Appendix D.

Let us now return to some further examples of simply supported beams. We consider the case of a uniformly loaded beam before that of a beam subjected to a concentrated force, because the latter poses greater difficulties, associated with the discontinuities introduced by such a loading.

EXAMPLE 6.4 Simply supported beam with a uniformly distributed force

Figure 6.8 shows a uniform beam of length L simply supported at its ends and with a uniformly distributed force (including its own weight) of intensity w per unit length. Find the maximum absolute deflection of the beam, and the slopes at its ends.

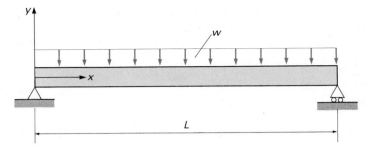

Fig. 6.8.

We found the shear force and bending moment distributions for this problem in Example 5.2. Equation (5.8) defines the bending moment distribution as

$$M(x) = \frac{w}{2}(Lx - x^2)$$

and the moment–curvature relationship is

$$EI\frac{d^2v}{dx^2} = -\frac{wx^2}{2} + \frac{wLx}{2}$$

Repeated integration gives the general solutions for slope and deflection as

$$EI\frac{dv}{dx} = -\frac{wx^3}{6} + \frac{wLx^2}{4} + A \qquad\qquad (6.17)$$

and

$$EIv = -\frac{wx^4}{24} + \frac{wLx^3}{12} + Ax + B \qquad\qquad (6.18)$$

where A and B are constants of integration. The boundary conditions for this problem are those of zero deflection at the beam supports. At the left-hand support, $x = 0$ and $v = 0$, values which may be substituted into equation (6.18) to give $B = 0$. At the right-hand support, $x = L$ and $v = 0$, giving

$$0 = -\frac{wL^4}{24} + \frac{wL^4}{12} + AL$$

from which

$$A = -\frac{wL^3}{24}$$

Substituting this expression for A into equation (6.17), we obtain the distribution of slope as

$$\frac{dv}{dx} = \frac{w}{EI}\left(-\frac{x^3}{6} + \frac{Lx^2}{4} - \frac{L^3}{24}\right)$$

Now, as we would expect from the symmetry of the problem, the slope is zero at the center of the beam, where $x = L/2$, which is therefore also the position of the (mathematical) maximum absolute deflection. The slopes at the ends of the beam are of magnitude

$$\text{end slope} = \frac{wL^3}{24EI} \tag{6.19}$$

but different signs. Also, the deflection distribution is given by equation (6.18) as

$$v(x) = \frac{w}{EI}\left(-\frac{x^4}{24} + \frac{Lx^3}{12} - \frac{L^3 x}{24}\right)$$

and the maximum absolute deflection, at $x = L/2$, is

$$v_{max} = v\left(\frac{L}{2}\right) = -\frac{5wL^4}{384EI} \tag{6.20}$$

EXAMPLE 6.5 Simply supported beam with a concentrated central force

Figure 6.9 shows a uniform simply supported beam of length L and negligible weight, with a downward vertical concentrated force of magnitude F applied at its center. Find the maximum absolute deflection of the beam, and the slopes at its ends.

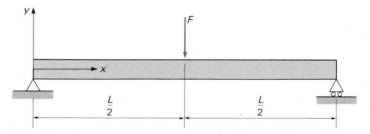

Fig. 6.9.

We found the shear force and bending moment distributions for this problem in Example 5.1. Equation (5.5) defines the bending moment distribution *in the left-hand half of the beam* as

$$M(x) = \frac{Fx}{2} \quad \text{for} \quad x < \frac{L}{2}$$

whereas equation (5.6) defines the distribution in the right-hand half. Because the problem is symmetrical about the center of the beam, we need only consider one half. The moment–curvature relationship for the left-hand half is

$$EI\frac{d^2v}{dx^2} = \frac{Fx}{2} \quad \text{for} \quad x < \frac{L}{2}$$

Repeated integration of this equation gives the general solutions for slope and deflection

$$EI\frac{dv}{dx} = \frac{Fx^2}{4} + A \tag{6.21}$$

and

$$EIv = \frac{Fx^3}{12} + Ax + B \tag{6.22}$$

where A and B are constants of integration. One boundary condition for this problem is that of zero deflection at the left-hand support, or $v = 0$ at $x = 0$, which when substituted into equation (6.22) gives $B = 0$. The second actual boundary condition is that of zero deflection at the right-hand support. But, our equations are only valid for the left-hand half of the beam. Instead, we must make use of the assumed symmetry to set the slope to zero at the center of the beam. Applying this condition to equation (6.21) gives

$$A = -\frac{F}{4}\left(\frac{L}{2}\right)^2 = -\frac{FL^2}{16}$$

and the slope at the left-hand end of the beam is

$$\left(\frac{dv}{dx}\right)_{x=0} = -\frac{FL^2}{16\,EI} \tag{6.23}$$

The negative sign indicates that the slope is in the downward direction at this end. From symmetry, the slope at the right-hand end will be of the same magnitude, but opposite sign. The maximum absolute deflection, at the center of the beam where the slope is zero, is given by equation (6.22) as

$$v_{max} = v\left(\frac{L}{2}\right) = -\frac{FL^3}{48\,EI} \tag{6.24}$$

Results for a number of commonly occurring types of loading applied to simply supported beams are summarized in Appendix D.

6.2.2 Use of step functions

Although Example 6.5 involved a discontinuity in the form of the bending moment distribution due to the presence of a concentrated force acting on the beam, we were able to effectively avoid the discontinuity by taking advantage of the symmetry of the problem. Now let us consider the case of a concentrated force applied at an arbitrary point on the beam, as shown in Fig. 6.10.

Firstly, we can use equations of equilibrium of forces and moments to find the reactions at the support points as

$$R_A = \frac{Fb}{L} \quad \text{and} \quad R_B = \frac{Fa}{L} \tag{6.25}$$

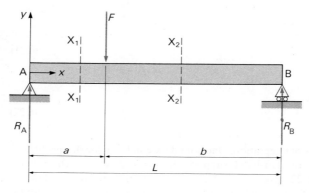

Fig. 6.10.

Considering a typical section $X_1 X_1$ to the left of the concentrated force F as shown, the bending moment there can be found as the sum of the external moments applied to the beam to the left of the section as

$$M(x) = R_A x = \frac{Fbx}{L} \quad \text{for} \quad x \leqslant a$$

where, as usual, x is the distance from the left-hand end of the beam. The moment–curvature relationship is

$$EI \frac{d^2 v}{dx^2} = \frac{Fbx}{L} \quad \text{for} \quad x \leqslant a$$

and repeated integration gives the general solutions for slope and deflection of the beam to the left of the applied force as

$$EI \frac{dv}{dx} = \frac{Fbx^2}{2L} + A_1 \tag{6.26}$$

and

$$EIv = \frac{Fbx^3}{6L} + A_1 x + B_1 \tag{6.27}$$

where A_1 and B_1 are constants of integration.

Similarly, for a typical section $X_2 X_2$ to the right of the concentrated force F as shown

$$EI \frac{d^2 v}{dx^2} = M(x) = R_B(L - x) = \frac{Fa}{L}(L - x) \quad \text{for} \quad x \geqslant a$$

which on integration gives

$$EI \frac{dv}{dx} = \frac{Fa}{L} \left(Lx - \frac{x^2}{2} \right) + A_2 \tag{6.28}$$

and

$$EIv = \frac{Fa}{L}\left(\frac{Lx^2}{2} - \frac{x^3}{6}\right) + A_2 x + B_2 \qquad (6.29)$$

where A_2 and B_2 are additional integration constants, which are not equal to A_1 and B_1.

It must be emphasized that equations (6.26) and (6.27) are only applicable in the region $x \leqslant a$, and equations (6.28) and (6.29) are only applicable in the region $x \geqslant a$. We must now find *four* boundary conditions which will allow us to determine the four constants of integration. The first two of these are the straightforward ones of zero deflection at the supports

1   $v = 0$ at $x = 0$, which must be applied to equation (6.27), giving $B_1 = 0$.
2   $v = 0$ at $x = L$, which must be applied to equation (6.29), giving an equation linking A_2 and B_2

$$A_2 L + B_2 + \frac{FaL^2}{3} = 0 \qquad (6.30)$$

The other two conditions are derived from the compatibility requirement that the two parts of the beam must meet under the concentrated applied force, so that both the deflection and slope are continuous there.

3   Slope is continuous at $x = a$, and equating the slopes defined there by equations (6.26) and (6.28) gives

$$\frac{Fba^2}{2L} + A_1 = \frac{Fa}{L}\left(La - \frac{a^2}{2}\right) + A_2 \qquad (6.31)$$

4   Deflection is continuous at $x = a$, and equating the deflections defined there by equations (6.27) and (6.29) gives

$$\frac{Fba^3}{6L} + A_1 a = \frac{Fa}{L}\left(\frac{La^2}{2} - \frac{a^3}{6}\right) + A_2 a + B_2 \qquad (6.32)$$

Now, equations (6.30), (6.31) and (6.32) provide three linear algebraic equations for finding the three unknowns A_1, A_2 and B_2, but which are relatively laborious to solve.

Even this relatively simple problem of a single concentrated force acting on a beam gives rise to considerable algebraic complexity. Several concentrated forces would be correspondingly worse to deal with (two integration constants and two continuity equations being added for each force). What causes the difficulty is the discontinuity in the form of function defining the bending moment distribution at the concentrated force, which obliges us to use two distinct expressions for moment over the two regions of the beam. Similar difficulties are experienced at any other loading features which cause changes in the form of the bending moment distribution, for example concentrated couples and abrupt changes in the intensities of distributed lateral forces. If we could devise a way of defining the bending moment distribution with a *single* function which is valid for the whole of the beam, then the analysis would be more straightforward. This is possible if we introduce suitable *step functions* (which are sometimes referred to as *singularity functions*).

Let us define a step function, $f(x)$

$$f(x) = \langle x - a \rangle^n \tag{6.33}$$

where a is a constant, and the index n is either zero or any positive number. The use of brackets $\langle$ and $\rangle$ rather than the usual parentheses (and) is quite deliberate, to indicate that the function has the following properties.

1 If $x < a$ then $f(x) = 0$
2 If $x \geqslant a$ then $f(x) = (x - a)^n$
3 The function can be integrated with respect to its independent variable as follows

$$\int \langle x - a \rangle^n \, dx = \frac{\langle x - a \rangle^{n+1}}{n+1} + \text{(a constant)} \tag{6.34}$$

Properties 1 and 2 mean that when the expression within the $\langle \ \rangle$ brackets is negative the function is zero, but that once the expression becomes positive (or zero) the function behaves just as though the brackets are the usual parentheses. In other words, the function appears abruptly or is 'switched on' as the expression within the $\langle \ \rangle$ brackets passes through zero. It is important to recognize that this remains true for even integer values of n, when the function $f(x)$ itself is positive for all values of x. Property 3 means that the step function can be integrated like a normal algebraic expression of this type, but that the resulting integral is itself a step function, with the step occurring at the same value of x. We should note that the bracketed expression must be integrated in its entirety rather than be expanded and integrated term by term. For example

$$\int \langle x - a \rangle \, dx \neq \left\langle \frac{x^2}{2} - ax \right\rangle + \text{(a constant)}$$

Figure 6.11 shows examples of the use of step functions to represent bending moment distributions. In Fig. 6.11a, a concentrated couple of magnitude C is applied to a beam at a distance a from its left-hand end. For simplicity, let us suppose that this end is free, with no other external loads applied to the region shown. To the left of the applied couple, where $x < a$, there is no bending moment in the beam, while to the right the bending moment is of constant magnitude C. Such a distribution can be represented with the aid of the step function with $n = 0$. In Fig. 6.11b, a concentrated force of magnitude F is applied to the beam at a distance a from its left-hand end. Again, to the left of the applied force there is no bending moment, while to the right it is of magnitude $F(x - a)$, a distribution which can be represented with the aid of the step function with $n = 1$. In Fig. 6.11c, a uniformly distributed force of intensity w per unit length is applied to the beam, but only starts at a distance a from its left-hand end. With no bending moment for $x < a$, and a hogging moment of

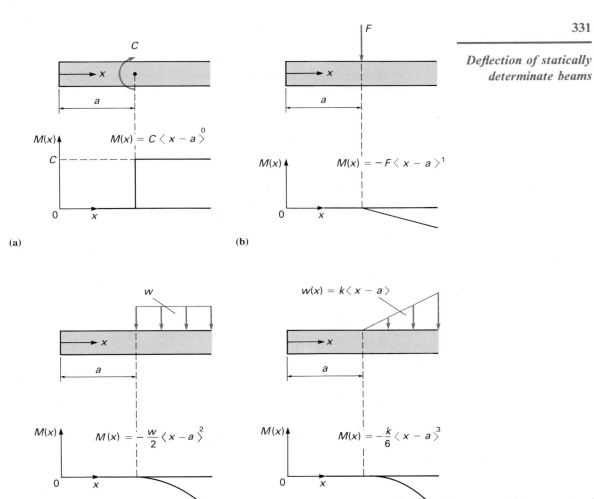

Fig. 6.11.

magnitude $w(x - a)^2/2$ for $x \geq a$, the distribution can be represented with the aid of the step function with $n = 2$. Similarly, Fig. 6.11d shows the case of a linearly varying distributed force which starts at $x = a$, involving the step function with $n = 3$. Other more complicated forms of variation of distributed forces can also be represented with the aid of step functions.

A distributed force which terminates before the right-hand end of a beam can also be accommodated. Consider, for example, the case shown in Fig. 6.12a of a uniformly distributed downward force of intensity w applied to the beam between $x = a$ and $x = b$. Using the step function expression shown in Fig. 6.11c, we can 'switch on' the effect of the force at $x = a$. To switch it off again at $x = b$, we effectively apply a uniformly distributed upward force of the same intensity from this point onwards. This can be done using a very similar step function, as shown in Fig. 6.12b.

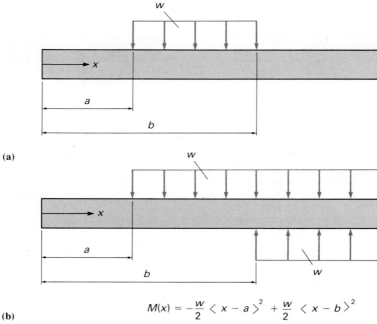

(a)

(b)

$$M(x) = -\frac{w}{2} \langle x - a \rangle^2 + \frac{w}{2} \langle x - b \rangle^2$$

Fig. 6.12.

We are now ready to consider some practical examples of the use of step functions. In these examples, we first establish an expression involving step functions for the bending moment distribution *which is valid for the entire beam*. Using the moment–curvature equation, we then integrate to find the distributions of slope and deflection.

EXAMPLE 6.6 Simply supported beam with a concentrated force

Figure 6.13 shows a uniform simply supported beam of length L and negligible weight, with a downward vertical concentrated force of magnitude F applied at a distance a from its left-hand end. Find the deflection under the force, the deflection at the center of the beam, and the slopes at its ends.

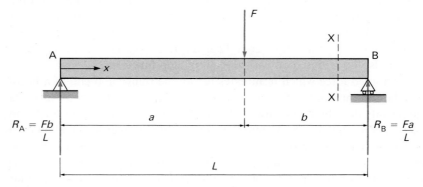

Fig. 6.13.

This problem we have already considered in Fig. 6.10, but without the aid of step functions. In order to establish an expression for the bending moment distribution which is valid for the entire beam, we consider the section XX as shown, at a distance of x along the beam. This section is close to the right-hand end of the beam, and certainly to the right of any concentrated applied forces (other than the support reaction at the extreme right-hand end), concentrated couples or changes in the form of variation of distributed forces. The bending moment there is found by summing the external moments acting on the beam to the left of the section as

$$M(x) = R_A x - F\langle x - a\rangle$$

where $R_A = Fb/L$ is the reaction force at the left-hand support. The step function is used to include the moment of force F only when $x \geqslant a$, just as in Fig. 6.11b. The moment–curvature relationship for the whole beam is

$$EI\frac{d^2v}{dx^2} = M(x) = \frac{Fbx}{L} - F\langle x - a\rangle$$

where EI is the constant flexural rigidity of the beam. Repeated integration gives the general solutions for slope and deflection as

$$EI\frac{dv}{dx} = \frac{Fbx^2}{2L} - \frac{F}{2}\langle x - a\rangle^2 + A \qquad (6.35)$$

and

$$EIv = \frac{Fbx^3}{6L} - \frac{F}{6}\langle x - a\rangle^3 + Ax + B \qquad (6.36)$$

where A and B are constants of integration.

The boundary conditions for this problem are those of zero deflection at the simple supports. Substituting $v = 0$ at $x = 0$ into equation (6.36), and noting that since $x < a$ the step function is then zero, we find that $B = 0$. Also, substituting $v = 0$ at $x = L$, we have

$$0 = \frac{FbL^2}{6} - \frac{F}{6}(L - a)^3 + AL$$

with the step function term now present. Since $(L - a) = b$, constant A may be obtained from this equation as

$$A = -\frac{Fb}{6L}(L^2 - b^2)$$

The deflection under the applied force is found by setting $x = a$ in equation (6.36) as

$$v(a) = \frac{1}{EI}\left[\frac{Fba^3}{6L} - \frac{Fba}{6L}(L^2 - b^2)\right]$$

which, since $L = a + b$, reduces to

$$v(a) = -\frac{Fa^2b^2}{3EIL} \qquad (6.37)$$

Similarly, the deflection at the center of the beam is found by setting $x = L/2$ as

$$v\left(\frac{L}{2}\right) = \frac{1}{EI}\left[\frac{Fb}{6L}\left(\frac{L}{2}\right)^3 - \frac{F}{6}\left\langle\frac{L}{2} - a\right\rangle^3 - \frac{Fb}{6L}(L^2 - b^2)\frac{L}{2}\right]$$

Simplification of this expression depends on the relative magnitudes of a and $L/2$. Let us first assume that $a > L/2$, so that the step function term is zero (the quantity within the $\langle \quad \rangle$ brackets is negative), giving

$$v\left(\frac{L}{2}\right) = \frac{1}{EI}\left[\frac{FbL^2}{48} - \frac{Fb}{12}(L^2 - b^2)\right] = -\frac{Fb}{48EI}(3L^2 - 4b^2) \tag{6.38}$$

If, on the other hand, we assume that $a < L/2$, so that the step function term is not zero, we obtain after rather more manipulation

$$v\left(\frac{L}{2}\right) = -\frac{Fa}{48EI}(3L^2 - 4a^2)$$

This is identical in form to equation (6.38), but with a in place of b: in each case, the shorter of the two lengths appears. This symmetry of form is to be expected because the two cases can be effectively interchanged by viewing the problem from either the front or back.

We should note that, in general, neither equation (6.37) nor equation (6.38) gives the maximum absolute deflection. If this were required, we would have to find its position as that of the point of zero slope. The only situation where they do both give the maximum absolute deflection is where $a = b = L/2$, and

$$v(a) = v\left(\frac{L}{2}\right) = -\frac{FL^3}{48EI}$$

which, as we should expect, is identical to the result obtained in equation (6.24) for a concentrated central force on the beam.

Finally, we may obtain expressions for the slopes at the ends of the beam from equation (6.35) by setting $x = 0$ and $x = L$, respectively.

$$\left(\frac{dv}{dx}\right)_{x=0} = \frac{A}{EI} = -\frac{Fb}{6EIL}(L^2 - b^2) \tag{6.39}$$

and

$$\left(\frac{dv}{dx}\right)_{x=L} = \frac{1}{EI}\left[\frac{FbL}{2} - \frac{F}{2}(L - a)^2 + A\right]$$

which reduces to

$$\left(\frac{dv}{dx}\right)_{x=L} = +\frac{Fa}{6EIL}(L^2 - a^2) \tag{6.40}$$

Note again the similar forms of equations (6.39) and (6.40). The change of sign merely reflects the fact that the beam slopes downward (negative slope) at its left-hand end, and upward (positive slope) at its right-hand end. Figure 6.14 shows the form of the deflected shape of the beam. When $a = b = L/2$ and the beam is centrally loaded,

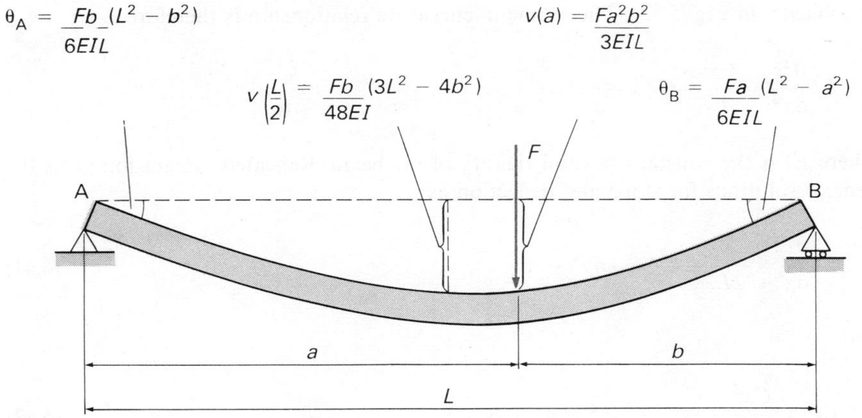

$$\theta_A = \frac{Fb}{6EIL}(L^2 - b^2)$$

$$v(a) = \frac{Fa^2b^2}{3EIL}$$

$$v\left(\frac{L}{2}\right) = \frac{Fb}{48EI}(3L^2 - 4b^2)$$

$$\theta_B = \frac{Fa}{6EIL}(L^2 - a^2)$$

Fig. 6.14.

equation (6.39) reduces to

$$\left(\frac{dv}{dx}\right)_{x=0} = -\frac{FL^2}{16EI}$$

which is identical to equation (6.23). For this symmetrical case, equation (6.40) gives an end slope of the same magnitude, but opposite sign.

EXAMPLE 6.7 Simply supported beam with a concentrated couple

Figure 6.15 shows a uniform simply supported beam of length L and negligible weight, with a counterclockwise concentrated couple of magnitude C applied at a distance a from its left-hand end. Find the slope and deflection of the beam at the applied couple, and the greatest absolute deflection.

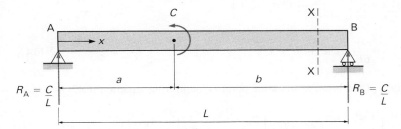

Fig. 6.15.

We have already determined the shear force and bending moment distributions for this problem in Example 5.6. Considering the section XX close to the right-hand end of the beam, at a distance x from the left-hand end, we establish an expression for the bending moment distribution which is valid for the entire beam as

$$M(x) = R_A x - C \langle x - a \rangle^0$$

where $R_A = C/L$ is the reaction force at the left-hand support. This distribution is shown graphically in Fig. 5.32c. The moment–curvature relationship is therefore

$$EI\frac{d^2v}{dx^2} = \frac{Cx}{L} - C\langle x - a\rangle^0$$

where EI is the constant flexural rigidity of the beam. Repeated integration gives the general solutions for slope and deflection as

$$EI\frac{dv}{dx} = \frac{Cx^2}{2L} - C\langle x - a\rangle^1 + A \tag{6.41}$$

and

$$EIv = \frac{Cx^3}{6L} - \frac{C}{2}\langle x - a\rangle^2 + Ax + B \tag{6.42}$$

where A and B are constants of integration.

The boundary conditions are those of zero deflection at the simple supports. Substituting $v = 0$ at $x = 0$ into equation (6.42), we find that $B = 0$. Also, substituting $v = 0$ at $x = L$, we have

$$0 = \frac{CL^2}{6} - \frac{C}{2}(L - a)^2 + AL$$

Since $(L - a) = b$, this equation gives the constant A as

$$A = \frac{C}{6L}(3b^2 - L^2)$$

The slope of the beam at the applied couple is found by setting $x = a$ in equation (6.41) as

$$\left(\frac{dv}{dx}\right)_{x=a} = \frac{1}{EI}\left[\frac{Ca^2}{2L} + \frac{C}{6L}(3b^2 - L^2)\right]$$

$$= \frac{C}{6EIL}(3a^2 + 3b^2 - L^2) \tag{6.43}$$

The deflection of the beam at the same point is found from equation (6.42) as

$$v(a) = \frac{1}{EI}\left[\frac{Ca^3}{6L} + \frac{Ca}{6L}(3b^2 - L^2)\right]$$

$$= \frac{Cab}{3EIL}(L - 2a) \tag{6.44}$$

If the couple is applied at the center of the beam, so that $a = L/2$, then the point of application does not deflect. A point of application not at the center is deflected either upward or downward according to whether a is less than or greater than $L/2$.

Now, we also need to find the greatest absolute deflection. Since the ends of the beam do not deflect, the point of greatest deflection must also be a mathematical maximum (in magnitude), and occurs where the slope of the beam is zero. From equation (6.41), the required value of x is given by

$$\frac{Cx^2}{2L} - C\langle x - a \rangle + \frac{C}{6L}(3b^2 - L^2) = 0 \qquad (6.45)$$

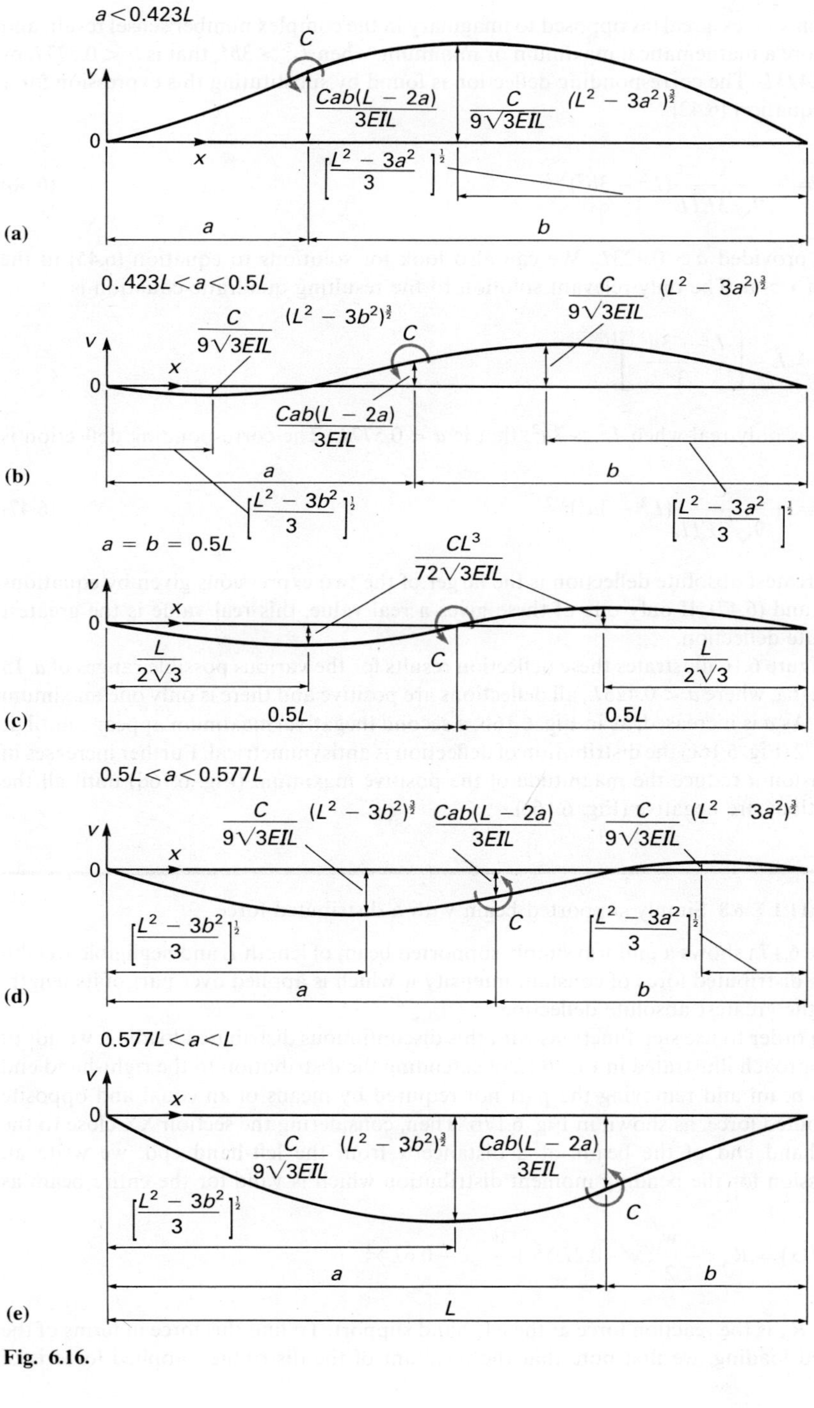

Fig. 6.16.

We first look for solutions in the region $x < a$, where the step function is zero, and we find

$$x = \left[\frac{L^2 - 3b^2}{3} \right]^{1/2}$$

This only gives a real (as opposed to imaginary in the complex number sense) result, and therefore a mathematical maximum or minimum, when $L^2 > 3b^2$, that is $b < 0.577L$ or $a > 0.423L$. The corresponding deflection is found by substituting this expression for x into equation (6.42)

$$v = -\frac{C}{9\sqrt{3}EIL}(L^2 - 3b^2)^{3/2} \tag{6.46}$$

again provided $a > 0.423L$. We can also look for solutions to equation (6.45) in the region $x > a$. The only relevant solution to the resulting quadratic equation is

$$x = L - \left[\frac{L^2 - 3a^2}{3} \right]^{1/2}$$

which is only real when $L^2 > 3a^2$, that is $a < 0.577L$. The corresponding deflection is

$$v = +\frac{C}{9\sqrt{3}EIL}(L^2 - 3a^2)^{3/2} \tag{6.47}$$

The greatest absolute deflection is the larger of the two expressions given by equations (6.46) and (6.47). If only one of these gives a real value, this real value is the greatest absolute deflection.

Figure 6.16 illustrates these deflection results for the various possible ranges of a. In Fig. 6.16a, where $a < 0.423L$, all deflections are positive and there is only one maximum point. As a is increased, as in Fig. 6.16b, a second (negative) maximum appears, until at $a = L/2$ (Fig. 6.16c) the distribution of deflection is antisymmetrical. Further increases in dimension a reduce the magnitude of the positive maximum (Fig. 6.16d) until all the deflections are negative (Fig. 6.16e).

EXAMPLE 6.8 Simply supported beam with a distributed force

Figure 6.17a shows a uniform simply supported beam of length L and negligible weight, with a distributed force of constant intensity w which is applied over part of its length. Find the greatest absolute deflection.

In order to use step functions with this discontinuous distributed loading, we adopt the approach illustrated in Fig. 6.12 of extending the distribution to the right-hand end of the beam and removing the part not required by means of an equal and opposite distributed force, as shown in Fig. 6.17b. Then, considering the section XX close to the right-hand end of the beam, at a distance x from the left-hand end, we write an expression for the bending moment distribution which is valid for the entire beam as

$$M(x) = R_A x - \frac{w}{2}\langle x - 0.2L \rangle^2 + \frac{w}{2}\langle x - 0.6L \rangle^2$$

where R_A is the reaction force at the left-hand support. To find this force in terms of the applied loading, we first note that the resultant of the distributed applied force is of

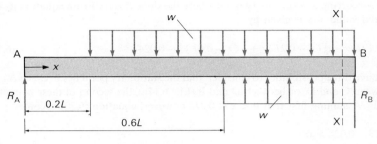

(a)

(b)

Fig. 6.17.

magnitude $0.4wL$, and acts downwards on the beam at a distance of $0.6L$ from the right-hand end **B**. Therefore, taking moments about this end, we have

$$R_A L - (0.4wL) \times (0.6L) = 0$$

and

$$R_A = 0.24wL$$

The moment–curvature relationship is therefore

$$EI \frac{d^2v}{dx^2} = M(x) = 0.24wLx - \frac{w}{2} \langle x - 0.2L \rangle^2 + \frac{w}{2} \langle x - 0.6L \rangle^2$$

where EI is the constant flexural rigidity of the beam, and this equation may be integrated twice to give the general solutions for slope and deflection as

$$EI \frac{dv}{dx} = 0.12wLx^2 - \frac{w}{6} \langle x - 0.2L \rangle^3 + \frac{w}{6} \langle x - 0.6L \rangle^3 + A \qquad (6.48)$$

and

$$EIv = 0.04wLx^3 - \frac{w}{24} \langle x - 0.2L \rangle^4 + \frac{w}{24} \langle x - 0.6L \rangle^4 + Ax + B \qquad (6.49)$$

where A and B are constants of integration. The boundary conditions are those of zero deflection at the simple supports. Substituting $v = 0$ at $x = 0$ into equation (6.49), we find that $B = 0$. Also, substituting $v = 0$ at $x = L$, we have

$$0 = 0.04wL^4 - \frac{w}{24}(0.8L)^4 + \frac{w}{24}(0.4L)^4 + AL$$

from which

$$A = -0.024wL^3$$

In order to find the greatest absolute deflection, we note first that since the ends of the beam do not deflect the required value must be a mathematical maximum or minimum, occurring at a stationary point where the slope is zero. From equation (6.48), the required value of x is given by

$$0 = 0.12Lx^2 - \tfrac{1}{6}\langle x - 0.2L\rangle^3 + \tfrac{1}{6}\langle x - 0.6L\rangle^3 - 0.024L^3 \tag{6.50}$$

Now, we have three possibilities to consider: that the stationary point lies between A and C, or between C and D, or between D and B (Fig. 6.17a), the second of these being the most likely. Assuming first that $0 < x < 0.2L$, however, equation (6.50) becomes

$$0.12X^2 - 0.024 = 0$$

or

$$X = \pm 0.447$$

where $X = x/L$ is the value of x nondimensionalized with respect to beam length L. Since neither of these values of $x = \pm 0.447L$ is within the assumed range, there is no stationary point in the deflection distribution between A and C. Now, assuming that $0.2 < X < 0.6$, equation (6.50) becomes

$$0.12X^2 - \tfrac{1}{6}(X - 0.2)^3 - 0.024 = 0 \tag{6.51}$$

which is a cubic equation in X. Although there are ways of solving such equations analytically, they are laborious. Much more convenient are computer methods which can find numerical solutions to a wide range of algebraic equations arising in engineering problems. Such methods are discussed in Appendix E, where a computer program named ROOT is used to find a solution to equation (6.51), in the required range, of

$$X = 0.4801$$

Hence, the required greatest absolute deflection is found by substituting this value into equation (6.49)

$$v = \frac{wL^4}{EI}(0.04 \times 0.4801^3 - \tfrac{1}{24}(0.4801 - 0.2)^4 - 0.024 \times 0.4801)$$

$$= -7.35 \times 10^{-3}\frac{wL^4}{EI}$$

Had we also explored the range $0.6 < X < 1.0$ between points D and B, we would not have found another stationary point.

6.2.3 Solution by superposition

In Chapter 5 we used superposition to construct shear force and bending moment diagrams for beams by combining relatively simple standard cases. Provided the stress–strain behavior of the beam material remains linearly elastic, we can do the same for slopes and deflections, often with a considerable saving in the amount of analysis involved. Let us consider two examples of the application of superposition to finding deflections of statically determinate beams: Examples 6.12 and 6.14 demonstrate its application to statically indeterminate problems. The results for standard cases given in Appendix D are particularly useful for this purpose.

EXAMPLE 6.9 Use of superposition

Figure 6.18 shows a uniform beam of length L and weight W simply supported at its ends and carrying a concentrated vertical force of $3W$ applied at its center. Find the greatest absolute deflection.

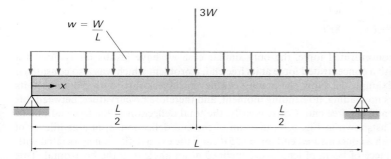

Fig. 6.18.

Symmetry of the problem means that the greatest absolute deflection must occur at the center of the beam. The loading is a combination of a uniformly distributed force due to the weight of beam, of intensity $w = W/L$ per unit length, and a concentrated force at the center of magnitude $F = 3W$. Expressions for central deflections for both of these cases are given in Appendix D, and were derived in detail in Examples 6.4 and 6.5. Using equations (6.20) and (6.24), the central deflection is given by

$$v\left(\frac{L}{2}\right) = -\frac{5wL^4}{384EI} - \frac{FL^3}{48EI}$$

$$= -\frac{5WL^3}{384EI} - \frac{3WL^3}{48EI}$$

$$= -\frac{29WL^3}{384EI}$$

EXAMPLE 6.10 Use of superposition

Figure 6.19 shows a uniform cantilever beam of length L and weight W with a concentrated vertical force of $5W$ applied at its center. Find the deflection of the free end of the beam.

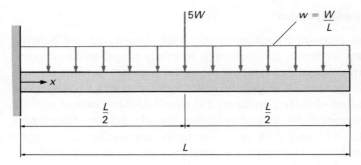

Fig. 6.19.

The loading is a combination of a uniformly distributed force due to the weight of beam, of intensity $w = W/L$ per unit length, and a concentrated force at the center of magnitude $F = 5W$. Let the end deflections due to each of the loads acting alone be v_1 and v_2, respectively. Deflection v_1 due to the uniformly distributed force is given in Appendix D, and was derived in detail in Example 6.3, equations (6.16), as

$$v_1 = -\frac{wL^4}{8EI} = -\frac{wL^3}{8EI}$$

As for the concentrated force, the only similar case we have results for is that of a cantilever with a concentrated force at its free end, a problem which was also considered in detail in Example 6.2. Now, the present cantilever, subject only to the $5W$ force at its center, suffers no loading, no bending moment, and therefore no curvature, between this center point and the free end. Consequently, the total deflection of the free end due to this force is, as shown in Fig. 6.20, the sum of the end deflection, δ, of a cantilever of length $L/2$ with a concentrated end force of $5W$, and the extra deflection associated with the straight piece of beam of length $L/2$ inclined at an angle θ to the horizontal. This angle is equal to the slope of the loaded half of the cantilever at the point of application of the concentrated force. From equations (6.14)

$$\delta = -\frac{(5W)}{3EI}\left(\frac{L}{2}\right)^3 = -\frac{5WL^3}{24EI}$$

and

$$\theta = -\frac{(5W)}{2EI}\left(\frac{L}{2}\right)^2 = -\frac{5WL^2}{8EI}$$

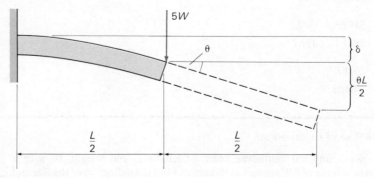

Fig. 6.20.

The total end deflection, v_2, due to the $5W$ force is therefore

$$v_2 = \delta + \frac{\theta L}{2} = -\frac{25WL^3}{48EI}$$

and due to both forms of loading is

$$v = v_1 + v_2 = -\frac{WL^3}{8EI} - \frac{25\,WL^3}{48EI}$$

$$= -\frac{31\,WL^3}{48EI}$$

6.2.4 Beams of varying cross section

So far in this chapter, we have been concerned with the deflection of beams of uniform cross section. If the dimensions of the cross section vary, the methods of analysis are still essentially the same, but we must use not equation (6.5) as the moment–curvature relationship, but the more general form given by equation (6.4), which is

$$\frac{d^2v}{dx^2} = \frac{M(x)}{EI(x)} \tag{6.52}$$

The presence of a variable second moment of area in the denominator of the right-hand side of this equation, where the form of variation may be relatively complicated, makes repeated integration by analytical methods much more difficult. Indeed, it may well be necessary to resort to numerical computer methods. Let us consider, however, a case where analytical solution is possible.

EXAMPLE 6.11 Tapered cantilever

Figure 6.21 shows a steel cantilever beam with a concentrated vertical force of magnitude $F = 10\,\text{kN}$ (2.2 kip) at its free end. The cross section of the beam is rectangular with a constant width, b, of 0.1 m (4 in), and depth, h, which tapers linearly from $h_0 = 0.3\,\text{m}$ (12 in) at the built-in end to $h_0/2$ at the free end, over a length of $L = 5\,\text{m}$ (16 ft). Find the deflection of the free end produced by the force F.

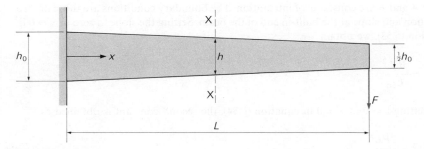

Fig. 6.21.

Note that, because we are only seeking the deflection due to F, we do not need to consider the deflection due to the weight of the beam, which would be significant in this case. Although we are given numerical values for the beam dimensions and loading, let us first work through the problem in symbols. If we consider the typical cross section marked XX, at a distance x from the built-in end of the beam and where the depth of the beam is h, the bending moment there is

$$M(x) = -F(L-x)$$

Since the variation of beam depth is linear, we can express h as a function of x as follows

$$h = h_0\left(1 - \frac{x}{2L}\right)$$

and the second moment of area of the beam cross section about its neutral axis as

$$I = \frac{bh^3}{12} = \frac{bh_0^3}{12}\left(1 - \frac{x}{2L}\right)^3 = I_0\left(1 - \frac{x}{2L}\right)^3$$

where I_0 is the second moment of area at the built-in end of the beam.

Moment–curvature equation (6.52) therefore becomes

$$\frac{d^2v}{dx^2} = -\frac{F}{EI_0}(L-x)\left(1 - \frac{x}{2L}\right)^{-3}$$

which is much more difficult to integrate analytically than the forms of equations we derived for beams of uniform cross section. We can, however, express it in partial fraction form as

$$\frac{d^2v}{dx^2} = -\frac{FL}{EI_0}\left[2\left(1 - \frac{x}{2L}\right)^{-2} - \left(1 - \frac{x}{2L}\right)^{-3}\right]$$

which can now be integrated to find first the slope

$$\frac{dv}{dx} = -\frac{FL^2}{EI_0}\left[4\left(1 - \frac{x}{2L}\right)^{-1} - \left(1 - \frac{x}{2L}\right)^{-2}\right] + A \qquad (6.53)$$

and then the deflection

$$v = -\frac{FL^3}{EI_0}\left[-8\ln\left(1 - \frac{x}{2L}\right) - 2\left(1 - \frac{x}{2L}\right)^{-1}\right] + Ax + B \qquad (6.54)$$

where A and B are constants of integration. The boundary conditions are those of zero deflection and slope at the built-in end of the beam. Setting the slope to zero at $x = 0$ in equation (6.53), we obtain

$$A = \frac{3FL^2}{EI_0}$$

and, setting $v = 0$ at $x = 0$ in equation (6.54), the second constant is obtained as

$$B = -\frac{2FL^3}{EI_0}$$

The distribution of deflection is therefore given by

$$v(x) = -\frac{FL^3}{EI_0}\left[-8\ln\left(1 - \frac{x}{2L}\right) - 2\left(1 - \frac{x}{2L}\right)^{-1} - \frac{3x}{L} + 2\right]$$

and the deflection of the free end of the cantilever is

$$v(L) = -\frac{FL^3}{EI_0}[-8\ln 0.5 - 5] = -0.5452\frac{FL^3}{EI_0}$$

SI units

In the present problem, $F = 10^4$ N, $L = 5$ m, $E = 207 \times 10^9$ N/m^2 (Appendix A), and

$$I_0 = \frac{bh_0^3}{12} = \frac{0.1 \times 0.3^3}{12} = 2.25 \times 10^{-4}\ \text{m}^4$$

giving

$$v(L) = -\frac{0.5452 \times 10^4 \times 5^3}{207 \times 10^9 \times 2.25 \times 10^{-4}} = -0.01463\ \text{m}$$

$$= -14.6\ \text{mm}$$

the negative sign indicating a downward deflection.

US customary units

In the present problem, $F = 2200$ lb, $L = 16$ ft $= 192$ in, $E = 30 \times 10^6$ psi (Appendix A), and

$$I_0 = \frac{bh_0^3}{12} = \frac{4 \times 12^3}{12} = 576\ \text{in}^4$$

giving

$$v(L) = -\frac{0.5452 \times 2200 \times 192^3}{30 \times 10^6 \times 576}$$

$$= -0.491\ \text{in}$$

the negative sign indicating a downward deflection.

6.3 Deflection of statically indeterminate beams

In Section 5.1.1, equation (5.1), we established that for a beam to be statically determinate is must have three, and only three, unknown support reactions (forces or moments) applied to it. While a roller support involves just one unknown reaction, in general a pinned support involves two (although the axial force along the beam is often zero), and a built-in support involves three (two forces and one moment). The three unknowns can be found using three equations of equilibrium for the beam as a whole, without taking account of its deformation. If the supports are such that the total number of unknown reactions exceeds three, then the beam is statically indeterminate, and we must analyze the deflections in order to find even the forces and moments. As in the case of pin-jointed structures in Section 2.1.1, the difference between the number of unknown reactions and the number of equilibrium equations (three) gives the number of redundancies for the system.

If a beam is statically indeterminate and deflections have to be taken into account, we may have to consider forms of support in addition to the pinned, roller and built-in types we have already met. In particular, we may be concerned with flexible supports which, under the loads applied to them do deform elastically, if only by relatively small amounts. Indeed, in practice all supports, even very massive steel or concrete structures, have at least some small flexibility, because no material is perfectly rigid. Figure 6.22a shows a

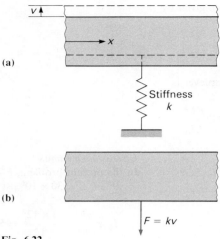

(a)

(b)

Fig. 6.22.

model for a flexible support, in the form of a spring located between the beam and a rigid surface. As in the case of pinned and roller supports, we assume that the spring makes contact with the beam at a single point. What we need to know to characterize the behavior of the support is its stiffness, k, which is the ratio between the reaction force it provides and the deflection of the beam at the support. If this deflection is v (upward positive according to our sign convention), the spring representing the support is extended by an amount v, and the reaction force provided by the spring is, as shown in Fig. 6.22b

$$F = kv \tag{6.55}$$

in the downward direction. In this case, we assume that the spring provides no reaction force along the beam, also that there is no force in the spring when the beam is in its undeflected state.

Rotational stiffness of supports may also be significant. Figure 6.23a shows the end of a beam embedded in a support which, rather than being perfectly rigid may have some flexibility, not only in the vertical direction but also in the rotational sense. Suppose that we know the rotational stiffness, λ, of the support, which is the ratio between the reaction moment it provides and the rotation of the beam. If the cross section of the beam where it enters the support rotates through an angle θ as shown (which, provided it is small, is equal to the slope of the beam there), the reaction moment at the end of the beam is, as illustrated in Fig. 6.23b

$$M = \lambda\theta = \lambda\frac{dv}{dx} \tag{6.56}$$

The moment always opposes the rotation: in this case, a clockwise moment is associated with a counterclockwise rotation. Supports with rotational stiffness need not be confined to the ends of beams.

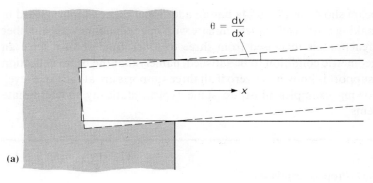

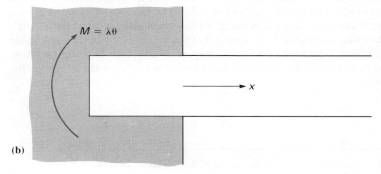

Fig. 6.23.

Our general approach to analyzing statically indeterminate beam problems is similar to the one we used to find the deflections of statically determinate beams. In other words, we use the moment–curvature relationship to set up a second-order differential equation for beam deflection, which is then solved with the appropriate boundary conditions to impose the geometry of deformation of the problem. The main difference is that for statically indeterminate beams we cannot determine all of the support reaction forces and moments before we start to analyze deflections. The redundant reactions can, however, be found when all of the boundary conditions for slope and deflection are applied. Figure 6.24 illustrates an example of what is involved. The simply

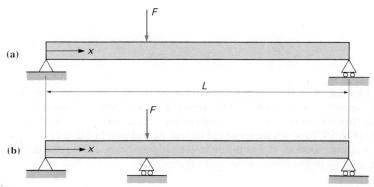

Fig. 6.24.

supported beam shown in Fig. 6.24a has an additional roller support added in Fig. 6.24b, making it statically indeterminate with one redundancy (the number of support reactions is increased from three to four). But, we also have an additional geometric condition to be satisfied, namely that the beam deflection at the extra support is known – as zero if all three supports are at the same level.

The following examples illustrate some typical statically indeterminate beam problems.

EXAMPLE 6.12 Propped cantilever

The uniform cantilever beam of length L shown in Fig. 6.25a is rigidly built in at one end and rests at its center on a flexible support of stiffness k. The second moment of area about the neutral axis of the beam in bending in the vertical plane is I, and the Young's modulus of the beam material is E. Assuming that the weight of the beam is negligible and that it just makes contact with the flexible support initially, find the vertical reaction force provided by this support when a downward force F is applied to the free end of the beam as shown.

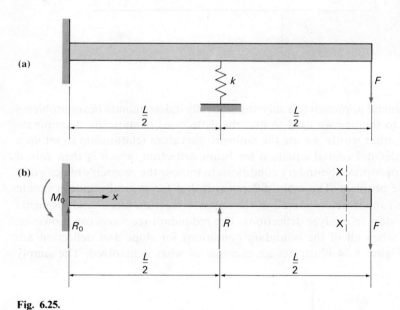

Fig. 6.25.

There are four unknown support reactions (three for the built-in support, and one for the flexible support), which means that the problem is statically indeterminate with one redundancy. It is convenient to take this redundancy as the reaction force, R, at the flexible support (Fig. 6.25b) which we are asked to find. Also shown in Fig. 6.25b are the reaction force, R_0, and moment, M_0, at the built-in end of the beam. For equilibrium of external forces applied to the beam in the vertical direction

$$R_0 + R - F = 0$$

and for equilibrium of moments about the left-hand end of the beam

$$FL - M_0 - \frac{RL}{2} = 0$$

These two equations can be used to express R_0 and M_0 in terms of F and the unknown R as

$$R_0 = F - R$$

and

$$M_0 = FL - \frac{RL}{2}$$

The equation of equilibrium of external forces applied to the beam in the horizontal direction provides no useful information (since there are no forces in this direction).

Considering the section XX in Fig. 6.25b close to the right-hand end of the beam, at a distance x from the left-hand end, we write an expression for the bending moment distribution which is valid for the entire beam as

$$M(x) = -M_0 + R_0 x + R\left\langle x - \frac{L}{2} \right\rangle$$

The moment–curvature relationship is therefore

$$EI\frac{d^2v}{dx^2} = M(x) = -FL + \frac{RL}{2} + (F - R)x + R\left\langle x - \frac{L}{2} \right\rangle$$

which still involves the unknown R. Integrating twice with respect to x gives the general solutions for slope and deflection as

$$EI\frac{dv}{dx} = \left(\frac{RL}{2} - FL\right)x + (F - R)\frac{x^2}{2} + \frac{R}{2}\left\langle x - \frac{L}{2} \right\rangle^2 + A \tag{6.57}$$

and

$$EIv = \left(\frac{RL}{2} - FL\right)\frac{x^2}{2} + (F - R)\frac{x^3}{6} + \frac{R}{6}\left\langle x - \frac{L}{2} \right\rangle^3 + Ax + B \tag{6.58}$$

where A and B are constants of integration. The two boundary conditions at the rigidly built-in end of the beam are those of zero deflection and zero slope at $x = 0$. Applying these to equations (6.57) and (6.58), we find that

$$A = B = 0$$

But there is a third condition, connecting reaction force and deflection at the flexible support. This takes the form of equation (6.55), with $-R$ in place of F (the negative sign is necessary because we have assumed that R acts upward on the beam)

$$v\left(\frac{L}{2}\right) = -\frac{R}{k} = \frac{1}{EI}\left[\left(\frac{RL}{2} - FL\right)\frac{L^2}{8} + (F - R)\frac{L^3}{48}\right]$$

from which R can be found

$$-\frac{R}{k} = \frac{L^3}{48EI}(2R - 5F)$$

$$R = \frac{5F}{2 + \dfrac{48EI}{kL^3}} \qquad (6.59)$$

From this general expression for the reaction at the flexible support, we can obtain results for limiting values of the support stiffness. For example, if the support is very rigid, $k \to \infty$ and $R \to 2.5F$. Also, if it is very flexible, $k \to 0$ and $R \to 0$ and the beam becomes a simple cantilever.

It is worth noting that this problem can be solved much more concisely by the use of superposition. Since we are only interested in what happens at the flexible support, we need only consider the right-hand half of the beam to the extent that it applies a shear force and a bending moment to the left-hand half. In other words, we may consider the cantilever of length $L/2$ shown in Fig. 6.26, subjected at its free end to the upward reaction force, R, a downward shear force F and a hogging bending moment of magnitude $FL/2$. The last two of these loads are due to the applied end force, F (Fig. 6.25). Using two of the formulae given in Appendix D for cantilevers, the deflections at the free end of this beam due to a total downward end force of $(F - R)$ and the bending moment of $-FL/2$ are

$$-\frac{(F - R)}{3EI}\left(\frac{L}{2}\right)^3 \quad \text{and} \quad -\frac{1}{2EI}\left(\frac{FL}{2}\right)\left(\frac{L}{2}\right)^2$$

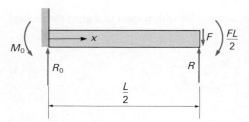

Fig. 6.26.

respectively, giving a total end deflection of

$$v = \frac{L^3}{48EI}(2R - 5F)$$

Equating this to the upward deflection, $-R/k$, of the flexible support at this point, we again obtain equation (6.59).

EXAMPLE 6.13 Built-in beam with a concentrated force

Figure 6.27 shows a uniform beam of length L and negligible weight, with a downward vertical concentrated force of magnitude F applied at a distance a from its left-hand end.

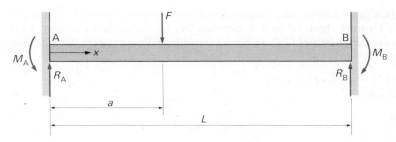

Fig. 6.27.

The ends of the beam are rigidly built-in to vertical walls, and are at the same height. Find the deflection under the applied force.

This problem is similar to Example 6.6, except that here the ends of the beam are built-in rather than simply supported. There are six unknown support reactions (three for each built-in support), which means that the problem is statically indeterminate with three redundancies. If we ignore axial loading of the beam, then we can eliminate two of the unknown reactions from our analysis, but we also lose one equilibrium equation. We are left with four unknowns and two equilibrium equations, implying two redundancies. Let us take these redundancies as the reaction force, R_A, and moment, M_A, at the left-hand end, A, of the beam (Fig. 6.27). While the reactions at end B can be expressed in terms of R_A and M_A by applying two equilibrium equations for the beam as a whole, these are not required explicitly in the present analysis.

The moment–curvature relationship for this problem is given by

$$EI\frac{d^2v}{dx^2} = M(x) = R_A x - M_A - F\langle x - a\rangle$$

where EI is the constant flexural rigidity of the beam. Integrating this equation twice with respect to x, we obtain

$$EI\frac{dv}{dx} = R_A\frac{x^2}{2} - M_A x - \frac{F}{2}\langle x-a\rangle^2 + A \tag{6.60}$$

and

$$EIv = R_A\frac{x^3}{6} - M_A\frac{x^2}{2} - \frac{F}{6}\langle x-a\rangle^3 + Ax + B \tag{6.61}$$

where A and B are constants of integration. The boundary conditions at the built-in left-hand end of the beam are zero deflection and slope, at $x = 0$. When substituted into equations (6.60) and (6.61), these give $A = B = 0$. The same zero deflection and slope boundary conditions apply at the built-in right-hand end of the beam, which allow the unknowns R_A and M_A to be found.

Setting slope equal to zero at $x = L$ in equation (6.60), we have

$$0 = R_A\frac{L^2}{2} - M_A L - \frac{F}{2}(L-a)^2 \tag{6.62}$$

and setting deflection equal to zero at $x = L$ in equation (6.61) gives

$$0 = R_A\frac{L^3}{6} - M_A\frac{L^2}{2} - \frac{F}{6}(L-a)^3 \tag{6.63}$$

Equations (6.62) and (6.63) are simultaneous linear algebraic equations for R_A and M_A in terms of F. To solve them, let us first eliminate M_A by multiplying equation (6.62) through by a common factor of $L/2$, and then subtracting it from equation (6.63), to give

$$0 = R_A L^3 \left(\frac{1}{6} - \frac{1}{4} \right) - F(L-a)^2 \left(\frac{L-a}{6} - \frac{L}{4} \right)$$

from which

$$R_A = \frac{F}{L^3}(L-a)^2(L+2a) \tag{6.64}$$

Substituting this expression for R_A into equation (6.62), we obtain the following definition for M_A

$$M_A = \frac{Fa}{L^2}(L-a)^2 \tag{6.65}$$

We have now solved the problem to the extent that we have found the unknown reactions. It only remains to extract the particular information we require. Substituting the expressions for R_A and M_A into equation (6.61), and setting $x = a$, we obtain the deflection under the applied force as

$$v(a) = \frac{1}{EI} \left[\frac{F}{L^3}(L-a)^2(L+2a)\frac{a^3}{6} - \frac{Fa}{L^2}(L-a)^2\frac{a^2}{2} \right]$$

$$= -\frac{F(L-a)^3 a^3}{3EIL^3} \tag{6.66}$$

The results of this example of a built-in beam are of considerable practical significance. Let us focus on the particular case of the concentrated force located at the center of the beam, with $a = L/2$. From equation (6.64), the support reaction force at A is then

$$R_A = \frac{F}{L^3}\left(\frac{L}{2} \right)^2 2L = \frac{F}{2}$$

With the applied force at the center of the beam, the system is symmetrical, and each of the supports carries half of this load. From equation (6.65), the moment at the built-in end of the beam at A is

$$M_A = \frac{F}{L^2}\left(\frac{L}{2} \right)\left(\frac{L}{2} \right)^2 = \frac{FL}{8}$$

and from equation (6.66), the deflection under the applied force, which is also the greatest absolute deflection, is

$$v\left(\frac{L}{2} \right) = -\frac{F}{3EIL^3}\left(\frac{L}{2} \right)^6 = -\frac{FL^3}{192EI}$$

These results can be compared with those obtained in Example 6.5, for a simply supported beam subjected to the same form of loading. Figure 6.28 shows the shear force and bending moment diagrams for these two cases, together with the greatest absolute deflections (at the centers of the beams). It is interesting to note that the shear force distributions are identical. Also, the bending moment distributions are of the same shape, but in the case of the built-in beam shifted vertically downward by an amount $FL/8$, the magnitude of the moments applied to its ends. This shift is important because it has the effect of halving the magnitude of the greatest absolute bending moment. Finally, the greatest absolute deflection of the built-in beam is only a quarter of that of the simply supported beam (equation (6.24)). Put in other words, the effect of building in the beam is to retain the same greatest shear stresses, to halve the greatest bending stresses, and to increase the bending stiffness by a factor of four. Similar effects are also obtained for other forms of loading. In many practical situations, therefore, built-in beams appear to offer significant advantages over simply supported ones.

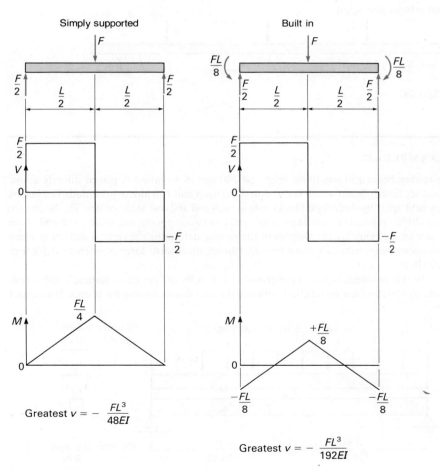

Fig. 6.28.

These conclusions must be treated with caution, however, for several reasons. Firstly, truly rigid built-in end conditions are difficult to achieve and maintain. Even when a beam is inserted in, say, a massive block of concrete, the stiffness of the block is not infinite, and in time some local crumbling of the concrete immediately surrounding the beam may further reduce the effective stiffness. Secondly, accurate alignment of the built-in ends is important, since relatively small differences in height or slope can cause significant bending moments and stresses. Thirdly, even if the system is accurately aligned initially, later settlement of one support relative to the other again can cause significant bending moments and stresses.

Although built-in beams should be designed with caution, one way to take advantage of the greater strength and stiffness of such beams is to make beams which rest on a number of supports continuous. For example, each span of the continuous beam shown in Fig. 6.29, which rests on simple supports, is effectively a built-in beam, since both the slope and deflection at each support are zero. The resistance to rotation is provided by the adjacent span rather than the support, and this resistance will not change significantly with time. A continuous beam is preferable to having separate simply supported beams covering each span.

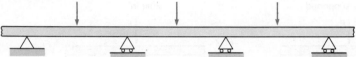

Fig. 6.29.

EXAMPLE 6.14

A timber beam 150 mm (6 in) deep and 200 mm (8 in) broad is placed directly above another timber beam which is 200 mm (8 in) deep and 150 mm (6 in) broad. The beams are held apart by three rigid blocks, one at each end and one at the center. The beams are 6 m (20 ft) long and the whole assembly rests on two supports, one under each end of the lower beam. Ignoring the weights of the beams, determine the central deflection when the upper beam carries a uniformly distributed downward force of intensity 1.5 kN/m (100 lb/ft).

In this problem, which is illustrated in Fig. 6.30, the system is statically indeterminate by virtue of the connection between the two simply supported beams. We cannot

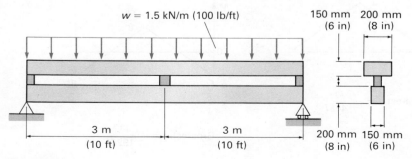

Fig. 6.30.

find from equilibrium considerations alone the force in the rigid block which connects the centers of the beams, but we do know that these center points deflect by the same amount. Let the compressive force in the block be F, and let the common central deflection be v_c. Figure 6.31 shows the two beams separated, with the forces acting on them. We can use superposition to find the central deflections which result from these forces.

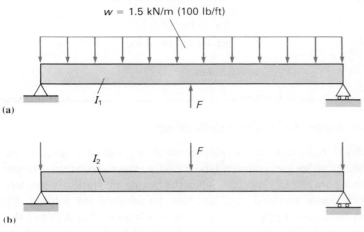

Fig. 6.31.

For the upper beam, Appendix D (also equations (6.20) and (6.24)) give the total central deflection of a simply supported beam relative to its supports due to a uniformly distributed downward force of intensity w per unit length, and an upward concentrated central force of magnitude F as

$$v_c = -\frac{5wL^4}{384EI_1} + \frac{FL^3}{48EI_1}$$

where L is the length of the beam, E is the Young's modulus for timber in the direction along the beam, and I_1 is the second moment of area about its neutral axis. For the lower beam, to which is applied only the downward concentrated central force F, the same central deflection is given by

$$v_c = -\frac{FL^3}{48EI_2}$$

where I_2 is the second moment of area of the lower beam. Eliminating F from these two expressions for the central deflection, we obtain

$$v_c = -\frac{5wL^4}{384E(I_1 + I_2)}$$

SI units
From the given numerical data we have $w = 1.5$ kN/m, $L = 6$ m, $E = 12$ GN/m^2 (Appendix A). Also

$$I_1 = \frac{0.2 \times 0.15^3}{12} = 5.625 \times 10^{-5} \text{ m}^4$$

US customary units
From the given numerical data we have $w = 100$ lb/ft, $L = 20$ ft, $E = 1.7 \times 10^6$ psi (Appendix A). Also

$$I_1 = \frac{8 \times 6^3}{12} = 144 \text{ in}^4$$

SI units
and

$$I_2 = \frac{0.15 \times 0.2^3}{12} = 1.0 \times 10^{-4} \text{ m}^4$$

and the common central deflection is

$$v_c = -\frac{5 \times 1.5 \times 10^3 \times 6^4}{384 \times 12 \times 10^9 \times 1.5625 \times 10^{-4}} \text{ m}$$

$$= -13.5 \text{ mm}$$

US customary units
and

$$I_2 = \frac{6 \times 8^3}{12} = 256 \text{ in}^4$$

and the common central deflection is

$$v_c = -\frac{5 \times 100 \times 20^4 \times 12^3}{384 \times 1.7 \times 10^6 \times 400}$$

$$= -0.529 \text{ in}$$

6.4 Computer method for beam deflections

Although we have been able to develop manual methods for finding the deflections of beams, these are relatively laborious for even the relatively simple problems we have considered, especially when the beams are statically indeterminate. The techniques involved can, however, be adapted for solution by computer. As in the cases of pin-jointed structures (Chapters 2 and 4) and the equilibrium analysis of statically determinate beams (Chapter 5), it is appropriate for us to formulate our analysis as a finite element method.

6.4.1 Finite element analysis

The development of a finite element method for beam deflections follows closely that described in Section 5.2.3 for finding shear force and bending moment distributions. We start by dividing the beam into a number of elements arranged along its length. Figure 6.32 shows a typical element, numbered *m* and of length L_m. We can treat this element as being joined to its immediate neighbors via its nodal points *i* and *j*. The way in which we divide the beam into elements is not always obvious. But, since we intend to consider equilibrium conditions at the nodes, a minimum requirement is for element interfaces to occur at the following:
(a) the supports,
(b) points of application of concentrated lateral forces,
(c) points of application of concentrated couples,
(d) points at which there are abrupt changes in the intensity of distributed lateral forces.

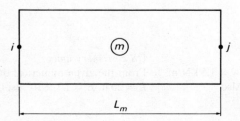

Fig. 6.32.

Whether it is necessary for any further subdivision depends on whether the beam is of uniform cross section, on how distributed lateral forces vary along the beam, and on the form of such variations which can be accurately modeled by the type of elements we decide to use.

In fact, we will use elements over each of which the cross section is uniform and any distributed lateral force is of constant intensity. This does not prevent us analyzing beams with continuously varying cross sections and distributed forces, but the solutions obtained are only approximate, and a larger number of elements is required to obtain results accurate enough for engineering purposes. We first met this idea of element refinement to improve accuracy in Section 5.2.3. The main advantage of choosing such simple elements is to keep the analysis and computer program as straightforward as possible.

Following our usual sign conventions, we show the constant distributed lateral force, w, the shear forces, V, and bending moments, M, acting on the typical element as in Fig. 6.33. In the cases of shear forces and bending moments, we assign subscripts i and j to indicate which nodes, and therefore which ends, of the element they refer to: we assume that nodes i and j are at the left and right-hand ends, respectively. Note the use of a superscript on the nodal point shear forces and bending moments to indicate the element number. This is because shear forces and bending moments may take different values on either side of a nodal point, and undergo step changes there due to a concentrated force or couple applied at the node. We use global coordinate X to define axial position along the beam, with some convenient origin relative to the beam as a whole (typically, but not necessarily, its left-hand end). We also introduce a local coordinate, x, which is parallel to X, but with its origin at the left-hand node, i, of the element.

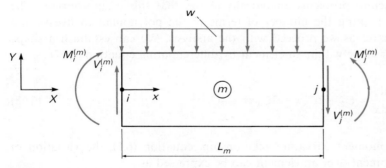

Fig. 6.33.

We next choose the variables we wish to have as the primary unknowns in the analysis. In other words, those variables we intend to solve for in the first instance, and from which other unknowns may then be obtained as required. In the analysis of statically determinate beams (Section 5.2.3), we solved for shear forces and bending moments at the nodes, and reactions at the supports, but in that case we had no alternative. Now we can choose either load variables or displacement variables, and in fact we choose the latter, in particular the beam deflections and slopes at the nodal points. These are shown for the typical element in Fig. 6.34, the symbol θ being used for the slope, dv/dx, which is an

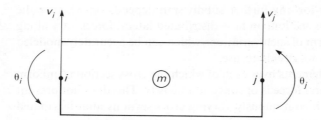

Fig. 6.34.

angular displacement. As the analysis is developed, we will see that this choice of variables results in a natural balance between the number of unknowns and the number of equations for finding them: there are two displacement variables at each node, and two equations of equilibrium at each node.

Having chosen nodal point deflections and slopes as the unknowns, we must now choose suitable approximations to define the variations of deflection and slope within the elements. For this purpose, we introduce a *shape function* for deflection as follows

$$v(x) = C_1 + C_2 x + C_3 x^2 + C_4 x^3 + C_5 x^4 \tag{6.67}$$

which describes how deflection varies with the local axial coordinate within any individual element. The parameters C_1 to C_5 are constants for any particular element, but may change from element to element. It is customary to choose shape functions which are polynomial in form, and our experience of solving beam deflection problems analytically shows that this is appropriate. The reason for limiting the number of terms in the polynomial to five should become clearer as we proceed with the analysis. We can establish a shape function for slope within an element by differentiating equation (6.67) as

$$\theta(x) = \frac{\mathrm{d}v}{\mathrm{d}x} = C_2 + 2C_3 x + 3C_4 x^2 + 4C_5 x^3 \tag{6.68}$$

Using the moment–curvature relationship, equation (6.5), the variation of bending moment over an element can be expressed as

$$M(x) = EI \frac{\mathrm{d}^2 v}{\mathrm{d}x^2} = EI(2C_3 + 6C_4 x + 12C_5 x^2) \tag{6.69}$$

Further, introducing the differential equation (5.11) linking shear force and bending moment, the variation of shear force within an element is given by

$$V(x) = \frac{\mathrm{d}M}{\mathrm{d}x} = EI(6C_4 + 24C_5 x) \tag{6.70}$$

and, finally, the differential equation (5.10) linking the intensity of the lateral distributed force on the beam to the shear force gives this intensity as

$$w = -\frac{dV}{dx} = -24EIC_5 \qquad (6.71)$$

We now see that the consequence of limiting the number of terms in the polynomial shape function (6.67) to five, involving not more than the fourth power of the local coordinate, is to accommodate only a constant lateral force intensity over an element. However, this is precisely the type of uniformly loaded element we initially chose to use. If we wished to allow the force intensity to vary over any element, we would have to use a shape function with more polynomial terms.

Considering the typical element, numbered m, shown in Fig. 6.33, equation (6.70) gives the shear forces at nodes $i(x = 0)$ and $j(x = L_m)$ as

$$V_i^{(m)} = 6EIC_4 \quad \text{and} \quad V_j^{(m)} = 6EIC_4 + 24EIC_5 L_m$$

while equation (6.69) gives the bending moments as

$$M_i^{(m)} = 2EIC_3 \quad \text{and} \quad M_j^{(m)} = 2EIC_3 + 6EIC_4 L_m + 12EIC_5 L_m^2$$

When we come to consider equilibrium at the nodes, it will be convenient to work with forces and moments of couples associated with the elements connected there, which are all in the same direction and sense, respectively. For forces we choose as positive the upward vertical direction defined by global coordinate Y, and for moments of couples we choose the counterclockwise sense as positive. Figure 6.35 shows the forces, Q, and couples, N, *acting on the element* in these directions at the nodes. Comparing Figs 6.33 and 6.35, we can see that there are some simple relationships between these new variables and the shear forces and bending moments, as follows

$$Q_i^{(m)} = V_i^{(m)}, \quad Q_j^{(m)} = -V_j^{(m)}, \quad N_i^{(m)} = -M_i^{(m)}, \quad N_j^{(m)} = M_j^{(m)} \qquad (6.72)$$

Consequently, we can link these forces and couples to the shape function

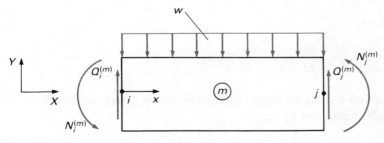

Fig. 6.35.

parameters by the following matrix relationship

$$[f]_m = \begin{bmatrix} Q_i^{(m)} \\ N_i^{(m)} \\ Q_j^{(m)} \\ N_j^{(m)} \end{bmatrix} = EI \begin{bmatrix} 0 & 6 & 0 \\ -2 & 0 & 0 \\ 0 & -6 & -24L_m \\ 2 & 6L_m & 12L_m^2 \end{bmatrix} \begin{bmatrix} C_3 \\ C_4 \\ C_5 \end{bmatrix} \tag{6.73}$$

where $[f]_m$ is the *element load vector* containing the four nodal point forces and couples arranged in the usual order. That is, values for the first node first, followed by those for the second node, in each case taking the force before the couple.

Now we need to find expressions in terms of nodal point deflections and slopes for the shape function parameters, C_1 to C_5, and in particular the last three of these, which appear in equation (6.73). The constant C_5 for a particular element can be determined directly from equation (6.71) as

$$C_5 = -\frac{w}{24EI} \tag{6.74}$$

where w is the intensity of lateral load on that element. If w varies over the element, we can use its value at the center of the element as the average value. The other constants can be found by using the fact that the shape functions given by equations (6.67) and (6.68) must define the nodal point values of deflection and slope, as follows.

$$v_i = v(0) = C_1 \tag{6.75}$$

$$\theta_i = \theta(0) = C_2 \tag{6.76}$$

$$v_j = v(L_m) = C_1 + C_2 L_m + C_3 L_m^2 + C_4 L_m^3 + C_5 L_m^4 \tag{6.77}$$

$$\theta_j = \theta(L_m) = C_2 + 2C_3 L_m + 3C_4 L_m^2 + 4C_5 L_m^3 \tag{6.78}$$

While equations (6.75) and (6.76) give C_1 and C_2 directly, C_3 and C_4 must be found from equations (6.77) and (6.78). To find C_3, we multiply equation (6.77) by 3 and subtract from it equation (6.78) multiplied by L_m

$$3v_j - \theta_j L_m = 3C_1 + 2C_2 L_m + C_3 L_m^2 - C_5 L_m^4$$

from which

$$C_3 = -\frac{3v_i}{L_m^2} - \frac{2\theta_i}{L_m} + \frac{3v_j}{L_m^2} - \frac{\theta_j}{L_m} - \frac{wL_m^2}{24EI} \tag{6.79}$$

Similarly, to find C_3, we multiply equation (6.77) by 2 and subtract from it equation (6.78) multiplied by L_m

$$2v_j - \theta_j L_m = 2C_1 + C_2 L_m - C_4 L_m^3 - 2C_5 L_m^4$$

from which

$$C_4 = \frac{2v_i}{L_m^3} + \frac{\theta_i}{L_m^2} - \frac{2v_j}{L_m^3} + \frac{\theta_j}{L_m^2} + \frac{wL_m}{12EI} \tag{6.80}$$

Substituting definitions (6.79), (6.80) and (6.74) into equation (6.73), we obtain the following relationship between loads and displacements for the element

$$\begin{bmatrix} Q_i^{(m)} \\ N_i^{(m)} \\ Q_j^{(m)} \\ N_j^{(m)} \end{bmatrix} = \frac{EI}{L_m^3} \begin{bmatrix} 12 & 6L_m & -12 & 6L_m \\ 6L_m & 4L_m^2 & -6L_m & 2L_m^2 \\ -12 & -6L_m & 12 & -6L_m \\ 6L_m & 2L_m^2 & -6L_m & 4L_m^2 \end{bmatrix} \begin{bmatrix} v_i \\ \theta_i \\ v_j \\ \theta_j \end{bmatrix}$$
$$- \begin{bmatrix} -wL_m/2 \\ -wL_m^2/12 \\ -wL_m/2 \\ wL_m^2/12 \end{bmatrix} \tag{6.81}$$

which takes the form

$$[f]_m = [k]_m[\delta]_m - [W]_m \tag{6.82}$$

where $[\delta]_m$ is the *element displacement vector* (containing the four nodal point displacement variables arranged in the same order as the force terms in equation (6.73)), $[k]_m$ is the *element stiffness matrix*, and $[W]_m$ is a vector of external load terms associated with the distributed lateral force on the beam. In fact, the components of $[W]_m$ are the forces and couples created at the nodes of the element by the presence of the distributed force, assuming the element to be built-in to the adjacent elements, as shown in Fig. 6.36.

It is worth noting that the first and third rows, also the first and third columns, of coefficients in the element stiffness matrix in equations (6.81) are proportional to each other. This means that the determinant of the matrix is zero, and the matrix is *singular* (a condition which is discussed in Appendix B).

Having established the form of relationship between the externally applied loads and the corresponding displacements for a typical element, we can apply this relationship to each of the elements of the beam. We now need to assemble all the resulting element stiffness information to establish the behavior of the whole beam. To do this, we take account of both the geometry of deformation and the equilibrium of forces and moments of couples acting at the nodes. The geometry of deformation of a beam is such that both beam deflection and slope are continuous functions of position along the beam. In terms of the symbols we have chosen to use, deflection v_i and slope θ_i at node i are the same for both elements which meet there.

We now write equilibrium equations for the forces and moments acting at each of the nodes of the beam. The first of these equates the external lateral

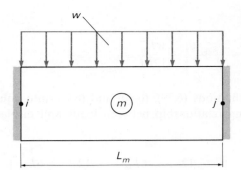

(a)

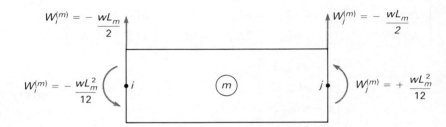

(b)

Fig. 6.36.

force to the sum of the lateral forces transmitted to the elements at that node

$$\Sigma Q_i^{(m)} = - F_i \qquad (6.83)$$

and the other defines a similar balance for the moment of the external couple and the moments of couples transmitted to the elements

$$\Sigma N_i^{(m)} = C_i \qquad (6.84)$$

In these equations, F_i and C_i are the concentrated external lateral force and concentrated external couple, respectively, applied at node i, either or both of which may be zero. The negative sign in the force equation is due to the external forces following the beam sign convention of positive downward, opposite to that of the Q forces. In the moment equation, all couples are assumed to be positive in the counterclockwise direction. Note that the element force and moment summations are carried out over only the two elements connected together at the particular node, and over only one element in the case of a node at an end of the beam. Also note that if node i is a support point, then F_i includes the unknown reaction force there, and C_i may include an unknown reaction moment.

   Force $Q_i^{(m)}$ and couple $N_i^{(m)}$ applied to element m are terms in the element load vector $[f]_m$ which, according to equation (6.81), can be expressed as sums of products of stiffnesses and displacements. While the stiffnesses are known functions of the element geometry and physical properties, the displacements concerned are the as yet unknown deflections and slopes at the nodes of the element. When the summations required by equations (6.83) and (6.84) are complete for node i, we can expect to have a pair of algebraic equations which are linear in certain displacements. These are the deflection and slope of node i itself, together with those of the immediately adjacent nodes which are directly connected to node i by elements.

The process of assembling equilibrium equations for all the nodes of the beam can be represented symbolically as

$$\Sigma[f]_m = [F]$$

where $[F]$ is the vector of external concentrated loads (forces and couples) applied at the nodes. Using equation (6.82), this becomes

$$\Sigma[k]_m[\delta]_m = [F] + \Sigma[W]_m$$

or

$$[K][\delta] = [F]^* \qquad\qquad\qquad (6.85)$$

The vector $[\delta]$ contains the unknown displacements, the square overall stiffness matrix $[K]$ contains known stiffnesses, while vector $[F]^*$ contains the applied concentrated forces and couples, together with the contributions at the nodes of the distributed forces. Matrix equation (6.85) represents a set of $2n$ linear algebraic equations (where n is the number of nodes along the beam) for the $2n$ displacements (two per node). Although we are free to arrange the variables in $[\delta]$ and $[F]^*$ in any order, it is convenient to choose an order similar to that used for element displacements and loads, namely to take the variables of the first node (node 1) first, followed by those of the second node (node 2), and so on for all the other nodes along the beam, in each case taking the linear displacement or force before the rotational displacement (slope) or couple. The details of the assembly process are similar to those explained in Section 4.1.1 for the finite element analysis of pin-jointed structures, and can be followed in the computer program below.

At first sight we may be tempted to think that we can solve equations (6.85) as they stand. We do, after all, have $2n$ linear algebraic equations involving $2n$ unknowns. If we attempted to solve them, however, we would find that the overall stiffness matrix is, like the individual element stiffness matrices, singular. The physical significance of a singular stiffness matrix is that we have not yet taken into account the conditions which must be applied at the beam supports, conditions which are part of the imposed geometry of deformation. Without these conditions, the beam as a whole is free to move (it is a mechanism). In Section 5.2.3, we imposed support conditions on statically determinate beams by including as unknowns the reaction forces and moments at the relevant nodes. Such reactions certainly exist, for both statically determinate and statically indeterminate beams, but since we are now choosing to use displacements as the unknowns it is necessary to express the conditions at the supports in terms of these displacements. There are at least three types of support to be considered, namely simple, built-in and flexible. The heights of simple supports are known, and in many problems are all the same, and we can take this as the datum for zero deflection. If their heights vary, we must choose a convenient datum, such as the height of one of them. The deflection of a nodal point at a typical simple support, here numbered i, is therefore known. This condition is readily implemented in equations (6.85) by replacing the $(2i-1)$th equation (which is the force equilibrium equation for node i) by

$$v_i = v_i' \tag{6.86}$$

where v_i' is the known deflection at the support. At a built-in support, not only is the deflection known but also the slope, a condition which is implemented in equations (6.85) by replacing the $2i$th equation (which is the moment equilibrium equation for node i) by

$$\theta_i = \theta_i' \tag{6.87}$$

where θ_i' is the known slope at the support.

At a flexible support with lateral stiffness, equation (6.55) can be adapted to the present sign convention to give the relationship between the deflection and the reaction force as

$$F_i = -k_i v_i \tag{6.88}$$

where k_i is the lateral stiffness of the support at node i. The negative sign appears because force and deflection are now defined as positive in the same direction, and the reaction force at a flexible support always opposes the movement of the beam. With a flexible support, we do not know in advance either the deflection or the reaction, merely the linear relationship between them. Therefore, rather than replacing the $(2i-1)$th equation in set (6.85) by equation (6.88), we use the latter to define the external concentrated force term on the right-hand side of the equation. But, because we know the force only in terms of the unknown v_i, the product $k_i v_i$ must be taken from the right-hand side to the left-hand side of the equation. In other words, the coefficient on the leading diagonal of the overall stiffness matrix, K_{pp}, where $p = 2i - 1$, is replaced by $(K_{pp} + k_i)$, and the stiffness of the support adds to the stiffness of the beam. We can treat a flexible support with rotational stiffness in very much the same way. Equation (6.56) can be adapted to give the relationship between rotation (slope) and the reaction moment as

$$C_i = -\lambda_i \theta_i \tag{6.89}$$

where λ_i is the rotational stiffness of the support at node i, a relationship which can be used to modify the $2i$th equation in set (6.85), adding λ_i to K_{pp}, where $p = 2i$.

Equations (6.85) are now ready to be solved and, once solved, we have the deflections and slopes at all the nodal points. From these can be found the shear forces and bending moments in the elements. This finite element method is equally applicable to statically determinate and statically indeterminate beams.

6.4.2 Computer program

Program SIBEAM (standing for Statically Indeterminate **BEAM** analysis) provides a FORTRAN implementation of the above analysis. As far as possible it follows in terms of FORTRAN variable names and general layout the programs SDPINJ and SIPINJ for the analysis of pin-jointed structures (Sections 2.1.3 and 4.1.2), and program SDBEAM for the analysis of statically determinate beams (Section 5.2.3).

```
      PROGRAM  SIBEAM
C
C PROGRAM TO FIND THE DEFLECTION, SLOPE, SHEAR FORCE AND BENDING
C MOMENT DISTRIBUTIONS ALONG A STATICALLY INDETERMINATE (OR
C STATICALLY DETERMINATE) BEAM, SUBJECT TO LATERAL FORCES AND
C CONCENTRATED COUPLES.
C
      DIMENSION   X(51),NPI(50),NPJ(50),SMA(50),OSTIFF(102,103),
     1            DELTA(102),F(51),C(51),W(50),NODE(51),NELEM(50),
     2            VS(51),THETAS(51),VSTIFF(51),RSTIFF(51),
     3            ESTIFF(50,4,4),WLOAD(50,4),V(51),THETA(51),
     4            ELOAD(4),XVMAX(50),VMAX(50),SF(50,2),BM(50,2),
     5            XBMMAX(50),BMMAX(50)
      COMMON /CONSTS/   C2,C3,C4,C5
      NNPMAX=51
      NELMAX=50
      PI=4.*ATAN(1.)
      OPEN(5,FILE='DATA')
      OPEN(6,FILE='RESULTS')
      WRITE(6,61)
   61 FORMAT('STATICALLY INDETERMINATE OR DETERMINATE BEAM')
C
C INPUT AND TEST THE NUMBER OF NODAL POINTS.
      READ(5,*) NNP
      IF(NNP.GT.NNPMAX) THEN
        WRITE(6,62) NNP
   62   FORMAT(/ 'NUMBER OF NODES = ',I4,' TOO LARGE - STOP')
        STOP
      END IF
C
C INPUT AND OUTPUT THE AXIAL COORDINATES OF THE NODES, WHICH MUST
C BE ARRANGED IN ORDER FROM LEFT TO RIGHT ALONG THE BEAM.
      READ(5,*) (X(I),I=1,NNP)
      WRITE(6,63) (I,X(I),I=1,NNP)
   63 FORMAT(/ 'COORDINATES OF THE NODES'
     1 / 4('NODE      X       ') / (4(I3,E12.4,2X)))
C
C DEFINE THE NUMBER OF ELEMENTS.
      NEL=NNP-1
C
C DEFINE THE NODES CONNECTED BY THE ELEMENTS, BOTH NODES AND ELEMENTS
C BEING NUMBERED FROM LEFT TO RIGHT ALONG THE BEAM.  ALSO, INITIALIZE
C THE DISTRIBUTED FORCES ON THE ELEMENTS.
      DO 1 M=1,NEL
      NPI(M)=M
      NPJ(M)=M+1
    1 W(M)=0.
C
C INPUT AND OUTPUT THE YOUNGS MODULUS OF THE BEAM MATERIAL AND SECOND
C MOMENT OF AREA ABOUT THE NEUTRAL AXIS OF THE BEAM CROSS SECTION - A
C ZERO (OR NEGATIVE) VALUE OF SECOND MOMENT IMPLIES THAT THE CROSS
C SECTION VARIES, AND VALUES OF SECOND MOMENT OF AREA MUST BE INPUT
C FOR THE INDIVIDUAL ELEMENTS.
      READ(5,*) E,SMAC
      WRITE(6,631) E
  631 FORMAT(/ 'YOUNGS MODULUS OF THE BEAM MATERIAL = ',E12.4)
      IF(SMAC.GT.0.) THEN
        WRITE(6,632) SMAC
  632   FORMAT(/ 'SECOND MOMENT OF AREA OF THE BEAM CROSS SECTION =
     1          E12.4)
        DO 101 M=1,NEL
```

```
101     SMA(M)=SMAC
        ELSE
        READ(5,*) (SMA(M),M=1,NEL)
        WRITE(6,633) (M,SMA(M),M=1,NEL)
633     FORMAT(/ 'SECOND MOMENTS OF AREA OF THE ELEMENTS'
     1          / 4('ELEM      SMA      ') / (4(I3,E12.4,2X)))
        END IF
C
C  INITIALIZE THE OVERALL STIFFNESS COEFFICIENTS AND EXTERNAL FORCES
C  AND COUPLES CONCENTRATED AT THE NODES.
        NEQN=2*NNP
        DO 2 IROW=1,NEQN
        DO 2 ICOL=1,NEQN
2       OSTIFF(IROW,ICOL)=0.
        DO 3 I=1,NNP
        F(I)=0.
3       C(I)=0.
C
C  INPUT AND TEST THE NUMBER OF NODES AT WHICH EXTERNAL CONCENTRATED
C  LATERAL FORCES ARE APPLIED.
        READ(5,*) NNPF
        IF(NNPF.GT.NNPMAX) THEN
        WRITE(6,64) NNPF
64      FORMAT(/ 'NUMBER OF NODES WITH CONCENTRATED FORCES = ',I4,
     1          ' TOO LARGE - STOP')
        STOP
        END IF
C
C  INPUT AND OUTPUT THE FORCES AT THESE NODES (DOWNWARDS POSITIVE),
C  AND CONVERT TO UPWARDS POSITIVE SIGN CONVENTION.
        IF(NNPF.GT.0) THEN
        READ(5,*) (NODE(K),F(NODE(K)),K=1,NNPF)
        WRITE(6,65) (NODE(K),F(NODE(K)),K=1,NNPF)
65      FORMAT(/ 'CONCENTRATED (DOWNWARD) LATERAL FORCES'
     1          / ' NODE',12X,'F' / (I5,5X,E15.4))
        DO 31 K=1,NNPF
        I=NODE(K)
31      F(I)=-F(I)
        END IF
C
C  INPUT AND TEST THE NUMBER OF NODES AT WHICH EXTERNAL CONCENTRATED
C  COUPLES ARE APPLIED.
        READ(5,*) NNPC
        IF(NNPC.GT.NNPMAX) THEN
        WRITE(6,66) NNPC
66      FORMAT(/ 'NUMBER OF NODES WITH CONCENTRATED COUPLES = ',I4,
     1          ' TOO LARGE - STOP')
        STOP
        END IF
C
C  INPUT AND OUTPUT THE COUPLES AT THESE NODES (COUNTERCLOCKWISE
C  POSITIVE).
        IF(NNPC.GT.0) THEN
        READ(5,*) (NODE(K),C(NODE(K)),K=1,NNPC)
        WRITE(6,67) (NODE(K),C(NODE(K)),K=1,NNPC)
67      FORMAT(/ 'CONCENTRATED (COUNTERCLOCKWISE) COUPLES'
     1          / ' NODE',12X,'C' / (I5,5X,E15.4))
        END IF
C
C  INPUT AND TEST THE NUMBER OF ELEMENTS OVER WHICH EXTERNAL
C  DISTRIBUTED LATERAL FORCES ARE APPLIED.
        READ(5,*) NELW
        IF(NELW.GT.NELMAX) THEN
        WRITE(6,68) NELW
68      FORMAT(/ 'NUMBER OF ELEMENTS WITH DISTRIBUTED FORCES = ',I4,
     1          ' TOO LARGE - STOP')
        STOP
        END IF
C
C  INPUT AND OUTPUT THE FORCE INTENSITIES ON THESE ELEMENTS (DOWNWARDS
C  POSITIVE).
        IF(NELW.GT.0) THEN
        READ(5,*) (NELEM(K),W(NELEM(K)),K=1,NELW)
        WRITE(6,69) (NELEM(K),W(NELEM(K)),K=1,NELW)
69      FORMAT(/ 'DISTRIBUTED (DOWNWARD) LATERAL FORCES'
```

```
      1          / ' ELEMENT',12X,'W' / (I5,5X,E15.4))
           END IF
C
C  INPUT AND TEST THE NUMBERS OF NODES AT WHICH THE BEAM IS SIMPLY
C  SUPPORTED (NNPSS), BUILT IN (NNPBI), AND HELD ON FLEXIBLE
C  SUPPORTS (NNPFL).
           READ(5,*) NNPSS,NNPBI,NNPFL
           NNPT=NNPSS+NNPBI+NNPFL
           IF(NNPT.GT.NNPMAX) THEN
             WRITE(6,70) NNPT
  70         FORMAT(/ 'NUMBER OF NODES AT WHICH BEAM IS SUPPORTED = ',I4,
      1             ' TOO LARGE - STOP')
             STOP
           END IF
C
C  INPUT AND OUTPUT THE NODES AT WHICH THE BEAM IS SIMPLY SUPPORTED,
C  TOGETHER WITH THE DEFLECTIONS (UPWARDS POSITIVE) THERE.
           IF(NNPSS.GT.0) THEN
             READ(5,*) (NODE(K),VS(NODE(K)),K=1,NNPSS)
             WRITE(6,71) (NODE(K),VS(NODE(K)),K=1,NNPSS)
  71         FORMAT(/ 'NODES AT WHICH BEAM IS SIMPLY SUPPORTED' /
      1             ' NODE      DEFLECTION'  / (I5,E15.4))
           END IF
C
C  INPUT AND OUTPUT THE NODES AT WHICH THE BEAM IS BUILT IN, TOGETHER
C  WITH THE DEFLECTIONS (UPWARDS POSITIVE) AND SLOPES (IN DEGREES
C  MEASURED COUNTERCLOCKWISE FROM THE POSITIVE X-AXIS) THERE.
           IF(NNPBI.GT.0) THEN
             READ(5,*) (NODE(NNPSS+K),VS(NODE(NNPSS+K)),
      1             THETAS(NODE(NNPSS+K)),K=1,NNPBI)
             WRITE(6,72) (NODE(NNPSS+K),VS(NODE(NNPSS+K)),
      1             THETAS(NODE(NNPSS+K)),K=1,NNPBI)
  72         FORMAT(/ 'NODES AT WHICH BEAM IS BUILT IN' /
      1             ' NODE      DEFLECTION     SLOPE (DEG)' / (I5,2E15.4))
           END IF
C
C  INPUT AND OUTPUT THE NODES AT WHICH THE BEAM IS HELD ON FLEXIBLE
C  SUPPORTS, AND THE STIFFNESSES OF THESE SUPPORTS (ROTATIONAL
C  STIFFNESSES IN UNITS OF MOMENT PER RADIAN).
           IF(NNPFL.GT.0) THEN
             READ(5,*) (NODE(NNPSS+NNPBI+K),VSTIFF(NODE(NNPSS+NNPBI+K)),
      1             RSTIFF(NODE(NNPSS+NNPBI+K)),K=1,NNPFL)
             WRITE(6,721) (NODE(NNPSS+NNPBI+K),VSTIFF(NODE(NNPSS+NNPBI+K)),
      1             RSTIFF(NODE(NNPSS+NNPBI+K)),K=1,NNPFL)
  721        FORMAT(/ 'NODES AND STIFFNESSES OF THE FLEXIBLE SUPPORTS' /
      1             ' NODE          VERT. STIFF.          ROT. STIFF. ' /
      2             (I5,2(5X,E15.4)))
           END IF
C
C  COUNT THE NUMBER OF NON-ZERO STIFFNESSES OF FLEXIBLE SUPPORTS.
           NSTIFF=0
           IF(NNPFL.GT.0) THEN
             DO 102 K=1,NNPFL
             I=NODE(NNPSS+NNPBI+K)
             IF(VSTIFF(I).GT.0.) NSTIFF=NSTIFF+1
  102        IF(RSTIFF(I).GT.0.) NSTIFF=NSTIFF+1
           END IF
C
C  TEST THE NUMBER OF UNKNOWN SUPPORT REACTION FORCES AND MOMENTS.
           NUNK=NNPSS+2*NNPBI+NSTIFF
           IF(NUNK.GT.2) WRITE(6,73)
  73       FORMAT(/ 'THE BEAM IS STATICALLY INDETERMINATE')
           IF(NUNK.EQ.2) WRITE(6,731)
  731      FORMAT(/ 'THE BEAM IS STATICALLY DETERMINATE')
           IF(NUNK.LT.2) THEN
             WRITE(6,74)
  74         FORMAT(/ 'THE BEAM IS A MECHANISM - STOP')
             STOP
           END IF
C
C  SET UP THE LINEAR EQUATIONS, FIRST CONSIDERING EACH ELEMENT IN TURN.
           DO 7 M=1,NEL
           I=NPI(M)
           J=NPJ(M)
           ELENGT=X(J)-X(I)
```

```
C
C   TEST FOR NEGATIVE ELEMENT LENGTH (WHICH RESULTS FROM NODAL
C   COORDINATES NOT BEING ENTERED IN ORDER FROM LEFT TO RIGHT ALONG THE
C   BEAM.
        IF(ELENGT.LT.0.) THEN
          WRITE(6,75) M,I,J
   75     FORMAT(/ 'ELEMENT ',I2,' WITH NODES ',I2,' AND ',I2,
      1          ' HAS NEGATIVE LENGTH - STOP' /
      2          '(NODAL COORDINATES NOT ENTERED IN ORDER FROM LEFT TO'
      3          ' RIGHT ALONG THE BEAM)')
          STOP
        END IF
C
C   COMPUTE THE ELEMENT STIFFNESS MATRIX.
        FACT=E*SMA(M)/ELENGT**3
        ESTIFF(M,1,1)=FACT*12.
        ESTIFF(M,1,2)=FACT*6.*ELENGT
        ESTIFF(M,1,3)=-ESTIFF(M,1,1)
        ESTIFF(M,1,4)=ESTIFF(M,1,2)
        ESTIFF(M,2,1)=ESTIFF(M,1,2)
        ESTIFF(M,2,2)=FACT*4.*ELENGT**2
        ESTIFF(M,2,3)=-ESTIFF(M,2,1)
        ESTIFF(M,2,4)=ESTIFF(M,2,2)/2.
        DO 4 IC=1,4
   4    ESTIFF(M,3,IC)=-ESTIFF(M,1,IC)
        DO 5 IC=1,3
   5    ESTIFF(M,4,IC)=ESTIFF(M,IC,4)
        ESTIFF(M,4,4)=ESTIFF(M,2,2)
C
C   ADD EACH ELEMENT STIFFNESS TO THE OVERALL STIFFNESS MATRIX.
        DO 6 IR=1,4
        DO 6 IC=1,4
        IF(IR.LE.2) IROW=2*(I-1)+IR
        IF(IR.GE.3) IROW=2*(J-1)+IR-2
        IF(IC.LE.2) ICOL=2*(I-1)+IC
        IF(IC.GE.3) ICOL=2*(J-1)+IC-2
   6    OSTIFF(IROW,ICOL)=OSTIFF(IROW,ICOL)+ESTIFF(M,IR,IC)
C
C   DEFINE THE FORCES AND COUPLES AT THE NODES DUE TO THE
C   DISTRIBUTED FORCE ON THE ELEMENT, AND ADD THEM TO THE
C   CONCENTRATED FORCES AND COUPLES.
        WLOAD(M,1)=-W(M)*ELENGT/2.
        WLOAD(M,2)=-W(M)*ELENGT**2/12.
        WLOAD(M,3)=WLOAD(M,1)
        WLOAD(M,4)=-WLOAD(M,2)
        F(I)=F(I)+WLOAD(M,1)
        C(I)=C(I)+WLOAD(M,2)
        F(J)=F(J)+WLOAD(M,3)
   7    C(J)=C(J)+WLOAD(M,4)
C
C   STORE THE EXTERNAL FORCES AND COUPLES APPLIED TO THE NODES.
        DO 8 I=1,NNP
        OSTIFF(2*I-1,NEQN+1)=F(I)
   8    OSTIFF(2*I,NEQN+1)=C(I)
C
C   IMPOSE THE GIVEN DEFLECTIONS AT THE SIMPLE SUPPORTS.
        IF(NNPSS.GT.0) THEN
          DO 10 K=1,NNPSS
          I=NODE(K)
          DO 9 J=1,NEQN
   9      OSTIFF(2*I-1,J)=0.
          OSTIFF(2*I-1,2*I-1)=1.
   10     OSTIFF(2*I-1,NEQN+1)=VS(I)
        END IF
C
C   IMPOSE THE GIVEN DEFLECTIONS AND SLOPES AT THE BUILT-IN SUPPORTS,
C   CONVERTING SLOPES FROM DEGREES TO RADIANS.
        IF(NNPBI.GT.0) THEN
          DO 12 K=1,NNPBI
          I=NODE(NNPSS+K)
          DO 11 J=1,NEQN
          OSTIFF(2*I-1,J)=0.
   11     OSTIFF(2*I,J)=0.
          OSTIFF(2*I-1,2*I-1)=1.
```

```
              OSTIFF(2*I,2*I)=1.
              OSTIFF(2*I-1,NEQN+1)=VS(I)
   12         OSTIFF(2*I,NEQN+1)=THETAS(I)*PI/180.
              END IF
C
C   IMPOSE THE STIFFNESS RELATIONSHIPS AT THE FLEXIBLE SUPPORTS.
          IF(NNPFL.GT.0) THEN
              DO 13 K=1,NNPFL
              I=NODE(NNPSS+NNPBI+K)
              OSTIFF(2*I-1,2*I-1)=OSTIFF(2*I-1,2*I-1)+VSTIFF(I)
   13         OSTIFF(2*I,2*I)=OSTIFF(2*I,2*I)+RSTIFF(I)
              END IF
C
C   SOLVE THE LINEAR ALGEBRAIC EQUATIONS BY GAUSSIAN ELIMINATION.
          CALL   SOLVE(OSTIFF,DELTA,NEQN,2*NNPMAX,2*NNPMAX+1,IFLAG)
C
C   STOP IF A UNIT VALUE OF THE ILL-CONDITIONING FLAG IS DETECTED.
          IF(IFLAG.EQ.1) THEN
              WRITE(6,76)
   76         FORMAT(/ 'EQUATIONS ARE ILL-CONDITIONED - STOP')
              STOP
              END IF
C
C   STORE DEFLECTIONS AND SLOPES IN CONVENIENT ARRAYS.
          DO 14 I=1,NNP
          V(I)=DELTA(2*I-1)
   14     THETA(I)=DELTA(2*I)
C
C   OUTPUT THE COMPUTED DEFLECTIONS AND SLOPES AT THE NODAL POINTS.
          WRITE(6,761) (I,V(I),THETA(I),I=1,NNP)
  761     FORMAT(/ 'COMPUTED DEFLECTIONS AND SLOPES AT THE NODES' /
         1 2(' NODE           V               SLOPE      ') / 2(I4,2E15.4,2X))
C
C   SEARCH FOR MATHEMATICAL MAXIMUM OR MINIMUM VALUES OF DEFLECTION
C   WITHIN ELEMENTS (WHERE SLOPE ZERO).
          NVMAX=0
          DO 15 M=1,NEL
          I=NPI(M)
          J=NPJ(M)
          ELENGT=X(J)-X(I)
          C1=V(I)
          C2=THETA(I)
          C3=-3.*V(I)/ELENGT**2-2.*THETA(I)/ELENGT+3.*V(J)/ELENGT**2
         1   -THETA(J)/ELENGT-W(M)*ELENGT**2/(24.*E*SMA(M))
          C4=2.*V(I)/ELENGT**3+THETA(I)/ELENGT**2-2.*V(J)/ELENGT**3
         1   +THETA(J)/ELENGT**2+W(M)*ELENGT/(12.*E*SMA(M))
          C5=-W(M)/(24.*E*SMA(M))
C
C   SUBDIVIDE THE ELEMENT INTO 100 STEPS FOR SEARCH PURPOSES.
          DO 15 ISTEP=1,100
          X1=0.01*ELENGT*REAL(ISTEP-1)
          X2=X1+0.01*ELENGT
          THETA1=C2+2.*C3*X1+3.*C4*X1**2+4.*C5*X1**3
          THETA2=C2+2.*C3*X2+3.*C4*X2**2+4.*C5*X2**3
          IF(THETA1*THETA2.LE.0.) THEN
              NVMAX=NVMAX+1
              CALL   BISECT(X1,X2,1.0E-6*ELENGT,XLOCAL,IFLAG)
              XVMAX(NVMAX)=X(I)+XLOCAL
              VMAX(NVMAX)=C1+C2*XLOCAL+C3*XLOCAL**2+C4*XLOCAL**3+C5*XLOCAL**4
              END IF
   15     CONTINUE
C
C   OUTPUT MATHEMATICAL MAXIMUM OR MINIMUM VALUES OF DEFLECTION.
          IF(NVMAX.GT.0) THEN
              WRITE(6,762) (XVMAX(IVMAX),VMAX(IVMAX),IVMAX=1,NVMAX)
  762         FORMAT(/ 'MATHEMATICAL MAXIMUM OR MINIMUM VALUES OF DEFLECTION'
         1            / ' AXIAL POSITION      DEFLECTION ' /
         2              (2E15.4))
              END IF
C
C   FIND GREATEST VALUE OF DEFLECTION.
          GRVP=0.
          XGRVP=0.
          GRVN=0.
          XGRVN=0.
```

```
               IF(NVMAX.GT.0) THEN
                  DO 16 IVMAX=1,NVMAX
                  IF(VMAX(IVMAX).GT.GRVP) THEN
                     GRVP=VMAX(IVMAX)
                     XGRVP=XVMAX(IVMAX)
                  END IF
                  IF(VMAX(IVMAX).LT.GRVN) THEN
                     GRVN=VMAX(IVMAX)
                     XGRVN=XVMAX(IVMAX)
                  END IF
      16         CONTINUE
               END IF
               DO 17 M=1,NEL
               DO 17 IEND=1,2
               IF(IEND.EQ.1) INODE=NPI(M)
               IF(IEND.EQ.2) INODE=NPJ(M)
               IF(V(INODE).GT.GRVP) THEN
                  GRVP=V(INODE)
                  XGRVP=X(INODE)
               END IF
               IF(V(INODE).LT.GRVN) THEN
                  GRVN=V(INODE)
                  XGRVN=X(INODE)
               END IF
      17       CONTINUE
      C
      C  OUTPUT THE GREATEST VALUES OF DEFLECTION.
               WRITE(6,763) GRVP,XGRVP,GRVN,XGRVN
      763      FORMAT(
           1     /'GREATEST UPWARD DEFLECTION   =',E12.4,'   AT X =',E12.4
           2     /'GREATEST DOWNWARD DEFLECTION =',E12.4,'   AT X =',E12.4)
      C
      C  COMPUTE SHEAR FORCES AND BENDING MOMENTS.
               DO 19 M=1,NEL
               I=NPI(M)
               J=NPJ(M)
               DO 18 IR=1,4
      18       ELOAD(IR)=ESTIFF(M,IR,1)*V(I)+ESTIFF(M,IR,2)*THETA(I)
           1            +ESTIFF(M,IR,3)*V(J)+ESTIFF(M,IR,4)*THETA(J)-WLOAD(M,IR)
               SF(M,1)=ELOAD(1)
               BM(M,1)=-ELOAD(2)
               SF(M,2)=-ELOAD(3)
      19       BM(M,2)=ELOAD(4)
      C
      C  OUTPUT THE COMPUTED SHEAR FORCES AND BENDING MOMENTS AT THE NODES
      C  OF EACH ELEMENT.
               WRITE(6,77)  (M,NPI(M),NPJ(M),SF(M,1),SF(M,2),BM(M,1),BM(M,2),
           1                 M=1,NEL)
      77       FORMAT(/ 'COMPUTED SHEAR FORCES AND BENDING MOMENTS AT THE'
           1            ' NODES OF EACH ELEMENT' /
           2            ' ELEM   NODES       SF AT I       SF AT J   '
           3            '    BM AT I       BM AT J   ' / (3I4,4E15.4))
      C
      C  SEARCH FOR MATHEMATICAL MAXIMUM OR MINIMUM VALUES OF BENDING
      C  MOMENT WITHIN ELEMENTS (WHERE SHEAR FORCE ZERO).
               NBMMAX=0
               DO 20 M=1,NEL
               IF(SF(M,1)*SF(M,2).LE.0.) THEN
                  NBMMAX=NBMMAX+1
                  I=NPI(M)
                  J=NPJ(M)
                  IF(SF(M,1)*SF(M,2).LT.0.) THEN
                     XLOCAL=SF(M,1)/W(M)
                  ELSE
                     IF(SF(M,1).EQ.0.) XLOCAL=0.
                     IF(SF(M,2).EQ.0.) XLOCAL=X(J)-X(I)
                  END IF
                  XBMMAX(NBMMAX)=X(I)+XLOCAL
                  BMMAX(NBMMAX)=BM(M,1)+SF(M,1)*XLOCAL-0.5*W(M)*XLOCAL**2
               END IF
      20       CONTINUE
      C
      C  OUTPUT MATHEMATICAL MAXIMUM OR MINIMUM VALUES OF BENDING MOMENT.
               IF(NBMMAX.GT.0) THEN
                  WRITE(6,82) (XBMMAX(IBMMAX),BMMAX(IBMMAX),IBMMAX=1,NBMMAX)
```

```
82      FORMAT(/ 'MATHEMATICAL MAXIMUM OR MINIMUM VALUES OF BENDING'
    1              ' MOMENT' / ' AXIAL POSITION        MOMENT ' /
    2              (2E15.4))
        END IF
C
C   FIND GREATEST VALUES OF SHEAR FORCE AND BENDING MOMENT.
        GRSF=0.
        XGRSF=0.
        GRBMS=0.
        XGRBMS=0.
        GRBMH=0.
        XGRBMH=0.
        IF(NBMMAX.GT.0) THEN
          DO 21 IBMMAX=1,NBMMAX
          IF(BMMAX(IBMMAX).GT.GRBMS) THEN
            GRBMS=BMMAX(IBMMAX)
            XGRBMS=XBMMAX(IBMMAX)
          END IF
          IF(BMMAX(IBMMAX).LT.GRBMH) THEN
            GRBMH=BMMAX(IBMMAX)
            XGRBMH=XBMMAX(IBMMAX)
          END IF
21        CONTINUE
        END IF
        DO 22 M=1,NEL
        DO 22 IEND=1,2
        IF(IEND.EQ.1) INODE=NPI(M)
        IF(IEND.EQ.2) INODE=NPJ(M)
        IF(ABS(SF(M,IEND)).GT.GRSF) THEN
          GRSF=ABS(SF(M,IEND))
          XGRSF=X(INODE)
        END IF
        IF(BM(M,IEND).GT.GRBMS) THEN
          GRBMS=BM(M,IEND)
          XGRBMS=X(INODE)
        END IF
        IF(BM(M,IEND).LT.GRBMH) THEN
          GRBMH=BM(M,IEND)
          XGRBMH=X(INODE)
        END IF
22      CONTINUE
C
C   OUTPUT THE GREATEST VALUES OF SHEAR FORCE AND BENDING MOMENT.
        WRITE(6,83) GRSF,XGRSF,GRBMS,XGRBMS,GRBMH,XGRBMH
83      FORMAT(/ 'GREATEST SHEAR FORCE =',E12.4,'   AT X =',E12.4
    1   /'GREATEST SAGGING BENDING MOMENT =',E12.4,'   AT X =',E12.4
    2   /'GREATEST HOGGING BENDING MOMENT =',E12.4,'   AT X =',E12.4)
        STOP
        END

        FUNCTION  FN(X,IOTEST)
C
C   FUNCTION SUBPROGRAM TO DEFINE THE CUBIC FUNCTION FOR SLOPE WITHIN
C   AN ELEMENT.
        COMMON /CONSTS/  C2,C3,C4,C5
        FN=C2+2.*C3*X+3.*C4*X**2+4.*C5*X**3
        RETURN
        END
```

The following list provides the definitions of the FORTRAN variables used,
arranged in alphabetical order.

Variable	Type	Definition
BM	real	Array storing the bending moments at the nodes of the elements (first subscript defines the element number, second subscript the node number within the element)
BMMAX	real	Array storing mathematical maximum and minimum values of bending moment within elements
C	real	Array storing the external concentrated couples (counterclockwise positive) acting on the beam at the nodes

Variable	Type	Definition
C1 to C5	real	Shape function constants C_1 to C_5
DELTA	real	Array storing the deflections and slopes at the nodal points
E	real	Young's modulus of the beam material
ELENGT	real	Element length
ELOAD	real	Array storing components of the element load vector $[f]_m$
ESTIFF	real	Array storing the stiffness coefficients for all the elements (first subscript defines the element number, second and third subscripts the row and column numbers of the element stiffness matrix)
F	real	Array storing the external concentrated forces (downward positive) acting on the beam at the nodes
FACT	real	Common factor in the element stiffness coefficients
GRBMH	real	Greatest value of bending moment (hogging)
GRBMS	real	Greatest value of bending moment (sagging)
GRSF	real	Greatest value of shear force
GRVP	real	Greatest value of deflection (positive)
GRVN	real	Greatest value of deflection (negative)
I	integer	Nodal point number (sometimes first node of an element)
IBMMAX	integer	Counter for mathematical maximum or minimum values of bending moments
IC	integer	Column number in array ESTIFF
ICOL	integer	Column number in array OSTIFF
IEND	integer	Counter for the (two) ends of an element
IFLAG	integer	Flag for a singular or very ill-conditioned coefficient matrix (normally returned as zero by SOLVE, one if the ill-conditioning test is satisfied)
INODE	integer	Node number
IR	integer	Row number in array ESTIFF
IROW	integer	Row number in array OSTIFF
ISTEP	integer	Counter for steps along an element when searching for maximum or minimum deflections
IVMAX	integer	Counter for mathematical maximum or minimum values of deflection
J	integer	Nodal point number (second node of an element)
K	integer	A counter (used in READ and WRITE statements)
M	integer	Element number
NBMMAX	integer	Number of mathematical maximum and minimum values of bending moment within elements
NEL	integer	Number of elements
NELEM	integer	Array storing element numbers
NELMAX	integer	Maximum number of elements permitted by the array dimensions
NELW	integer	Number of elements over which external distributed forces act
NEQN	integer	Number of equations to be solved
NNP	integer	Number of nodal points
NNPC	integer	Number of nodes at which external concentrated couples are applied
NNPF	integer	Number of nodes at which external concentrated forces are applied
NNPMAX	integer	Maximum number of nodal points permitted by the array dimensions
NNPBI	integer	Number of nodes at which the beam is built in
NNPFL	integer	Number of nodes at which the beam is supported on flexible supports
NNPSS	integer	Number of nodes at which the beam is simply supported
NNPT	integer	Total number of nodes at which the beam is supported
NODE	integer	Array storing node numbers
NPI	integer	Array storing the numbers of the first nodes of the elements
NPJ	integer	Array storing the numbers of the second nodes of the elements
NSTIFF	integer	Number of non-zero stiffnesses of flexible supports
NUNK	integer	Total number of unknown support reaction forces and moments
NVMAX	integer	Number of mathematical maximum and minimum values of deflection within elements
OSTIFF	real	Array storing coefficients of the overall stiffness matrix $[K]$, extended to include external force vector $[F]^*$ as its last column
PI	real	The mathematical constant π

Variable	Type	Definition
RSTIFF	real	Array storing the rotational stiffnesses of the flexible supports (in units of moment per radian)
SF	real	Array storing the shear forces at the nodes of the elements (first subscript defines the element number, second subscript the node number within the element)
SMA	real	Array storing the second moments of area of the elements
SMAC	real	Constant second moment of area for a uniform beam
THETA	real	Array storing the slopes (in radians) at the nodal points
THETAS	real	Array storing the slopes (in degrees) at the support points
THETA1	real	Slope at the beginning of a step within an element
THETA2	real	Slope at the end of a step within an element
V	real	Array storing the deflections at the nodal points
VMAX	real	Array storing mathematical maximum and minimum values of deflection within elements
VS	real	Array storing the deflections at the support points
VSTIFF	real	Array storing the vertical stiffnesses of the flexible supports
W	real	Array storing the intensities of the external distributed forces (downward positive) acting on the elements
WLOAD	real	Array storing the external load terms $[W]_m$ associated with the distributed lateral forces, for all the elements (first subscript defines the element number, second subscript the row number)
X	real	Array storing X global coordinates of the nodes
X1	real	Local coordinate at the beginning of a step within an element
X2	real	Local coordinate at the end of a step within an element
XBMMAX	real	Array storing global coordinates of mathematical maximum and minimum values of bending moment within elements
XGRBMH	real	Position (X coordinate) of the point of greatest hogging bending moment
XGRBMS	real	Position (X coordinate) of the point of greatest sagging bending moment
XGRSF	real	Position (X coordinate) of the point of greatest shear force
XGRVN	real	Position (X coordinate) of the point of greatest negative deflection
XGRVP	real	Position (X coordinate) of the point of greatest positive deflection
XLOCAL	real	Local coordinate, x, within an element
XVMAX	real	Array storing global coordinates of mathematical maximum and minimum values of deflection within elements

The arrays used in SIBEAM are dimensioned in such a way as to allow a beam with up to 50 elements and 51 nodes to be analyzed, implying that up to 102 equations (twice the number of nodes) may have to be solved. Note that array OSTIFF, which stores the overall stiffness matrix $[K]$ extended to include vector $[F]^*$ is allowed to have up to 103 columns. The array dimensions can of course be changed, provided the values of NNPMAX and NELMAX defined in the next two statements are also adjusted.

The program reads information from a file named DATA, and writes onto a file named RESULTS. All READ statements call for free format input data. After writing out a heading, the program first reads the number of nodal points along the beam to be analyzed, and then tests that this does not exceed the maximum allowed by the array dimensions. We should note that, as in programs SDPINJ, SIPINJ and SDBEAM, the checking of input data in the program is comparatively rudimentary, for the same reasons. Some of the more obvious data errors are trapped, and in all cases input data is immediately written out to allow it to be checked.

The program reads in and tests the number of nodes, followed by the global axial coordinates of the nodes, X, for node 1, then node 2 and so on. It is assumed that the nodes are numbered consecutively from left to right along the beam, and the axial coordinates must reflect this ordering (if they do not, execution will fail at a later point in the program). Although we do not have to make this assumption, and the program could be coded to accept an arbitrary numbering system, there is little to be gained with a geometrically simple system such as a beam. By assuming we know the node numbering system, it is then not necessary to enter as data the nodes of the elements, which is a significant saving of labor. The program also numbers the elements from left to right along the beam, which means that the element numbered m has as its nodal points the nodes m and $m+1$, as shown by Fig. 5.40. At the same time, the distributed forces on the elements are set to zero.

The Young's modulus of the beam material and the second moment of area of the beam cross section about its neutral axis are then read in. A nonzero value of the latter implies that the second moment of area is constant along the beam (the beam is uniform). If a zero (or negative) value is read for the second moment, then the beam is not uniform and numerical values must be supplied for each of the elements, taken in order from left to right.

All coefficients of the overall stiffness matrix and the applied concentrated forces and couples at the nodes are then set to zero in preparation for the subsequent assembly process. The number of nodes at which external concentrated forces are applied is read in, followed by the numbers of the nodes concerned together with the corresponding forces, taking the downward direction as positive. The signs of these forces are then reversed to conform with the sign convention used in the finite element analysis. Then, the number of nodes at which external concentrated couples are applied is read in, followed by the numbers of the nodes concerned together with the corresponding couples, taking the counterclockwise direction as positive. The final possible form of loading is that due to external distributed forces, and the number of elements over which such forces are applied is read in, followed by the numbers of the elements concerned together with the corresponding force intensities (force per unit length), again taking the downward direction as positive.

Next, the numbers of nodes at which the beam is simply supported, built-in and attached to flexible supports are read in. If there are some simple supports, the numbers of the nodes forming them are read, together with the corresponding deflections, taking the upward direction as positive. Any convenient common datum can be used for the deflections, typically the height of one of the supports. If there are some built-in supports, the numbers of the nodes at which they occur are read, together with the corresponding deflections and slopes, taking the counterclockwise direction as positive. If there are some flexible supports, the numbers of the nodes at which they occur are read, together with the corresponding vertical and rotational stiffnesses.

Note that, as in program SDBEAM, no distinction is made between pinned and rolling simple supports, because the program does not accept axial loading of the beam so that the distinction is unnecessary. It also means that the formal test for whether a beam is statically determinate, equation (5.1), must be modified, since both reaction force components and the equilibrium equation in the axial direction have been eliminated. The program calculates the total number of support reaction forces and moments: one for a simple support, two for a built-in support, and one or two for a flexible support depending on the number of nonzero stiffnesses. The beam is statically indeterminate, statically determinate or a mechanism according to whether this number is greater than, equal to or less than two. Only if the beam is not a mechanism is the analysis allowed to proceed.

The coefficients of the overall stiffness matrix in equations (6.85) are assembled by considering each of the elements in turn. Our typical element, shown in Fig. 6.32 and numbered m, has first and second nodes i and j, respectively, which in the program are recovered from arrays NPI and NPJ and stored in variables I and J. The length of the element is found as the distance between nodes i and j. If this length is negative, the global coordinates of the nodes must have been entered in an incorrect order, and execution is halted.

The coefficients of the element stiffness matrix are computed according to equations (6.81), and stored in array ESTIFF for later use in finding shear forces and bending moments from computed deflections and slopes. Each of these coefficients is added to the appropriate term in the overall stiffness matrix, according to the numbers of the nodes involved. The assembly process is very similar to that used in program SIPINJ for the finite element analysis of pin-jointed structures in Section 4.1.1. The nodal point forces and moments due to any distributed force acting on the element, contained in vector $[W]_m$ and defined by equations (6.81), are then computed and added to the vector $[F]$ of concentrated forces and couples to give $[F]^*$ (equation (6.85)). The components of this vector are then transferred to the last column of the extended overall stiffness matrix.

At each of the simple supports, the program implements the changes to the overall stiffness matrix required by equation (6.86) to impose prescribed values of deflection there. Similarly, at built-in supports, the changes required by equations (6.86) and (6.87) are made (noting that the units of prescribed slopes must be converted from degrees to radians). Also, at flexible supports, the stiffness relationships defined by equations (6.88) and (6.89) are incorporated by adding the vertical and rotational stiffnesses to the appropriate self stiffnesses (coefficients on the leading diagonal of the overall stiffness matrix) of the corresponding nodes.

The equations can now be solved, using subroutine SOLVE which is described in detail in Appendix B. If a unit value of IFLAG is returned to SIBEAM, this means that SOLVE has detected a singular or very ill-conditioned coefficient matrix. This is only likely to happen in the present program if the beam is inadequately constrained to prevent it moving as a rigid body. For example, if the two simple supports of a statically determinate beam are inadvertently defined at the same point, the beam could then rotate about this point. Execution would then be terminated. Otherwise, the computed deflections and slopes at the nodes are transferred to arrays V and THETA, and are then written out. The program then searches for mathematical maximum or minimum values of deflection, examining each element in turn. Within each element, the deflection is a quartic function of position (equation (6.67)), so that there may be more than one maximum or minimum point. This possibility is dealt with in a rather unsophisticated way, by dividing the element into 100 steps of equal length and, for each step, testing whether the slope passes through zero between the two ends of the step, or is zero at either one of them. These conditions can be detected by the product of the slopes at the ends being either less than or equal to zero, and the local coordinate of the required point can be found with the aid of subroutine BISECT for solving a nonlinear algebraic equation, which is described in detail in Appendix E. This subroutine finds the value of local coordinate x which makes the cubic function given in equation (6.68) zero. This expression is defined in the function subprogram FN which immediately follows the main program for SIBEAM. The shape function constants C_2 to C_5 which are required for this purpose are made available to FN via the labeled COMMON block CONSTS. The global coordinate of a point of maximum or minimum deflection is found by adding its local coordinate to the global coordinate of node i, and the actual value of the deflection is found with the aid of equation (6.67). The greatest positive and negative values of deflection are found, first by searching the mathematical maximum and minimum values, and then the values at the nodal points. These greatest values are then written out.

The shear forces and bending moments at the nodes of each of the elements are computed from the displacements, first using equations (6.81) to determine the element load vector, and then equations (6.72) to modify the signs. Treatment of shear forces and bending moments is then identical to that described for program SDBEAM: the values are written out, mathematical maximum and minimum values are determined, followed by the greatest values. Note that the searching process for the latter involves first setting the greatest values to zero and the corresponding positions (X coordinates) to zero. This can mean that if, for example, there are no hogging bending moments along the beam the initial settings would be unchanged and the result for the greatest hogging moment would be shown as zero at $X = 0$.

No units are specified for either the input data or the computed results, with the exception of prescribed slopes at built-in supports, which must be supplied in degrees, and rotational stiffnesses of flexible supports, which must be supplied in units of moment per radian. We may use any consistent set of force and length units.

6.4.3 Practical applications

Let us now examine how program SIBEAM can be used to analyze beam problems with the aid of some typical examples. The first of these, which is statically determinate, has already been subjected to both manual and computer methods for finding distributions of shear force and bending moment in Examples 5.4 and 5.9. We are now able to find the deflections and slopes as well.

EXAMPLE 6.15

Consider again the beam shown in Fig. 5.26, which has a uniform cross section with a second moment of area about its neutral axis of 220 in⁴, and is made of steel. Ignoring the weight of the beam, use program SIBEAM to find the distributions of beam deflection, slope, shear force and bending moment in US customary units.

The beam is redrawn in Fig. 6.37 with the nodes and elements shown numbered in the orders assumed by the program. The choice of nodes and elements is determined by the positions of the supports, and the abrupt changes in the distributed lateral forces. Element numbers are shown ringed, while node numbers are unringed. It is convenient to define the origin for global coordinate X at node 1 as shown. Figure 6.38 shows the file of input data used to set up this problem for solution by program SIBEAM, ft and kip units being used for length and force, respectively. In these units, the Young's modulus and second moment of area are 4.32×10^6 kip/ft^2 and 0.010 61 ft^4.

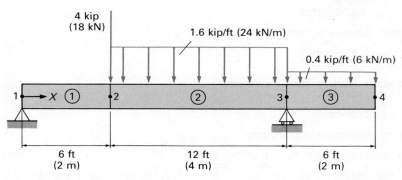

Fig. 6.37.

```
4                             (Number of nodes)
0.  6.  18.  24.              (Node coordinates)
4.32E6  0.01061     (Young's modulus and second moment of area)
1                   (Number of concentrated lateral forces)
2  4.                 (Node number and concentrated force)
0                      (Number of concentrated couples)
2            (Number of elements with distributed lateral forces)
2  1.6,  3  0.4         (Element numbers and force intensities)
2  0  0       (Numbers of simple, built-in and flexible supports)
1  0.,  3  0.   (Node numbers and deflections of the simple supports)
```

Fig. 6.38.

The file has been annotated to indicate what each line of data represents. The file RESULTS produced by the program is shown in Fig. 6.39. As we would expect, the computed shear forces and bending moments are all identical to those obtained by manual methods in Example 5.4, which are plotted in Fig. 5.27(US). They are also identical to the results given by program SDBEAM in Example 5.9, Fig. 5.43. The only minor exceptions are the bending moment at node 1 and the shear force and bending moment at node 4, which should be zero, but are in fact of the order of 10^6 times smaller than typical values elsewhere along the beam. These very small nonzero values, which are only noticeable because the results are presented in exponent format, are due to the fact that shear forces and bending moments are derived from displacements in the program, rather than computed directly as the primary variables. The computations involved are carried out with numbers which retain only a finite number of significant digits.

The computed deflections are plotted in Fig. 6.40. The greatest upward deflection of $+0.498$ in (0.041 52 ft) occurs at the right-hand end of the beam (but is not a mathematical maximum), while the greatest downward deflection of -0.501 in ($-0.041\,78$ ft) occurs 9.024 ft along the beam from the left-hand end, and does represent a local maximum (in magnitude).

```
STATICALLY INDETERMINATE OR DETERMINATE BEAM

COORDINATES OF THE NODES
NODE    X          NODE    X          NODE    X          NODE    X
 1  0.0000E+00      2  0.6000E+01      3  0.1800E+02      4  0.2400E+02

YOUNGS MODULUS OF THE BEAM MATERIAL =    0.4320E+07

SECOND MOMENT OF AREA OF THE BEAM CROSS SECTION =    0.1061E-01

CONCENTRATED (DOWNWARD) LATERAL FORCES
  NODE         F
    2         0.4000E+01

DISTRIBUTED (DOWNWARD) LATERAL FORCES
  ELEMENT        W
    2         0.1600E+01
    3         0.4000E+00

NODES AT WHICH BEAM IS SIMPLY SUPPORTED
  NODE     DEFLECTION
    1      0.0000E+00
    3      0.0000E+00

THE BEAM IS STATICALLY DETERMINATE

COMPUTED DEFLECTIONS AND SLOPES AT THE NODES
  NODE     V          SLOPE      NODE     V          SLOPE
    1   0.0000E+00  -0.7139E-02    2  -0.3602E-01  -0.3735E-02
    3   0.0000E+00   0.7156E-02    4   0.4152E-01   0.6842E-02

MATHEMATICAL MAXIMUM OR MINIMUM VALUES OF DEFLECTION
  AXIAL POSITION    DEFLECTION
    0.9024E+01      -0.4178E-01

GREATEST UPWARD DEFLECTION   =  0.4152E-01   AT X =  0.2400E+02
GREATEST DOWNWARD DEFLECTION = -0.4178E-01   AT X =  0.9024E+01

COMPUTED SHEAR FORCES AND BENDING MOMENTS AT THE NODES OF EACH ELEMENT
  ELEM  NODES    SF AT I      SF AT J      BM AT I      BM AT J
    1    1   2  0.8667E+01   0.8667E+01   0.7629E-05   0.5200E+02
    2    2   3  0.4667E+01  -0.1453E+02   0.5200E+02  -0.7200E+01
    3    3   4  0.2400E+01  -0.1454E-04  -0.7200E+01  -0.1824E-04

MATHEMATICAL MAXIMUM OR MINIMUM VALUES OF BENDING MOMENT
  AXIAL POSITION       MOMENT
    0.8917E+01        0.5881E+02
    0.2400E+02       -0.2575E-04

GREATEST SHEAR FORCE =  0.1453E+02   AT X =  0.1800E+02
GREATEST SAGGING BENDING MOMENT =  0.5881E+02   AT X =  0.8917E+01
GREATEST HOGGING BENDING MOMENT = -0.7200E+01   AT X =  0.1800E+02
```

Fig. 6.39.

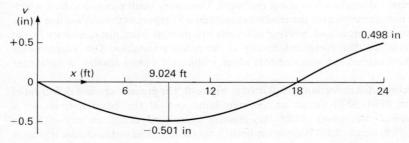

Fig. 6.40.

EXAMPLE 6.16   379

Figure 6.41 shows an initially straight and unstressed beam 12 m (39 ft) long which is built in horizontally at each end, A and F, and rests on simple supports at B and D, which are 3 m (10 ft) and 8 m (26 ft), respectively, from A. As a result of subsequent movement of the supports, end F is 4 mm (0.16 in) below A and is rotated clockwise through an angle of 0.5°, the support at B is 5 mm (0.2 in) below A, but the one at D is at the same level as A. The beam has a constant flexural rigidity of 30×10^6 Nm² $(1.0 \times 10^{10}$ lb in²) weighs 2 kN/m (140 lb/ft) and is subjected to two concentrated vertical forces, one of 30 kN (6.7 kip) at C, which is 5 m (16 ft) from A, and another of 25 kN (5.6 kip) at E, which is 9 m (30 ft) from A. Find the greatest absolute shear force, the greatest sagging and hogging bending moments and the greatest upward and downward deflections of the beam.

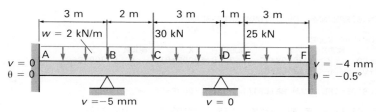

Fig. 6.41.

This statically indeterminate problem is considerably more complicated than any we have solved by manual methods, but is more typical of situations met in practice. Ignoring axial forces and the axial equilibrium equation, there are six unknown support reaction forces and moments (two at each end of the beam, and one at each simple support), and only two equilibrium equations, implying four redundancies.

The beam is redrawn in Fig. 6.42 with nodes and elements shown numbered in the orders assumed by the program. The choice of nodes and elements is determined by the positions of the supports, and the points of application of the concentrated lateral forces. It is convenient to define the origin for global coordinate X at node 1 as shown. Figure 6.43 shows the file of input data used to set up this problem for solution by program SIBEAM, the units of length and force being meters and newtons, respectively. Note that, since any combination of Young's modulus and second moment of area having a product equal to 30×10^6 Nm² is acceptable, we can choose E as 30×10^6 N/m² and I as 1 m⁴.

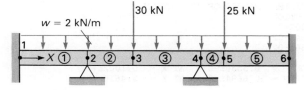

Fig. 6.42.

```
6
0.  3.  5.  8.  9.  12.
30.0E6  1.
2
3  30.0E3,  5  25.0E3
0
5
1 2.0E3, 2  2.0E3, 3  2.0E3, 4  2.0E3, 5  2.0E3
2  2  0
2  -0.005, 4  0.
1  0.  0., 6  -0.004  -0.5
```

Fig. 6.43.

The file RESULTS produced by the program is shown in Fig. 6.44. The required greatest absolute value of shear force is 53.1 kN (12 kip) at support F, and the greatest sagging and hogging bending moments are + 50.9 kNm (38 kip-ft) under the applied force at C, and − 183.5 kNm (− 130 kip-ft) at support F. The greatest upward and

```
STATICALLY INDETERMINATE OR DETERMINATE BEAM

COORDINATES OF THE NODES
NODE     X         NODE     X         NODE     X         NODE     X
   1  0.0000E+00      2  0.3000E+01      3  0.5000E+01      4  0.8000E+01
   5  0.9000E+01      6  0.1200E+02

YOUNGS MODULUS OF THE BEAM MATERIAL =   0.3000E+08

SECOND MOMENT OF AREA OF THE BEAM CROSS SECTION =   0.1000E+01

CONCENTRATED (DOWNWARD) LATERAL FORCES
  NODE           F
     3       0.3000E+05
     5       0.2500E+05

DISTRIBUTED (DOWNWARD) LATERAL FORCES
 ELEMENT          W
     1       0.2000E+04
     2       0.2000E+04
     3       0.2000E+04
     4       0.2000E+04
     5       0.2000E+04

NODES AT WHICH BEAM IS SIMPLY SUPPORTED
  NODE     DEFLECTION
     2    -0.5000E-02
     4     0.0000E+00

NODES AT WHICH BEAM IS BUILT IN
  NODE    DEFLECTION     SLOPE (DEG)
     1    0.0000E+00     0.0000E+00
     6   -0.4000E-02    -0.5000E+00

THE BEAM IS STATICALLY INDETERMINATE

COMPUTED DEFLECTIONS AND SLOPES AT THE NODES
NODE      V           SLOPE      NODE      V           SLOPE
   1  0.0000E+00   0.0000E+00      2  -0.5000E-02  -0.1887E-02
   3 -0.6576E-02   0.6217E-03      4   0.0000E+00   0.2711E-02
   5  0.2394E-02   0.1960E-02      6  -0.4000E-02  -0.8727E-02

MATHEMATICAL MAXIMUM OR MINIMUM VALUES OF DEFLECTION
 AXIAL POSITION       DEFLECTION
    0.0000E+00       0.0000E+00
    0.4616E+01      -0.6694E-02
    0.1002E+02       0.3536E-02

GREATEST UPWARD DEFLECTION   =  0.3536E-02   AT X =  0.1002E+02
GREATEST DOWNWARD DEFLECTION = -0.6694E-02   AT X =  0.4616E+01
```

Fig. 6.44 (cont.)

```
COMPUTED SHEAR FORCES AND BENDING MOMENTS AT THE NODES OF EACH ELEMENT
ELEM  NODES       SF AT I       SF AT J       BM AT I       BM AT J
  1    1    2    0.3192E+05    0.2592E+05   -0.6375E+05    0.2300E+05
  2    2    3    0.1597E+05    0.1197E+05    0.2300E+05    0.5094E+05
  3    3    4   -0.1803E+05   -0.2403E+05    0.5094E+05   -0.1216E+05
  4    4    5   -0.2008E+05   -0.2208E+05   -0.1216E+05   -0.3324E+05
  5    5    6   -0.4708E+05   -0.5308E+05   -0.3324E+05   -0.1835E+06

GREATEST SHEAR FORCE =  0.5308E+05   AT X =  0.1200E+02
GREATEST SAGGING BENDING MOMENT =  0.5094E+05   AT X =  0.5000E+01
GREATEST HOGGING BENDING MOMENT = -0.1835E+06   AT X =  0.1200E+02
```

Fig. 6.44.

downward deflections are $+3.54$ mm (0.14 in) at 10.0 m (33 ft) along the beam, and -6.69 mm (-0.26 in) at 4.62 m (15 ft) along the beam. Note that, due to the difference in heights of the supports at A and B, there is no mathematical maximum or minimum deflection between them.

EXAMPLE 6.17

Consider again the tapered steel cantilever beam shown in Fig. 6.21, whose end deflection due to the concentrated force of 10 kN applied there was found analytically in Example 6.11. Use program SIBEAM to check this result.

There are two main types of problems which cannot be solved exactly using the present finite element method: cases where distributed forces vary continuously along the beam, and those where the cross section of the beam varies continuously. We considered a problem involving a distributed force of varying intensity in Example 5.10, using program SDBEAM which employs similar elements over each of which the distributed force must be of constant intensity. The present example concerns a beam with a varying second moment of area. Figure 6.45 shows the beam divided into five elements of equal length. Node 1 is again chosen as the origin for global coordinate X.

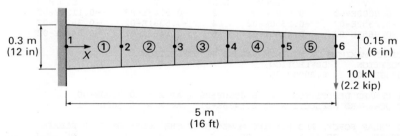

Fig. 6.45.

Figure 6.46 shows the file of input data used to set up this problem for solution by program SIBEAM, the units of length and force being meters and newtons. We must assume a constant second moment of area over each element, and we therefore take the

```
6
0.  1.0  2.0  3.0  4.0  5.0
207.0E9  0.
1.929E-4 1.382E-4 9.492E-5 6.179E-5 3.743E-5
1
6  10.0E3
0
0
0  1  0
1  0.  0.
```

Fig. 6.46.

value at the center of each element as the value applicable to the entire element. For example, the center of element 1 is at 0.5 m along the beam, where the breadth is 0.1 m, the depth is 0.285 m and the second moment of area is

$$I = \frac{0.1 \times 0.285^3}{12} = 1.929 \times 10^{-4} \, \text{m}^4$$

The file RESULTS produced by the program is shown in Fig. 6.47. The computed deflection at the free end of the cantilever is − 14.88 mm, which differs from the exact value of − 14.63 mm by only about 1.7%.

```
STATICALLY INDETERMINATE OR DETERMINATE BEAM

COORDINATES OF THE NODES
NODE    X         NODE    X         NODE    X         NODE    X
  1  0.0000E+00     2  0.1000E+01     3  0.2000E+01     4  0.3000E+01
  5  0.4000E+01     6  0.5000E+01

YOUNGS MODULUS OF THE BEAM MATERIAL =    0.2070E+12

SECOND MOMENTS OF AREA OF THE ELEMENTS
ELEM    SMA       ELEM    SMA       ELEM    SMA       ELEM    SMA
  1  0.1929E-03     2  0.1382E-03     3  0.9492E-04     4  0.6179E-04
  5  0.3743E-04

CONCENTRATED (DOWNWARD) LATERAL FORCES
  NODE       F
    6         0.1000E+05

NODES AT WHICH BEAM IS BUILT IN
  NODE   DEFLECTION    SLOPE (DEG)
    1     0.0000E+00     0.0000E+00

THE BEAM IS STATICALLY DETERMINATE

COMPUTED DEFLECTIONS AND SLOPES AT THE NODES
NODE      V           SLOPE      NODE      V           SLOPE
  1   0.0000E+00    0.0000E+00     2  -0.5843E-03   -0.1127E-02
  3  -0.2352E-02   -0.2350E-02     4  -0.5381E-02   -0.3623E-02
  5  -0.9655E-02   -0.4795E-02     6  -0.1488E-01   -0.5441E-02

MATHEMATICAL MAXIMUM OR MINIMUM VALUES OF DEFLECTION
  AXIAL POSITION       DEFLECTION
    0.0000E+00         0.0000E+00

GREATEST UPWARD DEFLECTION   =   0.0000E+00   AT X =   0.0000E+00
GREATEST DOWNWARD DEFLECTION = -0.1488E-01   AT X =   0.5000E+01

COMPUTED SHEAR FORCES AND BENDING MOMENTS AT THE NODES OF EACH ELEMENT
  ELEM  NODES     SF AT I        SF AT J        BM AT I        BM AT J
    1    1   2  0.1000E+05     0.1000E+05    -0.5000E+05    -0.4000E+05
    2    2   3  0.1000E+05     0.1000E+05    -0.4000E+05    -0.3000E+05
    3    3   4  0.1000E+05     0.1000E+05    -0.3000E+05    -0.2000E+05
    4    4   5  0.1000E+05     0.1000E+05    -0.2000E+05    -0.1000E+05
    5    5   6  0.1000E+05     0.1000E+05    -0.1000E+05    -0.4688E-01

GREATEST SHEAR FORCE =  0.1000E+05   AT X =  0.3000E+01
GREATEST SAGGING BENDING MOMENT =  0.0000E+00   AT X =  0.0000E+00
GREATEST HOGGING BENDING MOMENT = -0.5000E+05   AT X =  0.0000E+00
```

Fig. 6.47.

Although the result obtained using 5 elements is good, we can expect to improve it by using more elements. Table 6.1 shows the computed end deflections obtained using 5, 10 and 20 elements along the beam, in each case using elements of constant length. Using not more than 10 elements, accuracies of better than 1% can be obtained, which are adequate for most practical purposes. The progressive improvement of the results with element refinement is typical of approximate finite element methods. If we did not have the exact solution for comparison, we would refine the elements until we were satisfied that there was no further significant change in the results.

Table 6.1 Effect of element refinement on computed end deflection

Number of elements	End deflection (mm)	% error
5	−14.88	1.7
10	−14.69	0.4
20	−14.64	0.07
exact	−14.63	—

We see from these three examples that the main use for a computer program such as SIBEAM is for problems where all the details of the beam geometry and loading are defined numerically. It offers particular advantages for more complicated problems (with large numbers of applied forces and couples, and many supports), where manual methods become increasingly laborious. It is not suitable, however, for problems where results are required in the form of general algebraic expressions for deflection, slope, shear force or bending moment.

Problems

6.1 A 50 mm by 225 mm timber beam is used as a floor joist, and is simply supported at its ends. If the span is 6 m, find the greatest deflection of the beam when subjected to a uniformly distributed force (including its own weight) of intensity 940 N/m along its entire length.

6.2 Repeat Problem 6.1 assuming the beam is built in at each end.

6.3 For the beam defined in Problem 6.1, calculate the greatest absolute deflection of the beam due to the addition of a concentrated force of 2.8 kN at its center.

6.4 Repeat Problem 6.3 assuming the beam is built in at each end.

6.5 to 6.7 Each of the simple cantilever beams shown in Figs P6.5 to P6.7 has a constant flexural rigidity *EI*. Ignoring the weights of the beams, find expressions for the deflections and slopes at the free ends, and compare these with the results given in Appendix D.

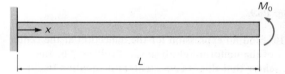

Fig. P6.5.

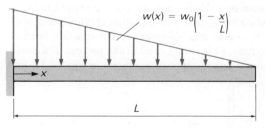

Fig. P6.6.

6.8 and 6.9 The simply supported light beams shown in Figs P6.8 and P6.9 each have a constant flexural rigidity *EI*. Find expressions for the slopes at the left-hand ends of the beams.

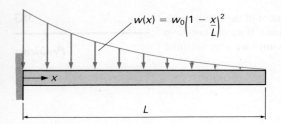

$$w(x) = w_0\left(1 - \frac{x}{L}\right)^2$$

Fig. P6.7.

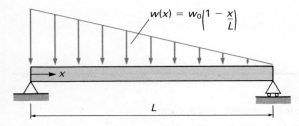

$$w(x) = w_0\left(1 - \frac{x}{L}\right)$$

Fig. P6.8.

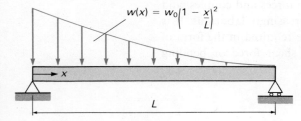

$$w(x) = w_0\left(1 - \frac{x}{L}\right)^2$$

Fig. P6.9.

6.10 Find an expression for the deflection at the center of the uniform light beam in Problem 5.6, and show that this is consistent with equation (6.20) when $a = 0$.

6.11 Find an expression for the deflection at the center of the uniform light beam in Problem 5.7, and show that this is consistent with equation (6.24) when $b = 0$, and with equation (6.20) when $b = L$.

6.12 A vertical mast of length L and constant flexural rigidity EI is firmly fixed in the ground. If a horizontal force is applied as shown in Fig. P6.12, find

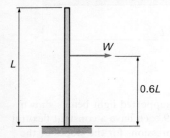

Fig. P6.12.

the horizontal deflection at the tip of the mast using the step function method.

6.13 Use superposition to solve Problem 6.12.

6.14 The uniform simply supported light beam shown in Fig. P6.14 has a flexural rigidity EI. Find the greatest absolute deflection with the aid of program ROOT given in Appendix E.

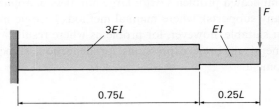

Fig. P6.14.

6.15 Find an expression for the end deflection of the stepped cantilever shown in Fig. P6.15 produced by the force F.

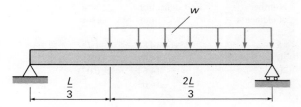

Fig. P6.15.

6.16 Figure P6.16 shows a solid cantilever beam of length L with a diameter which tapers linearly from d_0 at the built-in end to $d_0/3$ at the free end. If the free end is subjected to a concentrated lateral force of magnitude F and the material of the beam has a Young's modulus of E, find an expression for the deflection of the free end produced by force F.

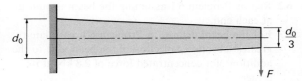

Fig. P6.16.

6.17 A uniform cantilever beam carries a force uniformly distributed over its length. If a prop is placed under the free end to hold it at the same level as the fixed end, find the proportion of the total load taken by this prop.

6.18 A beam which is rigidly built in at each end is of length L, has a constant flexural rigidity EI and carries a uniformly distributed force of intensity w per unit length (including the weight of the beam). Find expressions for the greatest absolute bending moment, and the deflection at the center of the beam, and compare the latter with the result given in Appendix D.

6.19 In Problem 6.18 the height of one of the supports is reduced by an amount δ due to settlement, the ends of the beam remaining horizontal. Find an expression for the new greatest absolute bending moment.

6.20 The horizontal uniform beam of length $3L$ shown in Fig. P6.20 is rigidly built in at A and rests on a simple support at B, which is at the same height as A. Assuming the weight of the beam is negligible, determine the reaction force at support B using the step function method.

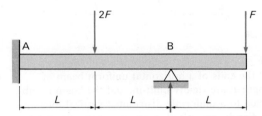

Fig. P6.20.

6.21 Solve Problem 6.20 by the superposition method.

6.22 The two cantilevers shown in Fig. P6.22 each have the same constant flexural rigidity EI, and are pin-jointed together at their ends. Find an expression for the deflection of the pin joint due to the applied force F.

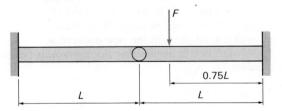

Fig. P6.22.

6.23 The continuous light beam shown in Fig. P6.23 has a constant flexural rigidity EI and rests on three simple supports at A, B and C which are initially at the same level. Obtain expressions for the reaction force at support B and the greatest absolute bending moment in the beam. Find the effect

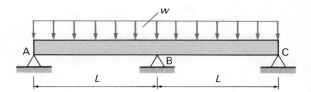

Fig. P6.23.

on these results of (a) raising, and (b) lowering support B by 10% of the amount by which the center of the beam would deflect if the support were removed.

6.24 A horizontal steel pipeline connects two pressure vessels, each of which may be considered rigid. The pipe is 6 m long and 250 mm outside diameter with a wall thickness of 10 mm. At a point 2 m from one end is a valve which may be treated as a concentrated mass of 85 kg. Taking account of the weight of the pipe as well as of the valve, find the position at which the greatest stress due to bending occurs and the magnitude of this stress.

6.25 The uniform light beam shown in Fig. P6.25 is built in at both ends A and C. It carries a uniformly distributed force of intensity 2 kN/m between the end A and a point B, 2 m from A, and a concentrated force of 10 kN at point B. Determine the reaction forces and moments at each end and the position and magnitude of the greatest absolute bending moment.

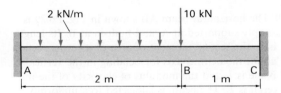

Fig. P6.25.

6.26 A beam which is rigidly built in at each end is of length L, has a constant flexural rigidity EI and is subject to a counterclockwise couple of magnitude C applied at its center. Find an expression for the greatest absolute deflection of the beam produced by this couple.

6.27 The uniform light cantilever shown in Fig. P6.27 has a flexural rigidity EI. The simple and built-in supports at A and B, respectively, are at the same height. The loading consists of a concentrated force W at the midpoint, and a total force W uniformly distributed between the built-in end and the midpoint. Find the position and magnitude of the greatest absolute deflection.

385

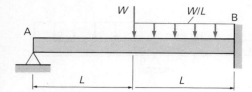

Fig. P6.27.

6.28 The steel beam of solid circular cross section shown in Fig. P6.28 has a diameter of 6 mm. The beam is simply supported at B, while at A the support allows no movement in the vertical direction and applies a bending moment proportional to the slope of the beam at that point: the stiffness of the rotational spring being 70 Nm/rad. Find the position and magnitude of the greatest absolute deflection of the beam produced by the applied force of 50 N.

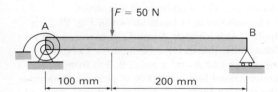

Fig. P6.28.

6.29 The horizontal beam AB shown in Fig. P6.29 is simply supported at A, and built-in at B. Its length is L, the second moment of area of its cross section about the neutral axis for bending in the vertical plane is I, and the modulus of elasticity of the material is E. The beam is subjected to a uniformly distributed force (including its own weight) in the downward vertical direction of intensity w per unit length. Obtain expressions for: (i) the location and magnitude of the greatest hogging bending moment, (ii) the location and magnitude of the

greatest sagging bending moment. The support at A is then raised a distance h to reduce the bending moment at B to zero. Determine this distance.

6.30 The uniform elastic beam of length $2L$ shown in Fig. P6.30 is built in at A, and simply supported at B and C, which are both at the same level as A. It is subject to a uniformly distributed force (including its own weight) of intensity w per unit length. Show that the vertical support reaction forces at A, B and C are in the ratios $13:32:11$, and determine the bending moment at A.

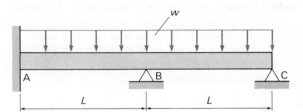

Fig. P6.30.

6.31 If the ends of a horizontal uniform beam of length L are rigidly built-in, and the beam is subjected to a uniformly distributed vertical force (including its weight) of intensity w per unit length, show that the maximum bending moment is $wL^2/12$. A horizontal steel pipe of outside diameter 100 mm and 5 mm wall thickness conveys a liquid of density 1100 kg/m³ in a chemical plant at a pressure of 8 MN/m². The pipe is supported by being clamped to pillars at 6 meter intervals as shown in Fig. P6.31. Each clamped support may be assumed to apply a built-in restraint to bending of the pipe over a length of 0.5 m, but negligible axial restraint. To improve access to the plant it is desirable to remove one pillar. If the maximum allowable axial stress in the pipe material is 90 MN/m², determine whether this removal is permissible.

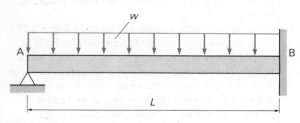

Fig. P6.29.

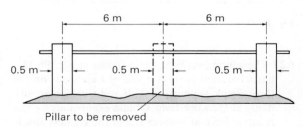

Fig. P6.31.

SI METRIC UNITS – COMPUTER METHOD OF SOLUTION

In each of the following problems, use program SIBEAM to obtain the solutions.

6.32 to 6.38 For the steel beams shown in Figs P5.11 to P5.16, and P5.30 confirm the shear force and bending moment distributions obtained previously, and find the greatest absolute deflection in each case. Take the second moment of area for each beam as 10^{-4} m^4 (and ignore the beam weight).

6.39 Solve Problem 5.34, and find the greatest upward and downward deflections of the beam.

6.40 and 6.41 Solve Problems 6.24 and 6.28.

6.42 Check the results of Example 6.7 for the case where $a = 0.45L$.

6.43 The flexural rigidity of the light beam shown in Fig. P6.43 is 2 MNm^2. Find the greatest absolute shear force, the greatest sagging and hogging bending moments, and the greatest upward and downward deflections of the beam.

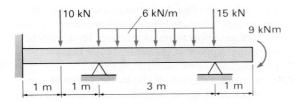

Fig. P6.43.

6.44 Find the effect on the results of Problem 6.43 of the two simple supports being flexible, with vertical stiffnesses of 10 MN/m.

6.45 The light beam shown in Fig. P6.45, which has a flexural rigidity of 0.55 MNm^2, rests on flexible supports, having the lateral and rotational stiffnesses shown. Find the greatest absolute bending moment. There are no loads in the supports before the external forces are applied.

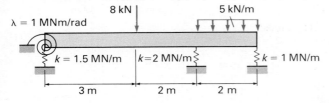

Fig. P6.45.

6.46 For the continuous uniform light beam shown in Fig. P6.46, find the reaction forces at the five supports as proportions of the total load on the beam. The supports are all at the same level.

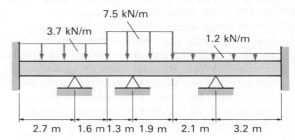

Fig. P6.46.

6.47 Using five elements of equal length, find the horizontal deflection of the tip of the mast in Problem 5.56.

US CUSTOMARY UNITS – MANUAL METHODS OF SOLUTION

6.48 A 2 in by 9 in timber beam is used as a floor joist, and is simply supported at its ends. If the span is 15 ft, find the greatest deflection of the beam when subjected to a uniformly distributed force (including its own weight) of intensity 60 lb/ft along its entire length.

6.49 Repeat Problem 6.48 assuming the beam is built in at each end.

6.50 For the beam defined in Problem 6.48, calculate the greatest absolute deflection of the beam due to the addition of a concentrated force of 600 lb at its center.

6.51 Repeat Problem 6.50 assuming the beam is built in at each end.

6.52 to 6.68 Solve Problems 6.5 to 6.21.

6.69 If the pipe in Problem 5.89 is simply supported at the free end at the same height as the horizontal fixed end, find the greatest bending stress in the pipe.

6.70 to 6.72 Solve Problems 6.22, 6.23 and 6.26.

6.73 The uniform beam shown in Fig. P6.73 is a brass strip 1.2 in broad and 0.4 in deep, and is rigidly

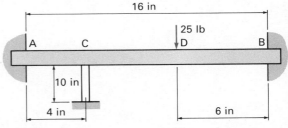

Fig. P6.73.

387

clamped at its ends A and B. It is also supported
at C on an aluminum tube of outer diameter 1 in
and wall thickness 0.04 in. Find the deflection of
the beam at D, due to the 25 lb force applied there.

6.74 to 6.76 Solve Problems 6.27, 6.29 and 6.30.

6.77 In the device shown in Fig. P6.77, the two steel
beams are rigidly built in at one end and are connected to each other by a freely pivoted link at the
other end. A force W, in the plane of the diagram
and perpendicular to the length of the beams, can
be applied to the center of this link. Find the displacement of the link from its unloaded position
when a force W of 50 lb is applied.

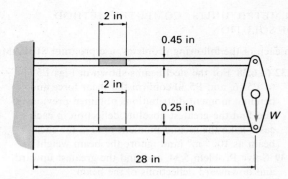

Fig. P6.77.

US CUSTOMARY UNITS –
COMPUTER METHOD OF SOLUTION

In each of the following problems, use program SIBEAM to obtain the solutions.

6.78 to 6.84 For the steel beams shown in Figs P5.67
to P5.72, and P5.85 confirm the shear force and
bending moment distributions obtained previously,
and find the greatest absolute deflection in each
case. Take the second moment of area for each
beam as 250 in⁴ (and ignore the beam weight).

6.85 Solve Problem 6.73.

6.86 Check the results of Example 6.7 for the case
where $a = 0.45L$.

6.87 For the continuous light beam shown in
Fig. P6.87, which has a flexural rigidity of

1.2×10^9 lb in², find the position and magnitude of
the greatest absolute bending moment when (a) all
the simple supports are at the same level, and (b)
when the support at the center of the beam has
been displaced downward by 0.3 in.

6.88 The stepped steel shaft shown in Fig. P6.88 can
be regarded as being built in at each end and supported by two bearings at the same level which act
as simple supports. Ignoring the weight of the
shaft, find the position and magnitude of the
greatest bending stress in the shaft.

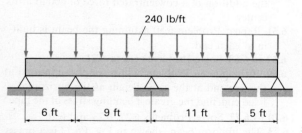

Fig. P6.87.

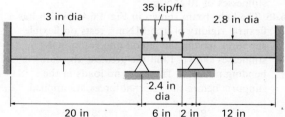

Fig. P6.88.

7
Torsion

In the bending of beams, we were concerned with relatively long slender structural members subject to lateral loading which gave rise to bending moments in planes (the planes of bending) passing through the axes of the members. We now turn to problems where such members are subjected to moments (of couples) in planes normal to their axes. In other words, couples about the axes which tend to cause twisting, putting the members into *torsion*. The couples are referred to as *torsional couples* or *torques*. Despite the differences in the forms of loading, we will find that there are a number of similarities between bending and torsion, including, for example, linear variations of stresses and strains with position. Let us first review some engineering examples of torsion.

Perhaps the most familiar example of torsion is that of a rotating circular shaft which transmits mechanical power. Figure 7.1 shows a typical car transmission system where power from the engine is transmitted to the rear wheels via a drive shaft and rear axle arrangement. The drive shaft in particular is subjected to torsional loading. The same is true of propeller shafts in ships, and indeed any rotating shafts linking power sources to the devices which absorb the power. Not all torsion problems involve rotating machinery,

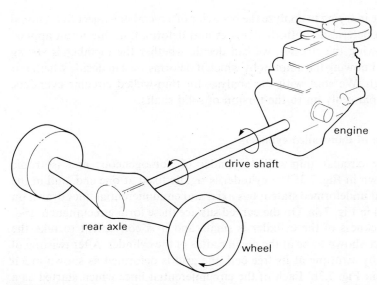

Fig. 7.1.

however. For example, some types of vehicle suspension systems employ torsional springs. Indeed, even coil springs are really curved members in torsion, as illustrated in Fig. 7.2.

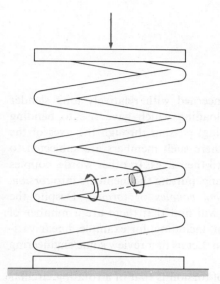

Fig. 7.2.

Although noncircular and curved members are sometimes subjected to torsion, straight circular shafts are the most common, and may be either solid or hollow in cross section. It is on these that we will focus our attention.

7.1 Torsion of shafts

Our objective in trying to analyze the behavior of a member subject to torsional loading is to determine both the stresses and deformation due to an applied torque. Knowing the stresses, we can decide whether the member is strong enough, and knowing the amount by which it deforms we can decide whether it is stiff enough. We start with the analysis for thin-walled circular cylinders, which leads naturally on to the torsion of solid shafts.

7.1.1 Torsion of thin-walled cylinders

Consider the circular thin-walled cylinder of homogeneous and isotropic material shown in Fig. 7.3. The cylinder is rigidly fixed at one end, and in the unloaded and undeformed state it has axial and circumferential lines drawn on its surface, as in Fig. 7.3a. On the curved surface these form a rectangular grid. The wall thickness of the cylinder is small, and it is convenient to take the curved surface shown to be at the mean radius of the cylinder. After twisting of the cylinder by a torque at its free end, the grid is deformed as shown much exaggerated in Fig. 7.3b. Each of the circumferential lines which started as a circle around the cylinder remains as such, but is rotated about the axis of the

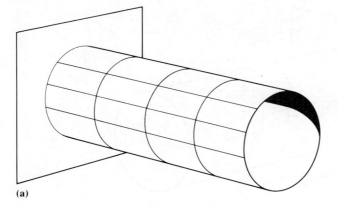

(a)

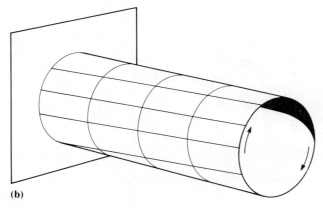

(b)

Fig. 7.3.

cylinder relative to the fixed end, the amount of rotation increasing with the axial distance from this end. The reason for this simple form of deformation is the symmetry about the axis of the cylinder of the geometry, material properties and loading. Since the cylinder is uniform in the axial direction, the amount of relative rotation of cross sections at a given axial distance apart must be the same at all positions along the cylinder. In other words, the amount of rotation is directly proportional to the axial distance from the fixed end of the cylinder.

Each of the rectangles drawn on the surface of the cylinder in Fig. 7.3a becomes a parallelogram in Fig. 7.3b. Let us consider the typical small rectangle ABCD shown in Fig. 7.4a, which after twisting of the cylinder becomes the parallelogram A′B′C′D′. Figure 7.4b shows the curved surface of the cylinder unrolled onto a flat plane. The small thin element of material represented by ABCD is both displaced bodily in the circumferential direction and deformed. This deformation is of the form illustrated in Fig. 1.23: the element is subjected to shearing. For this to happen, there must be equal and opposite shear stresses, say τ, applied to sides A′B′ and C′D′ as shown. As we have already seen in the case of shear stresses in bending (Section 5.3.2), and will

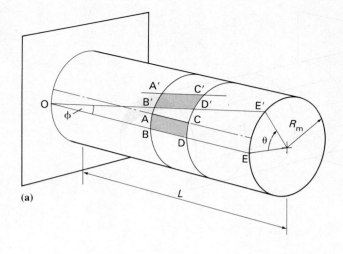

(a)

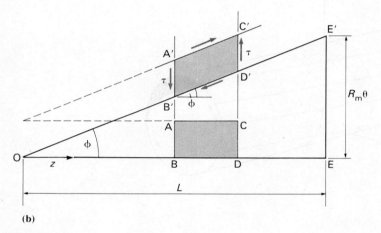

(b)

Fig. 7.4.

prove for general situations in Chapter 9, the existence of shear stresses on sides A′B′ and C′D′ of the element implies that there must also be *complementary shear stresses* (of equal magnitude) on sides A′C′ and B′D′ for the element to be in equilibrium.

Since the element ABCD is subjected only to shearing, it suffers no normal strains in either the axial, circumferential or radial (through-thickness) directions. This means that the cylinder, which is made up of a large number of elements similar to ABCD, does not change in length, diameter or thickness when loaded in torsion.

Let us define the angle through which sides AC and BD rotate in Fig. 7.4b to become A′C′ and B′D′ in the deformed state as ϕ. Now according to equation (1.13), this means that the shear strain, γ, in the element is

$$\gamma = \tan \phi \tag{7.1}$$

We can relate ϕ to the overall deformation of the cylinder. Suppose that the free end of the cylinder, which is at an axial distance L from the fixed end, is rotated through an angle θ in the plane of the cross section of the cylinder as shown in Fig. 7.4a. The point E at this end, which lies on the axial line passing through point O at the fixed end, and points B and D on element ABCD, is rotated to point E′. The distance between points E and E′ measured along the surface of the cylinder is $R_m\theta$, where R_m, which is the mean radius of the cylinder, is the radius of this surface. Therefore, from the geometry of right-angled triangle OEE′ in Fig. 7.4b, which defines the geometry of deformation of the problem, $\tan\phi = R_m\theta/L$, and

$$\gamma = \frac{R_m\theta}{L} \tag{7.2}$$

Introducing stress–strain equation (3.7), we can define the shear stress in the wall of the cylinder as

$$\tau = G\gamma = \frac{GR_m\theta}{L} \tag{7.3}$$

where G is the shear modulus of the material of the cylinder. So far we have not defined how the torque which creates this stress is applied to the cylinder. In practice the free end, like the fixed end, would most likely be attached to some relatively rigid plate or other component which is rotated about the axis of the cylinder, giving rise to uniform loading around the cylinder. If the loading on the free end is not so uniform around the circumference, however, the effects of the nonuniformity are limited to a relatively small region close to the end, and our analysis remains applicable to most of the length of the cylinder.

We assume that the shear stress, τ, is uniform around the circumference of the cylinder at any typical cross section. We also assume that it is constant through the thickness of the cylinder wall, on the grounds that this thickness is small compared to the cylinder radius, that is $t \ll R_m$ (say, $t < 0.1\, R_m$). Considering the typical cross section of the cylinder shown in Fig. 7.5, the resultant shear force, δF, applied to the small shaded element of the cylinder wall which subtends a small angle $\delta\psi$ at the cylinder axis is the product of the shear stress and the area over which it acts

$$\delta F = \tau R_m \delta\psi\, t$$

where $R_m\delta\psi$ is the mean circumferential length of the shaded element. The moment of this force, which acts in the direction tangential to the cylinder at the mid thickness of the element, about the axis of the cylinder is $\delta F R_m$. Hence, the total moment or torque for the entire cross section is found by summing over all elements around the circumference

$$T = \Sigma\delta F R_m = \Sigma\tau R_m^2\delta\psi\, t = \int_0^{2\pi} \tau R_m^2 t\, d\psi = 2\pi\tau R_m^2 t$$

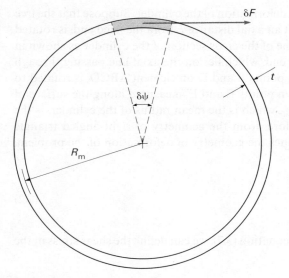

Fig. 7.5.

and for moments applied to the cylinder to be in equilibrium, this torque must be equal to the externally applied torque. The shear stress can therefore be expressed in terms of the applied torque as

$$\tau = \frac{T}{2\pi R_{\mathrm{m}}^2 t} \qquad (7.4)$$

and we can combine equations (7.3) and (7.4) to give

$$\frac{T}{2\pi R_{\mathrm{m}}^3 t} = \frac{G\theta}{L} = \frac{\tau}{R_{\mathrm{m}}} \qquad (7.5)$$

which summarizes the results of our analysis of the torsion of a thin-walled cylinder.

EXAMPLE 7.1

A hollow brass cylinder having a mean diameter of 50 mm (2 in), a wall thickness of 3 mm (0.12 in) and a length of 400 mm (16 in) is subjected to torsional loading which results in a mean shear stress in the wall of the cylinder of 80 MN/m² (12 ksi). Find the torque applied to the cylinder and the relative rotation of its two ends.

This example involves the straightforward application of equations (7.5). From the data given, $R_{\mathrm{m}} = 25$ mm, $t = 3$ mm, $L = 400$ mm and $\tau = 80$ MN/m². Also, the value of shear modulus for brass is $G = 38.3$ GN/m² (Appendix A). The required torque is

$$T = 2\pi R_{\mathrm{m}}^2 t\tau = 2\pi \times 0.025^2 \times 0.003 \times 80 \times 10^6 = \underline{942\ \mathrm{Nm}}\ (690\ \mathrm{lb\ ft})$$

and the relative twist of the ends of the cylinder is

$$\theta = \frac{\tau L}{GR_m} = \frac{80 \times 10^6 \times 0.4}{38.3 \times 10^9 \times 0.025} = 0.0334 \text{ radians} = \underline{1.91}^\circ$$

It is also worth noting that the shear strain is $\gamma = \tau/G = 7.77 \times 10^{-4}$ and, according to equation (7.1), the angle ψ in Fig. 7.4 is some 0.044°, a very small angle.

7.1.2 Torsion of solid circular shafts

Having analyzed the torsional behavior of thin-walled cylinders, we now seek to do the same for solid circular shafts. In view of the fact that thin-walled cylinders in torsion do not change in length, diameter or thickness, we might anticipate that a solid cylinder would behave as a set of concentric cylinders. This is indeed the case, but let us try to demonstrate it from a somewhat different point of view.

Consider the solid circular shaft of homogeneous and isotropic material shown in Fig. 7.6. The shaft is rigidly fixed at one end, and in the undeformed state has a rectangular grid of axial and circumferential lines drawn on its curved surface, as in Fig. 7.6a. After applying a torque to the shaft, this grid is deformed as shown exaggerated in Fig. 7.6b. Symmetry about the axis of the shaft ensures that each of the circumferential lines which started as circle around the cylinder remains as such, but is rotated about the shaft axis relative to the fixed end by an amount which is directly proportional to the distance from the fixed end. So far, the behavior is just like that of the thin-walled cylinder shown in Fig. 7.3, for the same reasons.

Now consider a typical shaft cross section, such as the one at the exposed end in Fig. 7.6, and in particular the typical radius OP in this cross section. What we need to determine is how this radial line deforms: whether it remains straight or becomes curved in the plane of the cross section as in Fig. 7.6b. Also, we need to know whether the cross section itself, which was flat in the undeformed state, remains flat or *warps* to form either a convex or concave surface (which is symmetrical about the axis) as shown in Figs 7.7a and b. To answer these questions, let us consider the shaft subject to torsional loading shown in Fig. 7.8a, and imagine that a thin disk is cut from it in the undeformed state by planes at right angles to its axis. The axial positioning of the disk is not important, provided it is not close to any local end effects. Let us label the front and rear surfaces of the disk as viewed in Fig. 7.8a as S_1 and S_2, respectively. Again as viewed, while S_1 is subject to clockwise torsional loading, S_2 is subject to balancing counterclockwise loading. Now suppose we view the disk at right angles to surface S_1 as in Fig. 7.8b, and from one side as in Fig. 7.8c. These views also show possible deformed shapes of an initially straight radius and the initially flat surfaces S_1 and S_2. Suppose we now take the views shown in Figs 7.8b and c, and rotate them through 180° about, say, the vertical axis, to give Figs 7.8d and e. Because the disk is thin, there can be no significant changes through its thickness of the mode of deformation, and the deformed shapes of the straight radius and flat surfaces must be reversed as shown. But, the

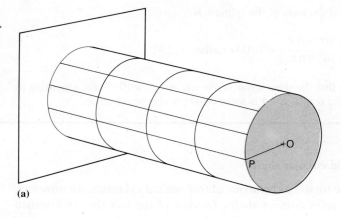

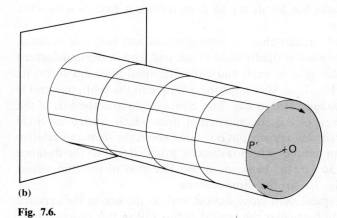

Fig. 7.6.

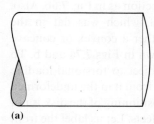

Fig. 7.7.

torsional loading on surface S_2, which was in the counterclockwise direction as viewed in Fig. 7.8a, is now clockwise in Fig. 7.8d, due to the rotation of the disk. So, in terms of geometry and loading, Figs 7.8b and d are identical. Therefore, there can be no difference in the deformations: the radius shown must remain straight, and surfaces S_1 and S_2 remain flat.

We conclude that the geometry of deformation of a homogeneous isotropic solid circular shaft in torsion is such that

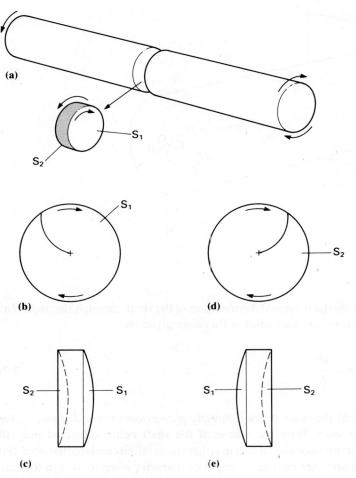

(a)

(b)

(d)

(c)

(e)

Fig. 7.8.

1   all radii remain straight as the shaft deforms, and
2   cross sections of the shaft remain plane.

At the curved outer surface of the solid shaft, the shear strain can be found, just as in Fig. 7.4 for a thin-walled cylinder, from the geometry of triangle NPP′ in Fig. 7.9 as

$$\gamma_{\mathrm{o}} = \tan \phi_{\mathrm{o}} = \frac{R_{\mathrm{o}}\theta}{L} \tag{7.6}$$

where θ is again the angle of twist over an axial length L of shaft of outer radius R_{o}. Now let us consider shear strain not at the outer surface of the shaft but at a cylindrical surface within the shaft at a smaller radius of r. Part of the shaft is cut away in Fig. 7.9 to show this surface, which intersects the exposed shaft cross section over the arc QQ′. Since all radii remain straight, the point Q is

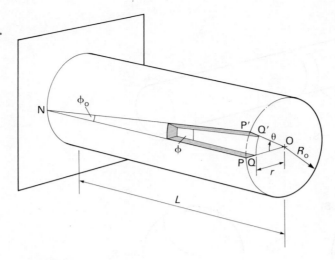

Fig. 7.9.

displaced to Q′ by the torsional deformation of the shaft, through the angle θ at O. The shear strain at this radius is therefore given by

$$\gamma(r) = \tan \phi = \frac{r\theta}{L} \qquad (7.7)$$

which shows that the *shear strain is directly proportional to the distance r from the axis of the shaft*. With the whole of the shaft being subjected only to shearing, it suffers no normal strains in either the axial, circumferential or radial directions, and does not change in length or diameter when loaded in torsion.

Introducing stress–strain equation (3.7), the corresponding shear stress is

$$\tau(r) = G\gamma(r) = \frac{Gr\theta}{L} \qquad (7.8)$$

where G is the shear modulus: the *shear stress is also directly proportional to the distance r from the axis of the shaft*, as shown in Fig. 7.10, but does not vary in the circumferential direction. The maximum shear stress, τ_o, occurs at the outer surface of the shaft, where $r = R_o$. We need now to determine the resultant torque associated with this shear stress distribution. Consider the cross-sectional view of the shaft shown in Fig. 7.11, with a narrow annular ring of material shown shaded. The radius of the ring is r, and its radial width δr. The shear stress acting on this ring element can be regarded as constant, and given by equation (7.8). The moment or torque applied to the element can therefore be found as the product of the shear stress, the area of the element, and its distance from the axis of rotation

$$\delta T = \tau \times (2\pi r \delta r) \times r = 2\pi \tau r^2 \, \delta r$$

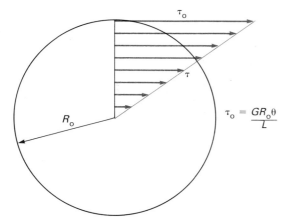

Fig. 7.10.

$$\tau_o = \frac{GR_o \theta}{L}$$

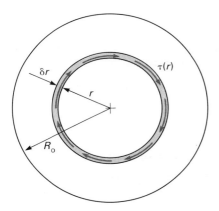

Fig. 7.11.

and the total moment, which for the shaft to be in equilibrium must equal the applied torque, is found by summing over all the ring elements forming the cross section

$$T = \Sigma 2\pi\tau r^2 \delta r = \int_0^{R_o} 2\pi\tau r^2 \, dr$$

Using equation (7.8) to define the variation of shear stress with radius, this expression becomes

$$T = \frac{G\theta}{L} \int_0^{R_o} 2\pi r^3 \, dr = \frac{G\theta J}{L} \tag{7.9}$$

where J is the *polar second moment of area* of the shaft cross section about its axis. Polar second moments of area are discussed in Appendix C, where an expression for J is derived in equation (C.16) as

$$J = \frac{\pi D_o^4}{32} = \frac{\pi R_o^4}{2} \qquad (7.10)$$

where $D_o = 2R_o$ is the outer diameter of the shaft. We note that, for the reasons given in Appendix C, this polar second moment of area has twice the magnitude of the bending second moment of area for a beam of circular cross section (equation (5.40)).

Combining equations (7.8) and (7.9), we have

$$\frac{T}{J} = \frac{G\theta}{L} = \frac{\tau}{r} \qquad (7.11)$$

which summarizes the results of our analysis of the torsion of a solid circular shaft. We note the similarity of form not only with equations (7.5) for the torsion of a thin-walled cylinder, but also with equations (5.33) for bending, which are repeated here

$$\frac{M}{I} = \frac{E}{R} = \frac{\sigma_x}{-y} \qquad (5.33)$$

In each case, an applied moment divided by a second moment of area is equated to first a modulus of elasticity divided by a characteristic length of the system concerned (the angle θ is dimensionless) and then to a stress divided by a distance from the axis at which the stress is zero.

EXAMPLE 7.2

A solid brass shaft having a diameter of 50 mm (2 in) and a length of 400 mm (16 in) is subjected to torsional loading which results in a maximum shear stress in the shaft of 80 MN/m² (12 ksi). Find the torque applied to the shaft and the relative rotation of its two ends.

This example is a repetition of Example 7.1, but with a solid rather than a hollow thin-walled circular cylindrical shaft. From the given data, $D_o = 50$ mm, $R_o = 25$ mm, $L = 400$ mm, $\tau_o = 80$ MN/m² and $G = 38.3$ GN/m² (Appendix A). The polar second moment of area is, from equation (7.10)

$$J = \frac{\pi D_o^4}{32} = \frac{\pi \times 0.050^4}{32} = 6.136 \times 10^{-7} \text{ m}^4$$

and the required torque is given by equations (7.11) as

$$T = \frac{J\tau_o}{R_o} = \frac{6.136 \times 10^{-7} \times 80 \times 10^6}{0.025} = \underline{1960 \text{ Nm}} \text{ (1400 lb ft)}$$

which compares with 942 Nm for the thin hollow cylinder. The relative twist of the ends of the shaft is

$$\theta = \frac{\tau_o L}{G R_o} = \frac{80 \times 10^6 \times 0.4}{38.3 \times 10^9 \times 0.025} = 0.0334 \text{ radians} = \underline{1.91^\circ}$$

We note that, for the same nominal overall dimensions and maximum shear stress, the angle of twist of the hollow and solid shafts is the same. The solid shaft can, however, transmit a substantially greater torque. In Example 7.3 we will examine further the relative merits of solid and hollow shafts.

7.1.3 Torsion of hollow circular shafts

For the torsion of solid circular shafts, we used symmetry arguments to conclude that radii remain straight and cross sections remain plane as the shafts deform. We can apply exactly the same arguments to hollow thick-walled circular shafts. Indeed, following through the same analysis, we arrive again at equation (7.11), the only change being that instead of the integral for torque in equation (7.9) being from a lower limit of $r = 0$ it is now from $r = R_i$, the inner radius of the hollow shaft. In other words, the polar second moment of area is

$$J = \int_{R_i}^{R_o} 2\pi r^3 \, dr = \frac{\pi}{2} (R_o^4 - R_i^4) = \frac{\pi}{32} (D_o^4 - D_i^4) \tag{7.12}$$

where $D_i = 2R_i$ is the inner diameter of the shaft. This expression is the difference between the polar second moments of area of the outer and inner circles forming the cross section. The radial distribution of shear stress remains linear, with stress proportional to radius, as illustrated in Fig. 7.12.

A special case of a hollow circular shaft is the thin-walled cylinder we considered in Section 7.1.1. To demonstrate that this is true, we need to show that equation (7.11) reduces to (7.5) under the right circumstances: namely that the definition for J in equation (7.12) reduces to the expression $2\pi R_m^3 t$ when $R_o \approx R_i \approx R_m$ and $t = R_o - R_i$. Now

$$J = \frac{\pi}{2} (R_o^4 - R_i^4) = \frac{\pi}{2} (R_o^2 + R_i^2)(R_o + R_i)(R_o - R_i)$$

$$\approx \frac{\pi}{2} (2R_m^2)(2R_m) t = 2\pi R_m^3 t$$

which is the result we need. Comparing the approximate expression $J_{approx} = 2\pi R_m^3 t$ with the exact expression, J_{exact}, given by equation (7.12), we obtain for a range of values of the ratio t/R_o:

t/R_o	0.05	0.1	0.2	0.3
J_{approx}/J_{exact}	0.999	0.997	0.988	0.970

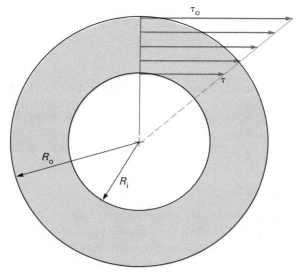

Fig. 7.12.

The error involved in using the approximate expression for a thin-walled cylindrical shaft is at most 1.2% for wall thicknesses up to 20% of the shaft radius. In practice, however, it is no more difficult to use the exact form and equation (7.11) than the approximate form of equation (7.5).

EXAMPLE 7.3

Compare the weights of hollow and solid circular shafts of the same length and material transmitting the same torque and subject to the same maximum allowable shear stress.

In a solid shaft, only the material near its outer surface carries high shear stresses, while the material near the axis carries very low ones. In a hollow shaft, this central region is eliminated, thereby making more efficient use of material and reducing the weight of the shaft. From equations (7.11), we have

$$\frac{T}{\tau_o} = \frac{J}{R_o}$$

where τ_o is the maximum shear stress at the outer radius R_o, an equation which is so far valid for either a solid or a hollow shaft. Since the torque and maximum stress are the same for both, it follows that the ratio of second moment of area to outer radius must also be the same for both.

For the solid shaft of outer radius R_s and second moment of area J_s shown in Fig. 7.13a

$$\frac{J_s}{R_s} = \frac{\pi R_s^3}{2}$$

while for the hollow shaft with a ratio of inner to outer radius of $n = R_i/R_h$ and second moment of area J_h shown in Fig. 7.13b

$$\frac{J_h}{R_h} = \frac{\pi}{2R_h}(R_h^4 - R_i^4) = \frac{\pi R_h^3}{2}(1 - n^4)$$

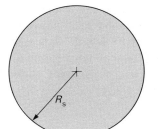

(a)

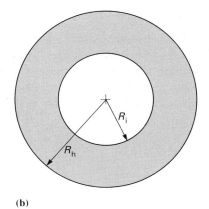

(b)

Fig. 7.13.

For these expressions for the two shafts to be identical

$$R_s^3 = R_h^3(1 - n^4)$$

or

$$R_s = R_h(1 - n^4)^{1/3}$$

The weights of the two shafts, W_s and W_h, are proportional to their cross-sectional areas and therefore

$$\frac{W_h}{W_s} = \frac{\pi(R_h^2 - R_i^2)}{\pi R_s^2} = \frac{R_h^2}{R_s^2}(1 - n^2)$$

which, given the relationship between the outer radii of the two shafts, becomes

$$\frac{W_h}{W_s} = \frac{(1 - n^2)}{(1 - n^4)^{2/3}}$$

Table 7.1 shows values of this weight ratio for some typical values of the hollow shaft radius ratio, n, together with the corresponding ratios of outer radii of the hollow and solid shafts. We note that the use of hollow shafts offers substantial weight savings – from just over 20% at $n = 0.5$ to over 60% at $n = 0.9$. At the upper end of this range, however, the hollow shaft must be substantially larger than the equivalent solid shaft. Although lighter in weight, it is less compact, which in some practical applications can be a significant disadvantage.

Table 7.1 Weight and radius ratios for hollow and solid shafts

n	W_h/W_s	R_h/R_s
0.5	0.78	1.02
0.6	0.70	1.05
0.7	0.61	1.10
0.8	0.51	1.19
0.9	0.39	1.43

7.1.4 Nonuniform circular shafts

Under certain circumstances, shafts may be used which are not uniform in cross section. There may be abrupt changes, as in the stepped solid shaft shown in Fig. 7.14a, or the shaft which is hollow over part of its length shown in Fig. 7.14b. Alternatively, the change may be gradual if the shaft (either solid or hollow) is tapered. Let us consider first a shaft with a step change in section.

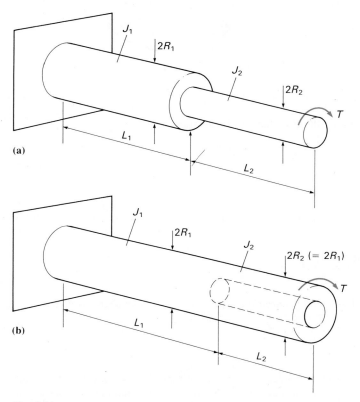

Fig. 7.14.

In the case of either of the shafts shown in Fig. 7.14, we can regard one end as being fixed, and the other end as subject to a torque of magnitude T. We can also regard the parts of the shaft on either side of the step change as separate uniform shafts of lengths L_1 and L_2, polar second moments of area J_1 and J_2, and outer radii R_1 and R_2, respectively. While equilibrium of the shaft requires that the torque T be the same throughout both parts, the geometry of deformation is such that the total angle of twist of the free end, θ, is the sum of the angles, θ_1 and θ_2, over each of the two parts. Applying equation (7.11) to each part, we have

$$\frac{T}{J_1} = \frac{G\theta_1}{L_1} = \frac{\tau_1}{R_1}$$

and

$$\frac{T}{J_2} = \frac{G\theta_2}{L_2} = \frac{\tau_2}{R_2}$$

where G is the common shear modulus and τ_1 and τ_2 are the maximum shear stresses at the outer surfaces of the two parts. We may obtain expressions for these stresses as

$$\tau_1 = \frac{TR_1}{J_1} \quad \text{and} \quad \tau_2 = \frac{TR_2}{J_2} \tag{7.13}$$

and for the total angle of twist as

$$\theta = \theta_1 + \theta_2 = \frac{T}{G}\left(\frac{L_1}{J_1} + \frac{L_2}{J_2}\right) \tag{7.14}$$

We should note that in the regions of the interfaces between the two parts of the shafts, the distributions of stresses and strains are actually much more complicated than those given by the present simple theory of torsion. Equation (7.14) gives only an approximation for the total angle of twist, but for a shaft which is reasonably long compared with its diameter the approximation is a good one. In the case of the stresses we should be aware that the abrupt change of section gives rise to a local stress concentration. Such effects were discussed in Section 1.3.3.

EXAMPLE 7.4

The drive shaft from an engine shown in Fig. 7.15 is made from a 1.0 m (40 in) length of solid round steel bar having an external diameter of 25 mm (1 in), together with a 0.9 m (36 in) length of steel tube having an external diameter of 31 mm (1.24 in) and a nominal internal diameter of 25 mm (1 in). The tube is shrunk onto the solid bar over a length of 150 mm (6 in). Determine the greatest shear stress in the shaft and the relative angular rotation of its ends when the engine is supplying a torque of 250 Nm (180 lb ft). Assume no relative movement between the contacting surfaces of the bar and tube.

   Although this example involves two abrupt changes of section, the same approach can be used.

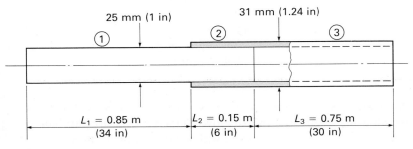

Fig. 7.15.

SI units

The shaft can be divided into three parts: (i) a solid bar of length $L_1 = 0.85$ m and radius $R_1 = 12.5$ mm, (ii) a solid bar of length $L_2 = 0.15$ m and radius $R_2 = 15.5$ mm, and (iii) a hollow tube of length $L_3 = 0.75$ m and radii $R_1 = 12.5$ mm and $R_2 = 15.5$ mm. The polar second moments of area are

$$J_1 = \frac{\pi \times 0.0125^4}{2} = 3.835 \times 10^{-8} \text{ m}^4$$

$$J_2 = \frac{\pi \times 0.0155^4}{2} = 9.067 \times 10^{-8} \text{ m}^4$$

$$J_3 = J_2 - J_1 = 5.232 \times 10^{-8} \text{ m}^4$$

Using equations (7.13), we find the maximum shear stresses in each of the three parts, for the given torque of $T = 250$ Nm, as

$$\tau_1 = \frac{TR_1}{J_1} = \frac{250 \times 0.0125}{3.835 \times 10^{-8}} \text{ N/m}^2 = 81.5 \text{ MN/m}^2$$

$$\tau_2 = \frac{TR_2}{J_2} = \frac{250 \times 0.0155}{9.067 \times 10^{-8}} \text{ N/m}^2 = 42.7 \text{ MN/m}^2$$

$$\tau_3 = \frac{TR_2}{J_3} = \frac{250 \times 0.0155}{5.232 \times 10^{-8}} \text{ N/m}^2 = 74.1 \text{ MN/m}^2$$

The greatest shear stress of 81.5 MN/m² occurs at the outer surface of the solid 25 mm diameter bar.

Applying equation (7.14) to the present shaft with two changes of section, and taking the value of the shear modulus for steel as $G = 79.6$ GN/m² (Appendix A), we obtain the total angle of twist as

$$\theta = \frac{T}{G}\left(\frac{L_1}{J_1} + \frac{L_2}{J_2} + \frac{L_3}{J_3}\right) = \frac{250}{79.6 \times 10^9}$$

$$\times \left(\frac{0.85}{3.835 \times 10^{-8}} + \frac{0.15}{9.067 \times 10^{-8}}\right.$$

$$\left. + \frac{0.75}{5.232 \times 10^{-8}}\right)$$

$$= 0.120 \text{ radians} = \underline{6.87°}$$

US customary units

The shaft can be divided into three parts: (i) a solid bar of length $L_1 = 34$ in and radius $R_1 = 0.5$ in, (ii) a solid bar of length $L_2 = 6$ in and radius $R_2 = 0.62$ in, and (iii) a hollow tube of length $L_3 = 30$ in and radii $R_1 = 0.5$ in and $R_2 = 0.62$ in. The polar second moments of area are

$$J_1 = \frac{\pi \times 0.5^4}{2} = 0.098\,17 \text{ in}^4$$

$$J_2 = \frac{\pi \times 0.62^4}{2} = 0.2321 \text{ in}^4$$

$$J_3 = J_2 - J_1 = 0.1339 \text{ in}^4$$

Using equations (7.13), we find the maximum shear stresses in each of the three parts, for the given torque of $T = 180$ lb ft, as

$$\tau_1 = \frac{TR_1}{J_1} = \frac{180 \times 12 \times 0.5}{0.098\,17} \text{ psi} = 11.0 \text{ ksi}$$

$$\tau_2 = \frac{TR_2}{J_2} = \frac{180 \times 12 \times 0.62}{0.2321} \text{ psi} = 5.77 \text{ ksi}$$

$$\tau_3 = \frac{TR_2}{J_3} = \frac{180 \times 12 \times 0.62}{0.1339} \text{ psi} = 10.0 \text{ ksi}$$

The greatest shear stress of 11.0 ksi occurs at the outer surface of the solid 1 in diameter bar.

Applying equation (7.14) to the present shaft with two changes of section, and taking the value of the shear modulus for steel as $G = 11.5 \times 10^6$ psi (Appendix A), we obtain the total angle of twist as

$$\theta = \frac{T}{G}\left(\frac{L_1}{J_1} + \frac{L_2}{J_2} + \frac{L_3}{J_3}\right) = \frac{180 \times 12}{11.5 \times 10^6}$$

$$\times \left(\frac{34}{0.098\,17} + \frac{6}{0.2321}\right.$$

$$\left. + \frac{30}{0.1339}\right)$$

$$= 0.112 \text{ radians} = \underline{6.42°}$$

The other main type of nonuniform shaft is one whose cross section varies continuously along its length. Figure 7.16 shows a tapered solid circular shaft of length L whose outer radius varies linearly with axial position, from R_1 to R_2. The angle of taper is relatively small, a condition which can be expressed

406

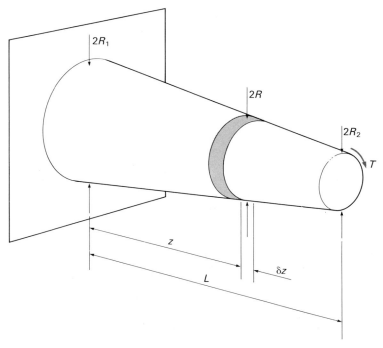

Fig. 7.16.

symbolically as $(R_1 - R_2) \ll L$. Consider a typical thin slice of the shaft at right angles to its axis. This takes the form of a circular disk at a distance z from the fixed end of the shaft and is of constant thickness δz. The disk is shown shaded in Fig. 7.16. Since the disk is thin, we can regard the outer radius, R, as constant, and apply equations (7.11)

$$\frac{T}{J(z)} = \frac{G\,\delta\theta}{\delta z} = \frac{\tau_o(z)}{R(z)}$$

where $J(z)$ is the local polar second moment of area, $\delta\theta$ is the relative angle of twist between the plane faces of the disk, and $\tau_o(z)$ is the local maximum shear stress at the curved surface of the disk. In the limit as $\delta z \to 0$, this becomes

$$\frac{T}{J(z)} = G\frac{\mathrm{d}\theta}{\mathrm{d}z} = \frac{\tau_o(z)}{R(z)} \qquad (7.15)$$

Since the radius of the shaft varies linearly with z, we can express this variation as

$$R(z) = R_1 - (R_1 - R_2)\frac{z}{L} = R_1(1 - \alpha z)$$

where

$$\alpha = \frac{(R_1 - R_2)}{R_1 L}$$

and the polar second moment of area is given as a function of z by

$$J(z) = \frac{\pi R^4}{2} = \frac{\pi R_1^4}{2}(1 - \alpha z)^4 = J_1(1 - \alpha z)^4$$

where J_1 is the value at $z = 0$, where $R = R_1$. Equations (7.15) therefore give the rate of variation of angle of twist with axial position as

$$\frac{d\theta}{dz} = \frac{T}{GJ_1}(1 - \alpha z)^{-4}$$

a differential equation which can be integrated to give the total angle of twist over the length L of the shaft as

$$\theta = \int_0^L \frac{T}{GJ_1}(1 - \alpha z)^{-4}\, dz = \frac{T}{GJ_1}\left[\frac{(1 - \alpha z)^{-3}}{3\alpha}\right]_0^L$$

$$\theta = \frac{T}{3\alpha GJ_1}\left[\frac{R_1^3}{R_2^3} - 1\right] \tag{7.16}$$

As for shear stresses, the maximum value at any shaft cross section varies with position according to equation (7.15), and the greatest value anywhere along the shaft occurs at the smaller of the two ends.

Any other form of continuous variation of shaft radius can be handled in a similar way, provided we can carry out the necessary integration. Even if this cannot be done analytically, it can always be done numerically with the aid of a computer.

7.1.5 Shafts of noncircular cross section

The methods we have developed for analyzing torsion of circular shafts are not applicable to noncircular sections. We must not calculate a polar second moment of area of such a section and use it in equations (7.11). The reason for this is that our conclusions about the geometry of deformation of a circular shaft, in particular that cross sections of the shaft remain plane, are no longer valid. In arriving at these conclusions we invoked arguments based on symmetry, notably the fact that since a circular shaft is symmetrical about its axis all points on the perimeter of any cross section must be displaced in exactly the same way. In a noncircular shaft, which is not symmetrical about its axis, this is not true. In fact, the deformation of a uniform noncircular shaft at a particular cross section is generally assumed to consist of a rotation of that cross section, together with a distribution over the section of *warping* displacement normal to it. More sophisticated methods of analysis are necessary, and are described in more advanced books on the theory of elasticity.

7.2 Statically determinate torsion problems

In Section 5.1.1 we established a condition for a beam problem to be statically determinate, based on the number of unknown reaction forces and moments at the beam supports. This number had to be three, equalling the number of independent equilibrium equations available. We can do much the same for torsion problems, although the number of equilibrium equations is now reduced to one, the equation of equilibrium of moments about the axis of a shaft. Therefore, to be statically determinate, a torsion problem should involve only one support which prevents, or at least restricts, rotation. For example, the shafts shown in Figs 7.3, 7.6, 7.14 and 7.16 are all statically determinate in that they are only fixed at one end: we were able to find the torques everywhere along them by considering equilibrium alone. Indeed, those shown in Figs 7.8 and 7.15 are also effectively statically determinate, because we are only interested in the angle of twist of one end relative to the other, which is equivalent to fixing one end and finding the rotation of the other.

One reason why in practice problems involving the torsion of shafts often have none of the shaft ends fixed is that the shafts rotate and transmit power, as for example in the car transmission shown in Fig. 7.1. We are only concerned here with such systems in dynamic equilibrium, rotating at constant speed. Since we may be interested in power and speed as well as torque, we need the relationship between power, torque and speed, which is

	Power	= Torque × Speed		(7.17)
SI metric units	(watts)	(Nm)	(radians/second)	
US customary units	(ft lb/second)	(lb ft)	(radians/second)	

Note that we must define rotational speed using the natural unit of angle, which is the radian (rather than, say, degrees or revolutions).

EXAMPLE 7.5

A hollow steel shaft transmits 400 kW (540 hp) at 1500 rev/min. If the maximum allowable shear stress is 85 MN/m² (12 ksi), and the ratio of outer to inner diameters of the shaft is 2 to 1, find the minimum outer diameter and the angle of twist over a 1 m (40 in) length of shaft.

To find the torque, we first express the speed in the required units as

$$\text{angular speed} = \frac{1500}{60} \text{ rev/s} = \frac{1500}{60} \times 2\pi = 157.1 \text{ rad/s}$$

and the torque is

$$T = \frac{\text{power}}{\text{speed}} = \frac{400 \times 10^3}{157.1} = 2546 \text{ Nm}$$

If R_o is the outer radius of the shaft, and the inner radius is half this, then equation (7.12) gives the polar second moment of area as

$$J = \frac{\pi}{2}\left[R_o^4 - \left(\frac{R_o}{2}\right)^4\right] = \frac{15}{32}\pi R_o^4$$

Rearranging equations (7.11) we have

$$\frac{J}{R_o} = \frac{T}{\tau_o}$$

where τ_o is the maximum shear stress at the outer surface of the shaft. If this is equal to the maximum allowable shear stress, then

$$\frac{15}{32}\pi R_o^3 = \frac{2546}{85 \times 10^6}\,\text{m}^3$$

from which we obtain the outer radius as $R_o = 0.0273$ m. In other words, the minimum outer diameter of the shaft necessary to prevent excessive shear stresses is <u>54.6 mm</u> (2.2 in).

The angle of twist per meter length of shaft can also be obtained with the aid of equations (7.11) as

$$\theta = \frac{L\tau_o}{GR_o} = \frac{1 \times 85 \times 10^6}{79.6 \times 10^9 \times 0.0273} = 0.0391 \text{ radians} = \underline{2.24°}$$

Since torsion problems are often concerned with rotating shafts transmitting power, they may also involve several shafts connected by gears. These have the effect not only of changing the speeds of the shafts but also the torques they transmit.

Consider the system shown in Fig. 7.17, which consists of two parallel solid shafts AB and CD connected via a pair of spur gears. It is convenient to think of end A of the first shaft as being fixed, even though the system may be rotating at

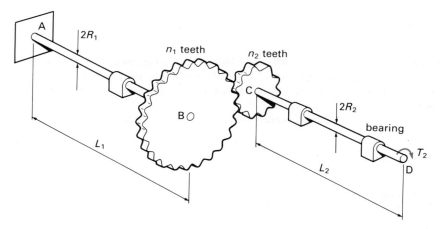

Fig. 7.17.

constant speed: we are interested in the relative rotations due to twisting of the shafts by the torques they carry. The first shaft is of radius R_1 and length L_1, and has mounted at its unconstrained end B a spur gear with n_1 teeth. This meshes with another gear with n_2 teeth on end C of the second shaft, which is of radius R_2 and length L_2 and is subjected to a torque of magnitude T_2 at its free end, at D. The thicknesses of the gear wheels are negligible compared with the lengths of the shafts. In practice, the shafts would be supported on bearings, typically as shown. We assume that these serve only as lateral supports, and do not provide any significant resistance to rotation of the shafts due to friction. With only one support, at A, imposing any rotational constraint, the system is statically determinate.

Figure 7.18a shows the two spur gears with their teeth intermeshed. It is convenient to think of them as plain circular disks whose outer surfaces just make contact. The circles representing these disks in Fig. 7.18a are the *pitch circles* for the gears concerned. Let their radii be r_1 and r_2, respectively. Now,

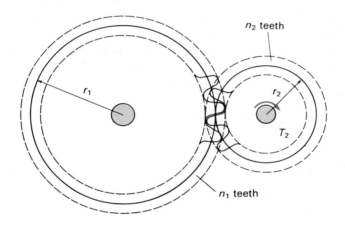

(a)

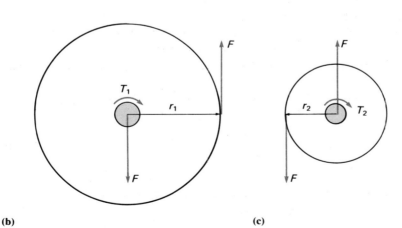

(b)   **(c)**

Fig. 7.18.

since the teeth of the two gears intermesh with each other, individual teeth must be of the same length or *pitch* measured along their pitch circles. Since this pitch is equal to the pitch circle circumference divided by the number of teeth, which is proportional to the pitch circle radius divided by the number of teeth, it follows that

$$\frac{r_1}{n_1} = \frac{r_2}{n_2} \tag{7.18}$$

Now consider the two gears, and the shafts to which they are attached, as separate free bodies, as shown in Fig. 7.18b and c. At the point of contact between the gear teeth, which is where the pitch circles meet in Fig. 7.18a, forces are transmitted between the gears. Let F be the component of force in the direction tangential to the pitch circles applied to both gears as shown. There may also be components of force in the common radial direction, but these have no moments about the shaft axes. For each shaft and gear to be in equilibrium, not only is a torque applied to the shaft, but a lateral force of magnitude F must also be applied to the shaft. This reaction force is supplied by the shaft bearings, and acts through the center of the shaft. The second shaft has a clockwise torque of magnitude T_2 applied at its free end at D. Therefore, for moment equilibrium of the free body in Fig. 7.18c

$$T_2 = Fr_2 \quad \text{from which} \quad F = \frac{T_2}{r_2}$$

Similarly, for the first shaft shown in Fig. 7.18b

$$T_1 = Fr_1 = T_2 \frac{r_1}{r_2} = T_2 \frac{n_1}{n_2} \tag{7.19}$$

where T_1 is the clockwise torque applied to the shaft at its fixed end. This result shows that the torques in the two shafts are directly proportional to the numbers of teeth on the gear wheels to which they are attached. The ratio of angular rotations of the shafts (speeds if they are rotating steadily) is the inverse of this, ensuring that mechanical energy is conserved (we have assumed that frictional effects, which would cause dissipation of energy as heat, are negligible).

From equation (7.11), the angles of twist of the two shafts are

$$\theta_1 = \frac{T_1 L_1}{GJ_1} \quad \text{and} \quad \theta_2 = \frac{T_2 L_2}{GJ_2} \tag{7.20}$$

In other words, θ_1 is the relative angle of twist between the two ends of the first shaft, while θ_2 provides the same information for the second shaft.

Now, since end A of the first shaft (Fig. 7.17) is assumed to be fixed, the angle of rotation of end B is $\theta_B = -\theta_1$ (the negative sign indicating counterclockwise rotation). This angle is transmitted through the gears to give an angle of rotation of end C of the second shaft of

$$\theta_C = -\theta_B \frac{n_1}{n_2} = \theta_1 \frac{n_1}{n_2}$$

in the clockwise direction. To this must be added the further clockwise rotation θ_2 of end D relative to end C to give the total clockwise rotation of D (relative to the fixed end A) of

$$\theta_D = \theta_C + \theta_2 = \theta_1 \frac{n_1}{n_2} + \theta_2 \qquad (7.21)$$

EXAMPLE 7.6

A motor drives end A of the solid steel shaft ABC shown in Fig. 7.19 at 1000 rev/min. The shaft transmits power to two machines; one is driven directly at end C of shaft ABC, while the other is driven at 400 rev/min through a pair of gears B and D and a second solid shaft DE. The thicknesses of the gear wheels are negligible compared with the lengths of the shafts. The machine at C consumes 66 kW (89 hp) and that at E 84 kW (113 hp), and energy losses in the gears and bearings are negligible. The shear stress in the shafting material must not exceed 75 MN/m² (11 ksi). Find the necessary diameters of the shafts, and the twist in the shafting between A and E, taking E as the datum.

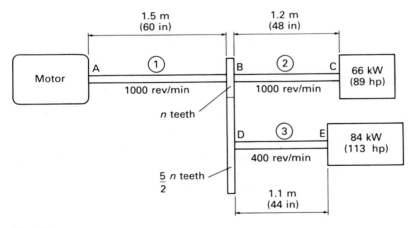

Fig. 7.19.

For convenience, let us number AB, BC and DE as shafts 1, 2 and 3. Their speeds are 104.7 rad/s (1000 rev/min), 104.7 rad/s, and 41.89 rad/s (400 rev/min), respectively. Since energy losses are negligible, the power transmitted from the motor to shaft 1 is the sum of the powers consumed by the machines at C and E, namely 150 kW (202 hp). Therefore, the torques in the three shafts are

SI units   *US customary units*

$$T_1 = \frac{150 \times 10^3}{104.7} = 1433 \text{ Nm}$$

$$T_1 = \frac{202 \times 550}{104.7} = 1061 \text{ lb ft}$$

SI units

$$T_2 = \frac{66 \times 10^3}{104.7} = 630.4 \text{ Nm}$$

$$T_3 = \frac{84 \times 10^3}{41.89} = 2005 \text{ Nm}$$

US customary units

$$T_2 = \frac{89 \times 550}{104.7} = 467.5 \text{ lb ft}$$

$$T_3 = \frac{113 \times 550}{41.89} = 1484 \text{ lb ft}$$

From equations (7.11) and (7.10)

$$\frac{T}{\tau_o} = \frac{J}{R_o} = \frac{\pi R_o^4}{2} \times \frac{1}{R_o} = \frac{\pi R_o^3}{2}$$

where τ_o is the maximum shear stress at the surface of a solid shaft of outer radius R_o when subjected to a torque of T. This gives the following expression for shaft radius

$$R_o = \left(\frac{2T}{\pi \tau_o}\right)^{1/3}$$

and the required diameters of the three shafts are

SI units

$$D_1 = 2\left(\frac{2 \times 1433}{\pi \times 75 \times 10^6}\right)^{1/3} = 0.0460 \text{ m} = \underline{46.0 \text{ mm}}$$

$$D_2 = 2\left(\frac{2 \times 630.4}{\pi \times 75 \times 10^6}\right)^{1/3} = 0.0350 \text{ m} = \underline{35.0 \text{ mm}}$$

$$D_3 = 2\left(\frac{2 \times 2005}{\pi \times 75 \times 10^6}\right)^{1/3} = 0.0514 \text{ m} = \underline{51.4 \text{ mm}}$$

US customary units

$$D_1 = 2\left(\frac{2 \times 1061 \times 12}{\pi \times 11 \times 10^3}\right)^{1/3} = \underline{1.81 \text{ in}}$$

$$D_2 = 2\left(\frac{2 \times 467.5 \times 12}{\pi \times 11 \times 10^3}\right)^{1/3} = \underline{1.37 \text{ in}}$$

$$D_3 = 2\left(\frac{2 \times 1484 \times 12}{\pi \times 11 \times 10^3}\right)^{1/3} = \underline{2.02 \text{ in}}$$

Also from equation (7.11), we can obtain an expression for the twist over a length L of shaft with a given maximum shear stress as

$$\theta = \frac{\tau_o L}{G R_o} = \frac{2\tau_o L}{G D}$$

where D is the shaft diameter. Taking the value of shear modulus for steel as 79.6 GN/m^2 (Appendix A), the angles of twist of shafts 1 and 3 are therefore

$$\theta_1 = \frac{2\tau_o L_1}{G D_1} = \frac{2 \times 75 \times 10^6 \times 1.5}{79.6 \times 10^9 \times 0.0460} = 0.0614 \text{ radians}$$

and

$$\theta_3 = \frac{2\tau_o L_3}{G D_3} = \frac{2 \times 75 \times 10^6 \times 1.1}{79.6 \times 10^9 \times 0.0514} = 0.0403 \text{ radians}$$

Also from equation (7.11), we can obtain an expression for the twist over a length L of shaft with a given maximum shear stress as

$$\theta = \frac{\tau_o L}{G R_o} = \frac{2\tau_o L}{G D}$$

where D is the shaft diameter. Taking the value of shear modulus for steel as 11.5×10^6 psi (Appendix A), the angles of twist of shafts 1 and 3 are therefore

$$\theta_1 = \frac{2\tau_o L_1}{G D_1} = \frac{2 \times 11 \times 10^3 \times 60}{11.5 \times 10^6 \times 1.81} = 0.0634 \text{ radians}$$

and

$$\theta_3 = \frac{2\tau_o L_3}{G D_3} = \frac{2 \times 11 \times 10^3 \times 44}{11.5 \times 10^6 \times 2.02} = 0.0417 \text{ radians}$$

414

The gears at B and D are such that B turns at 2.5 times (1000/400) faster than D. In other words, if there are n teeth on gear B there are $2.5n$ teeth on D, and we may adapt equation (7.21) to give the twist in the shafting at A relative to that at E as

$$\theta_A - \theta_E = \theta_1 + 2.5\theta_3 = 0.162 \text{ radians} = \underline{9.28°}$$

The gears at B and D are such that B turns at 2.5 times (1000/400) faster than D. In other words, if there are n teeth on gear B there are $2.5n$ teeth on D, and we may adapt equation (7.21) to give the twist in the shafting at A relative to that at E as

$$\theta_A - \theta_E = \theta_1 + 2.5\theta_3 = 0.168 \text{ radians} = \underline{9.63°}$$

7.3 Statically indeterminate torsion problems

Statically indeterminate torsion problems involve two or more supports which prevent or restrict rotation of the shaft or shafts involved. As in the case of systems of components under normal stress considered in Section 4.2.1, statically indeterminate torsion problems can involve shafts either in series or parallel with each other. Again, we can choose to solve such problems by either equilibrium or compatibility methods.

EXAMPLE 7.7

Figure 7.20 shows a uniform shaft of length L fixed at each end. A clockwise torque of magnitude T is applied to a thin circular disk attached to the shaft at a distance of $L/4$ from one end, and a counterclockwise torque of magnitude T is applied to a similar disk at the center of the shaft. Find the torques at the ends of the shaft, and the distribution of angular displacement along it.

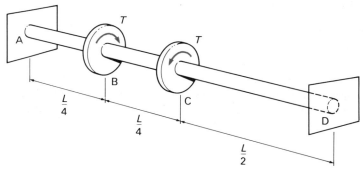

Fig. 7.20.

   With two fixed supports for the shaft, the problem is statically indeterminate. It also effectively involves three distinct sections of shaft in series, namely AB, BC and CD, where A and D are at the ends of the shaft, and B and C are at the cross sections where the external torques are applied. Figure 7.21 shows a free body diagram for the shaft, with torques applied at each of the four positions marked A, B, C and D, those at supports A and D being unknown. In other words, there are two torque variables, T_A and T_D. The distribution of angular displacement is defined once we know the rotations at points A, B, C and D (the distributions between them being linear for a uniform shaft). With the ends of the shaft being fixed, the rotations of A and D are both zero.

415

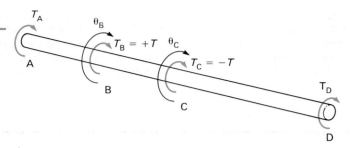

Fig. 7.21.

Consequently, there are two displacement variables, θ_B and θ_C, which we can assume to be positive in the clockwise direction, as shown.

With two torque variables and two displacement variables, the choice between an equilibrium method and a compatibility method of analysis is not clear-cut. The latter is, however, somewhat more convenient, if only because there is a simple moment equilibrium equation for the shaft

$$T_A + T_B + T_C + T_D = 0$$

where the torques are all assumed to be positive in the clockwise direction. With $T_B = -T_C = T$, this becomes

$$T_A + T_D = 0$$

which effectively reduces the number of torque variables to one, say T_A.

In Fig. 7.21 we see that the torque in AB is T_A, which in the direction shown tends to rotate A clockwise relative to B. Using equations (7.11), the twist over this section of shaft, which is of length $L/4$, is therefore

$$\theta_B - \theta_A = -\frac{T_A}{GJ} \times \frac{L}{4}$$

where G is the shear modulus of the shaft material and J is its polar second moment of area. Similarly, considering the external torques applied to the shaft to the left of a typical cross section through BC, the torque there is $T_A + T \, (= T_A + T_B)$ tending to rotate B clockwise relative to C, and the twist is

$$\theta_C - \theta_B = -\frac{(T_A + T)}{GJ} \times \frac{L}{4}$$

Finally, for CD the torque is $T_A \, (= T_A + T_B + T_C)$ tending to rotate C clockwise relative to D, and the twist is

$$\theta_D - \theta_C = -\frac{T_A}{GJ} \times \frac{L}{2}$$

Adding these three expressions for angular displacements together, we obtain

$$\theta_D - \theta_A = -\frac{L}{GJ}\left[\frac{T_A}{4} + \frac{(T_A + T)}{4} + \frac{T_A}{2}\right]$$

Since the ends of the shaft are fixed, both θ_A and θ_D are zero, which can only occur if

$$T_A + (T_A + T) + 2T_A = 0$$

or

$$T_A = -\frac{T}{4} = -T_D$$

In other words, the torque at A is actually in the counterclockwise direction, while the torque at D is clockwise, and both are of magnitude $T/4$.

We have now solved the problem to the extent that we have found the torques at the ends of the shaft. To find the distribution of angular displacement along the shaft, we first find the displacements at B and C as

$$\theta_B = \theta_A + \frac{TL}{16\,GJ} = \frac{TL}{16\,GJ}$$

and

$$\theta_C = \theta_B - \frac{3TL}{16\,GJ} = -\frac{TL}{8\,GJ}$$

and plot the distribution as shown in Fig. 7.22.

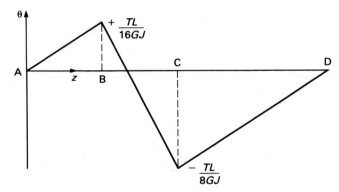

Fig. 7.22.

We could also have solved the problem by superposition. One way to do this is to imagine the shaft being fixed only at A, and to determine the displacement distributions along the shaft due to the torques $+T$, $-T$ and $+T_D$ applied separately at B, C and D, respectively, as shown in Figs 7.23a, b and c. Superimposing these three distributions, the resulting displacement at D is

$$\theta_D = \frac{L}{GJ}\left(\frac{T}{4} - \frac{T}{2} + T_D\right)$$

and for this to be zero, we require that

$$T_D = \frac{T}{4}$$

as before. The combined displacement distribution can then be plotted as in Fig. 7.22.

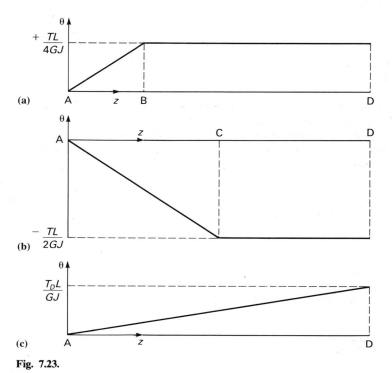

Fig. 7.23.

EXAMPLE 7.8

The three solid steel shafts AB, CD and EF shown in Fig. 7.24 are all of length 1.2 m (48 in). AB has a diameter of 25 mm (1 in), is fixed at A and carries a gear wheel having 30 teeth at B. CD has a diameter of 35 mm (1.4 in), is fixed at C and carries a gear wheel having 40 teeth at D. The gear wheels on AB and CD mesh with a gear wheel, having 20 teeth, fixed to the end E of shaft EF, which has a diameter of 50 mm (2 in). The thicknesses of the gear wheels are negligible compared with the lengths of the shafts. If a torque T of 260 Nm (190 lb ft) is applied at F, find the shaft which is most highly stressed and the rotation of end F. Friction at the bearings can be neglected.

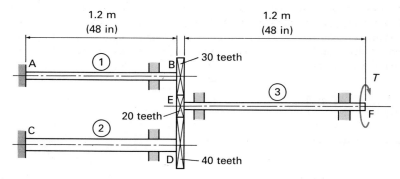

Fig. 7.24.

With two fixed supports for the shafts, the problem is statically indeterminate. It also involves two shafts, AB and CD, in parallel, the two connected in series with the third shaft EF. Counting unknowns, there are two unknown torques (in AB and CD), but really only one unknown rotation, that of gear wheel E (those of B and D follow directly from the ratios of numbers of teeth). An equilibrium method of analysis is therefore appropriate to use.

For convenience, let us number AB, CD and EF as shafts 1, 2 and 3, as shown. Also, let the rotation of wheel E when the torque is applied at F be θ_E in the same direction as the torque. Given the numbers of teeth on the wheels, the rotations of B and D are $\theta_B = 2\theta_E/3$ and $\theta_D = \theta_E/2$, both in the opposite direction to θ_E. The torques in shafts 1 and 2 are therefore

$$T_1 = \frac{G\theta_B J_1}{L} = \frac{2G\theta_E J_1}{3L}$$

and

$$T_2 = \frac{G\theta_D J_2}{L} = \frac{G\theta_E J_2}{2L}$$

where L is the common length of the shafts, G is the shear modulus for steel, and J_1 and J_2 are the polar second moments of area of shafts 1 and 2. The numerical values for all three shafts, with diameters of 25, 35 and 50 mm, are

$$J_1 = \frac{\pi}{32} \times 0.025^4 = 3.835 \times 10^{-8} \text{ m}^4$$

$$J_2 = \frac{\pi}{32} \times 0.035^4 = 1.473 \times 10^{-7} \text{ m}^4$$

$$J_3 = \frac{\pi}{32} \times 0.050^4 = 6.136 \times 10^{-7} \text{ m}^4$$

Figure 7.25a shows the end view of the three gear wheels B, E and D, looking from F. Figure 7.25b shows them separated as free bodies with the torques and tooth contact force components tangential to their pitch circles shown, but not the bearing reaction forces on the shafts to which they are attached. For moment equilibrium of gears B and D

$$T_1 = F_1 r_B \quad \text{and} \quad T_2 = F_2 r_D$$

where r_B and r_D are the pitch circle radii of gears B and D, and F_1 and F_2 are the tangential force components at the tooth contacts between B and E, and D and E, respectively. For moment equilibrium of gear E, with a pitch circle radius of r_E

$$T = F_1 r_E + F_2 r_E = T_1 \frac{r_E}{r_B} + T_2 \frac{r_E}{r_D}$$

and, since pitch circle radii are proportional to the numbers of teeth and T_1 and T_2 have already been defined in terms of θ_E

$$T = \tfrac{2}{3}T_1 + \tfrac{1}{2}T_2 = \frac{G\theta_E}{L}\left(\tfrac{4}{9}J_1 + \tfrac{1}{4}J_2\right)$$

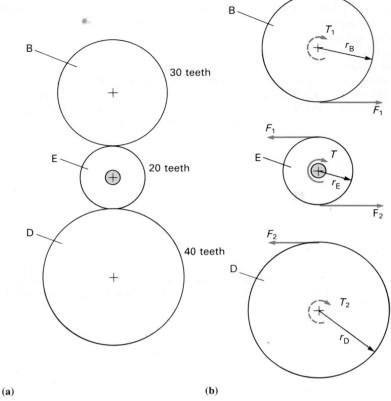

(a)   **(b)**

Fig. 7.25.

Substituting numerical values

$$260 = \frac{79.6 \times 10^9}{1.2} \left(\tfrac{4}{9} \times 3.835 + \tfrac{1}{4} \times 14.73\right) \times 10^{-8}\,\theta_E$$

from which

$$\theta_E = 0.07276 \text{ radians}$$

Consequently

$$\theta_B = \tfrac{2}{3}\theta_E = 0.04851 \text{ radians}$$

$$\theta_D = \tfrac{1}{2}\theta_E = 0.03638 \text{ radians}$$

The maximum shear stresses at the outer surfaces of each of the three shafts, of diameters D_1, D_2 and D_3, are

$$\tau_1 = \frac{G\theta_B D_1}{2L} = \frac{79.6 \times 10^9 \times 0.048\,51 \times 0.025}{2 \times 1.2} \text{ N/m}^2 = 40.2 \text{ MN/m}^2$$

$$\tau_2 = \frac{G\theta_D D_2}{2L} = \frac{79.6 \times 10^9 \times 0.036\,38 \times 0.035}{2 \times 1.2}\,\text{N/m}^2 = 42.2\,\text{MN/m}^2$$

$$\tau_3 = \frac{TD_3}{2J_3} = \frac{260 \times 0.050}{2 \times 6.136 \times 10^{-7}}\,\text{N/m}^2 = 10.6\,\text{MN/m}^2$$

The most highly stressed shaft is therefore the second one, namely CD, with a maximum shear stress of 42.2 MN/m² (6.1 ksi).

Finally, the rotation of end F is

$$\theta_F = \theta_E + \frac{TL}{GJ_3} = 0.072\,76 + \frac{260 \times 1.2}{79.6 \times 10^9 \times 6.136 \times 10^{-7}}$$

$$= 0.079\,15 \text{ radians} = \underline{4.53°}$$

7.4 Combined bending and torsion

In many practical situations, structural members may be subjected to both bending and torsion. Provided the material behavior remains linearly elastic, however, we can superimpose the deformations and the stresses due to the two effects. In the case of stresses, this may lead to rather more complex states of stress than we have met so far, and which we will not consider in detail until Chapter 9. To complete this chapter on torsion, let us consider two typical examples of combined bending and torsion.

EXAMPLE 7.9

Figure 7.26 shows a solid circular rod ABC of length 2L and diameter D which is built in horizontally at A, is bent through a right angle in the horizontal plane at its midpoint B, and supports a vertical concentrated force of F at C. If the material of the rod has a Young's modulus of E and a Poisson's ratio of v, find an expression for the vertical displacement of point C.

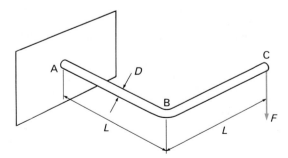

Fig. 7.26.

We can consider separately the behaviors of the two halves of the rod, AB and BC. While AB is effectively a cantilever beam subject also to twisting about its axis, BC is a cantilever with its constrained end not rigidly built in but supported flexibly in both the

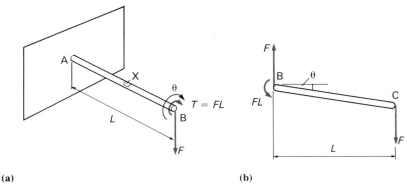

(a)   **(b)**

Fig. 7.27.

vertical direction and rotational sense. Figure 7.27 shows AB and BC separated, with the external forces and couples acting on them. We note that, in the case of AB, the state of stress at the typical surface element shown shaded and marked X is a combination of a tensile normal stress due to the bending caused by the force F at B, and a shear stress due to the twisting caused by the torque $T = FL$, also acting at B.

Let the angle of twist of AB at B be θ, and let the vertical displacement of B be v_B (positive in the upward vertical direction according to the bending sign convention). From equations (7.11)

$$\theta = \frac{TL}{GJ} = \frac{FL^2}{GJ}$$

where J is the polar second moment of area of the rod cross section, and from Appendix D

$$v_B = -\frac{FL^3}{3EI}$$

where I is the second moment of area of the rod cross section about its neutral axis in bending.

Considering now part BC of the rod, Fig. 7.27b, the vertical displacement of C is made up of three contributions: (i) the displacement of B which affects equally the whole of BC, (ii) the displacement of C relative to B due to the rotation θ at B with BC remaining straight, and (iii) the displacement of C relative to B due to bending. Expressed in symbols

$$v_C = v_B - \theta L - \frac{FL^3}{3EI}$$

$$= -\frac{2FL^3}{3EI} - \frac{FL^3}{GJ}$$

Now

$$J = 2I = \frac{\pi D^4}{32}$$

and, from equation (3.11), the elastic properties of the material are connected by

$$E = 2G(1 + v)$$

The expression for the displacement at C therefore becomes

$$v_C = -FL^3\left[\frac{2}{3EI} + \frac{2(1+v)}{E}\frac{1}{2I}\right] = -\frac{FL^3}{EI}\left(\frac{5}{3} + v\right)$$

where the negative sign indicates a downward displacement.

EXAMPLE 7.10

A steel brace used to tighten car wheel nuts takes the form of a 'T' as shown in Fig. 7.28a. The socket at A is connected to a spherical boss at B of 50 mm (2 in) diameter by a shaft which is 300 mm (12 in) long and 20 mm (0.8 in) diameter. The handle CD, which is 600 mm (24 in) long, also takes the form of a circular rod of 20 mm (0.8 in) diameter, and passes through the boss at B at right angles to shaft AB, with the center of the rod coinciding with the center of the boss. In use, the loads applied to the handle are equal and opposite forces at right angles to the handle, as shown in Fig. 7.28b, which is a view in the direction parallel to the axis of shaft AB. These forces are applied, not at the ends of the handle, but at points E and F which are each at 50 mm (2 in) from the ends of the handle. The brace is applied to a nut which is already tight, allowing no further rotation of socket A, and forces of 350 N (80 lb) are applied at points E and F. Assuming that socket A and boss B are rigid, and that the handle can be treated as two cantilevers, each of length 275 mm (11 in) rigidly built in to the boss, calculate the maximum stresses in the shaft and in the handle, also the displacements of points E and F in the directions of the applied loads.

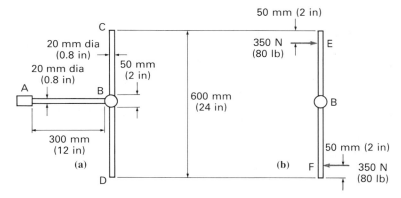

Fig. 7.28.

This is a more practical example of combined torsion and bending, and in some respects is simpler than Example 7.9. Because of the symmetrical loading applied to handle CD, it transmits only a torque and no lateral loading to shaft AB. Therefore, while CD is subject only to bending, AB is subject only to torsion.

SI units

With two equal and opposite forces of magnitude $P = 350$ N applied to CD at a distance of 500 mm apart, the torque in AB is

$$T = 350 \times 0.5 = 175 \text{ Nm}$$

The polar second moment of area of AB, which is of diameter $D = 20$ mm, is

$$J = \frac{\pi D^4}{32} = \frac{\pi \times 0.020^4}{32} = 1.571 \times 10^{-8} \text{ m}^4$$

and the maximum shear stress at the outer surface of the shaft is

$$\tau_0 = \frac{TD}{2J} = \frac{175 \times 0.020}{2 \times 1.571 \times 10^{-8}} \text{ N/m}^2 = \underline{111 \text{ MN/m}^2}$$

Since the length of the shaft is $L = 300$ mm, the angle of twist at B, relative to the fixed end at A, is

$$\theta_B = \frac{TL}{GJ} = \frac{175 \times 0.3}{79.6 \times 10^9 \times 1.571 \times 10^{-8}}$$

$$= 0.041\,98 \text{ radians}$$

As for the handle CD, with the same diameter ($D = 20$ mm) as the shaft AB, its second moment of area in bending is

$$I = \frac{J}{2} = 7.854 \times 10^{-9} \text{ m}^4$$

The greatest bending moment, which occurs where the handle enters the boss at B, a position which is 225 mm from one of the applied forces, is

$$M = 350 \times 0.225 = 78.75 \text{ Nm}$$

and the corresponding maximum bending stress is

$$\sigma = \frac{MD}{2I} = \frac{78.75 \times 0.020}{2 \times 7.854 \times 10^{-9}} \text{ N/m}^2 = \underline{100 \text{ MN/m}^2}$$

The displacement of, say, point E is made up of two contributions: (i) the displacement of E relative to B due to the rotation θ at B with the handle remaining straight, and (ii) the displacement of E relative to B due to bending of BE as a cantilever. Since this cantilever is of effective length $L_e = 225$ mm, and E is at a radius $R = 250$ mm from the axis of AB, the total displacement of E is (with the aid of Appendix D)

US customary units

With two equal and opposite forces of magnitude $P = 80$ lb applied to CD at a distance of 20 in apart, the torque in AB is

$$T = 80 \times 20 = 1600 \text{ lb in}$$

The polar second moment of area of AB, which is of diameter $D = 0.8$ in, is

$$J = \frac{\pi D^4}{32} = \frac{\pi \times 0.8^4}{32} = 0.040\,21 \text{ in}^4$$

and the maximum shear stress at the outer surface of the shaft is

$$\tau_0 = \frac{TD}{2J} = \frac{1600 \times 0.8}{2 \times 0.040\,21} \text{ psi} = \underline{15.9 \text{ ksi}}$$

Since the length of the shaft is $L = 12$ in, the angle of twist at B, relative to the fixed end at A, is

$$\theta_B = \frac{TL}{GJ} = \frac{1600 \times 12}{11.5 \times 10^6 \times 0.040\,21}$$

$$= 0.041\,52 \text{ radians}$$

As for the handle CD, with the same diameter ($D = 0.8$ in) as the shaft AB, its second moment of area in bending is

$$I = \frac{J}{2} = 0.020\,11 \text{ in}^4$$

The greatest bending moment, which occurs where the handle enters the boss at B, a position which is 9 in from one of the applied forces, is

$$M = 80 \times 9 = 720 \text{ lb in}$$

and the corresponding maximum bending stress is

$$\sigma = \frac{MD}{2I} = \frac{720 \times 0.8}{2 \times 0.020\,11} \text{ psi} = \underline{14.3 \text{ ksi}}$$

The displacement of, say, point E is made up of two contributions: (i) the displacement of E relative to B due to the rotation θ at B with the handle remaining straight, and (ii) the displacement of E relative to B due to bending of BE as a cantilever. Since this cantilever is of effective length $L_e = 9$ in, and E is at a radius $R = 10$ in from the axis of AB, the total displacement of E is (with the aid of Appendix D)

$$u_E = R\theta_B + \frac{PL_e^3}{3EI} = 0.250 \times 0.041\,98$$

$$+ \frac{350 \times 0.225^3}{3 \times 207 \times 10^9 \times 7.854 \times 10^{-9}}\ \text{m}$$

$$= 10.49 + 0.817\ \text{mm} = \underline{11.3\ \text{mm}}$$

in the horizontal direction in Fig. 7.28b. Note that the contribution due to twisting of the shaft is much greater than that due to bending of the handle.

$$u_E = R\theta_B + \frac{PL_e^3}{3EI} = 10 \times 0.041\,52$$

$$+ \frac{80 \times 9^3}{3 \times 30 \times 10^6 \times 0.020\,11}$$

$$= 0.4152 + 0.0322\ \text{in} = \underline{0.447\ \text{in}}$$

in the horizontal direction in Fig. 7.28b. Note that the contribution due to twisting of the shaft is much greater than that due to bending of the handle.

Problems

SI METRIC UNITS

7.1 A thin circular steel tube has an outer diameter of 75 mm and a wall thickness of 5 mm. If the maximum allowable mean shear stress in the tube wall is 100 MN/m², determine the maximum torque the tube can withstand.

7.2 A circular aluminum tube having an outer diameter of 120 mm, wall thickness of 8 mm and length 600 mm is attached to rigid plates at its ends. If one of these plates is fixed and the other rotated through an angle of 1° about the axis of the tube, find the torque required to do this, and the resulting mean shear stress in the aluminum.

7.3 Repeat Problem 7.2 for a solid bar of aluminum having the same external dimensions, in this case finding the maximum shear stress.

7.4 A vertical solid steel drilling rod of 100 mm diameter for an oil rig is driven by a torque at the top and against a resistance concentrated at the bottom end. If the maximum allowable shear stress in the rod is 80 MN/m² when drilling at a depth of 3000 m, calculate the angular rotation of one end of the rod relative to the other end.

7.5 A solid circular drive shaft is to be designed to transmit up to 40 kW at 1500 rev/min. If the maximum allowable shear stress is 65 MN/m², find the minimum diameter of the shaft.

7.6 A hollow circular steel shaft of length 1.5 m is to be designed to transmit a torque of 90 kNm, subject to a maximum allowable shear stress of 85 MN/m², and a maximum angle of twist of 1°. Find the inner and outer diameters of the shaft which just satisfies these requirements.

7.7 As part of a design modification to reduce the weight of a piece of equipment, a solid drive shaft is to be replaced by one which is hollow and made from the same material. If the same maximum torque has to be transmitted, and the outer diameter of the shaft cannot be increased by more than 15%, find the percentage reduction in weight of the shaft.

7.8 Repeat Problem 7.7 assuming that a different material is to be used for the hollow shaft, the new material having a 5% higher density and a 25% higher maximum allowable shear stress than the existing one.

7.9 Equation (7.16) defines the angle of twist for a tapered solid circular shaft. Show that this expression reduces to the correct form for a uniform shaft when $R_1 \rightarrow R_2$.

7.10 A solid steel shaft tapers linearly in diameter from 65 mm to 35 mm over a length of 200 mm. If the shaft is subjected to torsional loading which causes a relative twist of its ends of 0.2°, find the applied torque and the maximum shear stress in the steel.

7.11 Repeat Problem 7.10 assuming the shaft is not tapered but has a step change in diameter from 65 mm to 35 mm at its center.

7.12 The solid aluminum shaft shown in Fig. P7.12 has a linearly tapered section between two parallel sections. If the maximum allowable shear stress in the material is 70 MN/m², find the greatest torque the shaft can transmit, and the angle of twist between the ends of the shaft when this torque is applied.

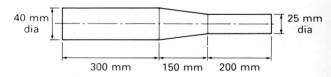

Fig. P7.12.

425

7.13 A solid shaft of 50 mm diameter is connected to a hollow shaft of 100 mm outside diameter by means of a flanged coupling as shown in Fig. P7.13, fitted with ten bolts equally spaced on a pitch circle diameter of 250 mm. If the maximum allowable shear stress is 50 MN/m² in the shafting and 25 MN/m² in the bolts, calculate the required diameter of the bolts and internal diameter of the hollow shaft so that both the shafts and the coupling have the same strength. Find the maximum power that can be transmitted by the shaft at a speed of 120 rev/min.

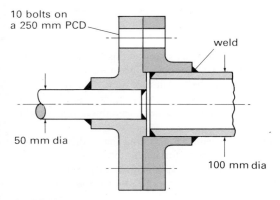

10 bolts on a 250 mm PCD

weld

50 mm dia

100 mm dia

Fig. P7.13.

7.14 An electric motor drives a load via two steel shafts and a step-down gearbox, as shown in Fig. P7.14. The gearbox has an input to output speed ratio of 100 to 1, and may be assumed to cause negligible power losses due to friction. Shaft 1, which connects motor and gearbox, is 3 m long and is solid, with a diameter of 10 mm. Shaft 2, which connects gearbox and load is 10 m long and is hollow, with an internal diameter of 40 mm and an external diameter of 50 mm. The maximum (stalled) torque exerted by the motor is 20 Nm. The normal resistance to rotation offered by the load is 1 kNm, but a mechanical failure causes the load to seize, and resist any further motion. Inertial effects are negligible. Determine the magnitudes of the maximum shear stresses in the two shafts once

motion has ceased, and the number of revolutions turned by the motor after seizure of the load.

7.15 A motor vehicle has its two rear wheels driven by a steel propeller shaft which is parallel to the length of the vehicle and drives two steel half-shafts through a right-angled gear giving a speed reduction of 3.9 to 1. Each half-shaft is solid, 25 mm in diameter and 0.8 m long, and the propeller shaft is 2.1 m long and hollow with an outer diameter of 25 mm. If the maximum allowable shear stress in the half-shafts is 90 MN/m², find the maximum torque that can be safely transmitted by each of them. Find the internal diameter of the propeller shaft which would produce the same maximum shear stress in it as in the half-shafts. If the rear wheels are locked and the maximum safe torque is applied to the front end of the propeller shaft, calculate how much this end will rotate relative to the ground.

7.16 An engine develops 60 kW. The power is transmitted through a gearbox having an efficiency of 83% to a hollow cylindrical steel output shaft rotating at 350 rev/min. The maximum allowable shear stress for the material of the shaft is 80 MN/m². Tubes for the shaft are available in all combinations of wall thickness and outer diameter starting with a wall thickness of 2 mm increasing in steps of 1 mm and with a minimum outer diameter of 30 mm increasing in steps of 5 mm. To avoid problems of buckling, the ratio of wall thickness to outer radius must not be less than 0.1. Find a suitable size for the shaft and calculate the angle of twist per meter of the chosen shaft.

7.17 Figure P7.17 shows a uniform circular shaft AB of length L fixed at both ends. Torques of $2T$, $6T$ and $3T$ are applied as shown. Find expressions for the greatest torque in the shaft and the greatest angular displacement. The polar second moment of area of the shaft is J, and the shear modulus of its material is G.

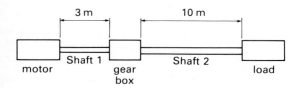

3 m

10 m

motor

Shaft 1

gear box

Shaft 2

load

Fig. P7.14.

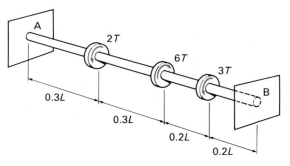

A

2T

6T

3T

B

0.3L

0.3L

0.2L

0.2L

Fig. P7.17.

7.18 Repeat Problem 7.17 assuming that the support at B is not rigid but has an effective torsional stiffness (ratio of torque to angular displacement) of magnitude μ.

7.19 A stepped shaft of circular cross section, 700 mm long, is made by fixing a solid steel section 20 mm in diameter and 300 mm long to a solid brass section 400 mm long. The two sections are coaxial, and cannot move relative to each other at the joint between them. The composite shaft is fixed at both ends, and a torque T is applied at the joint. Find the diameter of the brass section if the maximum shear stress in the steel is twice that in the brass, and find the value of T when the maximum shear stress in the steel is 80 MN/m².

7.20 In the mine winder system shown in Fig. P7.20, the 5 m diameter winding drum is connected to an electric motor by a solid steel shaft 360 mm in diameter and 1.2 m long. When the cage is lowered into the mine, there is 200 m of steel rope, having an effective cross-sectional area of 3000 mm², between the cage and the drum. The motor is used as a brake to prevent further rotation of the end of the shaft. If material with a mass of 9000 kg is then loaded into the cage, which itself has a mass of 6000 kg, determine by how much the cage is displaced downwards. The winding drum may be treated as rigid, and bending of the shaft neglected. Find also the resulting maximum shear stress in the shaft.

7.21 Figure P7.21 shows a solid circular rod ABCD of length $3L$ and diameter D which is built in hori-

zontally at A, is bent through a right angle in the horizontal plane at B, and again in the same plane at C. If a vertical concentrated force F is applied to end D, find an expression for the amount by which this end is displaced in the direction of the force.

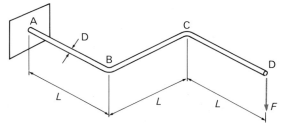

Fig. P7.21.

US CUSTOMARY UNITS

7.22 A thin circular steel tube has an outer diameter of 3 in and a wall thickness of 0.125 in. If the maximum allowable mean shear stress in the tube wall is 15 ksi, determine the maximum torque the tube can withstand.

7.23 A circular aluminum tube having an outer diameter of 5 in, wall thickness of 0.25 in and length 2 ft is attached to rigid plates at its ends. If one of these plates is fixed and the other rotated through an angle of 1° about the axis of the tube, find the torque required to do this, and the resulting mean shear stress in the aluminum.

7.24 Repeat Problem 7.23 for a solid bar of aluminum having the same external dimensions, in this case finding the maximum shear stress.

7.25 A solid circular steel shaft with a diameter of 1 in is used to transmit power from an electric motor. If the maximum allowable shear stress is 12 ksi, find the greatest power that can be transmitted at a speed of 1000 rev/min.

7.26 A copper cylinder with inner and outer diameters of 1.5 in and 2.5 in is subjected to torsional loading about its axis. If the maximum allowable shear stress in the copper is 7 ksi, find the maximum torque and the corresponding angle of twist per foot of cylinder.

7.27 and 7.28 Solve Problems 7.7 and 7.8.

7.29 Figure P7.29 shows a solid steel rod AB of diameter 0.6 in which is attached to and coaxial with the flanged brass cylinder CD which has internal and external diameters of 1.4 in and 1.6 in. If a torque of 35 lb ft is applied to the rod at B, find the maximum shear stresses in the rod and cylinder, and the angle of twist of B. Assume that end D of the cylinder is fixed, and that the only signifi-

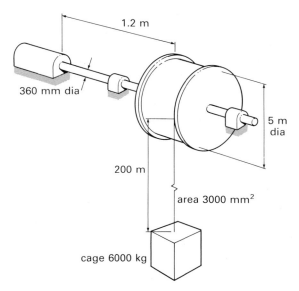

Fig. P7.20.

427

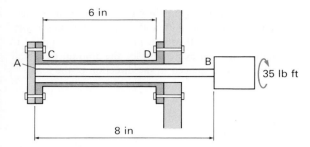

Fig. P7.29.

cant torsional deformations occur in the rod between A and B and in the cylinder between C and D.

7.30 Solve Problem 7.9.

7.31 A solid steel shaft tapers linearly in diameter from 2.5 in to 1.5 in over a length of 10 in. If the shaft is subjected to torsional loading which causes a relative twist of its ends of 0.2°, find the applied torque and the maximum shear stress in the steel.

7.32 Repeat Problem 7.31 assuming the shaft is not tapered but has a step change in diameter from 2.5 in to 1.5 in at its center.

7.33 A composite shaft consists of a solid steel rod, 3 in diameter, surrounded by a closely-fitting tube of bronze bonded to it. Find the outside diameter of the tube so that when a torque is applied to the composite shaft it will be shared equally by the two materials. Take the shear modulus for the bronze to be half that for steel. If the torque is 12 kip-ft, calculate the maximum shear stress in each material.

7.34 In the arrangement of two steel shafts shown in Fig. P7.34, shaft 1 is solid, 3 in diameter, and shaft 2 is hollow, with diameters of 4.8 and 3.5 in. The two shafts are connected by one pair of gear wheels, the wheel on shaft 2 having three times as many teeth as that on shaft 1. If end B of shaft 2 is held stationary while a torque of 1.8 kip-ft is applied to end A of shaft 1, find the greatest shear

stress in each shaft and the rotation of end A of shaft 1.

7.35 An aircraft control column contains an aileron transmission tube of aluminum alloy, whose length is 30 in; it is connected by bevel gears to the pilot's wheel, as in Fig. P7.35. The stiffness criterion for the tube is that under a maximum torque of 50 lb ft applied to the wheel, the rotation of the wheel shall not exceed 0.01 radians, assuming that the lower end of the transmission is prevented from rotating. If the tube has a mean diameter of 1.6 in and a thickness of 0.02 in, determine the gear ratio n_1/n_2 necessary to meet the design requirement.

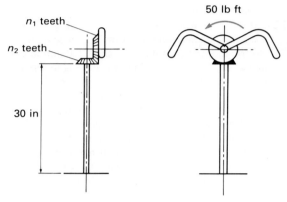

Fig. P7.35.

7.36 A solid circular steel rod of diameter 0.75 in and length 32 in is fixed at both ends. If a concentrated torque applied to the shaft at a position 14 in from one end causes an angular rotation there of 1°, find the magnitude of the torque.

7.37 and 7.38 Solve Problems 7.17 and 7.18.

7.39 Figure P7.39 shows a solid circular steel rod of diameter 0.6 in and length 20 in built in horizontally at one end and bent through a right angle in the horizontal plane at a distance of 8 in from the fixed end. If a downward vertical force of 12 lb is applied at the other end, find by how much this end is deflected downwards.

7.40 Solve Problem 7.21.

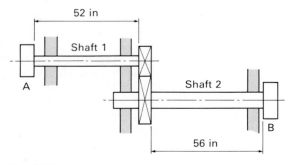

Fig. P7.34.

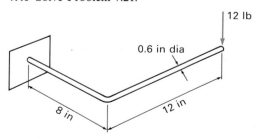

Fig. P7.39.

428

8

Instability and the Buckling of Struts and Columns

In all of the types of problems considered so far, we have assumed the deformation to be both progressive with increasing load and simple in form. For example, we assumed that a member in simple tension or compression becomes progressively longer or shorter, but remains straight. Similarly, in the case of torsion, we assumed that the angle of twist increases progressively with increasing torque, but that the shaft remains straight. In each of these examples, provided the material involved is linearly elastic the deformation is directly proportional to the applied loading.

Under some circumstances, however, our assumptions of progressive and simple deformation may no longer hold, and the member may become *unstable*, suffering gross deformations at relatively low levels of loading. Consider, for example, a solid circular cylinder placed between the platens of a testing machine and subjected to compressive loading, as in Fig. 8.1. If the cylinder has a relatively small length-to-diameter ratio, then it merely becomes slightly shorter and fatter as in Fig. 8.1a (for present purposes we assume the platens to be well lubricated, allowing the cylinder to expand freely at its ends). On the other hand, if the cylinder is in the form of a slender rod, at first it will be uniformly compressed, but at higher loads it is then more likely to *buckle* as in Fig. 8.1b, with the center of the rod being displaced laterally by a large amount. In fact the rod is now bending as a beam subjected to axial loading. Had the

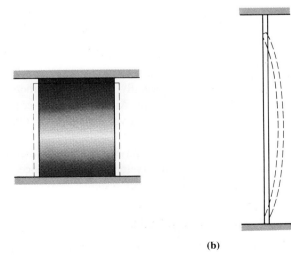

(a)   (b)

Fig. 8.1.

same two cylinders been subjected to tensile loading, both would have extended uniformly, with no tendency to bend.

The possibility of a structural member displaying unstable behavior well before the internal stresses reach their maximum allowable levels is of considerable practical importance, and must be anticipated if the member is to perform its function satisfactorily. In addition to the buckling of long slender members under axial compression, further examples of instability include the crushing and torsional buckling of thin cylinders illustrated in Fig. 8.2. While these two phenomena are relatively difficult to analyze theoretically, we can predict with reasonable accuracy the buckling of *struts* and *columns*.

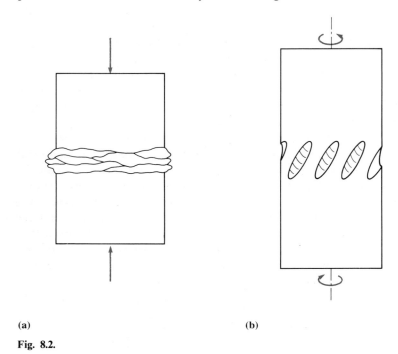

(a)   (b)

Fig. 8.2.

The terms *strut* and *column* (sometimes pillar) are widely used, often interchangeably, in the context of buckling of slender members. While the term strut is applied particularly to the members of pin-jointed structures which are in compression, the term column is used especially to describe a member in compression which has at least one end built-in. Although in the context of buildings a column is usually thought of as being a vertical member, this is not essential.

Before we attempt to analyze the buckling behavior of struts and columns, it is useful to review the types of equilibrium a system may experience, and to consider the stability of rather simpler mechanical systems.

8.1 Stable, neutral and unstable equilibrium

In order to illustrate types of equilibrium, let us first consider some situations encountered in the mechanics of rigid bodies which are shown in Fig. 8.3. In

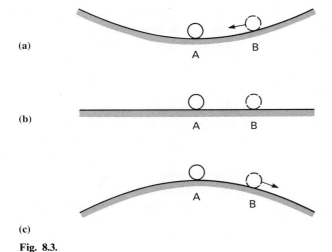

(a)

(b)

(c)

Fig. 8.3.

each case a sphere rests on a surface, which may be either concave, flat, or convex. If the surface is concave (Fig. 8.3a), and the sphere rests at the lowest point A, then if it is displaced a small distance to some other point B it will tend to return to A. It is said to be in *stable equilibrium* at A. If the surface is flat and horizontal (Fig. 8.3b), then when the sphere is displaced from A to B it remains there, neither returning to A nor moving further away: it is in *neutral equilibrium*. Finally, if the surface is convex (Fig. 8.3c), and the sphere rests at the highest point A, then if it displaced a small distance to some other point B it will tend to move away from A. It is said to be in *unstable equilibrium* at A. We note that it is not what happens at the equilibrium point A itself that determines the type of equilibrium, but what happens as the sphere is displaced slightly from this position. We will find this to be a common feature of stability problems.

Let us now consider the mechanical system shown in Fig. 8.4a, which brings us a little closer to the buckling of struts and columns. The weightless rigid rod

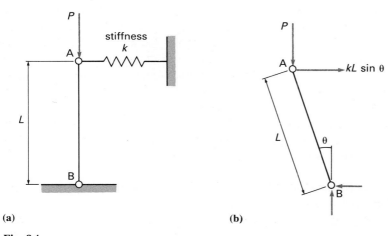

(a)   (b)

Fig. 8.4.

AB is pin-jointed to a fixed surface at B, and is held in the vertical position by a horizontal spring pin-jointed to the rod at A. The length of the rod is L, the stiffness of the spring is k, and the vertical force P applied at A puts the rod into compression. We assume initially that the system is perfect: when the rod is exactly vertical there is no deformation of the spring, and the line of action of force P is also exactly vertical.

Since we are interested in the stability of the system, we must consider its behavior when displaced by a small amount from its equilibrium position. Figure 8.4b shows a free body diagram for the rod displaced through a small angle θ from the vertical. In this position, the horizontal displacement of A is $L \sin \theta$, which means that the horizontal spring force acting on the rod at A is $kL \sin \theta$ (the product of the stiffness of the spring and its change in length) as shown. Also acting on the rod are the force P and reaction force components at the support at B (the resultant reaction force acts along AB). If the rod is in equilibrium in the position shown, we may take moments about B to give

$$(kL \sin \theta) \times (L \cos \theta) - P \times (L \sin \theta) = 0$$

or

$$P = kL \cos \theta$$

Provided the angle θ is small, $\cos \theta \approx 1$ and

$$P \approx kL \tag{8.1}$$

We need to consider carefully what this result means. If the rod is to be in equilibrium at some position displaced from the vertical, then the applied force, P, must be of the particular magnitude kL, *which does not depend on the amount of displacement* (provided this is small), and the rod is in neutral equilibrium. If $P < kL$, the rod cannot be in equilibrium when displaced from the vertical and must return to the vertical (the moment about B of the force P in Fig. 8.4b is less than that of the spring force): the rod remains in stable equilibrium in the vertical position. If, on the other hand, $P > kL$, the rod again cannot be in equilibrium when displaced from the vertical, but is forced away from the vertical (the moment about B of force P in Fig. 8.4b is greater than that of the spring force): the rod is in unstable equilibrium. These three types of equilibrium are equivalent to those illustrated for the rolling sphere in Fig. 8.3.

If we plot the magnitude of force P against angular displacement for the system, we obtain Fig. 8.5. Until P reaches the *critical* value $P_c = kL$ there is no displacement. At $P = P_c$ the system can take up any displaced position. The force cannot exceed the critical value without the system collapsing. The relationship between force and displacement is highly nonlinear, despite the fact that the spring, which is the only deformable part of the system, is linearly elastic.

Now let us consider what happens if the system is slightly imperfect, in that when the spring is undeformed the rod is at a very small angle θ_0 to the vertical, as in Fig. 8.6a. Figure 8.6b shows a free body diagram for the rod displaced through a small angle θ (where $\theta > \theta_0$) from the vertical. In this position, the

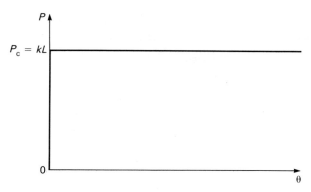

Fig. 8.5.

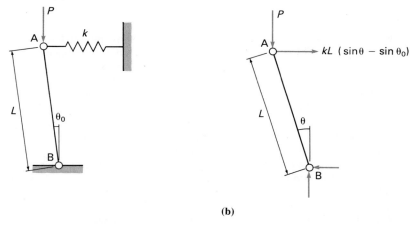

(a) (b)

Fig. 8.6.

horizontal displacement of A from its original position is $L(\sin \theta - \sin \theta_0)$, which means that the horizontal spring force acting on the rod at A is $kL(\sin \theta - \sin \theta_0)$. If the rod is in equilibrium in the position shown, we may take moments about B to give

$$(kL(\sin \theta - \sin \theta_0)) \times (L \cos \theta) - P \times (L \sin \theta) = 0$$

or

$$P = \frac{kL(\sin \theta - \sin \theta_0)\cos \theta}{\sin \theta} \approx kL\left(1 - \frac{\theta_0}{\theta}\right) \qquad (8.2)$$

provided θ is small (when $\sin \theta \approx \theta$, $\cos \theta \approx 1$). This relationship between force and displacement is plotted in Fig. 8.7, together with the result for the perfect system for comparison. The imperfection has the effect of removing the abrupt change in behavior at the critical value of force, and the system experiences significant deformations before this critical value is reached. The most important point, however, is that the magnitude of the critical force is unchanged: unbounded deformation still occurs at $P = P_c = kL$, and the system collapses.

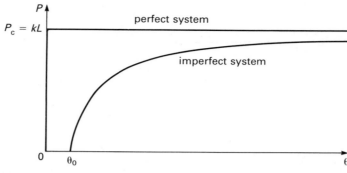

Fig. 8.7.

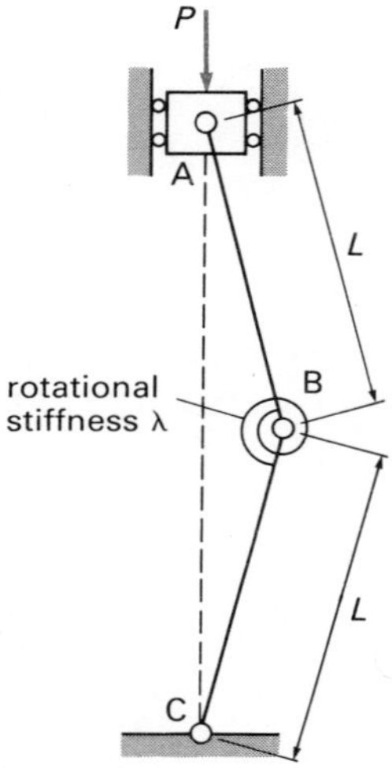

EXAMPLE 8.1

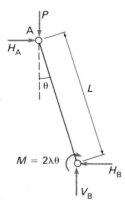

Fig. 8.8.

Figure 8.8 shows a mechanical system consisting of two pin-jointed weightless rigid members, AB and BC, each of length L, linked by a rotational spring of stiffness λ at B. The position of the pin joint C is fixed, but A is free to move vertically. If the spring is undeformed when both of the members are vertical, find the critical value of the compressive force P applied at A which just causes the system to collapse.

This example brings us closer to the buckling of struts. Figure 8.9 shows the free body diagram for AB when it is at a small angle θ to the vertical. If AB has rotated through angle θ, the geometry of deformation of the system is such that BC has also rotated through θ (in the opposite direction). In other words, the angle between the members has changed by an amount 2θ, and the moment applied to each of them by the spring is $2\lambda\theta$ (the product of rotational stiffness and relative angular displacement). In general, member AB would also have horizontal and vertical reaction force components H_B and V_B acting on it at B, and a horizontal reaction force H_A at A, in addition to the applied force P. In this particular case, however, symmetry of the system about the horizontal line through B means that there are no horizontal components of reaction between the members there, and $H_B = 0$. For member AB to be in equilibrium in the horizontal direction, this in turn means that $H_A = 0$. Therefore, for equilibrium about B of moments applied to AB

$$2\lambda\theta - PL\sin\theta = 0$$

and for small values of θ

$$P \approx \frac{2\lambda}{L}$$

This is the required critical value of force needed to produce a state of neutral equilibrium, in which the magnitude of the deformation is arbitrary.

Fig. 8.9.

The buckling of pin-ended struts is of particular concern in the design of pin-jointed structures. Indeed, in the structures of this type we considered in Chapters 2 and 4, any member which is in compression is a strut, and capable of buckling. We start by analyzing the behavior of a perfect strut subject only to axial compression, before considering the effects of geometrical imperfections and lateral loading.

8.2.1 The perfect pin-ended strut

Figure 8.10 shows a uniform pin-ended strut, typically forming part of a pin-jointed structure. It is perfect in the sense that it is perfectly straight, and that the compressive forces at its ends act through the centroid of every cross section. According to our analysis of the bending effects of axial loads in Section 5.4, this should mean that the member is subject only to uniform compressive stress, and no bending. Indeed, this is true for small values of the applied force, but as the force is increased the strut becomes unstable.

Fig. 8.10.

In order to study this unstable behavior, we must consider the strut in its deformed state, as modeled in Fig. 8.11. At this stage it is not important what form the deflected shape takes. The length of the strut between the centers of the pin joints is L, the second moment of area about the neutral axis of its cross section is I, and the material of which it is made has a Young's modulus of E. It is subjected to an axial compressive force of magnitude P. We can regard the position of one of the pinned ends as being fixed, and the other as being constrained to move freely along the line joining them. Suppose that at a distance x from the left-hand end the deflection (of the neutral surface) is $v(x)$, which is positive in the upward direction as shown. Therefore, at this position, the bending moment is $-Pv$, negative because it is a hogging moment. The bending deflections can be determined with the aid of the moment–curvature relationship, equation (6.5), which in this case is

$$EI \frac{d^2 v}{dx^2} = M(x) = -Pv \tag{8.3}$$

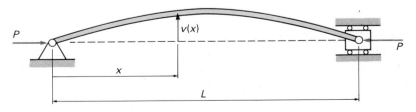

Fig. 8.11.

Such an equation is usually written in the form

$$\frac{d^2v}{dx^2} + n^2v = 0 \tag{8.4}$$

where

$$n^2 = \frac{P}{EI} \tag{8.5}$$

Now, equation (8.4) is a second-order linear ordinary differential equation with constant coefficients, but is less straightforward to solve than those we met in Chapter 6 for the bending of beams due to lateral loading. The difference is that now we have not only the second derivative of deflection appearing in the equation, but also the deflection itself. It is, however, a relatively common form of equation, other examples including the simple harmonic motion of vibrating systems (when time rather than distance is the independent variable). Although there are formal mathematical techniques for solving such equations, we can observe that a possible solution is

$$v = \sin nx$$

since

$$\frac{dv}{dx} = n \cos nx$$

and

$$\frac{d^2v}{dx^2} = -n^2 \sin nx = -n^2v$$

Also, we could multiply the sine function by any arbitrary constant and achieve the same result, or we could replace the sine by cosine. The general solution of the differential equation (8.4) is therefore given by

$$v = A \sin nx + B \cos nx \tag{8.6}$$

where A and B are arbitrary and independent *constants of integration*. Since we are solving a second-order differential equation, we expect there to be two such constants.

We find the values of constants A and B from the boundary conditions, which define the geometry of deformation of the particular problem. In this case, both ends of the strut are pinned and therefore cannot deflect. Substituting the condition $v = 0$ at $x = 0$ into equation (8.6) we obtain

$$0 = A \times 0 + B \times 1 \quad \text{or} \quad B = 0$$

and the cosine term is eliminated. Similarly, substituting $v = 0$ at $x = L$, we find

$$0 = A \sin nL$$

Now, there are two quite distinct ways in which this equation can be satisfied: either $A = 0$ or $\sin nL = 0$. If $A = 0$, there is no lateral deflection of the strut, which is the stable condition for the system. If $\sin nL = 0$, there are an infinite number of possible values of n, given by

$$nL = 0, \pi, 2\pi, 3\pi, \ldots. \qquad (8.7)$$

From the first of these, we obtain the trivial result $P = 0$, representing an unloaded and straight strut. The second one, however, gives with the aid of equation (8.5) for n

$$P = P_c = \frac{\pi^2 EI}{L^2} \qquad (8.8)$$

and a deflected shape defined by

$$v = A \sin \frac{\pi x}{L} \qquad (8.9)$$

The strut deflection follows a sinusoidal shape, with one half of a complete sine wave between $x = 0$ and $x = L$, the length of the strut, as in Fig. 8.11. The other notable feature of equation (8.9) is that the constant A remains undefined, which means that if the end force is given by equation (8.8) then the sinusoidal deflection is of arbitrary amplitude. In other words, a state of neutral equilibrium has been achieved, and the value of end force, P_c, given by equation (8.8) is a critical value necessary to cause instability. Indeed, it is referred to as the *Euler critical buckling force* for a pin-ended strut (after the Swiss mathematician Leonhard Euler, 1707–1783). Although it would appear that the deflection amplitude can be arbitrarily large, we must remember that the simple theory of bending we used to derive equation (8.9) assumes small deformations. An analysis of large amplitude buckling requires a more sophisticated treatment, but this is not necessary to predict the onset of buckling.

The results corresponding to the higher values of n in equation (8.7) are

$$P = \frac{4\pi^2 EI}{L^2} = 4P_c \qquad v = A \sin \frac{2\pi x}{L} \qquad (8.10)$$

and

$$P = \frac{9\pi^2 EI}{L^2} = 9P_c \qquad v = A \sin \frac{3\pi x}{L} \qquad (8.11)$$

and the sinusoidal deflected shapes are illustrated in Fig. 8.12. Under normal circumstances, such shapes cannot be achieved: they correspond to end forces much greater than the Euler critical one which itself is sufficient to cause buckling. The only exception to this is where a strut is subjected to lateral constraints forcing it into one of these forms. For example, if lateral deflection is prevented at the center of the strut, the deflected shape could be as in Fig. 8.12a.

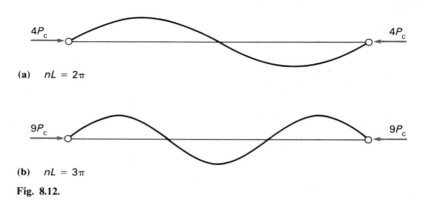

(a) $nL = 2\pi$

(b) $nL = 3\pi$

Fig. 8.12.

We may summarize our conclusions on the buckling of a perfect pin-ended strut as follows.

1 For $0 < P < P_c$, the Euler critical value given by equation (8.8), the condition $\sin nL = 0$ cannot be satisfied. Hence $A = 0$ and the strut is stable as a straight member in simple compression.

2 At $P = P_c$, a condition of neutral equilibrium is achieved: the strut can take up a deflected shape in the form of a half sine wave of arbitrary (small) amplitude.

3 For $P > P_c$, the strut is unstable and collapses, unless it happens to be laterally constrained in such a way as to compel it to take up a sinusoidal shape of shorter wavelength.

So far in our analysis, we have not identified the plane in which a particular strut will buckle, and therefore the neutral axis of bending about which we should take the second moment of area, I. If the strut is a circular rod or tube, the direction of buckling is arbitrary (unless the pin joints allow rotation in one plane only, which is then the plane of buckling). For any other cross-sectional shape, equation (8.8) shows that, because the critical buckling force is directly proportional to I, buckling will tend to occur in the plane of minimum I. As discussed in Section 5.3.4, this means that the neutral axis is one of the principal axes of the cross section.

EXAMPLE 8.2

An aluminum alloy member of a pin-jointed structure is 1.5 m (60 in) long and has a rectangular cross section 10 mm by 25 mm (0.4 in by 1 in). Find the compressive force at which it will buckle, and the corresponding compressive stress.

This example involves the straightforward application of equation (8.8) for the Euler critical buckling force. First we must find the minimum second moment of area of the cross section, which occurs about the axis through the center of the rectangular cross section parallel to the longer pair of sides.

SI units
With a cross section of breadth 0.025 m and depth 0.010 m

$$I = \frac{0.025 \times 0.010^3}{12} = 2.083 \times 10^{-9} \text{ m}^4$$

Since the length of the strut is $L = 1.5$ m, and the value of Young's modulus is $E = 68.9$ GN/m^2 (Appendix A), the critical buckling force is

$$P_c = \frac{\pi^2 EI}{L^2} = \frac{\pi^2 \times 68.9 \times 10^9 \times 2.083 \times 10^{-9}}{1.5^2}$$

$$= \underline{630 \text{ N}}$$

and the compressive stress is

$$\sigma = \frac{630}{0.025 \times 0.010} \text{ N/m}^2 = \underline{2.52 \text{ MN/m}^2}$$

which is very small in relation to the strength of the material.

US customary units
With a cross section of breadth 1 in and depth 0.4 in

$$I = \frac{1 \times 0.4^3}{12} = 5.333 \times 10^3 \text{ in}^4$$

Since the length of the strut is $L = 60$ in, and the value of Young's modulus is $E = 10 \times 10^6$ psi (Appendix A), the critical buckling force is

$$P_c = \frac{\pi^2 EI}{L^2} = \frac{\pi^2 \times 10 \times 10^6 \times 5.333 \times 10^{-3}}{60^2}$$

$$= \underline{146 \text{ lb}}$$

and the compressive stress is

$$\sigma = \frac{146}{1 \times 0.4} = \underline{365 \text{ psi}}$$

which is very small in relation to the strength of the material.

8.2.2 Eccentrically loaded struts

In arriving at the Euler critical buckling force, we assumed the strut concerned to be perfect, whereas real struts often have small but significant imperfections in either their geometries or the way in which they are loaded. One of the most common imperfections occurs when the resultant axial compressive force is applied not at the centroid of the strut cross section, but at some small distance from it, as in Fig. 8.13. The distance, e, of the line of action of the force P from the centroid, C, is referred to as the *eccentricity* of the loading. We can model this situation as shown in Fig. 8.14, and proceed with the analysis in much the same way as for a perfect strut.

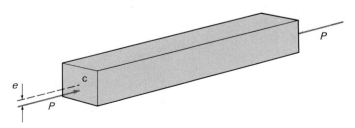

Fig. 8.13.

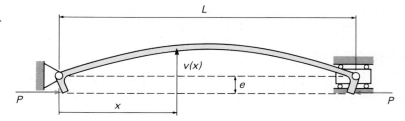

Fig. 8.14.

At a typical position at a distance x along the beam, the bending moment is $-P(v + e)$, and the moment–curvature relationship is

$$EI \frac{d^2v}{dx^2} = -P(v + e)$$

which may be rearranged as

$$\frac{d^2v}{dx^2} + n^2v = -n^2e \tag{8.12}$$

where $n^2 = P/EI$ is as defined in equation (8.5). While the left-hand side of this equation is identical to equation (8.4), the right-hand side is now not zero. The general mathematical approach to solving such a linear differential equation is to find the general solution as the sum of the *complementary function* and a *particular integral*. While the complementary function is the general solution to the *homogeneous* equation, which contains only the terms involving v or its derivatives (that is, setting the right-hand side of equation (8.12) to zero), a particular integral is any function which satisfies the full equation. In the case of equation (8.12), we already have the complementary function in equation (8.6), and a particular integral is $v = -e$. The general solution of equation (8.12) is therefore

$$v = A \sin nx + B \cos nx - e \tag{8.13}$$

containing as usual two constants of integration, A and B.

We find the values of A and B from the boundary conditions, $v = 0$ at $x = 0$, which gives

$$0 = A \times 0 + B \times 1 - e \quad \text{or} \quad B = e$$

and $v = 0$ at $x = L$, which gives

$$0 = A \sin nL + e \cos nL - e$$

or

$$A = \frac{e(1 - \cos nL)}{\sin nL}$$

Recalling that

$$1 - \cos nL = 2 \sin^2 \frac{nL}{2} \quad \text{and} \quad \sin nL = 2 \sin \frac{nL}{2} \cos \frac{nL}{2}$$

we have

$$A = \frac{2e \sin^2 \dfrac{nL}{2}}{2 \sin \dfrac{nL}{2} \cos \dfrac{nL}{2}} = e \tan \frac{nL}{2}$$

Substituting the expressions for A and B into equation (8.13), the deflected shape of the strut is given by

$$v = e \left(\tan \frac{nL}{2} \sin nx + \cos nx - 1 \right) \tag{8.14}$$

The value of the maximum deflection, which occurs at $x = L/2$, is

$$v_{\text{max}} = e \left(\tan \frac{nL}{2} \sin \frac{nL}{2} + \cos \frac{nL}{2} - 1 \right)$$

or

$$v_{\text{max}} = e \left(\frac{\sin^2 \dfrac{nL}{2} + \cos^2 \dfrac{nL}{2}}{\cos \dfrac{nL}{2}} - 1 \right) = e \left(\sec \frac{nL}{2} - 1 \right) \tag{8.15}$$

Substituting for n from equation (8.5) we have

$$v_{\text{max}} = e \left[\sec \left(\frac{L}{2} \sqrt{\frac{P}{EI}} \right) - 1 \right] \tag{8.16}$$

The strut collapses when this maximum deflection becomes infinite. The lowest value of force P at which this can occur is given by

$$\frac{L}{2} \sqrt{\frac{P}{EI}} = \frac{\pi}{2}$$

or

$$P = P_c = \frac{\pi^2 EI}{L^2} \tag{8.17}$$

which is the Euler critical buckling force we derived in equation (8.8). So, the eccentricity of the loading does not affect the critical force at which buckling occurs. It does, however, affect the deformation of the strut at lower loads. Equation (8.17) shows that the strut does deflect at all values of P: we can rewrite this equation in terms of the critical value of force as

$$v_{\max} = e\left[\sec\left(\frac{\pi}{2}\sqrt{\frac{P}{P_c}}\right) - 1\right] \tag{8.18}$$

Figure 8.15 shows this relationship between force and maximum deflection, together with the result for the perfect strut for comparison. The effect of reducing the eccentricity towards zero is shown by the dotted lines. The behavior is very similar to that of the rod and spring system represented in Fig. 8.7: again the imperfection removes the abrupt change at the critical value of force.

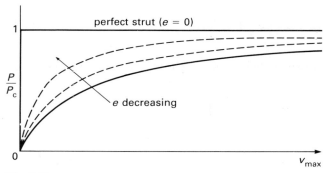

Fig. 8.15.

The fact that an eccentrically loaded strut experiences significant lateral deflections at all levels of axial force means that it also suffers bending stresses, not only due to the initial eccentricity of the force (which we saw how to deal with in Section 5.4) but also due to the additional deflections. This is in contrast to the other types of problems we have considered, where the deformations have negligible effects on geometry and loading. The maximum bending moment in an eccentrically loaded pin-ended strut occurs where the deflection is greatest, at its midpoint, and is equal to $P(v_{\max} + e)$, or

$$M_{\max} = Pe\,\sec\left(\frac{\pi}{2}\sqrt{\frac{P}{P_c}}\right)$$

Consequently, the maximum compressive stress, which also occurs at the cross section at the center of the strut, is

$$\sigma_{\max} = -\frac{P}{A} - c\frac{M_{\max}}{I} = -P\left[\frac{1}{A} + \frac{ce}{I}\sec\left(\frac{\pi}{2}\sqrt{\frac{P}{P_c}}\right)\right] \tag{8.19}$$

where A is the cross-sectional area of the strut and c is the distance of the greatest compressive stress from the neutral axis of the cross section. If we know the value of the axial force P, it is a straightforward matter to calculate the maximum stress. If we are given the maximum stress, however, it is more difficult to find the force, because P occurs both as a linear multiplier on the right-hand side of the equation, and within the secant function. It can, however, be found numerically with the aid of a computer

EXAMPLE 8.3

A member of a pin-jointed structure takes the form of a solid steel rod 2 m (6.6 ft) long and 25 mm (1 in) in diameter. If the eccentricity of the axial compressive force it carries is 5 mm (0.2 in), find the magnitudes of this force at which the member will buckle and at which the maximum allowable compressive stress of 100 MN/m² (15 ksi) is reached.

We first find the critical buckling force from the Euler formula, equation (8.17) or (8.8). To do this we need the second moment of area in bending for a rod of diameter $D = 25$ mm, which is

$$I = \frac{\pi D^4}{64} = \frac{\pi \times 0.025^4}{64} = 1.917 \times 10^{-8} \text{ m}^4$$

Hence

$$P_c = \frac{\pi^2 EI}{L^2} = \frac{\pi^2 \times 207 \times 10^9 \times 1.917 \times 10^{-8}}{2.0^2} \text{ N} = \underline{9.79 \text{ kN}} \text{ (2.2 kip)}$$

Rearranging equation (8.19), we have

$$\frac{A\sigma_{max}}{P_c} = \frac{P}{P_c} \left[1 + \frac{ceA}{I} \sec \left(\frac{\pi}{2} \sqrt{\frac{P}{P_c}} \right) \right]$$

which, with $e = 5$ mm, $c = 12.5$ mm, $\sigma_{max} = 100$ MN/m²,

$$A = \frac{\pi D^2}{4} = \frac{\pi \times 0.025^2}{4} = 4.909 \times 10^{-4} \text{ m}^2$$

$$\frac{A\sigma_{max}}{P_c} = \frac{4.909 \times 10^{-4} \times 100 \times 10^6}{9.79 \times 10^3} = 5.014$$

and

$$\frac{ceA}{I} = \frac{0.0125 \times 0.005 \times 4.909 \times 10^{-4}}{1.917 \times 10^{-8}} = 1.600$$

becomes

$$X \left[1 + 1.6 \sec \left(\frac{\pi}{2} \sqrt{X} \right) \right] - 5.014 = 0 \qquad (8.20)$$

where $X = P/P_c$. Methods for solving such a nonlinear algebraic equation are discussed in Appendix E, where a computer program named ROOT is used to find the value of the unknown X in the required range, $0 < X < 1$, of

$$X = 0.6956$$

From this value of X we can determine the force required to generate the maximum allowable stress as

$$P = XP_c = 0.6956 \times 9.79 = \underline{6.81 \text{ kN}} \ (1.5 \text{ kip})$$

8.2.3 Initially curved struts

Another form of imperfection of struts is lack of straightness before the loading is applied. Figure 8.16 shows a model of this situation with the strut in its deflected state, together with the dotted line indicating the initial position of the neutral surface. The form of this initial curved shape is arbitrary. A typical one, and one which is relatively straightforward to analyze, is given by

$$v_0(x) = a_0 \sin \frac{\pi x}{L}$$

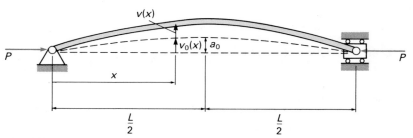

Fig. 8.16.

with a maximum deflection of a_0 at the center of the strut. When the axial load P is applied, its effect is to increase the deflection everywhere along the strut: at position x from $v_0(x)$ to $v_0(x) + v(x)$, giving a bending moment there of $-P(v_0(x) + v(x))$. The moment–curvature relationship is therefore

$$EI \frac{d^2v}{dx^2} = -P\left(v + a_0 \sin \frac{\pi x}{L} \right)$$

which may be rearranged as

$$\frac{d^2v}{dx^2} + n^2v = -n^2 a_0 \sin \frac{\pi x}{L} \tag{8.21}$$

where $n^2 = P/EI$ is as defined in equation (8.5). As in the case of an eccentrically loaded strut, the difference between this equation and equation (8.4) for the perfect strut lies in the form of the right-hand side. The complementary function forming one part of the solution is therefore given by equation (8.6), and a particular integral is

$$v^* = \frac{n^2 a_0 \sin \dfrac{\pi x}{L}}{\dfrac{\pi^2}{L^2} - n^2}$$

giving the general solution as

$$v = A \sin nx + B \cos nx + v^* \tag{8.22}$$

We find the values of integration constants A and B from the boundary conditions, $v = 0$ at $x = 0$, which gives

$$0 = A \times 0 + B \times 1 + 0 \quad \text{from which} \quad B = 0$$

and $v = 0$ at $x = L$, which gives

$$0 = A \sin nL + 0 \times \cos nL + 0 \quad \text{from which} \quad A \sin nL = 0$$

There are two possibilities: either (i) $\sin nL = 0$, in which case we obtain the Euler critical buckling force as in equations (8.7) and (8.8), or (ii) $A = 0$. In the latter case we are left with only the particular integral, which can be expressed as

$$v = \frac{a_0 \sin \dfrac{\pi x}{L}}{(P_c/P - 1)} = \frac{v_0}{(P_c/P - 1)} \tag{8.23}$$

where $P_c = \pi^2 EI/L^2$ is the Euler critical buckling force. The strut collapses when the deflection becomes infinite, which once again occurs when $P = P_c$. Figure 8.17 shows the relationship between force and maximum deflection (at $x = L/2$), together with the result for the perfect strut for comparison. The

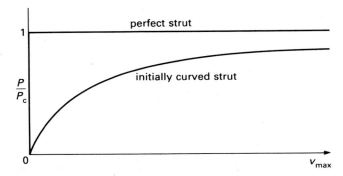

Fig. 8.17.

behavior is very similar in form to that of an eccentrically loaded strut shown in Fig. 8.15, also the rod and spring system represented in Fig. 8.7, with the imperfection removing the abrupt change at the critical value of force.

8.2.4 Effects of lateral loading

While imperfections in the form of eccentric loading or initial curvature do influence the deformation of an axially loaded strut, they do not affect the critical value of force which causes collapse. We now need to determine what effect lateral loading, in addition to the axial loading, has on the buckling behavior.

Figure 8.18 shows a pin-ended strut AB of length L subjected to an axial force P and a concentrated lateral force of F at its midpoint. The direction of this force, either upward or downward, is not important, provided we take the beam deflection to be in the same direction – which it certainly would be in practical situations. For convenience, we take both to be upward, the direction of positive deflection according to our sign convention for beam bending. Since the strut and its loading are symmetrical about its center, the reaction forces at the supports A and B are

$$R_A = R_B = \frac{F}{2}$$

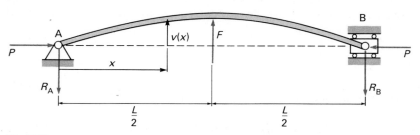

Fig. 8.18.

Also due to the symmetry, we need only consider one half, say $0 < x < L/2$, which avoids the use of a step function. In this range, the moment–curvature relationship is given by

$$EI \frac{d^2v}{dx^2} = -Pv - \frac{Fx}{2}$$

which may be rearranged as

$$\frac{d^2v}{dx^2} + n^2v = -\frac{Fx}{2EI}$$

where as usual $n^2 = P/EI$. The general solution, composed of complementary function (8.6) and a particular integral, is

$$v = A \sin nx + B \cos nx - \frac{Fx}{2EIn^2} \tag{8.24}$$

We find the values of the integration constants A and B from the boundary conditions, one of which is $v = 0$ at $x = 0$, which gives $B = 0$. The other is the symmetry condition $dv/dx = 0$ at $x = L/2$ (*not* $v = 0$ at $x = L$, which is outside the range of x for which the moment–curvature equation is valid). Differentiating equation (8.24) to obtain the general expression for the slope

$$\frac{dv}{dx} = An \cos nx - Bn \sin nx - \frac{F}{2P}$$

and applying the symmetry condition gives

$$0 = An \cos \frac{nL}{2} - \frac{F}{2P}$$

from which

$$A = \frac{F}{2Pn} \sec \frac{nL}{2}$$

The maximum deflection, at $x = L/2$, is therefore given by equation (8.24) as

$$v_{max} = \frac{F}{2Pn} \left(\tan \frac{nL}{2} - \frac{nL}{2} \right) \tag{8.25}$$

The strut collapses when this maximum deflection becomes infinite. The lowest value of force P at which this can occur is given by

$$\frac{nL}{2} = \frac{L}{2} \sqrt{\frac{P}{EI}} = \frac{\pi}{2}$$

from which

$$P = P_c = \frac{\pi^2 EI}{L^2} \tag{8.26}$$

which once again is the Euler critical buckling force. While the presence of a concentrated lateral force does not affect the final collapse condition for the strut, it does affect the deformation at lower levels of axial loading.

EXAMPLE 8.4

Show that a uniformly distributed lateral force applied over the length of a pin-ended strut does not affect the critical value of the axial force at which buckling occurs.

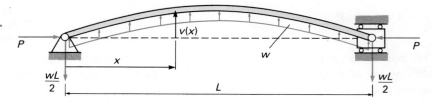

Fig. 8.19.

Figure 8.19 shows the strut subjected to an axial force P and a uniformly distributed lateral force of intensity w. Symmetry ensures that the reaction forces at the supports are each equal to half the total lateral load. The moment–curvature relationship, which is valid for the whole strut, is

$$EI\frac{d^2v}{dx^2} = -Pv - \frac{wLx}{2} + \frac{wx^2}{2}$$

which may be rearranged as

$$\frac{d^2v}{dx^2} + n^2v = -\frac{wLx}{2EI} + \frac{wx^2}{2EI}$$

where $n^2 = P/EI$. The general solution is

$$v = A \sin nx + B \cos nx + \frac{w}{2EIn^2}\left(x^2 - xL - \frac{2}{n^2}\right) \tag{8.27}$$

The constants A and B are found from the boundary conditions $v = 0$ at $x = 0$, which gives

$$0 = B + \frac{w}{2P}\left(-\frac{2}{n^2}\right)$$

from which $B = w/Pn^2$, and $v = 0$ at $x = L$, which gives

$$0 = A \sin nL + \frac{w}{Pn^2}\cos nL + \frac{w}{2P}\left(-\frac{2}{n^2}\right)$$

from which

$$A = \frac{w}{Pn^2}(\operatorname{cosec} nL - \cot nL) = \frac{w}{Pn^2}\tan\frac{nL}{2}$$

The maximum deflection, at $x = L/2$, is therefore given by equation (8.27) as

$$v_{max} = \frac{w}{Pn^2}\left(\tan\frac{nL}{2}\sin\frac{nL}{2} + \cos\frac{nL}{2}\right) + \frac{w}{2P}\left(\frac{L^2}{4} - \frac{L^2}{2} - \frac{2}{n^2}\right)$$

which simplifies to

$$v_{max} = \frac{w}{Pn^2}\left(\sec\frac{nL}{2} - 1\right) - \frac{wL^2}{8P} \tag{8.28}$$

The strut collapses when this maximum deflection becomes infinite. The lowest value of force P at which this can occur is given by

$$\frac{nL}{2} = \frac{L}{2} \sqrt{\frac{P}{EI}} = \frac{\pi}{2}$$

from which

$$P = P_c = \frac{\pi^2 EI}{L^2}$$

which is the Euler critical buckling force, and independent of the lateral loading.

8.3 Struts and columns with other end conditions

We have seen that the critical buckling force for a pin-ended strut is not affected by eccentric loading, initial curvature of the strut, or lateral loading. Also of interest, however, are the effects of changing the support conditions at the ends. In terms of practical applications, we are now moving away from pin-jointed structures towards, for example, columns used to support buildings or other structures. Figure 8.20 shows two such column arrangements, both built-in at their lower ends and respectively free (unsupported) and built-in at their upper ends.

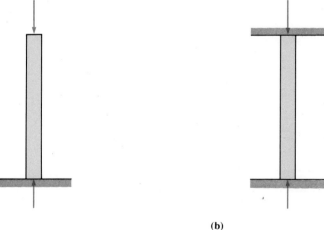

(a)   (b)

Fig. 8.20.

A number of types of end conditions are possible. In Section 5.1.1 we identified the three basic types of supports for statically determinate beams as pinned, roller and built-in. Indeed, we have again encountered both pinned and roller supports when considering pin-ended struts. In the case of roller supports, we represented them pictorially not as single-sided rollers (Fig. 5.5b) but as double-sided rollers (Figs 8.11, 8.14, 8.16, 8.18 and 8.19) in order to emphasize that they prevent lateral deflection in both directions. Apart from pinned or roller supports at both ends of a column, which we have already considered, the three main combinations of end conditions of practical interest are

1   built-in and free
2   built-in and pinned
3   built-in at both ends

The first of these is statically determinate, the other two statically indeterminate. In the case of a statically indeterminate column, we have the further possibility that the deflections and/or slopes at the supports need not be zero, also that the supports may have finite lateral and rotational stiffnesses. Let us first consider the three main combinations defined above.

8.3.1 Column with ends built-in and free

Following our usual practice of illustrating beams in the horizontal position, Fig. 8.21 shows a column of length L with one end rigidly built-in and the other free, subjected to an axial compressive force P. As usual, we show the column deflected from its initial straight condition, following a shape which has yet to be defined. Let δ be the lateral deflection at the free end of the column. Consequently, at a distance x from the fixed end the bending moment (found as the moment of the external force acting to the right of the cross section concerned) is $P(\delta - v)$, and the moment–curvature relationship is

$$EI\,\frac{\mathrm{d}^2v}{\mathrm{d}x^2} = P(\delta - v)$$

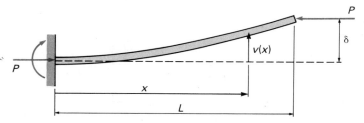

Fig. 8.21.

which may be rearranged as

$$\frac{\mathrm{d}^2v}{\mathrm{d}x^2} + n^2v = n^2\delta \tag{8.29}$$

where $n^2 = P/EI$. This equation is very similar in form to equation (8.12) for an eccentrically loaded pin-ended strut. The general solution for deflection is

$$v = A \sin nx + B \cos nx + \delta \tag{8.30}$$

from which the slope may be obtained as

$$\frac{dv}{dx} = An \cos nx - Bn \sin nx \qquad (8.31)$$

The boundary conditions for defining constants A and B are zero slope at $x = 0$, which substituted into equation (8.31) gives $A = 0$, and $v = 0$ at $x = 0$, which in equation (8.30) gives $B = -\delta$. Hence

$$v = \delta(1 - \cos nx)$$

When we introduced δ we merely defined it to be the deflection at the free end: we must now impose this requirement on the solution

$$\delta = \delta(1 - \cos nL) \quad \text{from which} \quad \delta \cos nL = 0$$

There are two quite distinct ways in which this equation can be satisfied: either $\delta = 0$ or $\cos nL = 0$. If $\delta = 0$, there is no lateral deflection of the column, which is the stable condition for the system. If $\cos nL = 0$, there are an infinite number of possible values of n, the smallest of which is

$$nL = L\sqrt{\frac{P}{EI}} = \frac{\pi}{2}$$

from which

$$P = \frac{\pi^2 EI}{4L^2} \qquad (8.32)$$

The deflected shape corresponding to this force is

$$v = \delta\left(1 - \cos\frac{\pi x}{2L}\right) \qquad (8.33)$$

The strut deflection follows a sinusoidal shape, with one quarter of a complete cycle between $x = 0$ and $x = L$, the length of the strut, as in Fig. 8.21. However, the amplitude, δ, of the deflected shape remains undefined. In other words, a state of neutral equilibrium has been achieved, and the value of end force given by equation (8.32) is a critical value necessary to cause instability.

The buckling behavior of the column is very similar to that of the perfect pin-ended strut we investigated in Section 8.2.1. The only significant differences are that the critical buckling force is reduced by a factor of four, and that the deflected shape now forms one quarter of a sine wave rather than one half. These two facts are closely related. To appreciate why this is so, we can think of the present column together with its mirror image in the wall at its built-in end, as shown in Fig. 8.22. The member so formed would be effectively a pin-ended

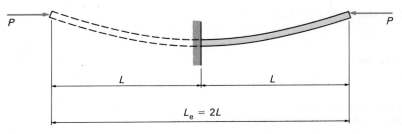

Fig. 8.22.

strut of length $2L$ taking the form of a half sine wave, and having a lowest critical buckling force according to equation (8.8) of

$$P_c = \frac{\pi^2 EI}{(2L)^2} = \frac{\pi^2 EI}{4L^2}$$

which is precisely the result we obtained in equation (8.32). Therefore, another way of expressing our present result is in terms of the *effective length of the equivalent pin-ended strut*, L_e, having the same critical buckling force, as

$$P_c = \frac{\pi^2 EI}{L_e^2} \quad \text{where} \quad L_e = 2L \tag{8.34}$$

8.3.2 Column with both ends built-in

Figure 8.23 shows a column of length L with both ends built-in and subjected to an axial compressive force P. With zero slope at each end, we can anticipate that the simplest form of deflected shape, corresponding to the lowest critical buckling force, follows one complete sine wave as shown (with peaks at each end, and a peak in the opposite direction at the center of the column). Since the deflected shape of a pin-ended strut represents only one half of a sine wave, we can expect the effective length in this case to be $L_e = L/2$. Now let us see whether a more detailed treatment confirms this result.

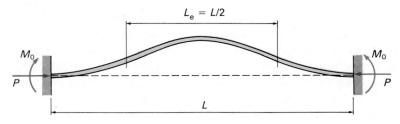

Fig. 8.23.

The problem is statically indeterminate and we must include in our analysis the reaction moments M_0 at the ends of the column as shown. The symmetry of the system and the absence of external lateral forces means, however, that there are no lateral reaction forces at the supports. The moment–curvature relationship can be expressed as

$$EI \frac{d^2v}{dx^2} = M_0 - Pv$$

which may be rearranged as

$$\frac{d^2v}{dx^2} + n^2 v = \frac{M_0}{EI}$$

where $n^2 = P/EI$. The general solution for deflection is

$$v = A \sin nx + B \cos nx + \frac{M_0}{P} \tag{8.35}$$

and for slope

$$\frac{dv}{dx} = An \cos nx - Bn \sin nx \tag{8.36}$$

We now use the boundary conditions to find the constants A and B and the moment M_0 at the supports. Firstly, the slope is zero at $x = 0$, which when substituted into equation (8.36) gives $A = 0$. Secondly, $v = 0$ at $x = 0$, which in equation (8.35) gives $B = -M_0/P$. Thirdly, $v = 0$ at $x = L$, which gives

$$\frac{M_0}{P}(1 - \cos nL) = 0$$

Either $M_0 = 0$ (which corresponds to the stable state of no deflection), or

$$\cos nL = 1 \quad \text{with} \quad nL = 0, 2\pi, 4\pi, \ldots \tag{8.37}$$

Incidentally, had we used the fourth boundary condition of zero slope at $x = L$, we would have found

$$Bn \sin nL = 0 \quad \text{with either} \quad nL = 0, \pi, 2\pi, 3\pi, \ldots$$

or $B = 0$ (no deflection) or $n = 0$ (no load). This result for nL gives all the solutions defined by equation (8.37), together with some (π, 3π and so on) which the latter does not permit.

Equation (8.37) is therefore the one we must use to find the critical buckling force, the lowest nonzero value of which is given by

$$nL = L\sqrt{\frac{P}{EI}} = 2\pi$$

and

$$P = \frac{4\pi^2 EI}{L^2} \tag{8.38}$$

We can express this result in terms of an effective length as

$$P_c = \frac{\pi^2 EI}{L_e^2} \quad \text{where} \quad L_e = \frac{L}{2} \tag{8.39}$$

As we anticipated, the effective length (of the equivalent pin-ended strut) is half the length of the column. Also, substituting the derived expressions for A, B and n into equation (8.35), we obtain the deflected shape of the column as

$$v = \frac{M_0}{P}\left(1 - \cos\frac{2\pi x}{L}\right) \tag{8.40}$$

which, as anticipated, follows one complete sine wave between $x = 0$ and $x = L$, as shown in Fig. 8.23.

8.3.3 Column with ends built-in and pinned

Figure 8.24 shows a column of length L with one end built-in, the other pinned, and subjected to an axial compressive force P. With such different conditions at each end, the deflected shape cannot follow a simple sine wave, and we cannot readily predict the effective length. We can, however, determine upper and lower bounds for it. From the examples we have considered so far, we see that the effect of increasing the degree of constraint at the ends of a column, from free to pinned to built-in, is to progressively reduce its effective length (and increase the critical buckling force), from $2L$ to L to $L/2$. Now, we can anticipate that the present column with one end built-in and the other pinned will have an effective length lying between those of columns with both ends built-in and both ends pinned. In other words, $L/2 < L_e < L$, and L_e probably lies near the middle of this range.

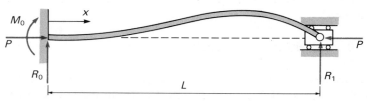

Fig. 8.24.

The problem is statically indeterminate and we must include in our analysis both the reaction moment M_0 and the reaction force R_0 at the built-in end of the column as shown. The reaction force R_1 at the pinned end can be found if required by considering equilibrium in the lateral direction ($R_1 = -R_0$). If the column is in equilibrium, we may take moments about the pinned end to give

$$M_0 + R_0 L = 0 \quad \text{from which} \quad R_0 = -\frac{M_0}{L}$$

The moment–curvature relationship can be expressed as

$$EI \frac{\mathrm{d}^2 v}{\mathrm{d}x^2} = M_0 + R_0 x - Pv$$

which may be rearranged as

$$\frac{\mathrm{d}^2 v}{\mathrm{d}x^2} + n^2 v = \frac{M_0}{EI}\left(1 - \frac{x}{L}\right)$$

where $n^2 = P/EI$. The general solution for deflection is

$$v = A \sin nx + B \cos nx + \frac{M_0}{P}\left(1 - \frac{x}{L}\right) \tag{8.41}$$

and for slope

$$\frac{\mathrm{d}v}{\mathrm{d}x} = An \cos nx - Bn \sin nx - \frac{M_0}{PL} \tag{8.42}$$

We now use the boundary conditions to find the constants A and B and the moment M_0. Firstly, the slope is zero at $x = 0$, which when substituted into equation (8.42) gives $A = M_0/nPL$. Secondly, $v = 0$ at $x = 0$, which in equation (8.41) gives $B = -M_0/P$. Thirdly, $v = 0$ at $x = L$, which gives

$$\frac{M_0}{P}\left(\frac{\sin nL}{nL} - \cos nL\right) = 0$$

Either $M_0 = 0$ (which corresponds to the stable state of no deflection), or

$$\tan nL = nL \tag{8.43}$$

Methods for solving such a nonlinear algebraic equation are discussed in Appendix E. In order to use the program named ROOT which is given there, we need to provide a range for the unknown nL in which we expect the required solution to be found. Figure 8.25 shows plots of the functions $y = nL$ and $y = \tan nL$: where the curves cross are the roots of equation (8.43). Apart from $nL = 0$ (which is the trivial case of no axial force and no deflection), the lowest such value is just less than $3\pi/2$ (that is, 4.712). A suitable range is therefore $4.0 < nL < 4.7$ (avoiding the infinite value of the tangent function at $3\pi/2$).

Program ROOT gives the required solution as $nL = 4.493$, from which we obtain the critical value of force as

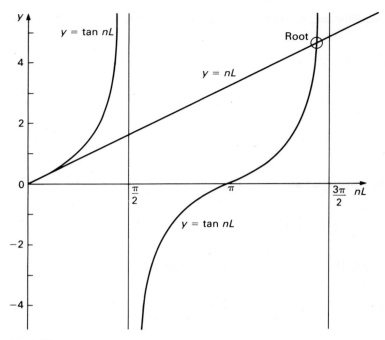

Fig. 8.25.

$$P_c = EIn^2 = EI\left(\frac{4.493}{L}\right)^2 = 20.19\frac{EI}{L^2} \tag{8.44}$$

We can express this result in terms of an effective length as

$$P_c = \frac{\pi^2 EI}{L_e^2} \quad \text{where} \quad L_e = 0.699L \approx 0.7L \tag{8.45}$$

As we anticipated, the effective length is approximately half-way between $L/2$ and L.

8.3.4 More general cases of column buckling

In the last three Sections, we determined expressions for the critical buckling forces of columns with the three main combinations of end conditions, and related these to the result for a pin-ended strut via effective lengths. While we assumed the columns to be initially perfectly straight and the end conditions to be perfect, with a pinned support imposing exactly zero lateral deflection, and a built-in support imposing exactly zero deflection and slope, such perfection is never achieved in practice. If we were to examine the effects of these imperfections we would find, just as we did for a pin-ended strut in Sections 8.2.2 and 8.2.3, that while the deformations prior to collapse are affected, the critical value of the axial force needed to cause collapse is not. The same is true of lateral loading.

One form of imperfection which can have a profound effect on the collapse condition is lack of perfect rigidity of the supports in the lateral direction and rotational sense. The following two examples deal with the particular cases of a column with ends built-in and pinned, with lateral flexibility at the pinned end, and one with ends built-in and free, with rotational flexibility at the built-in end.

EXAMPLE 8.5

Figure 8.26a shows a column of length L subjected to an axial compressive force P. One end is built-in, and the other is pinned, with the pinned joint being constrained in the lateral direction by a spring of stiffness k. There is no force in the spring when the column is straight. Derive an algebraic equation from which the critical value of force P to cause buckling can be found.

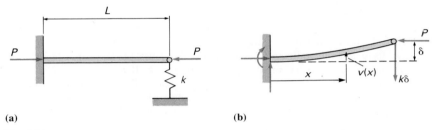

(a)   (b)

Fig. 8.26.

Although we cannot predict the effective length of the column with any degree of accuracy, we can anticipate that it will lie between that of the rigidly pinned case (Section 8.3.3) when k is infinite, and the free end case (Section 8.3.1) when k is zero. Figure 8.26b shows the forces and moments acting on the column. Let the lateral deflection at the pinned end be δ: the force in the spring is therefore $k\delta$. Considering the bending moment at a typical cross section of the strut at a distance x from the built-in end, we can write the moment–curvature relationship as

$$EI\frac{d^2v}{dx^2} = P(\delta - v) - k\delta(L - x)$$

or

$$\frac{d^2v}{dx^2} + n^2v = n^2\delta - \frac{k\delta}{EI}(L - x)$$

where $n^2 = P/EI$. The general solution for deflection is

$$v = A\sin nx + B\cos nx + \delta - \frac{k\delta}{P}(L - x) \tag{8.46}$$

and for slope

$$\frac{dv}{dx} = An\cos nx - Bn\sin nx + \frac{k\delta}{P} \tag{8.47}$$

The boundary condition of zero slope at $x = 0$ applied to equation (8.47) gives $A = -k\delta/Pn$, and that of $v = 0$ at $x = 0$ applied to equation (8.46) gives

$$B = \delta\left(\frac{kL}{P} - 1\right)$$

Also, $v = \delta$ at $x = L$, which in equation (8.46) gives

$$\delta = -\frac{k\delta}{Pn}\sin nL + \delta\left(\frac{kL}{P} - 1\right)\cos nL + \delta$$

from which, for $\delta \neq 0$

$$\tan nL = nL\left(1 - \frac{P}{kL}\right) \tag{8.48}$$

This nonlinear algebraic equation can be solved numerically to find P (which appears both explicitly and in the parameter n), just as we did for equation (8.43) in Section 8.3.3 (using the computer program given in Appendix E). To do this, we need numerical values for k, L and the flexural rigidity EI.

As we have already anticipated, equation (8.48) gives the appropriate limiting cases for very large and very small values of stiffness k. If $k \to \infty$, equation (8.48) reduces to equation (8.43) for a pinned end with no lateral flexibility, whose solution is $nL = 4.493 = 1.43\pi$. If $k \to 0$, equation (8.48) reduces to $\tan nL = -\infty$, $nL = \pi/2$ and P is given by equation (8.32) for a pinned end with no lateral constraint. For intermediate values of k, we can therefore expect that $0.5\pi < nL < 1.43\pi$.

EXAMPLE 8.6

Figure 8.27a shows a column of length L subjected to an axial compressive force P. One end is built-in, but with a rotational stiffness there of λ, and the other is free. Derive an algebraic equation from which the critical value of force P to cause buckling can be found.

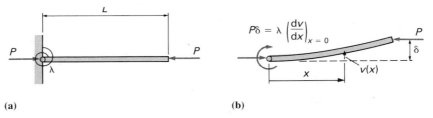

(a)　　　　　　　　　　　　　　　　(b)

Fig. 8.27.

Although we cannot predict the critical buckling force for the column with any degree of accuracy, we can anticipate that it will lie between that of the rigidly built-in case (Section 8.3.1) when λ is infinite, and that of a rigid member with a rotational spring at one pin-jointed end (similar to Example 8.1) when λ is small. Figure 8.27b shows the forces and moments acting on the column. Let the lateral deflection at the pinned end be δ. For the column to be in equilibrium, the couple applied by the spring, which is the product of λ and the slope of the column at the pinned end, must be equal to $P\delta$. The moment–curvature relationship and the general solution to the resulting differential equation are identical to those obtained in Section 8.3.1, especially equations (8.30) and (8.31)

$$v = A \sin nx + B \cos nx + \delta \tag{8.30}$$

$$\frac{\mathrm{d}v}{\mathrm{d}x} = An \cos nx - Bn \sin nx \tag{8.31}$$

The boundary condition of $v = 0$ at $x = 0$ gives $B = -\delta$, and that of slope at $x = 0$ being equal to $P\delta/\lambda$ gives $A = P\delta/\lambda n$. Also, since $v = \delta$ at $x = L$

$$\delta = \frac{P\delta}{\lambda n} \sin nL - \delta \cos nL + \delta$$

from which, for $\delta \neq 0$

$$\tan nL = \frac{\lambda n}{P} \tag{8.49}$$

Provided we have numerical values for λ, L and EI, this equation can be solved numerically to find P (using, for example, the computer program given in Appendix E).

As we have already anticipated, equation (8.49) gives the appropriate limiting cases for very large and very small values of stiffness λ. If $\lambda \to \infty$, $\tan nL \to \infty$, and $nL \to \pi/2$, which is the result for a rigidly built-in column (Section 8.3.1). If $\lambda \to 0$, $\lambda n/P$ and therefore $\tan nL$ are small, and

$$nL \approx \frac{\lambda n}{P} \quad \text{or} \quad P \approx \frac{\lambda}{L} \tag{8.50}$$

This result is similar to the one obtained in Example 8.1 for two rigid rods pinned together with a rotational spring between them. Indeed, the results for critical force to cause collapse differ only by a factor of two ($P = 2\lambda/L$ as compared with equation (8.50)). This factor arises because, for a given angle of rotation at the spring of an individual member, the couple generated in the spring in Fig. 8.8 is twice that in Fig. 8.27. Therefore, what the limiting case defined by equation (8.50) means physically is that when λ is very small, the stiffness of the column is so much greater than that of the spring that the system behaves as though the column is perfectly rigid. When λ is truly zero, equation (8.49) gives $\tan nL = 0$ and $nL = P = 0$: the column is pin-jointed at one end only, becoming a mechanism incapable of carrying any axial force. For intermediate values of λ, we can conclude that $0 < nL < 0.5\pi$.

8.3.5 Failure of real struts and columns

We have seen that the critical axial loads at which struts and columns buckle are influenced mainly by the support conditions at their ends, although imperfections and lateral loading do cause some deformation before critical conditions are reached. Also, we can relate the critical buckling force for any member to that of a simple pin-ended strut by means of an effective length. Equations (8.34), (8.39) and (8.45) all take the form of

$$P_c = \frac{\pi^2 EI}{L_e^2} \tag{8.51}$$

with differing interpretations for the effective length, L_e, in relation to the physical length of the particular column concerned. In place of the bending second moment of area, I, we can write Ar^2, where A is the area of the column cross section and r is its *radius of gyration* (see Appendix C, equation (C.13)). In the same way that I is the minimum second moment of area for the section, r is its minimum radius of gyration, about one of the principal axes of the cross section. Equation (8.51) becomes

$$P_c = \frac{\pi^2 EAr^2}{L_e^2}$$

and we can write the *critical buckling stress* (the mean compressive stress over the cross section when subjected to the critical buckling force) as

$$\sigma_c = \frac{P_c}{A} = \frac{\pi^2 E}{(L_e/r)^2} \tag{8.52}$$

where L_e/r is the *slenderness ratio*. Plotting critical stress against slenderness ratio, as in Fig. 8.28a, we obtain a curve which is sometimes referred to as the *Euler hyperbola*. For a very long slender strut or column, the critical stress is small, while for a short one it is high. At very short lengths, the critical stress according to the Euler hyperbola may be higher than the material can withstand in compression without yielding. This means that the member would fail not by buckling but by yielding, at yield stress σ_Y, and the upper part of the hyperbola which is shown as a dotted line is not of practical significance. Figure 8.28a in effect provides a failure criterion for struts and columns: whenever the mean stress in a member in compression lies above the envelope defined by the solid line AB and curve BC, failure will occur, the mechanism of failure being either yielding or elastic buckling according to the slenderness ratio. In practice, Fig. 8.28a provides a very idealized description of failure, particularly in the transition region from yielding to buckling at intermediate slenderness ratios. Figure 8.28b provides a more typical empirical representation of the behavior of real struts and columns, based on experimental data. This takes into account both imperfections and the effects of combined yielding and buckling. For short columns, it is reasonable to assume that failure occurs by yielding in uniform compression. While long columns fail by elastic buckling, it is advisable to use the Euler formula only in conjunction with substantial safety factors. For columns of intermediate slenderness ratio,

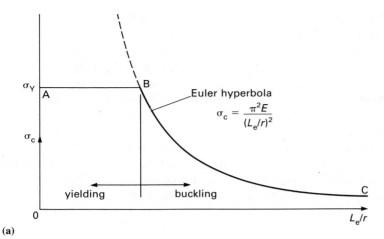

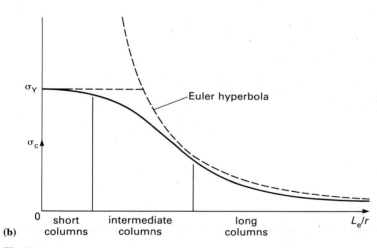

Fig. 8.28.

empirical formulae appropriate for different classes of materials, and which incorporate appropriate safety factors, are available in design codes and national Standards.

We are also now in a position to generalize the results of our analysis of eccentrically loaded pin-ended struts. Replacing I by Ar^2, and using equation (8.51) to define P_c, we can express equation (8.19) in the form

$$
\frac{P}{A} = \frac{|\sigma_{max}|}{1 + \dfrac{ce}{r^2} \sec\left(\dfrac{L_e}{2r}\sqrt{\dfrac{P}{EA}}\right)} \tag{8.53}
$$

This is often referred to as the *secant formula*. It defines the force per unit area which causes a particular maximum compressive stress, σ_{max}, in a column of effective slenderness ratio L_e/r, for a given value of the eccentricity ratio ce/r^2, e being the eccentricity of the axial force P, and c the distance of the greatest compressive stress from the neutral axis. Setting σ_{max} equal to the yield stress, σ_Y, Fig. 8.29 shows typical curves of P/A plotted against slenderness ratio for various values of the eccentricity ratio ce/r^2. These curves define the onset of yielding rather than buckling. For long slender columns, the effect of eccentric loading is relatively small, but for short columns it is much greater. The main difficulty in using these results in practice is that it is often difficult to estimate the eccentricity with sufficient accuracy. Another point to note is that the radius of gyration, r, must be for bending in the plane of the eccentricity. If this is not the minimum radius of gyration, then, despite the eccentricity, buckling in the other plane may represent the more critical condition.

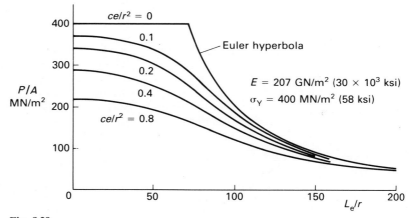

Fig. 8.29.

EXAMPLE 8.7

A member of a pin-jointed structure takes the form of a 50 mm (2 in) diameter steel rod. If the yield stress of the material is 300 MN/m² (44 ksi), find the length of rod at which failure by yielding and buckling according to the Euler formula is equally likely when the rod is in compression.

In this example, we are concerned with the point B in Fig. 8.28a where the Euler hyperbola BC meets the horizontal straight line AB representing the yield stress, which is where the critical buckling stress is equal to the yield stress. Therefore, from equation (8.52), the corresponding slenderness ratio is given by

SI units

$$\frac{L_e}{r} = \pi \sqrt{\frac{E}{\sigma_Y}} = \pi \sqrt{\frac{207 \times 10^9}{300 \times 10^6}} = 82.5$$

Now, since $Ar^2 = I$, where A is the area of the rod cross section and I is its second moment of area in bending

$$\left(\frac{\pi D^2}{4}\right)r^2 = \frac{\pi D^4}{64}$$

US customary units

$$\frac{L_e}{r} = \pi \sqrt{\frac{E}{\sigma_Y}} = \pi \sqrt{\frac{30 \times 10^3}{44}} = 82.0$$

Now, since $Ar^2 = I$, where A is the area of the rod cross section and I is its second moment of area in bending

$$\left(\frac{\pi D^2}{4}\right)r^2 = \frac{\pi D^4}{64}$$

where D is the diameter of the rod. From this equation $r = D/4 = 12.5$ mm, and

$$L_e = 82.5\,r = \underline{1.03 \text{ m}}$$

where D is the diameter of the rod. From this equation $r = D/4 = 0.5$ in, and

$$L_e = 82.0\,r = \underline{41.0 \text{ in}}$$

EXAMPLE 8.8

A vertical 5 m (16 ft) high column used in the structure of a building is made from an I section steel beam rigidly built-in to the foundations at its lower end. The I section is of depth 252 mm (10 in), breadth 203 mm (8 in), flange thickness 13.5 mm (0.54 in) and web thickness 8.0 mm (0.32 in). The maximum allowable compressive stress in the steel is 150 MN/m² (22 ksi) and the minimum safety factor on axial force against failure by buckling is 2.0. Find the maximum weight the column can support if (i) it is effectively free of constraint at its upper end, and (ii) it is rigidly built-in at this end, with no lateral deflection allowed.

The column cross section, which is shown in Fig. 8.30, is the same as the beam cross section we considered in Example 5.14, and found the second moment of area to be 85.6×10^6 mm⁴. But, that was for bending about axis AB, and is the maximum second moment of area, whereas we now need the minimum value, about axis XY. We find this by treating the cross section as three rectangular regions (the two flanges and the web), each of which is symmetrical about XY

$$I = 2 \times \left(\frac{13.5 \times 203^3}{12} \right) + \frac{225 \times 8^3}{12} = 18.8 \times 10^6 \text{ mm}^4$$

Similarly, the area of the cross section is given by

$$A = 2 \times (13.5 \times 203) + 225 \times 8 = 7280 \text{ mm}^2$$

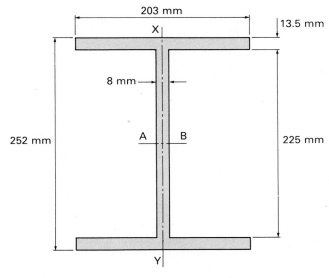

Fig. 8.30.

463

and the radius of gyration is

$$r = \sqrt{\frac{I}{A}} = \sqrt{\frac{18.8 \times 10^6}{7280}} = 50.8 \text{ mm}$$

If we first ignore the possibility of buckling, also the weight of the column, the maximum weight which can be supported is obtained when the maximum allowable compressive stress is uniformly distributed over the column cross section

$$P_1 = 150 \times 10^6 \times 7280 \times 10^{-6} \text{ N} = 1090 \text{ kN}$$

Considering buckling, in case (i) where the top of the column is free, the effective length of the column is twice its actual length. With $L_e = 10$ m, the slenderness ratio is $L_e/r = 197$, and from equation (8.52) the critical buckling force is

$$\frac{\pi^2 EA}{(L_e/r)^2} = \frac{\pi^2 \times 207 \times 10^9 \times 7280 \times 10^{-6}}{197^2} \text{ N} = 383 \text{ kN}$$

Applying the safety factor, the maximum weight that can be supported is $P_2 = 383/2.0$ = 192 kN. Since $P_2 < P_1$, in case (i) 192 kN (43 kip) is the required maximum weight.

In case (ii) where the top of the column is built-in, the effective length of the column is half its actual length. This reduces the slenderness ratio by a factor of four from case (i) to about 49, and therefore increases the maximum weight which can be supported before buckling occurs by a factor of 16 to $P_3 = 16 \times 192 = 3070$ kN. Since $P_3 > P_1$, in case (ii) 1090 kN (250 kip) is the required maximum weight.

Problems

SI METRIC UNITS

8.1 and 8.2 Find the critical value of force P for each of the systems of pin-jointed rigid members shown in Figs P8.1 and P8.2 at which collapse will just occur.

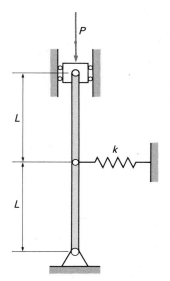

Fig. P8.2.

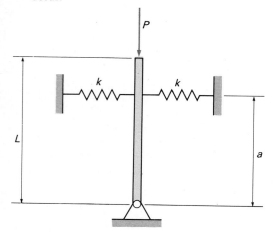

Fig. P8.1.

8.3 The rigid structure shown in Fig. P8.3 is supported on two similar springs, each of stiffness k. The constraint at A ensures that the structure is only free to move vertically and rotate about A. If the force F is small compared with $2ka^2/b$, obtain an expres-

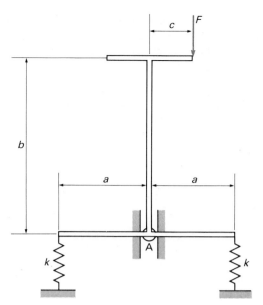

Fig. P8.3.

sion for the angle of rotation θ at A. Show that this expression no longer applies if F approaches the value $2ka^2/b$.

8.4 A tubular steel member of a pin-jointed structure is 2.5 m long and has an outer diameter of 85 mm and a wall thickness of 8 mm. Find the critical compressive force at which it will buckle, and the corresponding compressive stress.

8.5 A structural steel member with an effective length of 4.0 m is made from two angle sections welded together as shown in Fig. P8.5. Find the critical buckling force.

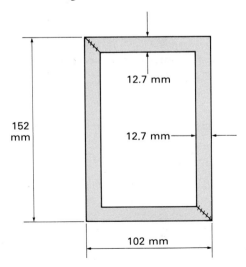

Fig. P8.5.

8.6 The members of the pin-jointed framework analyzed in Problem 2.3 take the form of hollow steel tubes of external diameter 60 mm and wall thickness 3 mm. All the members except AB and BC are of length 1.5 m. Find the maximum value of the force P if the structure is not to fail by buckling.

8.7 The members of the steel structure shown in Fig. P2.27 all have the same cross section. Find which one is the most likely to buckle, and the minimum second moment of area required to give a safety factor of 2.5 against buckling.

8.8 Repeat Problem 8.7 for the steel structure shown in Fig. P2.29.

8.9 The members of the pin-jointed structure defined in Problem 4.15 are to be hollow and tubular in cross section. The standard tubes available have a wall thickness equal to 10% of the outer diameter, and a range of outer diameters from 30 mm to 80 mm in steps of 10 mm. Choose a tube size to give a safety factor against buckling of at least 2.5.

8.10 The plane pin-jointed structure shown in Fig. P8.10 is used to support a vertical load W at a horizontal distance H from a vertical wall. The weights of the two members, AB and BC, can be neglected. The maximum allowable compressive force in the uniform strut BC, whose material has a Young's modulus of E, and whose cross section has a smallest second moment of area of I, is given by the Euler critical buckling force, P_c. The minimum acceptable value of I is required. First, show that the ratio of the maximum allowable to the actual compressive force, P, in BC is given by

$$\frac{P_c}{P} = \frac{\pi^2 EI}{H^2 W} \sin^2 \theta \cos \theta$$

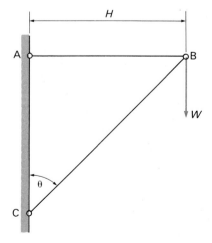

Fig. P8.10.

465

where θ is the angle BC makes with the vertical. Hence find the optimum value of θ for a given value of I. With this optimum value of θ, find an expression for the lowest value of I which strut BC can have and just support the load W without buckling.

8.11 A pin-ended strut has a moment M_0 applied to one end. Find the critical buckling force.

8.12 A uniform beam AB of length L with pin-jointed ends is subjected to an axial thrust P and a distributed lateral force which varies linearly in intensity from zero at end A to w per unit length at end B. Show that the maximum bending moment occurs at a distance x from A given by the solution of

$$\cos nx = \frac{\sin nL}{nL}$$

where $n^2 = P/EI$, E is the Young's modulus of the beam material, and I is the second moment of area of the beam cross section.

8.13 Obtain an expression for the critical buckling force for the system defined in Problem 8.12.

8.14 A member of a pin-jointed structure takes the form of a hollow steel tube 3 m long, 60 mm external diameter with a 6 mm wall thickness. If the eccentricity of the axial compressive force it carries is 15 mm, find both the magnitude of this force at which the member will buckle and the magnitude at which the maximum allowable compressive stress of 120 MN/m² is reached.

8.15 Find the effective length of the pin-ended strut AB shown in Fig. P8.15, which is constrained to have no lateral deflection at C, at a distance $a = L/2$ from A.

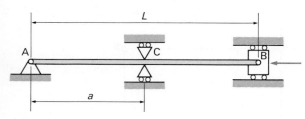

Fig. P8.15.

8.16 Repeat Problem 8.15 for a strut with two lateral constraints, at distances of $L/3$ from A and B.

8.17 A cantilever beam of flexural rigidity EI and length L is subjected to an axial compressive force P and also a uniformly distributed lateral force of intensity w. Obtain expressions for the deflection of

the free end of the cantilever and for the critical value of P at which elastic buckling occurs.

8.18 Find the effective length of a column which is rigidly built-in at one end and free to deflect laterally, but not to rotate, at the other end.

8.19 Find the effective length of a column which is rigidly built-in at both ends, and subjected to a uniformly distributed lateral force of intensity w per unit length.

8.20 A uniform column of length L is rigidly built-in at both ends and has a slightly curved shape which is defined by

$$v_0(x) = a_0\left(1 - \cos\frac{2\pi x}{L}\right)$$

where x is the distance from one end of the column, and a_0 is small compared with L. Find expressions for the deflected shape produced by an axial compressive force P, and the critical value of P at which buckling occurs.

8.21 A horizontal cantilever of uniform cross section and length L has a relevant flexural rigidity EI. It is subjected to an end force P whose line of action makes a small angle θ with the axis of the cantilever, as shown in Fig. P8.21. Obtain expressions for the lateral deflection of the free end of the cantilever and for the value of P at which elastic buckling would occur.

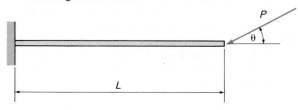

Fig. P8.21.

8.22 Find the effective length of the eccentrically loaded column shown in Fig. P8.22.

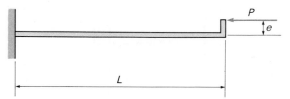

Fig. P8.22.

8.23 For the eccentrically loaded column shown in Fig. P8.22, determine the lateral end deflection when P is one quarter of its critical value.

8.24 For the eccentrically loaded column shown in Fig. P8.22, find a relationship between the axial force P and the maximum compressive stress, and show that this is consistent with equation (8.53).

8.25 The uniform elastic beam shown in Fig. P8.25 has a flexural rigidity of EI. The support at A can provide a vertical reaction either upward or downward. Find the reaction at A for the eccentric axial loading shown, assuming that P is less than its critical value.

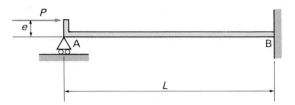

Fig. P8.25.

8.26 Explain the significance of the special case $P = kL$ in equation (8.48).

8.27 The column in Example 8.6 has a flexural rigidity of 3 MN/m^2 and is 7 m long. If the rotational stiffness at the built-in end is 0.2 MNm/rad, find the critical value of force P to cause buckling.

8.28 A uniform strut has length L and relevant flexural rigidity EI. At each end it is free to move axially, while lateral movement is prevented; at each end there also is a rotational spring, of stiffness λ. The strut is subjected to an axial compressive force P. Show that the smallest critical value of this force to cause buckling can be determined from the equation

$$\frac{2EI}{\lambda L} \phi + \tan \phi = 0$$

where $\phi^2 = PL^2/4EI$. Confirm that this gives the appropriate results for the limiting cases of very large and very small values of λ.

8.29 Figure P8.29 shows a pin-ended strut of length L subjected to an axial compressive force P. Lateral deflection at the center of the strut is restricted by a spring of stiffness k, which is unloaded when the strut is straight. Derive an algebraic equation from which the critical value of force P to cause elastic buckling can be found.

8.30 A cantilever in the form of a solid circular rod of length L and diameter D is rigidly built-in at one end and subject to an axial force at the other free end. If the yield stress of the material is σ_Y, derive an expression for the ratio of length to diameter of the cantilever at which failure by yielding and buckling according to the Euler formula occur at the same value of the end force. Evaluate this ratio when the material is steel, and $\sigma_Y = 350 \text{ MN/m}^2$.

8.31 A column with an effective length of 4.5 m is made from an I section steel beam. The I section, shown in Fig. P8.31, is of depth 210 mm, breadth 134 mm, flange thickness 10.2 mm and web thickness 6.4 mm. The maximum allowable compressive stress in the steel is 150 MN/m^2, and the minimum safety factor on axial force against failure by buckling is 2.5. Find the maximum weight the column can support (i) if the load is applied through the centroid C of the cross section, (ii) if it is applied at an eccentricity of 50 mm from the centroid along the axis AB, and (iii) if it is applied at the same eccentricity along axis XY.

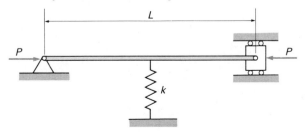

Fig. P8.29.

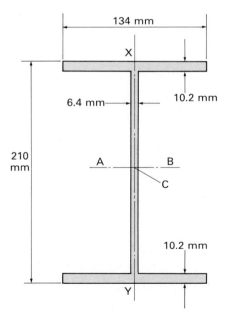

Fig. P8.31.

8.32 to 8.34 Solve Problems 8.1 to 8.3.

8.35 A tubular steel member of a pin-jointed structure is 10 ft long and has an outer diameter of 4 m and a wall thickness of 0.25 in. Find the critical compressive force at which it will buckle, and the corresponding compressive stress.

8.36 A piece of timber 9 ft long is required to form a temporary support for part of the roof of a building. The weight which must be supported is estimated to be 8 kip. Assuming the support can be treated as a light pin-ended strut, find the minimum dimensions of its square cross section to give a safety factor of 3.0 against buckling.

8.37 Repeat Problem 8.7 for the steel structure shown in Fig. P2.62.

8.38 The members of the pin-jointed structure defined in Problem 4.48 are to be solid and circular in cross section. Find the minimum safety factor against buckling.

8.39 The members of the pin-jointed structure defined in Problem 4.52 are to be solid and square in cross section. Find the minimum safety factor against buckling.

8.40 to 8.55 Solve Problems 8.10 to 8.13, and 8.15 to 8.26.

8.56 The column in Example 8.5 has a flexural rigidity of 5000 kip ft² and is 15 ft long. If the stiffness of the spring is 4 kip/ft, find the critical value of force P to cause buckling.

8.57 and 8.58 Solve Problems 8.28 and 8.29.

8.59 Solve Problem 8.30, taking the yield stress of steel as 50 ksi.

8.60 A 20 in diameter concrete pillar with an effective length of 25 ft is subjected to an eccentric axial compressive force with an eccentricity of 3 in. If the maximum allowable compressive stress is 1.5 ksi, find the maximum permissible value of the force.

9

Transformations of Stress and Strain

In many of the classes of problems considered so far, we have been concerned with only one type of stress and strain, either normal or shear, often acting in only one direction. For example, in pin-jointed structures we only had to consider uniaxial normal stresses and strains. In the bending of beams we worked mainly with normal bending stresses (and strains), although we did also consider shear stresses. Because these shear stresses proved to be small in magnitude, particularly in regions where the normal stresses were large, it was not necessary to consider the two types in combination. In the case of torsion, we were only concerned with shear stresses acting on the cross-sectional planes of twisted members. Although the analysis of thin-walled cylinders and spheres did involve biaxial normal stresses and strains, and some other problems in Chapter 3 were triaxial, none of them involved shear stresses and strains. The only situation we encountered where significant levels of normal and shear stresses occurred at the same location was in Example 7.9 involving combined bending and torsion – and there we were only concerned with superimposing displacements.

In many members of engineering systems, the states of stress and strain at points of interest are more *complex*, involving both normal and shear components in various directions. Consider, for example, the situation illustrated in Fig. 9.1a of a thin-walled cylinder subjected to both internal pressure and a torque. The stress components acting on a typical element ABCD of the cylinder wall are the axial stress, σ_z, the hoop stress, σ_θ, and the shear stress, τ, as shown in Fig. 9.1b. If we wish to determine whether the cylinder is strong enough, it is not sufficient to consider only the values of these stress components individually, because the combination of stresses may represent a more severe form of loading than any one component acting alone.

Figure 9.2a shows another example of combined stresses. The solid shaft driven by a belt and pulley is subject to both torsion and bending, and a typical element EFGH on the shaft surface experiences both an axial bending stress, σ_z, and a shear stress, τ, as shown in Fig. 9.2b. An important difference between this example and the one illustrated in Fig. 9.1 is that now one of the stress components, σ_z, varies with the position of the element on the shaft surface. If the dimensions of the element itself are large, the axial stress also varies significantly with position over the element. Although Fig. 9.2b shows an element of finite dimensions, we assume that it is so small that the stresses can be treated as constant. In other words, we are concerned with stresses acting effectively at a point, in the sense defined in Section 1.3. In Chapter 10 we will consider the variations of stresses from point to point within a solid body.

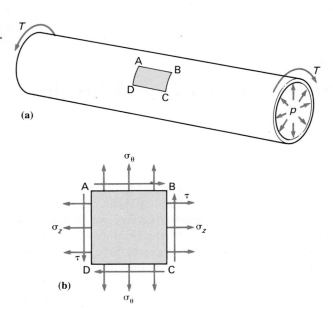

(a)

(b)

Fig. 9.1.

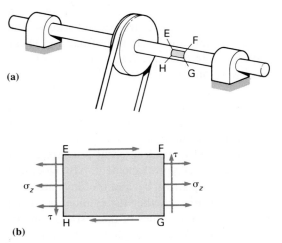

(a)

(b)

Fig. 9.2.

In both Figs 9.1b and 9.2b, the state of stress at a point in a body is defined in terms of normal and shear stress components acting on the faces of an infinitesimally small element at that point. This element is chosen with its sides parallel to the coordinate directions (hoop and axial) which are convenient for describing the problem. If we were to choose a different element orientation, then we might anticipate that the stresses acting on its sides would also be different from the ones shown, and possibly greater in magnitude. In order to investigate this possibility, we need to be able to transform given stress components to equivalent sets acting on other planes at the point of interest. This in turn allows us to relate material behavior under a complex state of stress to its strength characteristics measured in, say, a simple tension test, by means of a yield or fracture criterion. We must also be concerned with the transformation of strains at a point, not least because in practice we cannot measure stresses directly, but must deduce them from measured strains.

9.1 Transformation of stress

We first defined simple normal and shear stresses in Section 1.3. Before we can consider more complex states of stress, with components acting in two or three directions at a point, we must extend the notation we have used to represent stresses, particularly shear stresses.

9.1.1 Notation for stresses

We have already chosen to distinguish between normal and shear stresses by the use of the symbols σ and τ, respectively. In the case of normal stress, we have also introduced subscripts to indicate the direction of the stress. For example, in the (x, y, z) Cartesian and (r, θ, z) cylindrical polar coordinate systems, the normal stress components are

$$\sigma_x, \sigma_y, \sigma_z \quad \text{and} \quad \sigma_r, \sigma_\theta, \sigma_z$$

respectively, as shown in Fig. 9.3 acting on infinitesimally small elements of material. The sign convention defines tensile stresses as positive, compressive stresses as negative. Because the direction of each of the normal stress components shown is the same as the direction of the normal to the surface of the element on which it acts, a single subscript attached to the stress symbol is sufficient to define this direction. A stress acting in the direction of the outward normal to the surface is positive (tensile).

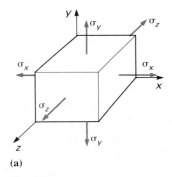 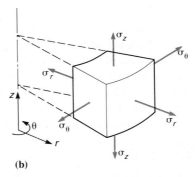

(a) (b)

Fig. 9.3.

With shear stress components the single subscript notation is not practical, because such stresses are in directions parallel to the surfaces on which they act. We therefore have two directions to specify, that of the normal to the surface and that of the stress itself. To do this, we attach two subscripts to the symbol, τ, for shear stress. For example, in the Cartesian and cylindrical polar coordinate systems the shear stress components are

$$\tau_{xy}, \tau_{yx}, \tau_{yz}, \tau_{zy}, \tau_{zx}, \tau_{xz} \quad \text{and} \quad \tau_{r\theta}, \tau_{\theta r}, \tau_{\theta z}, \tau_{z\theta}, \tau_{zr}, \tau_{rz}$$

respectively. Taking the first subscript to indicate the direction of the normal to the surface, and the second one the direction of the stress, these components act on the elements of material as shown in Fig. 9.4. The sign convention we adopt for a shear stress component such as τ_{xy} is that if it acts in the positive y direction on the surface whose outward normal is in the positive x direction, then it is itself positive. A general state of stress at a point involves all the normal stress components shown in Fig. 9.3, together with all the shear stress components shown in Fig. 9.4.

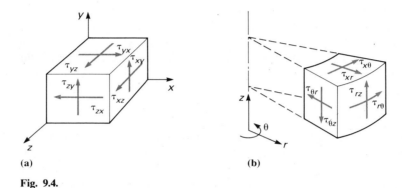

(a)   (b)

Fig. 9.4.

It is worth noting that the double subscript notation, with the same meanings for the subscripts, is sometimes extended to normal stress components. Then, for example, $\sigma_x, \sigma_y, \sigma_z$ become $\sigma_{xx}, \sigma_{yy}, \sigma_{zz}$. Also, all stresses may be represented by the same symbol, typically σ, with normal and shear components being distinguished by whether or not the subscripts are identical. Finally, the meaning of the subscripts is sometimes reversed, so that the first refers to the direction of the stress and the second the direction of the normal to the surface on which it acts. This only affects shear stresses.

In Section 5.3.2 where we considered shear stresses in beams, we found when we considered a small rectangular element of a beam (Fig. 5.59) that the *complementary shear stresses* on opposite pairs of faces of this element are equal in magnitude. The same was true for torsion of a thin-walled cylinder (Section 7.1.1, Fig. 7.4), although we did not attempt to prove the result. Let us now do so for the general case. Figure 9.5 shows an infinitesimally small rectangular element with sides of length δx and δy parallel to the x and y directions, respectively. Its thickness normal to the plane shown is δz, in the z direction. All nine normal and shear stress components may act on the element: only those in the x and y directions are shown. Note that, according to the sign convention, the shear stress τ_{xy} acting on the left-hand face of the element, whose normal is in the negative x direction, is positive in the downward or negative y direction. Similarly, the shear stress τ_{yx} acting on the bottom surface of the element is

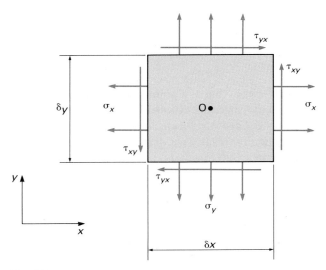

Fig. 9.5.

positive when acting from right to left. With the stresses as shown acting on the element, the resulting forces applied to it are in equilibrium in the x and y directions, which is therefore true of any other direction in the (x, y) plane. Although the other normal and shear stress components are not shown, their presence does not affect this conclusion. The weight of the element is ignored.

Since the element is a static piece of a solid body, the moments applied to it must also be in equilibrium, about any axis normal to the plane shown. Consider, for example, the axis through the point O at the center of the element. The resultant forces associated with normal stresses σ_x and σ_y acting on the sides of the element each pass through this axis, and therefore have no moment about it. The shear force on the right-hand face due to the shear stress τ_{xy} acting over an area of $\delta y\,\delta z$ is of magnitude $\tau_{xy}\,\delta y\,\delta z$, and with an equal and opposite shear force on the left-hand face forms a couple with a moment of $\tau_{xy}\,\delta x\,\delta y\,\delta z$ in the counterclockwise direction. Similarly, the shear stress τ_{yx} acting on the top and bottom surfaces gives rise to a clockwise couple of magnitude $\tau_{yx}\,\delta x\,\delta y\,\delta z$. Therefore, for the element to be in equilibrium

$$\tau_{xy}\,\delta x\,\delta y\,\delta z = \tau_{yx}\,\delta x\,\delta y\,\delta z \quad \text{from which} \quad \tau_{xy} = \tau_{yx}$$

In other words, the complementary shear stresses are equal in magnitude. The same form of relationship can be obtained for the other two pairs of shear stress components to give

$$\tau_{xy} = \tau_{yx}, \quad \tau_{yz} = \tau_{zy}, \quad \tau_{zx} = \tau_{xz} \tag{9.1}$$

Similarly, in the cylindrical polar system

$$\tau_{r\theta} = \tau_{\theta r}, \quad \tau_{\theta z} = \tau_{z\theta}, \quad \tau_{zr} = \tau_{rz} \tag{9.2}$$

Since we are concerned with stresses acting at a point, it is not important what form of coordinate system, such as Cartesian, cylindrical polar or spherical, we use, provided the directions so defined are locally *orthogonal* (meeting each other at right angles). Consequently, results we derive in this chapter for a Cartesian system are equally applicable to others.

9.1.2 Plane stress

Let us begin our study of transformation of stress by considering not the most general three-dimensional state of stress at a point, but a simpler two-dimensional state known as *plane stress*. It is two-dimensional in the sense that the stress components in one direction are all zero. For example, in Cartesian coordinates, if this direction is the z direction, then

$$\sigma_z = \tau_{yz} = \tau_{xz} = 0 \tag{9.3}$$

and from equation (9.1) it follows that τ_{zy} and τ_{zx} are also zero. Figure 9.6 illustrates the simplified state of stress acting on a small element of material. Not only are there no stresses acting in the z direction, there are also none acting on the front and back faces of the element. The choice of an element which is relatively thin in the z direction is intentional: typical practical examples of states of plane stress are to be found in thin members, such as plates and shells. If equations (9.3) do not hold, and the state of stress is three-dimensional, then the analysis which follows is still applicable, but incomplete.

We wish to determine the normal and shear stress components acting on the element after it has been rotated in the (x, y) plane through an angle θ as shown in Fig. 9.7, *given that the same state of stress still exists in the material at*

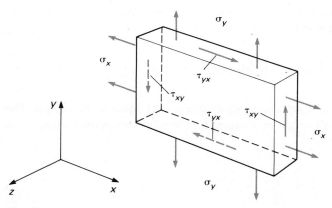

Fig. 9.6.

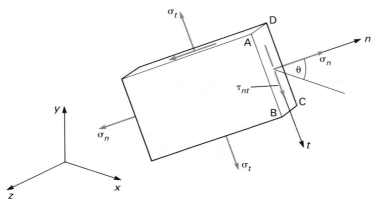

Fig. 9.7.

the point concerned. In particular, we wish to find the stresses acting on the typical face ABCD, whose normal is n, at an angle θ to the x axis measured in the counterclockwise sense in the (x, y) plane, and whose tangent is t at $(\theta - 90°)$ to the x axis. This choice of direction for t (down rather than up the sloping face shown) is made deliberately, to minimize subsequent difficulties with signs of shear stresses. According to our notation for stresses, the stress normal to the face is σ_n, and the shear stress parallel to it is τ_{nt}. To find these stresses, we examine the prismatic element of triangular cross section shown in Fig. 9.8, which has its three rectangular faces normal to the x, y and n directions. Since the state of stress is unchanged, the stress components acting on the first two of these faces are σ_x, τ_{xy} and σ_y, τ_{yx}, respectively, as on the original element in Fig. 9.6. The dimensions of the element in the x, y, z, and t directions are taken to be δx, δy, δz and δt as shown. For the element to be in equilibrium, the *forces*

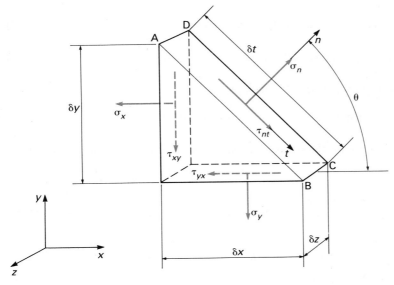

Fig. 9.8.

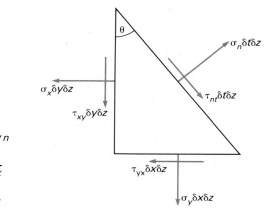

Fig. 9.9.

acting on it must be in equilibrium, and these forces are shown in Fig. 9.9. In the n direction

$$\sigma_n \delta t\, \delta z - \sigma_x \delta y\, \delta z \cos\theta - \tau_{xy} \delta y\, \delta z \sin\theta - \sigma_y \delta x\, \delta z \sin\theta - \tau_{yx} \delta x\, \delta z \cos\theta = 0$$

and in the t direction

$$\tau_{nt} \delta t\, \delta z - \sigma_x \delta y\, \delta z \sin\theta + \tau_{xy} \delta y\, \delta z \cos\theta + \sigma_y \delta x\, \delta z \cos\theta - \tau_{yx} \delta x\, \delta z \sin\theta = 0$$

Now, the length δz is a common factor in both of these equations and may therefore be eliminated. Also, the complementary shear stresses are equal in magnitude, and the geometry of the element is such that

$$\delta x = \delta t \sin\theta \quad \text{and} \quad \delta y = \delta t \cos\theta$$

We may therefore simplify the equilibrium equations to

$$\sigma_n = \sigma_x \cos^2\theta + \sigma_y \sin^2\theta + 2\tau_{xy} \sin\theta \cos\theta \tag{9.4}$$

$$\tau_{nt} = (\sigma_x - \sigma_y)\sin\theta \cos\theta - \tau_{xy}(\cos^2\theta - \sin^2\theta) \tag{9.5}$$

Using the trigonometric relationships

$$\sin 2\theta = 2\sin\theta \cos\theta \quad \text{and} \quad \cos 2\theta = \cos^2\theta - \sin^2\theta$$

these expressions for the stress components on the inclined plane can be written as

$$\sigma_n = \frac{\sigma_x + \sigma_y}{2} + \frac{\sigma_x - \sigma_y}{2}\cos 2\theta + \tau_{xy}\sin 2\theta \tag{9.6}$$

$$\tau_{nt} = \frac{\sigma_x - \sigma_y}{2}\sin 2\theta - \tau_{xy}\cos 2\theta \tag{9.7}$$

These are equilibrium equations for stresses at a point. They do not depend on material properties, and are equally valid for elastic and inelastic material behavior.

Equations (9.4) and (9.5), or (9.6) and (9.7) can be used to calculate the normal and shear stress components on any plane whose normal is inclined at an angle θ to the x direction. For example, if $\theta = 0$ then $\sigma_n = \sigma_x$ and $\tau_{nt} = -\tau_{xy}$ (directions n and t coincide with $+x$ and $-y$, respectively), and if $\theta = 90°$ then $\sigma_n = \sigma_y$ and $\tau_{nt} = +\tau_{xy}$ (directions n and t coincide with $+y$ and $+x$, respectively). Also, we can find the normal stress component σ_t in Fig. 9.7 by replacing θ by $\theta \pm 90°$ in equation (9.6) (recalling that $\sin(2\theta \pm 180°) = -\sin 2\theta$ and $\cos(2\theta \pm 180°) = -\cos 2\theta$) as

$$\sigma_t = \frac{\sigma_x + \sigma_y}{2} - \frac{\sigma_x - \sigma_y}{2} \cos 2\theta - \tau_{xy} \sin 2\theta \tag{9.8}$$

Adding equations (9.6) and (9.8), we obtain

$$\sigma_n + \sigma_t = \sigma_x + \sigma_y \tag{9.9}$$

which means that the sum of the two normal stress components acting on mutually perpendicular planes at a point in a state of plane stress is not affected by the orientation of these planes.

EXAMPLE 9.1

A prismatic bar of material is subjected to a uniform axial tensile stress of magnitude σ. Find the normal and shear stress components acting on a plane in the material whose normal is at an angle of 45° to the axis of the bar, as shown in Fig. 9.10.

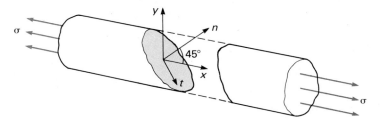

Fig. 9.10.

This is a straightforward example of the application of equations (9.6) and (9.7). If we take the x direction to be along the axis of the bar, then because the loading is uniaxial $\sigma_x = \sigma$, $\sigma_y = 0$ and $\sigma_{xy} = 0$. Also, at $\theta = 45°$, $\sin 2\theta = 1$, $\cos 2\theta = 0$, and

$$\sigma_n = \frac{\sigma}{2}, \qquad \sigma_{nt} = \frac{\sigma}{2} \tag{9.10}$$

Both the normal and shear stress components have magnitudes equal to half that of the applied stress. Even in a state of simple uniaxial tension, substantial shear stresses exist on certain planes. Indeed, we will show in Section 9.1.3 that the plane at $\theta = 45°$ carries the *maximum shear stress* in the material.

EXAMPLE 9.2

In a state of plane stress, the normal and shear stress components acting in the (x, y) plane are $\sigma_x = 57.8$ MN/m² (8.4 ksi), $\sigma_y = -35.9$ MN/m² $(-5.2$ ksi) and $\tau_{xy} = 29.3$ MN/m² (4.2 ksi). Find the normal and shear stress components acting on a plane whose normal is at 35° to the x direction, measured in the counterclockwise sense. With $\theta = 35°$, equations (9.6) and (9.7) give

$$\sigma_n = \frac{57.8 - 35.9}{2} + \frac{57.8 + 35.9}{2} \cos 70° + 29.3 \sin 70°$$

$$= 54.5 \text{ MN/m}^2 \quad (7.9 \text{ ksi})$$

$$\tau_{nt} = \frac{57.8 + 35.9}{2} \sin 70° - 29.3 \cos 70°$$

$$= 34.0 \text{ MN/m}^2 \quad (4.9 \text{ ksi})$$

EXAMPLE 9.3

A closed-ended circular cylindrical tube of diameter 250 mm (10 in) and wall thickness 5 mm (0.2 in) is subjected to an internal pressure of 3 MN/m² (440 psi), and a torque about its axis of 15 kNm (11 kip-ft). Find the normal and shear stress components acting in the wall of the cylinder on a plane whose normal is at 40° to the cylinder axis, as shown in Fig. 9.11.

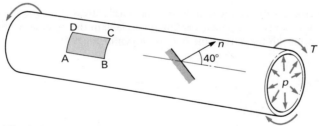

Fig. 9.11.

SI units

With a tube radius of $R = 0.125$ m and a wall thickness of $t = 0.005$ m, equation (2.18) gives the axial stress in the cylinder due to an internal pressure of $p = 3$ MN/m² as

$$\sigma_z = \frac{pR}{2t} = \frac{3 \times 0.125}{2 \times 0.005} = 37.5 \text{ MN/m}^2$$

and equation (2.16) gives the hoop stress as

$$\sigma_\theta = 2\sigma_z = 2 \times 37.5 = 75 \text{ MN/m}^2$$

The shear stress in the wall due to a torque $T = 15$ kNm is given by equation (7.4) as

US customary units

With a tube radius of $R = 5$ in and a wall thickness of $t = 0.2$ in, equation (2.18) gives the axial stress in the cylinder due to an internal pressure of $p = 440$ psi as

$$\sigma_z = \frac{pR}{2t} = \frac{440 \times 5}{2 \times 0.2} \text{ psi} = 5.50 \text{ ksi}$$

and equation (2.16) gives the hoop stress as

$$\sigma_\theta = 2\sigma_z = 2 \times 5.50 = 11.0 \text{ ksi}$$

The shear stress in the wall due to a torque $T = 11$ kip-ft is given by equation (7.4) as

$$\tau = \frac{T}{2\pi R^2 t} = \frac{15 \times 10^3}{2\pi \times 0.125^2 \times 0.005}\,\text{N/m}^2$$

$$= 30.56\,\text{MN/m}^2$$

The typical element ABCD of the tube wall which is shown shaded in Fig. 9.11 is shown enlarged in Fig. 9.12 with the stress components acting on it. The sides of this element are parallel to the axial and hoop directions, in which coordinate system the shear stress τ is the component $\tau_{z\theta}$. Adapting equations (9.6) and (9.7) to the change in coordinates, we obtain the required normal and shear stresses as

$$\tau = \frac{T}{2\pi R^2 t} = \frac{11 \times 12}{2\pi \times 5^2 \times 0.2}$$

$$= 4.202\,\text{ksi}$$

The typical element ABCD of the tube wall which is shown shaded in Fig. 9.11 is shown enlarged in Fig. 9.12 with the stress components acting on it. The sides of this element are parallel to the axial and hoop directions, in which coordinate system the shear stress τ is the component $\tau_{z\theta}$. Adapting equations (9.6) and (9.7) to the change in coordinates, we obtain the required normal and shear stresses as

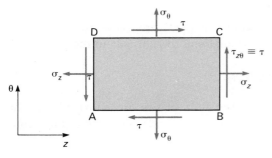

Fig. 9.12.

$$\sigma_n = \frac{37.5 + 75}{2} + \frac{37.5 - 75}{2}\cos 80° + 30.56\sin 80°$$

$$= \underline{83.1\,\text{MN/m}^2}$$

$$\tau_{nt} = \frac{37.5 - 75}{2}\sin 80° - 30.56\cos 80°$$

$$= \underline{-23.8\,\text{MN/m}^2}$$

$$\sigma_n = \frac{5.50 + 11.0}{2} + \frac{5.50 - 11.0}{2}\cos 80° + 4.202\sin 80°$$

$$= \underline{11.9\,\text{ksi}}$$

$$\tau_{nt} = \frac{5.50 - 11.0}{2}\sin 80° - 4.202\cos 80°$$

$$= \underline{-3.44\,\text{ksi}}$$

9.1.3 Principal stresses and maximum shear stresses

Equations (9.6) and (9.7) give normal and shear stress components as functions of angle for a state of plane stress, from which we can find the maximum and minimum values of these stresses at the point of interest. These are mathematical maximum and minimum values with respect to the direction of the plane considered. Differentiating equation (9.6) with respect to angle θ, we obtain

$$\frac{d\sigma_n}{d\theta} = -(\sigma_x - \sigma_y)\sin 2\theta + 2\tau_{xy}\cos 2\theta$$

479

and, if this derivative is zero, we obtain particular values of θ defined by

$$\tan 2\theta_p = \frac{2\tau_{xy}}{\sigma_x - \sigma_y} \tag{9.11}$$

The nature of the trigonometrical tangent function is such that there will always be a value of the angle $2\theta_p$ in the range $0 \leqslant 2\theta_p \leqslant 180°$ which satisfies this equation. Let this value be defined as

$$2\phi = \tan^{-1}\left(\frac{2\tau_{xy}}{\sigma_x - \sigma_y}\right) \tag{9.12}$$

Other angles which also satisfy equation (9.11) are $2\phi + 180°$, $2\phi + 360°$, $2\phi + 540°$, and so on. Therefore, the angles of the normals to the planes are

$$\theta_p = \phi,\ \phi + 90°,\ \phi + 180° \text{ and } \phi + 270° \tag{9.13}$$

which are illustrated in Fig. 9.13. The four planes are either parallel or orthogonal to each other, forming a rectangular element at the point concerned.

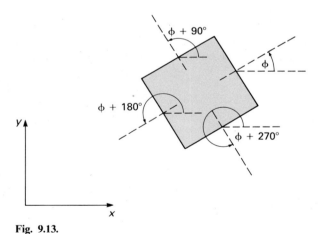

Fig. 9.13.

These planes are planes of maximum and minimum normal stress. We find the values of these stresses by using equation (9.11), together with Fig. 9.14 (which shows a right angled triangle as a geometric figure rather than an element of material) to define the sine and cosine of the angle $2\theta_p$ as

$$\sin 2\theta_p = \pm \frac{2\tau_{xy}}{\sqrt{[(\sigma_x - \sigma_y)^2 + 4\tau_{xy}^2]}} \tag{9.14}$$

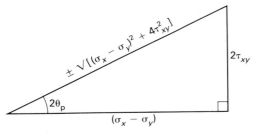

Fig. 9.14.

and

$$\cos 2\theta_p = \pm \frac{(\sigma_x - \sigma_y)}{\sqrt{[(\sigma_x - \sigma_y)^2 + 4\tau_{xy}^2]}} \qquad (9.15)$$

where the same sign is taken for each. Substituting these expressions into equation (9.6), we obtain

$$\sigma_{1,2} = \frac{\sigma_x + \sigma_y}{2} \pm \tfrac{1}{2}\sqrt{[(\sigma_x - \sigma_y)^2 + 4\tau_{xy}^2]} \qquad (9.16)$$

an equation which defines *two* normal stress components, σ_1 and σ_2. These are the maximum and minimum normal stresses, and are termed the *principal stresses* at the point concerned. The planes on which they act, known as the *principal planes*, have already been defined by equation (9.13): hence the use of the subscript 'p' on the plane angle θ. It is customary to assume that $\sigma_1 \geqslant \sigma_2$ in the mathematical sense, where negative quantities are regarded as less than positive quantities, regardless of their magnitudes. Then σ_1 is the *maximum principal stress*, given by the positive root in equation (9.16), and σ_2 is the *minimum principal stress*, given by the negative root. We note that

$$\sigma_1 + \sigma_2 = \sigma_x + \sigma_y \qquad (9.17)$$

which is consistent with equation (9.9).

Also of interest are the shear stresses on the principal planes. Substituting expressions (9.14) and (9.15) into equation (9.7) for shear stress, we find that $\tau_{nt} = 0$. In other words, the *principal planes are not only planes of maximum or minimum normal stress, but also planes of zero shear stress.* We may therefore represent the state of stress in terms of the principal stresses and their orientations relative to the chosen coordinates as in Fig. 9.15. Although the element of material shown in Fig. 9.15 has the same orientation as that shown in Fig. 9.13, it does not necessarily follow that the maximum principal stress, σ_1, acts on the planes whose normals are at ϕ and $\phi + 180°$ to the x axis: it could act on the other pair of planes. If it is necessary to know which is the plane of maximum principal stress, we can, however, substitute one of the values of θ_p

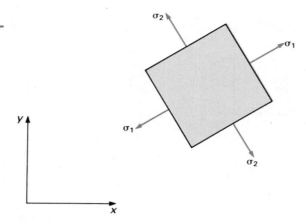

Fig. 9.15.

into equation (9.6) to find the principal stress acting on the chosen plane. The principal stresses and principal planes are properties only of the state of stress, and do not depend on the particular coordinate system used to define stress components.

Having found the maximum and minimum normal stresses at a point for a state of plane stress, we can do the same for shear stresses. Differentiating equation (9.7) with respect to angle θ, we obtain

$$\frac{d\tau_{nt}}{d\theta} = (\sigma_x - \sigma_y)\cos 2\theta + 2\tau_{xy}\sin 2\theta$$

and if this derivative is zero, we obtain values of θ defined by

$$\tan 2\theta_s = -\frac{(\sigma_x - \sigma_y)}{2\tau_{xy}} \tag{9.18}$$

Now, the product of the tangent functions defined by equations (9.11) and (9.18) is -1, which means that the angles (2θ) they represent differ by 90°. Therefore, the angles of the normals to the planes differ by 45°, and the *planes of maximum and minimum shear stress are at 45° to the principal planes* (the planes of maximum and minimum normal stress).

From equation (9.18), we obtain expressions for the sines and cosines of the angles $2\theta_s$ as

$$\sin 2\theta_s = \pm\frac{(\sigma_x - \sigma_y)}{\sqrt{[(\sigma_x - \sigma_y)^2 + 4\tau_{xy}^2]}} \tag{9.19}$$

and

$$\cos 2\theta_s = \pm\frac{2\tau_{xy}}{\sqrt{[(\sigma_x - \sigma_y)^2 + 4\tau_{xy}^2]}} \tag{9.20}$$

where opposite signs are taken for the two functions. Substituting these expressions into equation (9.7) we obtain

$$\tau_{1,2} = \pm \tfrac{1}{2}\sqrt{[(\sigma_x - \sigma_y)^2 + 4\tau_{xy}^2]} \tag{9.21}$$

where τ_1 and τ_2 are the maximum and minimum shear stresses (corresponding to the positive and negative roots, respectively) at the point of interest. These have the same magnitude, but are opposite in sign. Given the definition for the principal stresses in equation (9.16), we can express the maximum and minimum shear stresses as

$$\tau_1 = \tfrac{1}{2}(\sigma_1 - \sigma_2) \quad \text{and} \quad \tau_2 = -\tfrac{1}{2}(\sigma_1 - \sigma_2) \tag{9.22}$$

It should be emphasized that these are the maximum and minimum shear stresses *in the (x, y) plane considered*. Shear stresses of greater magnitude may exist in other planes, as we will see in Section 9.1.5.

Also of interest are the normal stresses on the planes of maximum and minimum shear stress. Substituting the expressions for sines and cosines defined by equations (9.19) and (9.20) into equation (9.6), and using equation (9.17), we find that

$$\sigma_n = \frac{\sigma_x + \sigma_y}{2} = \frac{\sigma_1 + \sigma_2}{2} \tag{9.23}$$

on all the planes. Figure 9.16 shows these normal stresses and the maximum shear stresses acting on an element of material, the normals to whose surfaces are at 45° to the directions of the principal stresses.

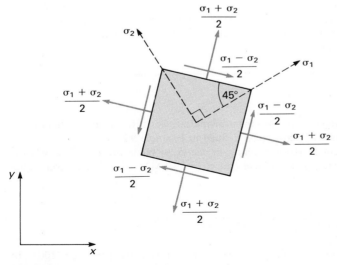

Fig. 9.16.

EXAMPLE 9.4

The hoop and axial stresses produced in a thin-walled circular cylinder by an internal pressure are 2σ and σ, respectively. Find the magnitudes and directions of the principal stresses and maximum shear stresses acting in the plane of the cylinder wall.

As shown in Fig. 9.17, a typical rectangular element of material with its sides parallel to the axial and hoop directions is subject only to the normal hoop and axial stresses. With no shear stresses on these sides, they must represent principal planes, and the hoop and axial stresses, 2σ and σ, are the maximum and minimum principal stresses, σ_1 and σ_2, respectively. From equations (9.22), the maximum shear stresses are of magnitude $(2\sigma - \sigma)/2$ or $\sigma/2$, and act on planes at $45°$ to the axial and hoop directions.

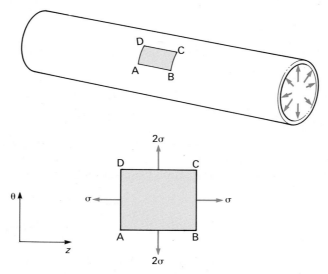

Fig. 9.17.

We note that many of the problems we met in earlier chapters were of this type, with one or two normal stresses acting on the chosen planes, but not shear stresses. These normal stresses were therefore principal stresses, and we were finding the maximum normal stresses in the material.

EXAMPLE 9.5

Find the magnitudes and directions of the principal stresses, also the maximum in-plane shear stress, for the state of plane stress defined in Example 9.2.

The normal and shear stresses acting in the (x, y) plane are $\sigma_x = 57.8 \text{ MN/m}^2$ (8.4 ksi), $\sigma_y = -35.9 \text{ MN/m}^2$ (-5.2 ksi) and $\tau_{xy} = 29.3 \text{ MN/m}^2$ (4.2 ksi). From equation (9.12)

$$\phi = \tfrac{1}{2}\tan^{-1}\left(\frac{2\tau_{xy}}{\sigma_x - \sigma_y}\right) = \tfrac{1}{2}\tan^{-1}\left(\frac{2 \times 29.3}{57.8 + 35.9}\right) = 16.0°$$

and, from equation (9.13), the principal planes have their normals at $16.0°$ and $106.0°$ to the positive x direction: the further planes with normals at $196.0°$ and $286.0°$ are parallel to these and carry the same principal stresses. The principal stresses are given by equation (9.16) as

$$\sigma_{1,2} = \frac{57.8 - 35.9}{2} \pm \tfrac{1}{2}\sqrt{[(57.8 + 35.9)^2 + 4 \times 29.3^2]}$$

$$= 10.95 \pm 55.26 \text{ MN/m}^2$$

that is

$$\sigma_1 = 66.2 \text{ MN/m}^2 \ (9.6 \text{ ksi}) \quad \text{and} \quad \sigma_2 = -44.3 \text{ MN/m}^2 \ (-6.4 \text{ ksi})$$

To find which principal plane each of these acts on, we substitute $\theta = 16°$ into equation (9.6)

$$\sigma_n = \frac{57.8 - 35.9}{2} + \frac{57.8 + 35.9}{2} \cos 32° + 29.3 \sin 32°$$

$$= 66.2 \text{ MN/m}^2$$

which shows that σ_1, the maximum principal stress, which is tensile, acts on the principal plane whose normal is at $16°$ to the x axis (also the one with a normal at $196°$ to the same axis). We could have anticipated this: the plane with a normal at $16°$ to the x axis is close to the (y, z) plane, on which the tensile stress σ_x acts. The other principal plane with a normal at $106°$ to the x axis, carrying the compressive principal stress, is close to the (z, x) plane, on which the compressive σ_y acts.

The maximum shear stress in the (x, y) plane is found with the aid of equations (9.22) as

$$\tau_1 = \tfrac{1}{2}(66.2 + 44.3) = 55.3 \text{ MN/m}^2 \ (8.0 \text{ ksi})$$

9.1.4 Mohr's circle for plane stress

It is possible to describe the state of plane stress at a point in pictorial form, which can help in the visualization and understanding of its properties. The form of representation we use is known as *Mohr's stress circle* (after the German engineer Otto Mohr, 1835–1918). Rearranging and squaring equations (9.6) and (9.7) for the normal and shear stresses on a plane whose normal is at an angle θ to the x direction, we obtain

$$\left[\sigma_n - \frac{(\sigma_x + \sigma_y)}{2}\right]^2 + \tau_{nt}^2 = \left[\frac{\sigma_x - \sigma_y}{2} \cos 2\theta + \tau_{xy} \sin 2\theta\right]^2$$

$$+ \left[\frac{\sigma_x - \sigma_y}{2} \sin 2\theta - \tau_{xy} \cos 2\theta\right]^2$$

$$= \frac{(\sigma_x - \sigma_y)^2}{4} + \tau_{xy}^2$$

Using equations (9.16) and (9.17), this result becomes

$$\left[\sigma_n - \frac{(\sigma_1 + \sigma_2)}{2}\right]^2 + \tau_{nt}^2 = \left[\frac{\sigma_1 - \sigma_2}{2}\right]^2 \tag{9.24}$$

If we now plot τ_{nt} against σ_n, as in Fig. 9.18, we obtain the locus represented by this equation as a circle of radius $(\sigma_1 - \sigma_2)/2$ with a center at $\sigma_n = (\sigma_1 + \sigma_2)/2$, $\tau_{nt} = 0$. Points A and B on this circle represent the principal stresses, while points D and E represent the maximum and minimum shear stresses. A typical point P on the circle gives the normal and shear stress components, σ_n and τ_{nt}, on a particular plane, whose normal is at an angle of θ' to the direction of the normal to the plane of maximum principal stress. We note that an angle subtended at the center of Mohr's circle by an arc connecting two points on the circle is twice the physical angle in the material between the directions of the normals associated with the two points. Points X and Y represent the stresses on the planes perpendicular to the x and y directions, with normal and shear stress components $(\sigma_x, -\tau_{xy})$ and $(\sigma_y, +\tau_{xy})$, respectively.

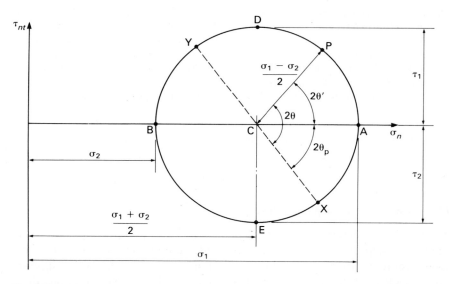

Fig. 9.18.

There are a number of features of Mohr's circle which serve to illustrate properties of states of plane stress we previously derived analytically. These include the following.

1 Complementary shear stresses (on planes 90° apart in the material, 180° apart on the circle) are equal in magnitude.
2 The principal planes are orthogonal: points A and B are 180° apart on the circle (90° apart in the material).
3 There are no shear stresses on the principal planes: points A and B lie on the normal stress axis.

4   The planes of maximum shear stress are 45° from the principal planes: points D and E are 90°, measured round the circle, from points A and B.
5   The maximum shear stresses are equal in magnitude, and given by equation (9.22): points D and E are the same distance (the radius of the circle) from the normal stress axis, which passes through the center of the circle.
6   The normal stresses on the planes of maximum shear stress are equal, and given by equation (9.23): points D and E both have a normal stress coordinate which is equal to the average of the two principal stresses.

Figure 9.18 is typical of Mohr's stress circles obtained for problems of practical interest, the only variations for different states of stress being in the size of the circle and in the position of its center relative to the shear stress axis. Some illustrations are provided by the example which follows. Mohr's stress circle was originally intended as an approximate graphical method for analyzing states of stress. Although it remains a useful aid to understanding, detailed analysis is now often more conveniently carried out using a computer. A computer method is given in Section 9.3, which deals with both stresses and strains at a point.

EXAMPLE 9.6

Draw the Mohr's stress circles for the following states of plane stress: (a) simple uniaxial tension, with a tensile stress of magnitude σ, (b) pure shear as in Fig. 9.19, with a shear stress of magnitude τ, (c) the state of biaxial tension examined in Example 9.4 for a thin-walled cylinder subjected to internal pressure, and (d) the state of stress analyzed in Examples 9.2 and 9.5.

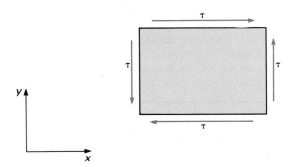

Fig. 9.19.

(a) For a state of simple uniaxial tension, the principal stresses are σ (the tensile stress) and zero, and Fig. 9.20a shows the Mohr's stress circle. The circle is tangential to the shear stress axis, and the maximum shear stress is of magnitude $\sigma/2$, acting on planes at 45° to the principal planes.
(b) For the state of pure shear shown in Fig. 9.19, equations (9.6) and (9.7) give

$$\sigma_n = \tau \sin 2\theta \quad \text{and} \quad \tau_{nt} = \tau \cos 2\theta$$

and a Mohr's stress circle as shown in Fig. 9.20b, with its center at the origin. We note that in this case, the planes of maximum shear stress (the faces of the element in

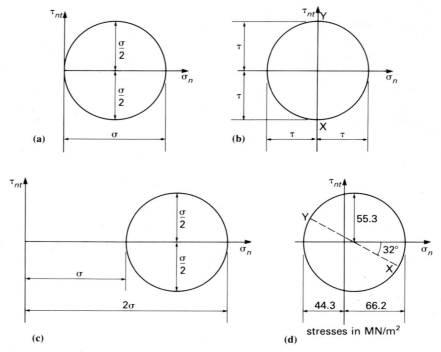

Fig. 9.20.

Fig. 9.19) are also planes of zero normal stress. Also, the principal stresses are $\pm\tau$, acting on planes at $45°$ to the faces of the element in Fig. 9.19.

(c) Figure 9.20c shows the Mohr's stress circle for the state of biaxial tension examined in Example 9.4, with principal stresses (the hoop and axial stresses in the cylinder) of 2σ and σ.

(d) Figure 9.20d shows the Mohr's stress circle for the state of stress analyzed in Examples 9.2 and 9.5, the points X and Y representing the stresses on the planes perpendicular to the x and y directions, respectively. Points on the circle to the left of the shear stress axis correspond to planes on which the normal stress is compressive.

9.1.5 Principal stresses and maximum shear stresses in a three-dimensional state of stress

Although the detailed analysis of a general three-dimensional state of stress at a point is beyond the scope of this book, there are some features we need to be aware of when considering two-dimensional stress states. Since there are two principal stresses for a two-dimensional state of stress, we can anticipate that there will be three in three dimensions. This is indeed the case. Also, the principal planes on which they act are all orthogonal to each other, and are free of shear stresses. In other words, we can always represent a general three-dimensional state of stress as in Fig. 9.21, with the three principal stresses, σ_1, σ_2, and σ_3 acting on the faces of a rectangular element which are free of shear stress. The orientation of the element within the material depends on the state of stress at the point of interest.

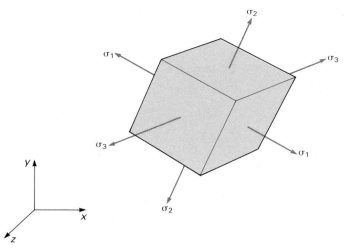

Fig. 9.21.

Now, when we started to consider two-dimensional states of stress, we used equation (9.3) to define a state of plane stress. That is, $\sigma_z = \tau_{yz} = \tau_{xz} = 0$ to give plane stress in the (x, y) plane, as illustrated in Fig. 9.22. The fact that the shear stresses τ_{yz} and τ_{xz} are zero means that τ_{zy} and τ_{zx} are also zero, and the face of the element shown normal to the z direction is free of shear stresses, *and is therefore a principal plane.* Since the normal stress on this plane is $\sigma_z = 0$, a state of plane stress is merely a special case of the general situation shown in Fig. 9.21, in which one of the principal stresses is zero. To ignore this principal stress, however, can lead to serious errors, particularly when calculating maximum shear stresses in the material, which is important when we come to consider yield criteria in Section 9.4.

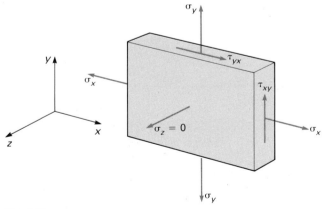

Fig. 9.22.

Suppose we take the normal stress σ_z as the third principal stress, σ_3. If we confine our attention to the (x, y) plane, which contains the principal stresses σ_1 and σ_2, we can draw a Mohr's stress circle as in Fig. 9.23a. On the other hand, if we considered only the plane containing principal stresses σ_2 and σ_3, we would

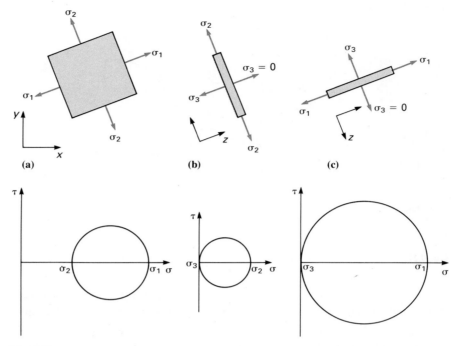

Fig. 9.23.

obtain the circle shown in Fig. 9.23b, which is tangential to the shear stress axis. Similarly, in the plane containing σ_3 and σ_1, the Mohr's stress circle is as shown in Fig. 9.23c, again tangential to the shear stress axis. If we combine the three circles on a single diagram, we obtain Fig. 9.24. We can see that there are maximum shear stresses associated with each of the three principal planes, namely

$$\frac{\sigma_1 - \sigma_2}{2}, \quad \frac{\sigma_2 - \sigma_3}{2} \quad \text{and} \quad \frac{\sigma_1 - \sigma_3}{2} \tag{9.25}$$

and the absolute maximum shear stress, τ_{max}, is

$$\tau_{max} = \frac{\sigma_1 - \sigma_3}{2} \tag{9.26}$$

which is greater than the maximum value, $(\sigma_1 - \sigma_2)/2$, found in the (x, y) plane. This conclusion, however, depends on the assumption inherent in the Mohr's stress circle drawn in Fig. 9.23a that the principal stresses in the (x, y) plane are of the same sign (both tensile). If one had been tensile and the other compressive, with the zero third principal stress lying between them, then the absolute maximum shear stress would have been found in the (x, y) plane.

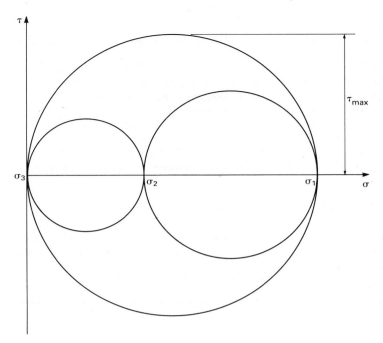

Fig. 9.24.

When analyzing a general three-dimensional state of stress it is customary to arrange the principal stresses such that

$$\sigma_1 \geqslant \sigma_2 \geqslant \sigma_3 \tag{9.27}$$

It then follows that the absolute maximum shear stress is given by equation (9.26). But it also follows that, in a state of plane stress, the principal stresses in the plane concerned are not necessarily σ_1 and σ_2. If we do define them as σ_1 and σ_2, with $\sigma_1 \geqslant \sigma_2$, as we have done so far in this chapter, then we must find the absolute maximum shear stress as either

$$\tau_{max} = \frac{\sigma_1 - \sigma_2}{2} \quad \text{or} \quad \frac{\sigma_1}{2} \quad \text{or} \quad -\frac{\sigma_2}{2} \tag{9.28}$$

according to which is the greatest in magnitude.

It should be noted that, although we have not attempted a formal proof for the general case, any three-dimensional state of stress can be represented on a Mohr stress diagram similar in form to Fig. 9.24, the three circles corresponding to the three principal planes, but with σ_3 not necessarily zero.

EXAMPLE 9.7

Find the absolute maximum shear stresses for the states of plane stress considered in Example 9.6.

(a) In a state of simple uniaxial tension (Fig. 9.20a), the three principal stresses are σ, 0 and 0. The absolute maximum shear stress is therefore $\sigma/2$.

(b) In a state of pure shear (Fig. 9.20b), the three principal stresses are $+\tau$, 0 and $-\tau$. The absolute maximum shear stress is therefore τ.

(c) In the state of biaxial tension represented in Fig. 9.20c, the three principal stresses are 2σ, σ and 0. The absolute maximum shear stress is therefore σ.

(d) In the state of stress represented in Fig. 9.20d, the three principal stresses are $+66.2 \text{ MN/m}^2$, 0 and -44.3 MN/m^2. The absolute maximum shear stress is therefore

$$\tau_{\text{max}} = \tfrac{1}{2}(66.2 + 44.3) = \underline{55.3 \text{ MN/m}^2} \ \ (8.0 \text{ ksi})$$

We note that only in the biaxial tension case (c) is the absolute maximum shear stress greater than the maximum shear stress in the plane of the nonzero stresses. This is because only in case (c) are the principal stresses in this plane of the same sign.

9.2 Transformation of strain

We first defined normal and shear strains in Section 1.3.5. Before we can consider more complex states of strain, with components acting in two or three directions at a point, we must extend the notation we have used to represent strains, particularly shear strains.

9.2.1 Notation for strains

We have already chosen to distinguish between normal and shear strains by the use of the symbols e and γ, respectively. In the case of normal strain, we have also introduced subscripts to indicate the direction of the strain. For example, in the (x, y, z) Cartesian coordinate system, the normal strains are e_x, e_y and e_z, respectively. These were defined as the relative changes in length in the coordinate directions. The sign convention for normal strains defines increases in length in the relevant directions as positive, and decreases as negative. In all the analysis which follows, it is assumed that strains are small in magnitude.

With shear strains the single subscript notation is not practical, because such strains involve displacements and lengths which are not in the same direction. In equation (1.13), we defined the shear strain applied to the initially rectangular element shown in Fig. 9.25a as equal to the small angle ϕ (in radians), where ϕ is the total angle through which each corner of the element is distorted into the form of a parallelogram. In the figure, the bottom edge of the element does not rotate, but in general this is not the case. For example, considering the element shown in Fig. 9.25b which is initially rectangular, with its edges parallel to the x and y coordinate axes, we define the shear strain of this element in the (x, y) plane, again as the total angle of distortion, as

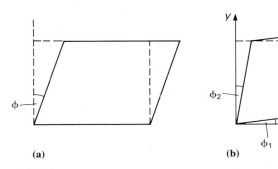

(a)   (b)

Fig. 9.25.

$$\gamma_{xy} = \phi_1 + \phi_2 \tag{9.29}$$

Similar definitions apply to elements in the (y, z) and (z, x) planes. As we will demonstrate below, the magnitude of the shear strain depends on the orientation in the (x, y) plane of the element chosen. The order of the subscripts is unimportant: γ_{xy} and γ_{yx} refer to the same physical quantity. The sign convention we adopt for a shear strain such as γ_{xy} is that if it represents a decrease in angle between the sides of an element of material lying along or parallel to the positive x and y axes, then it is positive. Alternatively, we can think of positive shear strain as being the form of distortion produced by positive shear stresses, as indicated by Fig. 9.26.

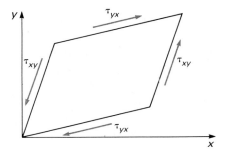

Fig. 9.26.

A link between shear strain and shear stress via the shear modulus for a linearly elastic material has already been established in equation (3.7). This can be generalized to a three-dimensional situation, for example in Cartesian coordinates, as

$$\tau_{xy} = G\gamma_{xy}, \quad \tau_{yz} = G\gamma_{yz}, \quad \tau_{zx} = G\gamma_{zx} \tag{9.30}$$

9.2.2 Plane strain

Let us begin our study of transformation of strain by considering the two-dimensional state known as *plane strain*. It is two-dimensional in the sense that the strain components in one direction are zero. For example, in Cartesian coordinates, if this direction is the z direction, then

$$e_z = \gamma_{yz} = \gamma_{xz} = 0 \tag{9.31}$$

and we can focus our attention on deformations in the (x, y) plane. In practice, states of plane strain occur in bodies which are relatively long and uniformly loaded in the out-of-plane direction: mirror symmetry about cross-sectional planes requires that there be no distortion in this direction. For example, a concrete dam forming a water reservoir is both long and approximately uniform in cross section, and uniformly loaded along its length: plane strain provides a good approximation to the conditions which exist over the cross section. Plane stress and plane strain are two-dimensional approximations to three-dimensional situations which are appropriate for bodies which are respectively very thin and very thick, relative to their in-plane dimensions. The states of stress and strain in bodies of intermediate thickness should be treated as three-dimensional.

If equations (9.31) do not hold, and the state of strain is three-dimensional, then the analysis which follows is still applicable, but incomplete. We should also be clear that plane strain is not the state of strain associated with plane stress, and vice versa. In Chapter 3 we saw that for a linearly elastic material obeying the generalized Hooke's Law, equations (3.6), a uniaxial or biaxial state of stress gives rise to a triaxial state of strain. In particular, according to equation (3.6c) a state of plane stress in the (x, y) plane can in general only give rise to zero normal strain in the z direction if Poisson's ratio is zero, which is not the case for real engineering materials.

Figure 9.27a shows an infinitesimally small rectangular element of material ABCD lying in the (x, y) plane with its sides parallel to the coordinate axes. The lengths of the sides of the element are chosen such that the diagonal AC is at an angle θ to the x axis. Let n and t be a new pair of coordinates, respectively along and perpendicular to this diagonal as shown, with the direction of t being consistent with that used for the analysis of stresses in Fig. 9.8. We wish to find

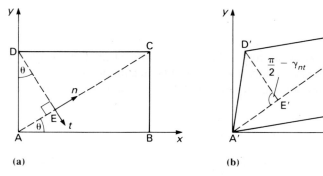

(a)   (b)

Fig. 9.27.

the normal strain, e_n, in the n direction and the shear strain, γ_{nt}, referred to the n and t directions when the element experiences strain components e_x, e_y, and γ_{xy} referred to the x and y directions. Figure 9.27b shows a much exaggerated view of the deformed element, with its corners displaced to the points A', B', C' and D'.

Bearing in mind that all strain components are in reality very small, the modified dimensions of the element are

$$C'D' = A'B' = AB(1 + e_x)$$

$$A'D' = B'C' = BC(1 + e_y)$$

$$A'C' = AC(1 + e_n)$$

and the angle between sides A'B' and B'C' is $(\pi/2 + \gamma_{xy})$ radians. Applying the cosine rule to triangle A'B'C'

$$(A'C')^2 = (A'B')^2 + (B'C')^2 - 2(A'B')(B'C') \cos (\pi/2 + \gamma_{xy})$$

which, since $\cos(\pi/2 + \gamma_{xy}) = -\sin \gamma_{xy}$, becomes

$$(AC)^2(1 + e_n)^2 = (AB)^2(1 + e_x)^2 + (BC)^2(1 + e_y)^2$$
$$+ 2(AB)(BC)(1 + e_x)(1 + e_y) \sin \gamma_{xy}$$

Because the strains are very small, terms involving products or higher powers of strains may be neglected, and this equation approximated by

$$(AC)^2(1 + 2e_n) = (AB)^2(1 + 2e_x) + (BC)^2(1 + 2e_y) + 2(AB)(BC)\gamma_{xy}$$

which, with $(AC)^2 = (AB)^2 + (BC)^2$, reduces to

$$(AC)^2 e_n = (AB)^2 e_x + (BC)^2 e_y + (AB)(BC)\gamma_{xy}$$

Dividing through by $(AC)^2$, and replacing AB/AC and BC/AC by $\cos \theta$ and $\sin \theta$, respectively, we obtain

$$e_n = e_x \cos^2 \theta + e_y \sin^2 \theta + \gamma_{xy} \sin \theta \cos \theta \tag{9.32}$$

which, using the trigonometric relationships

$$\sin 2\theta = 2 \sin\theta \cos\theta \quad \text{and} \quad \cos 2\theta = \cos^2 \theta - \sin^2 \theta$$

becomes

$$e_n = \frac{e_x + e_y}{2} + \frac{e_x - e_y}{2} \cos 2\theta + \frac{\gamma_{xy}}{2} \sin 2\theta \tag{9.33}$$

We note that equations (9.32) and (9.33) are very similar in form to equations (9.4) and (9.6) for stress transformation. Indeed, they are identical if we replace normal stress components by the corresponding normal strain components, and τ_{xy} by $\gamma_{xy}/2$. The factor of a half is a consequence of the way we have defined shear strains. Although it can be eliminated by changing the definition, this is rarely done in engineering practice.

Equation (9.32) or (9.33) can be used to find the normal strain in any direction, at an angle θ to the x axis. For example, the normal strain in the t direction is obtained by replacing θ by $\theta \pm 90°$ in equation (9.33) (recalling that $\sin(2\theta \pm 180°) = -\sin 2\theta$ and $\cos(2\theta \pm 180°) = -\cos 2\theta$) as

$$e_t = \frac{e_x + e_y}{2} - \frac{e_x - e_y}{2} \cos 2\theta - \frac{\gamma_{xy}}{2} \sin 2\theta \tag{9.34}$$

Adding equations (9.33) and (9.34), we obtain

$$e_n + e_t = e_x + e_y \tag{9.35}$$

which means that the sum of two normal strain components in mutually perpendicular directions at a point in a state of plane strain is not affected by the orientation of these directions.

The shear strain, γ_{nt}, referred to the n and t directions in Fig. 9.27 is the decrease in the angle between lines AE and ED. The modified lengths of the sides of triangle AED are

$$A'E' = AE(1 + e_n)$$

$$E'D' = ED(1 + e_t)$$

$$A'D' = AD(1 + e_y)$$

Applying the cosine rule

$$(A'D')^2 = (A'E')^2 + (E'D')^2 - 2(A'E')(E'D')\cos(\pi/2 - \gamma_{nt})$$

which, since $\cos(\pi/2 - \gamma_{xy}) = \sin \gamma_{xy}$, becomes

$$(AD)^2(1 + e_y)^2 = (AE)^2(1 + e_n)^2 + (ED)^2(1 + e_t)^2$$
$$- 2(AE)(ED)(1 + e_n)(1 + e_t) \sin \gamma_{nt}$$

Neglecting terms involving products or higher powers of strains, this equation becomes

$$(AD)^2(1 + 2e_y) = (AE)^2(1 + 2e_n) + (ED)^2(1 + 2e_t) - 2(AE)(ED)\gamma_{nt}$$

which, with $(AD)^2 = (AE)^2 + (ED)^2$, reduces to

$$(AD)^2 e_y = (AE)^2 e_n + (ED)^2 e_t - (AE)(ED)\gamma_{nt}$$

Dividing through by $(AD)^2$, and replacing AE/AD and ED/AD by $\sin\theta$ and $\cos\theta$, respectively, we obtain

$$e_y = e_n \sin^2\theta + e_t \cos^2\theta - \gamma_{nt}\sin\theta\cos\theta$$

Using equations (9.33) and (9.34) to define e_n and e_t in terms of e_x, e_y and γ_{xy}, this becomes

$$e_y = \frac{(e_x + e_y)}{2} - \frac{(e_x - e_y)}{2}\cos^2 2\theta - \frac{\gamma_{xy}}{2}\sin 2\theta\cos 2\theta - \gamma_{nt}\sin\theta\cos\theta$$

or

$$\frac{\gamma_{nt}}{2}\sin 2\theta = \frac{(e_x - e_y)}{2}\sin^2 2\theta - \frac{\gamma_{xy}}{2}\sin 2\theta\cos 2\theta$$

from which

$$\frac{\gamma_{nt}}{2} = \frac{(e_x - e_y)}{2}\sin 2\theta - \frac{\gamma_{xy}}{2}\cos 2\theta \qquad (9.36)$$

Once again, this is identical to the equivalent equation for stress transformations, equation (9.7), if we replace normal stresses by the corresponding normal strains, and shear stresses by half the corresponding shear strains.

EXAMPLE 9.8

The 10 mm by 10 mm square shown in Fig. 9.28 forms part of a body which is in a state of uniform plane strain. When the body is loaded, the square increases in width (in the x direction) by 5.24×10^{-3} mm, reduces in height (in the y direction) by 3.47×10^{-3} mm, and the angle at the bottom left-hand corner increases by 2.98×10^{-4} radians. Find the normal strain in the direction at $41°$ to the positive x axis.

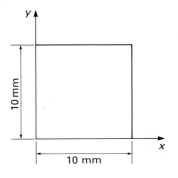
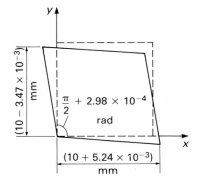

Fig. 9.28.

Because the state of strain is uniform, we can apply the equations developed for infinitesimally small elements of material at a point to an element of finite dimensions. The normal and shear strain components in the (x, y) coordinate system are

$$e_x = \frac{5.24 \times 10^{-3}}{10} = 5.24 \times 10^{-4}$$

$$e_y = -\frac{3.47 \times 10^{-3}}{10} = -3.47 \times 10^{-4}$$

and

$$\gamma_{xy} = -2.98 \times 10^{-4}$$

Note that, because the angle at the origin of coordinates is increased, the shear strain is negative. With these values of strain, and $\theta = 41°$, equation (9.33) gives

$$e_n = \frac{(5.24 - 3.47) \times 10^{-4}}{2} + \frac{(5.24 + 3.47) \times 10^{-4}}{2} \cos 82°$$

$$-\frac{2.98 \times 10^{-4}}{2} \sin 82° = 1.56 \times 10^{-6}$$

9.2.3 Principal strains and maximum shear strains

Just as we did for stresses at a point in a two-dimensional state of stress, we can find maximum and minimum (with respect to direction) principal strains for a two-dimensional state of strain at a point. In view of the close similarity between the transformation equations for stresses and strains (equations (9.6) and (9.7) for stresses, (9.33) and (9.36) for strains), we do not need to work through the analysis again in detail, but can adapt the results derived for stresses, replacing normal stresses by the corresponding normal strains, and shear stresses by half the corresponding shear strains.

From equations (9.11) and (9.16), the directions of the principal strains, which are normal to the principal planes, are given by

$$\tan 2\theta_p = \frac{\gamma_{xy}}{e_x - e_y} \tag{9.37}$$

and the principal strains themselves by

$$e_{1,2} = \frac{e_x + e_y}{2} \pm \frac{1}{2}\sqrt{[(e_x - e_y)^2 + \gamma_{xy}^2]} \tag{9.38}$$

The shear strains associated with the principal planes are zero. The relative orientations of the planes of principal stress and principal strain depend on the relationships between stress and strain for the material concerned. If the relationship between shear stress and shear strain is given by equation (3.7) ($\tau = G\gamma$, where G is the shear modulus of elasticity), and more generally by equations (9.30), then when the shear strain is zero so is the shear stress. Since we have shown that the planes of principal stress and planes of principal strain are also planes of zero shear stress and zero shear strain, respectively, they must coincide.

From equations (9.18) and (9.21), the planes of maximum and minimum shear strain are defined by

$$\tan 2\theta_s = -\frac{(e_x - e_y)}{\gamma_{xy}} \tag{9.39}$$

at 45° to the principal planes, and the strains themselves by

$$\frac{\gamma_{1,2}}{2} = \pm \tfrac{1}{2}\sqrt{[(e_x - e_y)^2 + \gamma_{xy}^2]} = \pm \frac{e_1 - e_2}{2} \tag{9.40}$$

From equation (9.28), taking into account strains not in the (x, y) plane, the absolute maximum shear strain is either

$$\gamma_{max} = e_1 - e_2 \quad \text{or} \quad e_1 \quad \text{or} \quad -e_2 \tag{9.41}$$

according to which is greatest in magnitude.

EXAMPLE 9.9

For the two-dimensional state of strain defined in Example 9.8, find the directions of the principal planes, the principal strains, the maximum in-plane shear strain and the absolute maximum shear strain.

The normal and shear strains in the (x, y) plane are $e_x = 5.24 \times 10^{-4}$, $e_y = -3.47 \times 10^{-4}$, $\gamma_{xy} = -2.98 \times 10^{-4}$. From equation (9.37)

$$\tan 2\theta_p = \frac{\gamma_{xy}}{e_x - e_y} = \frac{-2.98}{5.24 + 3.47} = -0.3421$$

and the only angle in the range $0 \leqslant 2\theta_p \leqslant 180°$ with this value of tangent is $2\theta_p = 161.1°$. Consequently, the angles relative to the x axis of the normals to the principal planes are

$$\theta_p = 80.6° \text{ and } 170.6° \text{ (also } 260.6° \text{ and } 350.6°)$$

The principal strains are given by equation (9.38) as

$$e_{1,2} = \left\{ \frac{5.24 - 3.47}{2} \pm \tfrac{1}{2}\sqrt{[(5.24 + 3.47)^2 + 2.98^2]} \right\} \times 10^{-4}$$

from which

$$e_1 = 5.49 \times 10^{-4} \quad \text{and} \quad e_2 = -3.72 \times 10^{-4}$$

To find which principal plane each of these acts on, we substitute $\theta = 80.6°$ into equation (9.33)

$$e_n = \left\{ \frac{5.24 - 3.47}{2} + \frac{5.24 + 3.47}{2} \cos 161.1° - \frac{2.98}{2} \sin 161.1° \right\} \times 10^{-4}$$

$$= -3.72 \times 10^{-4}$$

which shows that e_2, the minimum principal strain, acts on the principal plane whose normal is at $80.6°$ to the x axis (also the one at $260.6°$ to the same axis).

From equation (9.40), the magnitude of the maximum in-plane shear strain is

$$\gamma_1 = e_1 - e_2 = 9.21 \times 10^{-4}$$

and, because e_1 and e_2 are of opposite signs, this is also the absolute maximum shear strain.

9.2.4 Relationship between Young's modulus and shear modulus

In Chapter 3, equation (3.11), we stated without proof a relationship between the Young's modulus of elasticity, E, shear modulus of elasticity, G, and Poisson's ratio, v, for a material, which was

$$E = 2G(1 + v) \tag{3.11}$$

We are now in a position to derive this result. To do this, we can consider the small element of material shown in Fig. 9.29 subjected to pure shear under plane stress conditions: Figs 9.29a and c show the stresses and strains, respectively, in the (x, y) coordinate system. From equation (9.16), the principal stresses are

$$\sigma_1 = +\tau_{xy} \quad \text{and} \quad \sigma_2 = -\tau_{xy}$$

which act on the principal planes at $45°$ to the x and y axes, as shown in Fig. 9.29b. Similarly, from equation (9.38), the principal strains, in the directions at right angles to the same principal planes, are

$$e_1 = +\frac{\gamma_{xy}}{2} \quad \text{and} \quad e_2 = -\frac{\gamma_{xy}}{2}$$

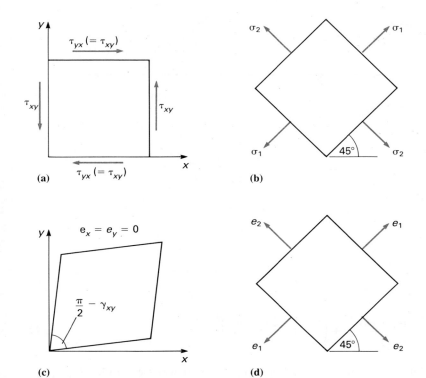

Fig. 9.29.

as shown in Fig. 9.29d. Now, Hooke's Law applied to the biaxial state of stress shown in Fig. 9.29b gives another expression for the maximum principal strain as

$$e_1 = \frac{1}{E}(\sigma_1 - v\sigma_2) = \frac{\tau_{xy}}{E}(1 + v)$$

For the two expressions for e_1 to be identical, we require that

$$E = 2\frac{\tau_{xy}}{\gamma_{xy}}(1 + v)$$

which, in view of equations (9.30) linking the shear stress and strain, becomes

$$E = 2G(1 + v)$$

9.2.5 Mohr's circle for plane strain

Since the transformation equations for plane strain are similar to those for plane stress, we can employ a similar form of pictorial representation. This is known as *Mohr's strain circle*. The strain equivalent of equation (9.24) is

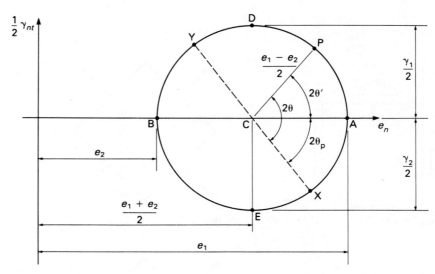

Fig. 9.30.

$$\left[e_n - \frac{(e_1 + e_2)}{2}\right]^2 + \left(\frac{\gamma_{nt}}{2}\right)^2 = \left[\frac{e_1 - e_2}{2}\right]^2 \tag{9.42}$$

If we plot $\gamma_{nt}/2$ against e_n, as in Fig. 9.30, we obtain a circle of radius $(e_1 - e_2)/2$ with a center at $e_n = (e_1 + e_2)/2$, $\gamma_{nt} = 0$. The main difference between this and Fig. 9.18 for stresses is the usual factor of a half attached to the shear strains. Points A and B on this circle represent the principal strains, while points D and E represent the maximum and minimum shear strains. A typical point P on the circle gives the normal strain and half the shear strain, e_n and $\gamma_{nt}/2$, associated with a particular plane, whose normal is at an angle of θ' to the direction of the normal to the plane of maximum principal stress. We note again that an angle subtended at the center of Mohr's circle by an arc connecting two points on the circle is twice the physical angle in the material between the directions of the normals associated with the two points. Points X and Y represent the strains associated with the x and y directions, with e_n and $\gamma_{nt}/2$ coordinates of $(e_x, -\gamma_{xy}/2)$ and $(e_y, +\gamma_{xy}/2)$, respectively.

EXAMPLE 9.10

Draw Mohr's strain circle for the state of strain analyzed in Examples 9.8 and 9.9.

Figure 9.31 shows the required circle, with principal strains of $+5.49 \times 10^{-4}$ and -3.72×10^{-4}. Points X and Y represent the strains associated with the x and y directions.

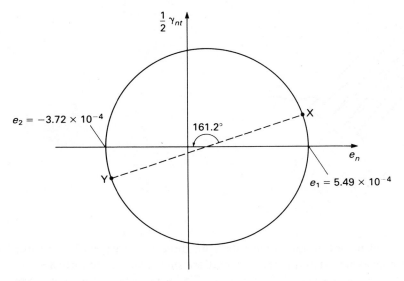

$e_2 = -3.72 \times 10^{-4}$

$161.2°$

$\frac{1}{2} \gamma_{nt}$

e_n

$e_1 = 5.49 \times 10^{-4}$

X

Y

Fig. 9.31.

9.2.6 Use of strain gages

Although we cannot measure stresses within a structural member, we can measure strains, and from them deduce the stresses. Even so, we can only measure strains on the surface. For example, we can mark points and lines on the surface and measure changes in their spacings and angles. In doing this we are of course only measuring average strains over the region concerned. Also, in view of the very small changes in the dimensions, it is difficult to achieve adequate accuracy in the measurements.

In practice, electrical *strain gages* provide a more accurate and convenient method of measuring strains. A typical gage is illustrated in Fig. 9.32. A length of thin wire is folded as shown and cemented to a thin strip of material such as paper or resin, which in turn is cemented to the surface where the strain is to be measured. As the wire is strained axially, its electrical resistance changes by an amount proportional to the strain, and this change can be measured. The constant of proportionality between strain and change in resistance can be found by calibrating the particular type of gage under known strain conditions, such as simple tension. The gage shown in Fig. 9.32 can measure normal strain in the local plane of the surface in the direction of the line PQ, which is parallel to the folds of wire. This strain is an average value for the region covered by the gage, rather than a value at any particular point. The gage is not sensitive to normal strain in the direction perpendicular to PQ, nor does it respond to shear strain. Therefore, in order to determine the state of strain at a particular small region of the surface, we usually need more than one strain gage. To define a general two-dimensional state of strain, we need to have *three* pieces of information, such as strain components e_x, e_y, and γ_{xy} referred to any convenient orthogonal coordinates x and y in the plane of the surface. We

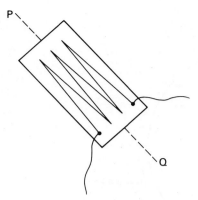

Fig. 9.32.

therefore need to obtain measurements from three strain gages. These three gages must be arranged at different orientations on the surface to form a *strain rosette*. Typical examples are shown in Fig. 9.33, where the gages are arranged at either 45° or 60° to each other.

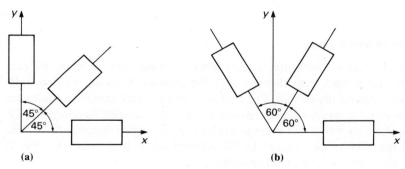

(a) (b)

Fig. 9.33.

Under circumstances where some of the properties of the state of strain are known, we can use fewer than three strain gages. For instance, we may know the directions of the principal strains. A good example is provided by a closed-ended thin-walled cylindrical pressure vessel, where the principal strains are in the hoop and axial directions, and can be measured by two gages attached to the outer surface. Indeed, since we also know the ratio between the principal strains (from the ratio between the principal stresses, and the elastic properties of the material), just one gage would be sufficient.

Knowing the orientations of the three gages forming a rosette, together with the in-plane normal strains they record, the state of strain at the region of the surface concerned can be found. Let us consider the general case shown in Fig. 9.34, where three strain gages numbered 1, 2 and 3 are arranged at angles of θ_1, θ_2 and θ_3 measured counterclockwise from some reference direction, which we can treat as the x axis. Now, although the conditions at a surface, on which there are no shear or normal stress components, are those of plane stress rather than plane strain, we can still use strain transformation equation (9.32) to

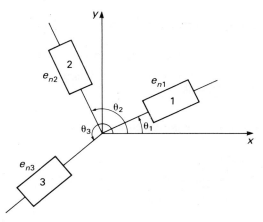

Fig. 9.34.

express the three measured normal strains in terms of strain components e_x, e_y, and γ_{xy} referred to the x and y coordinates, as

$$e_{n1} = e_x \cos^2 \theta_1 + e_y \sin^2 \theta_1 + \gamma_{xy} \sin \theta_1 \cos \theta_1 \qquad (9.43a)$$

$$e_{n2} = e_x \cos^2 \theta_2 + e_y \sin^2 \theta_2 + \gamma_{xy} \sin \theta_2 \cos \theta_2 \qquad (9.43b)$$

$$e_{n3} = e_x \cos^2 \theta_3 + e_y \sin^2 \theta_3 + \gamma_{xy} \sin \theta_3 \cos \theta_3 \qquad (9.43c)$$

This is a set of three simultaneous linear algebraic equations for the three unknowns e_x, e_y, and γ_{xy}, which is laborious to solve by manual methods, but straightforward by the computer method described below. It is also possible to obtain the required information graphically with the aid of Mohr's strain circle.

Program ROSETTE provides a numerical method for solving equations (9.43), using subroutine SOLVE which is described in detail in Appendix B. Having found the strain components in the (x, y) coordinate system, the program then finds the corresponding stresses with the aid of the elastic properties for the material concerned, also the principal strains and principal stresses. The latter calculations use subroutine POINT for the analysis of stresses and strains at a point, which is described in the next section.

```
      PROGRAM  ROSETTE
C
C  PROGRAM TO DETERMINE THE STATE OF STRAIN AND STRESS FROM
C  STRAIN ROSETTE MEASUREMENTS.
C
      DIMENSION  EN(3),A(3,4),EXY(3),THETA(3),SIGXY(3),EP(2),
     1           SIGP(2),THETAP(2)
      REAL NU
      OPEN(5,FILE='DATA')
      OPEN(6,FILE='RESULTS')
      PI=4.*ATAN(1.)
      WRITE(6,61)
   61 FORMAT('STRESSES AND STRAINS DERIVED FROM STRAIN ROSETTE'
     1        ' MEASUREMENTS')
```

```
C
C  INPUT THE YOUNGS MODULUS AND POISSONS RATIO OF THE MATERIAL.
      READ(5,*) E,NU
      WRITE(6,62) E,NU
   62 FORMAT(/ 'YOUNGS MODULUS = ',E12.4,10X,'POISSONS RATIO = ',F5.3)
C
C  INPUT THE ANGULAR ORIENTATIONS (IN DEGREES) OF THE THREE STRAIN
C  GAGES, MEASURED COUNTERCLOCKWISE FROM SOME REFERENCE DIRECTION
C  (WHICH WILL BE TREATED AS THE X  AXIS).
      READ(5,*) (THETA(I),I=1,3)
C
C  INPUT IN THE SAME ORDER THE NORMAL STRAINS MEASURED BY THESE GAGES.
      READ(5,*) (EN(I),I=1,3)
C
C  OUTPUT THE GAGE ORIENTATIONS AND MEASURED STRAINS.
      WRITE(6,63) (I,THETA(I),EN(I),I=1,3)
   63 FORMAT(/ 'ANGULAR ORIENTATIONS OF THE GAGES IN DEGREES MEASURED'
     1 / 'COUNTERCLOCKWISE FROM THE X AXIS, AND THE MEASURED STRAINS'
     2 / '    GAGE NO.    THETA         EN' / (I8,F14.2,E15.4))
C
C  SET UP THE LINEAR EQUATIONS.
      DO 1 I=1,3
      ANG=THETA(I)*PI/180.
      SINT=SIN(ANG)
      COST=COS(ANG)
      A(I,1)=COST**2
      A(I,2)=SINT**2
      A(I,3)=SINT*COST
    1 A(I,4)=EN(I)
C
C  SOLVE THE EQUATIONS.
      CALL   SOLVE(A,EXY,3,3,4,IFLAG)
C
C  A UNIT VALUE OF THE ILL-CONDITIONING FLAG IMPLIES DATA ERROR OR
C  GAGES UNSUITABLY ARRANGED.
      IF(IFLAG.EQ.1) THEN
         WRITE(6,64)
   64    FORMAT(/ 'DATA ERROR OR GAGES UNSUITABLY ARRANGED - STOP')
         STOP
      END IF
C
C  OUTPUT THE IN-PLANE STRAINS IN THE (X,Y) COORDINATE SYSTEM.
      WRITE(6,65) (EXY(I),I=1,3)
   65 FORMAT(/'COMPUTED IN-PLANE STRAINS IN THE (X,Y) COORDINATE SYSTEM'
     1 / 'EX = ',E12.4,5X,'EY = ',E12.4,5X,'GAMXY = ',E12.4)
C
C  CALCULATE AND OUTPUT THE CORRESPONDING IN-PLANE STRESSES.
      DENOM=1.-NU**2
      SIGXY(1)=E*(EXY(1)+NU*EXY(2))/DENOM
      SIGXY(2)=E*(EXY(2)+NU*EXY(1))/DENOM
      G=0.5*E/(1.+NU)
      SIGXY(3)=G*EXY(3)
      WRITE(6,66) (SIGXY(I),I=1,3)
   66 FORMAT(/ 'COMPUTED STRESSES IN THE (X,Y) COORDINATE SYSTEM'
     1 / 'SIGX = ',E12.4,5X,'SIGY = ',E12.4,5X,'TAUXY = ',E12.4)
C
C  ANALYZE THE STATE OF STRESS AND STRAIN AT THE POINT CONCERNED.
      CALL   POINT('STRAIN',EXY,THETAP,EP,ESMAX,AESMAX,SE)
      CALL   POINT('STRESS',SIGXY,THETAP,SIGP,SSMAX,ASSMAX,SE)
C
C  CALCULATE THE THIRD PRINCIPAL STRAIN (WHICH IS NOT ZERO), AND
C  CORRECT THE ABSOLUTE MAXIMUM SHEAR STRAIN.
      EP3=-NU*(SIGP(1)+SIGP(2))/E
      AESMAX=MAX(EP(1)-EP(2),ABS(EP(2)-EP3),ABS(EP3-EP(1)))
C
C  OUTPUT THE PRINCIPAL STRESSES AND STRAINS AND THE DIRECTIONS OF THE
C  NORMALS TO THE CORRESPONDING PRINCIPAL PLANES.
      WRITE(6,67) (I,SIGP(I),EP(I),THETAP(I),I=1,2)
   67 FORMAT(/ 'PRINCIPAL STRESSES AND STRAINS, AND DIRECTIONS OF THE'
     1         ' PRINCIPAL PLANES'
     2       / '(IN DEGREES MEASURED COUNTERCLOCKWISE FROM THE X AXIS)'
     3       / ' NO.      STRESS          STRAIN        DIRECTION'
     4       / (I3,2E15.4,5X,F8.1))
      WRITE(6,68) EP3
   68 FORMAT(' (3',' 0.       ',E15.4,'   OUT OF PLANE)')
```

```
C
C   OUTPUT THE MAXIMUM SHEAR STRESSES AND STRAINS.
        WRITE(6,69) SSMAX,ESMAX,ASSMAX,AESMAX
  69    FORMAT(/ 'MAXIMUM SHEAR STRESSES AND STRAINS'
       1        /  32X,'        STRESS         STRAIN'
       2        / 'MAXIMUM IN-PLANE SHEAR VALUES = ',2E15.4
       3        / 'ABSOLUTE MAXIMUM SHEAR VALUES = ',2E15.4)
C
C   OUTPUT THE VON MISES EQUIVALENT STRESS.
        WRITE(6,70) SE
  70    FORMAT(/ 'VON MISES EQUIVALENT STRESS = ',E12.4)
        STOP
        END
```

The following list provides the definitions of the FORTRAN variables used, arranged in alphabetical order.

Variable	Type	Definition
A	real	Array storing the coefficients of linear equations (9.43)
AESMAX	real	Absolute maximum shear strain
ANG	real	Angle θ of gage orientation in radians
ASSMAX	real	Absolute maximum shear stress
COST	real	Cosine of angle θ of gage orientation
DENOM	real	Denominator expression $(1 - v^2)$ in equations (9.45), explained below
E	real	Young's modulus
EN	real	Array storing the normal strains measured by the three gages
EP	real	Array storing the two in-plane principal strains
EP3	real	Third (out of plane) principal strain
ESMAX	real	Maximum in-plane shear strain
EXY	real	Array storing the strains e_x, e_y and γ_{xy}
G	real	Shear modulus
I	integer	A counter
IFLAG	integer	Flag for a singular or very ill-conditioned coefficient matrix (normally returned as zero by SOLVE, one if the ill-conditioning test is satisfied)
NU	real	Poisson's ratio
PI	real	The mathematical constant π
SE	real	Von Mises equivalent stress
SIGP	real	Array storing the two in-plane principal stresses
SIGXY	real	Array storing the stresses σ_x, σ_y and τ_{xy}
SINT	real	Sine of angle θ of gage orientation
SSMAX	real	Maximum in-plane shear stress
THETA	real	Array storing the angular orientations (in degrees) of the three gages
THETAP	real	Array storing the directions of the normals to the principal planes

The program reads information from a file named DATA, and writes onto a file named RESULTS. All READ statements call for free format input data. After writing out a heading, the program first reads the values of the Young's modulus and Poisson's ratio for the material concerned. It then reads the angular orientations in degrees of the three strain gages, measured counterclockwise from some convenient direction, which is subsequently treated as the x axis. The normal strains measured by these gages are then read. The order in which the gage angles and measured strains are arranged in the input file is arbitrary, provided the same order is used for both.

After writing out the input data, the program calculates and stores in array A the coefficients of the linear equations (9.43), in preparation for solution by subroutine SOLVE. This subroutine is only likely to detect an ill-conditioned set of equations if there is an error in the data (for instance, a duplicated gage angle), or if the gages themselves were unsuitably arranged (for example, at 180° apart, which is equivalent to being at the same angle). The solutions to the equations are the in-plane strains in the (x, y) coordinate system. After writing these out, the program proceeds to calculate the corresponding stresses. Since the conditions at the region of strain measurement are those of plane stress $(\sigma_z = 0)$, Hooke's Law gives

$$e_x = \frac{1}{E}(\sigma_x - v\sigma_y) \tag{9.44a}$$

$$e_y = \frac{1}{E}(\sigma_y - v\sigma_x) \tag{9.44b}$$

which can be rearranged to give stresses in terms of strains as

$$\sigma_x = \frac{E(e_x + ve_y)}{(1 - v^2)} \tag{9.45a}$$

$$\sigma_y = \frac{E(e_y + ve_x)}{(1 - v^2)} \tag{9.45b}$$

The shear stress τ_{xy} is obtained from equations (9.30), using the shear modulus defined by equation (3.11) in terms of the two given elastic constants.

After writing out these three stresses, the program then calls subroutine POINT twice to analyze first the state of strain and then the state of stress, to find the principal values, the principal planes and the maximum shear values. In fact, the principal stresses and maximum in-plane shear stresses could equally well be found from the corresponding strains using stress–strain equations similar to equations (9.45) and (9.30). Since subroutine POINT applied to strains assumes a state of plane strain, it is necessary to first find the third principal strain, e_3 or e_z, which is not zero in the present circumstances, using Hooke's Law

$$e_z = -\frac{v}{E}(\sigma_x + \sigma_y) = -\frac{v}{E}(\sigma_1 + \sigma_2) \tag{9.46}$$

Using this value, the program recalculates the absolute maximum shear strain, before writing out the magnitudes and directions of the principal stresses and strains, also the maximum shear stresses and strains. Finally, the von Mises equivalent stress, the purpose of which is explained in Section 9.4.1, is written out.

EXAMPLE 9.11

A 60° strain rosette is applied to part of a steel structure as shown in Fig. 9.35. The normal strain measurements obtained from the gages numbered 1, 2 and 3 are

$$e_{n1} = 0.32 \times 10^{-4}, \quad e_{n2} = 7.84 \times 10^{-4}, \quad e_{n3} = 2.64 \times 10^{-4}$$

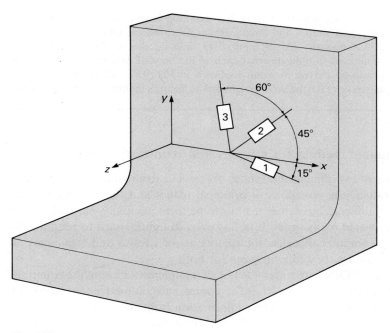

Fig. 9.35.

```
207.0E9     0.3                      (Young's modulus and Poisson's ratio)
-15.0      45.0     105.0     (Angles of orientation of the three gages)
0.32E-4    7.84E-4   2.64E-4     (Strains recorded by the three gages)
```

Fig. 9.36.

```
STRESSES AND STRAINS DERIVED FROM STRAIN ROSETTE MEASUREMENTS

YOUNGS MODULUS =  0.2070E+12          POISSONS RATIO =  .300

ANGULAR ORIENTATIONS OF THE GAGES IN DEGREES MEASURED
COUNTERCLOCKWISE FROM THE X AXIS, AND THE MEASURED STRAINS
     GAGE NO.     THETA        EN
        1        -15.00     0.3200E-04
        2         45.00     0.7840E-03
        3        105.00     0.2640E-03

COMPUTED IN-PLANE STRAINS IN THE (X,Y) COORDINATE SYSTEM
EX =  0.2261E-03     EY =  0.4939E-03     GAMXY =  0.8480E-03

COMPUTED STRESSES IN THE (X,Y) COORDINATE SYSTEM
SIGX =  0.8513E+08     SIGY =  0.1278E+09     TAUXY =  0.6751E+08

PRINCIPAL STRESSES AND STRAINS, AND DIRECTIONS OF THE PRINCIPAL PLANES
(IN DEGREES MEASURED COUNTERCLOCKWISE FROM THE X AXIS)
  NO.     STRESS        STRAIN          DIRECTION
   1    0.1773E+09    0.8047E-03          53.8
   2    0.3565E+08   -0.8465E-04         143.8
  (3    0.         -0.3086E-03     OUT OF PLANE)

MAXIMUM SHEAR STRESSES AND STRAINS
                                  STRESS         STRAIN
MAXIMUM IN-PLANE SHEAR VALUES =   0.7080E+08    0.8893E-03
ABSOLUTE MAXIMUM SHEAR VALUES =   0.8863E+08    0.1113E-02

VON MISES EQUIVALENT STRESS =   0.1624E+09
```

Fig. 9.37.

respectively. Use program ROSETTE to find the strain components e_x, e_y and γ_{xy}, the principal stresses and strains, and the maximum shear stresses and strains.

Figure 9.36 shows the file of input data defining this problem for program ROSETTE, annotated to indicate what each of the lines of data represents. The file RESULTS produced by the program is shown in Fig. 9.37. While the strains are dimensionless, stresses take the same dimensions as Young's modulus, in this case N/m².

9.3 Computer method for stresses and strains at a point

The analysis of two-dimensional states of stress or strain at a point to find the principal values, the directions of principal planes and the maximum shear values is not difficult, but rather tedious to perform manually, particularly if it has to be repeated many times. It is, however, straightforward to program for solution by computer. Because the treatments of stresses and strains are so similar, we can use the same program for both.

Subroutine POINT provides a FORTRAN implementation of the equations for analyzing either stresses or strains at a point, and was used in the last section as part of the process of extracting information from strain rosette measurements. In order to use POINT in the present context, we need a small main program to supply the necessary data and to output the computed results. This program is given the name ANAL2D.

```
      PROGRAM  ANAL2D
C
C  PROGRAM TO ANALYZE A TWO-DIMENSIONAL STATE OF STRESS OR STRAIN
C  AT A POINT.
C
      DIMENSION  SXY(3),S(2),THETAP(2)
      CHARACTER*6  CASE
      OPEN(5,FILE='DATA')
      OPEN(6,FILE='RESULTS')
C
C  INPUT 6 CHARACTER WORD 'STRESS' OR 'STRAIN' TO INDICATE THE TYPE
C  OF CASE TO BE TREATED.
      READ(5,*) CASE
      WRITE(6,61) CASE
  61  FORMAT('ANALYSIS OF A TWO-DIMENSIONAL STATE OF ',A6,' AT A POINT')
C
C  INPUT THE STRESSES/STRAINS IN THE (X,Y) COORDINATE SYSTEM,IN THE
C  ORDER OF X COMPONENT, Y COMPONENT, AND XY (SHEAR) COMPONENT.
      READ(5,*) (SXY(I),I=1,3)
      WRITE(6,62) (SXY(I),I=1,3)
  62  FORMAT(/ 'COMPONENTS IN THE (X,Y) COORDINATE SYSTEM'/
     1    'SX = ',E12.4,5X,'SY = ',E12.4,5X,'SXY = ',E12.4)
C
C  CARRY OUT THE ANALYSIS.
      CALL   POINT(CASE,SXY,THETAP,S,SHMAX,ASHMAX,SE)
C
C  OUTPUT THE PRINCIPAL VALUES AND THE DIRECTIONS OF THE NORMALS TO THE
C  CORRESPONDING PRINCIPAL PLANES.
      WRITE(6,63) (I,S(I),THETAP(I),I=1,2)
  63  FORMAT(/ 'PRINCIPAL VALUES AND DIRECTIONS OF THE PRINCIPAL PLANES'
     1       / '(IN DEGREES MEASURED COUNTERCLOCKWISE FROM THE X AXIS)'
     2       / ' NO.      VALUE      DIRECTION'
     3       / (I3,E15.4,F8.1))
C
C  OUTPUT THE MAXIMUM SHEAR VALUES.
      WRITE(6,64) SHMAX,ASHMAX
  64  FORMAT(/ 'MAXIMUM IN-PLANE SHEAR VALUE = ',E12.4
     1       / 'ABSOLUTE MAXIMUM SHEAR VALUE = ',E12.4)
```

```
C
C   FOR A CASE OF PLANE STRESS, OUTPUT THE VON MISES EQUIVALENT STRESS.
        IF(CASE.EQ.'STRESS') WRITE(6,65) SE
   65   FORMAT(/ 'VON MISES EQUIVALENT STRESS = ',E12.4)
        STOP
        END


        SUBROUTINE  POINT(CASE,SXY,THETAP,S,SHMAX,ASHMAX,SE)
C
C   SUBROUTINE TO ANALYZE THE STATE OF TWO-DIMENSIONAL STRESS OR
C   STRAIN AT A POINT, GIVEN TWO NORMAL COMPONENTS AND ONE SHEAR
C   COMPONENT IN A PARTICULAR ORTHOGONAL COORDINATE SYSTEM - HERE
C   LABELLED X AND Y.
C
        DIMENSION  SXY(3),S(2),THETAP(2)
        CHARACTER*6  CASE
        PI=4.*ATAN(1.)
C
C   IF THE CASE IS ONE OF TWO-DIMENSIONAL STRAIN, THE SHEAR TERM IS
C   HALVED.
        IF(CASE.EQ.'STRAIN') SXY(3)=0.5*SXY(3)
C
C   FIND THE ANGLES OF THE NORMALS TO THE PRINCIPAL PLANES.
        TWOPHI=ATAN2(2.*SXY(3),SXY(1)-SXY(2))
        IF(TWOPHI.LT.0.) TWOPHI=TWOPHI+PI
        THETAP(1)=0.5*TWOPHI
        THETAP(2)=THETAP(1)+0.5*PI
C
C   FIND THE PRINCIPAL STRESSES/STRAINS.
        DO 1 I=1,2
    1   S(I)=0.5*(SXY(1)+SXY(2))+0.5*(SXY(1)-SXY(2))*COS(2.*THETAP(I))
       1                        +SXY(3)*SIN(2.*THETAP(I))
C
C   ARRANGE THE FIRST PRINCIPAL VALUE TO BE GREATER THAN THE SECOND.
        IF(S(1).LT.S(2)) THEN
          TEMP=S(1)
          S(1)=S(2)
          S(2)=TEMP
          TEMP=THETAP(1)
          THETAP(1)=THETAP(2)
          THETAP(2)=TEMP
        END IF
C
C   FIND THE MAXIMUM IN-PLANE SHEAR STRESS/STRAIN.
        SHMAX=0.5*(S(1)-S(2))
C
C   FIND THE ABSOLUTE MAXIMUM SHEAR STRESS/STRAIN.
        ASHMAX=MAX(SHMAX,ABS(0.5*S(1)),ABS(0.5*S(2)))
C
C   FIND THE VON MISES EQUIVALENT STRESS.
        SE=0.
        IF(CASE.EQ.'STRESS') SE=SQRT(S(1)**2+S(2)**2-S(1)*S(2))
C
C   IF THE CASE IS ONE OF TWO-DIMENSIONAL STRAIN, THE SHEAR TERMS
C   ARE DOUBLED.
        IF(CASE.EQ.'STRAIN') THEN
          SXY(3)=2.*SXY(3)
          SHMAX=2.*SHMAX
          ASHMAX=2.*ASHMAX
        END IF
C
C   CONVERT ANGLES OF PRINCIPAL PLANES FROM RADIANS TO DEGREES.
        THETAP(1)=THETAP(1)*180./PI
        THETAP(2)=THETAP(2)*180./PI
        RETURN
        END
```

The following list provides the definitions of the FORTRAN variables used, arranged in alphabetical order.

Variable	Type	Definition
ASHMAX	real	Absolute maximum shear stress or strain
CASE	character	Label for the current case (either 'STRESS' for plane stress or 'STRAIN' for plane strain)
I	integer	A counter
PI	real	The mathematical constant π
S	real	Array storing the two in-plane principal stresses or strains
SE	real	Von Mises equivalent stress
SHMAX	real	Maximum in-plane shear stress or strain
SXY	real	Array storing the stresses or strains in the (x, y) coordinate system
TEMP	real	A temporary store used during reordering of principal values
THETAP	real	Array storing the directions of the normals to the principal planes
TWOPHI	real	A variable representing 2ϕ in equation (9.12)

Program ANAL2D reads information from a file named DATA, and writes onto a file named RESULTS. All READ statements call for free format input data. After reading the type of case to be considered (two-dimensional stress or strain), a heading is written out. The program then reads in and writes out the three components of stress or strain in the (x, y) coordinate system, before calling subroutine POINT to carry out the analysis. When control is returned to the main program, results in the form of magnitudes and directions of the principal stresses and strains, also the maximum shear values, are written out. If it is a case of plane stress, the von Mises equivalent stress, which is discussed in Section 9.4.1, is written out.

Subroutine POINT, which takes the Cartesian components of stress or strain as input data, first multiplies the shear term by a factor of a half if the case is one involving strains. This allows the same equations to be used for stresses and strains. The directions of the principal planes are found with the aid of equations (9.12) and (9.13). Rather than use equation (9.16) to find the principal values, which does not indicate directly which value is associated with which plane, the values are found by substituting each of the plane directions in turn into equation (9.6). If necessary, the principal values and the corresponding angles are reordered so that the larger principal stress or strain is taken first. The maximum in-plane and absolute maximum shear values are then calculated with the aid of equations (9.28), and in the case of plane stress the von Mises equivalent stress using equation (9.54). Finally, if the case is one of two-dimensional strain, all shear terms are doubled to give their correct numerical values, and in all cases the angles of the principal planes are converted from radians to degrees before being returned to the calling program.

EXAMPLE 9.12

Use program ANAL2D to check the results of Examples 9.5 and 9.7d.

The state of plane stress involved in these examples is given by $\sigma_x = 57.8$ MN/m^2 (8.4 ksi), $\sigma_y = -35.9$ MN/m^2 (-5.2 ksi) and $\tau_{xy} = 29.3$ MN/m^2 (4.2 ksi). Figure 9.38 shows the file of input data defining this problem for program ANAL2D, annotated to indicate what each of the lines of data represents.

```
STRESS                                    (Type of case)
57.8   -35.9   29.3        (Normal (x then y ) and shear components)
```

Fig. 9.38.

The file RESULTS produced by the program is shown in Fig. 9.39. While the strains are dimensionless, stresses take the same dimensions as those supplied in the data, in this case MN/m^2. The computed principal stresses, directions of principal planes and maximum shear stresses agree exactly with the values obtained manually.

```
ANALYSIS OF A TWO-DIMENSIONAL STATE OF STRESS AT A POINT

COMPONENTS IN THE (X,Y) COORDINATE SYSTEM
SX =   0.5780E+02     SY =  -0.3590E+02     SXY =   0.2930E+02

PRINCIPAL VALUES AND DIRECTIONS OF THE PRINCIPAL PLANES
(IN DEGREES MEASURED COUNTERCLOCKWISE FROM THE X AXIS)
  NO.      VALUE      DIRECTION
   1      0.6621E+02     16.0
   2     -0.4431E+02    106.0

MAXIMUM IN-PLANE SHEAR VALUE =    0.5526E+02
ABSOLUTE MAXIMUM SHEAR VALUE =    0.5526E+02

VON MISES EQUIVALENT STRESS =    0.9633E+02
```

Fig. 9.39.

EXAMPLE 9.13

Use program ANAL2D to check the results of Example 9.9.

The state of plane strain involved in this example is given by $e_x = 5.24 \times 10^{-4}$, $e_y = -3.47 \times 10^{-4}$ and $\gamma_{xy} = -2.98 \times 10^{-4}$, and Fig. 9.40 shows the file of input data. The file of computed results is shown in Fig. 9.41. The computed principal strains, directions of principal planes and maximum shear strains agree exactly with the values obtained manually.

```
STRAIN
5.24E-4     -3.47E-4     -2.98E-4
```

Fig. 9.40.

```
ANALYSIS OF A TWO-DIMENSIONAL STATE OF STRAIN AT A POINT

COMPONENTS IN THE (X,Y) COORDINATE SYSTEM
SX =   0.5240E-03     SY =  -0.3470E-03     SXY =  -0.2980E-03

PRINCIPAL VALUES AND DIRECTIONS OF THE PRINCIPAL PLANES
(IN DEGREES MEASURED COUNTERCLOCKWISE FROM THE X AXIS)
  NO.      VALUE      DIRECTION
   1      0.5488E-03    170.6
   2     -0.3718E-03     80.6

MAXIMUM IN-PLANE SHEAR VALUE =    0.9206E-03
ABSOLUTE MAXIMUM SHEAR VALUE =    0.9206E-03
```

Fig. 9.41.

9.4 Yield and fracture criteria

In Section 1.4 we saw that materials in uniaxial tension which exhibit linear elastic stress–strain behavior at moderate stress levels deviate from this behavior once a point on the stress–strain curve known as the *yield point* is reached. In other words, beyond a certain *yield stress* or *yield strain*, the behavior is no longer linear elastic, and the material is said to be *yielding* or behaving plastically. What happens subsequently depends on whether the material is ductile or brittle (typical stress–strain curves for the two types are shown in Fig. 1.25). Ductile materials, at least under static loading, undergo relatively large amounts of plastic deformation before final failure. Necking of specimens of the type shown in Fig. 1.24 is evidence of this deformation. Also, if the appearance of the exposed surfaces after failure at the neck is examined it is found that the separation of layers of material occurs not along the plane at right angles to the axis of the specimen but along a series of small planes inclined (typically at about 45°) to the axis. This suggests that yielding and failure are due to shear stresses causing slipping of the ductile material along oblique planes. On the other hand, brittle materials exhibit little or no plastic deformation before failure occurs, by fracturing along planes normal to the applied load. Under cyclically varying loading conditions all materials, even those which are ductile under steady load, tend to exhibit brittle failure at relatively low stress levels, due to the phenomenon of *fatigue*.

When designing a member of an engineering system we must avoid failure, and we often wish to avoid yielding of the material. Indeed, for the reasons discussed in Section 1.3.4, we often have to maintain a safety factor substantially greater than unity against either yielding or ultimate failure. In other words, we must ensure that the maximum stress anywhere in the member does not exceed some maximum allowable value. This maximum allowable value is some fraction of the yield stress, σ_Y, for a ductile material or some fraction of the ultimate (fracture) stress, σ_U, for a brittle material, where σ_Y or σ_U is obtained from a simple tension test. So long as the state of stress in the member to be designed is that of uniaxial tension the maximum stress in it is straightforward to define. Similarly, if the state of stress is simple shear, the shear stress can in principle be compared with the yield or ultimate stress measured in a shear test.

For the more complex states of stress which occur in most situations of practical interest, we would like to have a way of comparing the severity·of the combined effect of the several stress components with that of a simple stress state, usually simple tension. In other words, from a complex state of stress we wish to extract a single value of normal stress which can be compared with σ_Y or σ_U to determine whether the loading is excessive. A particular way of characterizing a complex state of stress by a single stress to test for failure is known as a *yield or fracture criterion*.

A yield or fracture criterion is a theoretical formula, which can only be verified by comparison with appropriate experimental results obtained under complex states of stress. Although a number of criteria have been developed over the past century or so, we will only consider those which have proved to be reliable and are currently in common use. Yield and fracture criteria are most conveniently expressed in terms of principal stresses, which completely determine a general state of stress, and which do not depend on the particular coordinate system chosen to describe it.

9.4.1 Yield criteria for ductile materials

Because experimental evidence suggests that the yielding of ductile materials is due to shear stresses causing slipping between layers of the material, we can anticipate that appropriate yield criteria will be based on shear stresses. Two such criteria are in common use.

Maximum shear stress (Tresca) criterion

This criterion is based on the idea that a ductile material will yield under a general state of stress when the absolute maximum shear stress is equal to the absolute maximum shear stress at the yield point in a simple tension test. It is often known as Tresca's criterion, after the French engineer Henri Tresca, (1814–1885).

As we saw in Example 9.7a, in a state of simple uniaxial tension with an applied stress of σ, the absolute maximum shear stress is $\sigma/2$. Consequently, the maximum shear stress yield criterion can be expressed symbolically as

$$\tau_{max} = \frac{\sigma_Y}{2} \tag{9.47}$$

In a general three-dimensional state of stress, the absolute maximum shear stress is given in terms of the maximum and minimum principal stresses, σ_1 and σ_3, by equation (9.26) as

$$\tau_{max} = \frac{\sigma_1 - \sigma_3}{2} \tag{9.48}$$

On the other hand, in a state of plane stress with principal stresses σ_1 and σ_2, the absolute maximum is given by equation (9.28) as the largest of

$$\tau_{max} = \left|\frac{\sigma_1 - \sigma_2}{2}\right| \quad \text{or} \quad \left|\frac{\sigma_1}{2}\right| \quad \text{or} \quad \left|\frac{\sigma_2}{2}\right| \tag{9.49}$$

Therefore, if σ_1 and σ_2 are of opposite signs

$$\sigma_1\sigma_2 < 0 \quad \text{and} \quad |\sigma_1 - \sigma_2| = \sigma_Y \tag{9.50a}$$

and if they are of the same sign

$$\sigma_1\sigma_2 > 0 \quad \text{and} \quad |\sigma_1| = \sigma_Y \quad \text{or} \quad |\sigma_2| = \sigma_Y \tag{9.50b}$$

These relationships are plotted with the principal stresses as the axes in Fig. 9.42. Lines CD and AF in the second and fourth quadrants are given by equation (9.50a), while AB, BC, DE and EF in the first and third quadrants are given by equations (9.50b). The closed hexagon ABCDEF is known as the *yield locus*. In other words, if a two-dimensional state of stress has principal stresses which when plotted define a point within the hexagon, the material has not yielded; if it is outside, yielding has occurred. We can identify the various points marked on the locus with different types of stress states. Point A represents simple tension in one direction, while C corresponds to simple tension at right angles to this. Similarly, points D and F represent simple compression, and points B and E equal biaxial tension and compression, respectively. Also, points P and Q represent states of simple shear (Example 9.6 showed that such states have principal stresses which are equal in magnitude to the applied shear stress, τ, but opposite in sign). We note that a consequence of assuming that yielding depends only on shear stresses is that the yield stress in simple compression must be equal in magnitude to that in simple tension.

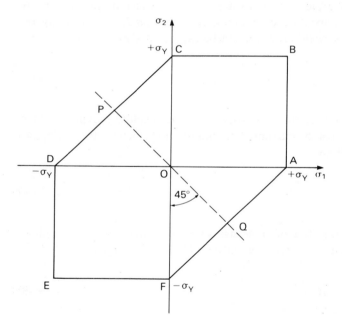

Fig. 9.42.

Shear strain energy (Von Mises) criterion

This criterion is often presented as being based on the idea that yielding of a ductile material under a general state of stress will occur when the density of shear strain energy is equal to the density of shear strain energy at the yield point in a simple tension test. It is often referred to as von Mises' criterion, after the German–American mathematician Richard von Mises (1883–1953). An equivalent, but much more straightforward, way of deriving it is to base yielding not on just the absolute maximum shear stress in the material (as in the Tresca criterion), but on the root mean square maximum shear stress, thereby taking into account the shear stresses on planes at right angles to that of the absolute maximum.

Using equation (9.25) to define the maximum shear stresses associated with the three principal planes, the root mean square maximum shear stress for a complex three-dimensional state of stress is

$$\tau_m = \sqrt{\left\{\frac{1}{3}\left[\left(\frac{\sigma_1 - \sigma_2}{2}\right)^2 + \left(\frac{\sigma_2 - \sigma_3}{2}\right)^2 + \left(\frac{\sigma_3 - \sigma_1}{2}\right)^2\right]\right\}} \qquad (9.51)$$

In simple uniaxial tension, with $\sigma_1 = \sigma_Y$, $\sigma_2 = 0$ and $\sigma_3 = 0$, this becomes

$$\tau'_m = \frac{\sigma_Y}{\sqrt{6}}$$

and we obtain the yield criterion by equating τ_m and τ'_m to give

$$(\sigma_1 - \sigma_2)^2 + (\sigma_2 - \sigma_3)^2 + (\sigma_3 - \sigma_1)^2 = 2\sigma_Y^2 \qquad (9.52)$$

Under plane stress conditions, with $\sigma_3 = 0$, this becomes

$$\sigma_1^2 + \sigma_2^2 - \sigma_1\sigma_2 = \sigma_Y^2 \qquad (9.53)$$

Another way of expressing the same result is to define a von Mises *equivalent stress* (sometimes referred to as an *effective stress*) as

$$\sigma_e = \sqrt{(\sigma_1^2 + \sigma_2^2 - \sigma_1\sigma_2)} \qquad (9.54)$$

and take yielding to occur when this normal stress is equal to the measured yield stress in simple tension. In other words, the equivalent stress is the stress in uniaxial tension which is equivalent to the complex state of stress according to the von Mises criterion of yielding. The equivalent stress is therefore a useful parameter with which to characterize a state of stress, which is why it is calculated in programs ROSETTE and ANAL2D in Sections 9.2.6 and 9.3.

If we plot the curve defined by equation (9.53) with axes of σ_1 and σ_2, we obtain Fig. 9.43. The shape of the yield locus is an ellipse, with the major and minor axes along the biaxial tension/compression line TT and pure shear line SS, respectively. Figure 9.44 shows the Tresca and von Mises yield loci, Figs 9.42 and 9.43, plotted on the same axes. We see that the points A, B, C, D, E and F at the corners of the Tresca locus also lie on the von Mises locus. This must be the case at points A and C, because both yield criteria use simple

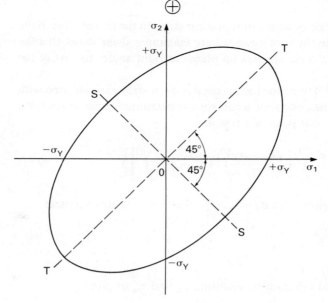

Fig. 9.43.

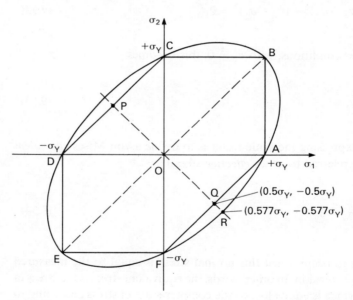

Fig. 9.44.

tension as a reference condition. We have seen already that the assumption that
yielding is a phenomenon controlled by shearing means that the yield stress in
simple compression is equal in magnitude to that in simple tension, which
explains why the loci are coincident at points D and F. They are also coincident
at points B and E representing biaxial tension and compression.

The most significant differences between the two criteria occur under
pure shear conditions. For example, point Q on the Tresca locus is defined

by $\sigma_1 = -\sigma_2 = \sigma_Y/2$, while point R on the von Mises locus is given by $\sigma_1 = -\sigma_2 = \sigma_Y/\sqrt{3} = 0.577\,\sigma_Y$, a difference in values of some 15%. A simple test for which of the two yield criteria is the more appropriate for a given material is therefore to compare the yield stresses measured in simple tension and pure shear, the latter in a torsion test on a thin-walled cylinder, for example. A ratio of shear to tensile yield stress around 0.5 suggests yielding according to Tresca, while one of about 0.577 favors von Mises. In practice, ratios for most ductile materials are closer to 0.577 than to 0.5. Measured yield behavior under more complex states of stress also tends to confirm the von Mises criterion as the more accurate. The Tresca maximum shear stress criterion nevertheless has a useful role to play. It is somewhat easier to use than von Mises, and has the important merit of being moderately conservative. Figure 9.44 shows that the Tresca locus either lies within or just touches the von Mises locus, so that it tends to predict the onset of yielding at stress levels somewhat below the actual ones. The difference, however, is not more than the 15% for the case of pure shear. A strategy sometimes used is to locate points of possible yield in a member with the aid of Tresca, and then to find the equivalent stress more accurately using von Mises.

It is instructive to extend the plotting of yield criteria to three-dimensional states of stress. Figure 9.45 does this for Tresca and von Mises, the two-dimensional loci becoming three-dimensional yield surfaces or envelopes. An envelope represents the interface between elastic states of stress inside and plastic outside, according to the particular criterion of yielding. The von Mises envelope is a circular cylinder with its geometric axis lying along the line $\sigma_1 = \sigma_2 = \sigma_3$ of equal principal stresses, which is equally inclined to the three

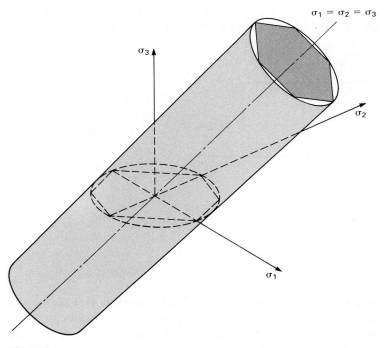

Fig. 9.45.

principal stress axes. The Tresca surface has the same geometric axis but has a cross section in the form of a regular hexagon just touching the von Mises cylinder at six positions around its circumference. The loci formed at the intersections of these surfaces with the (σ_1, σ_2) plane remain as shown in Fig. 9.44. The line $\sigma_1 = \sigma_2 = \sigma_3$ is of particular interest, because according to both criteria yielding never occurs, irrespective of the magnitude of the stresses. This is because with three equal principal stresses there are no shear stresses, and therefore no ductile yielding. The state of stress concerned is that which we described as hydrostatic in Section 3.1, where the absence of yielding was also mentioned.

9.4.2 Failure criteria for brittle materials

With experimental evidence suggesting that failure of brittle materials is due to normal stresses (particularly tensile stresses) causing fracture, we can anticipate that appropriate failure criteria will be based on normal stresses. Two such criteria are used.

Maximum principal stress (Rankine) criterion

This criterion assumes that failure of a brittle material under a general state of stress will occur when the magnitude of the greatest principal stress is equal to the ultimate or failure stress in simple tension. Under plane stress conditions

$$|\sigma_1| = \sigma_U \quad \text{or} \quad |\sigma_2| = \sigma_U \tag{9.55}$$

which can be plotted as in Fig. 9.46. We note that in the first and third quadrants, the locus is the same as that for the Tresca criterion.

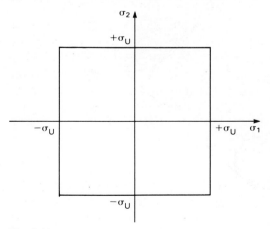

Fig. 9.46.

The main shortcoming of this criterion is that it predicts that failure will occur at the same stress levels in tension and compression, which does not agree with practical experience. Brittle materials such as cast iron and concrete are much stronger in compression than in tension. This is because microscopic cracks and flaws in the material tend to weaken it under tensile loading which opens them further, but have little effect when closed by compressive loading.

Mohr fracture criterion

This criterion allows for different ultimate strengths in tension and compression. In the first and third quadrants of the failure locus, shown in Fig. 9.47, the maximum principal stress theory is adopted, based on the ultimate strengths of the material in tension and compression, σ_{UT} and σ_{UC}, respectively. In the second and fourth quadrants, a form of maximum shear stress theory is used.

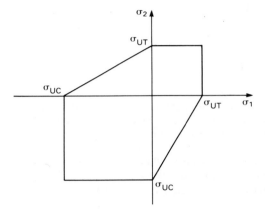

Fig. 9.47.

In addition to microscopic flaws in materials, there may also be macroscopic cracks in the members of engineering systems. The effects of such cracks are not measured by tensile tests on uncracked samples of material. If a crack is detected, it is necessary to investigate whether it will increase in size and lead to failure of the member involved. Such investigations are the province of the subject of *fracture mechanics*, which is beyond the scope of this book.

9.4.3 Examples

A failure criterion is applicable to a state of stress at a particular point of a structural member. In a practical situation it is necessary first to locate the most critical point. If this cannot be done by inspection, then we must calculate the maximum shear stress, von Mises equivalent stress or maximum principal stress (according to which criterion is to be used) at as many points as is necessary to find the most severe state of stress. Of the four examples which follow, the first two involve specified states of stress, the third a uniform state everywhere, and only the fourth requires the preliminary location of the critical point.

EXAMPLE 9.14

The state of plane stress shown in Fig. 9.48 occurs at a point in a steel structure. The yield stress of the steel is 320 MN/m² (46 ksi). Find the safety factors against yielding according to the Tresca and von Mises criteria.

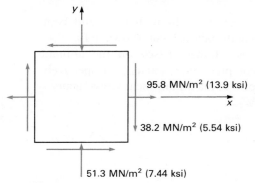

Fig. 9.48.

The state of stress defined in the (x, y) coordinate system is $\sigma_x = 95.8$ MN/m² (13.9 ksi), $\sigma_y = -51.3$ MN/m² (-7.44 ksi) and $\tau_{xy} = -38.2$ MN/m² (-5.54 ksi). Use of program ANAL2D (Section 9.3) gives the following relevant results:

Absolute maximum shear stress = 82.9 MN/m² (12.0 ksi)
Von Mises equivalent normal stress = 145.3 MN/m² (21.1 ksi)
(Principal stresses 105.1 and -60.6 MN/m², 15.2 and -8.8 ksi)

According to the Tresca criterion, equation (9.47), yielding would occur when the absolute maximum shear stress reaches half the tensile yield stress, in this case 160 MN/m² (23 ksi). The factor of safety against yielding is therefore

$$\text{Factor of safety (Tresca)} = \frac{160}{82.9} = \underline{1.93}$$

According to the von Mises criterion, equations (9.53) and (9.54), yielding would occur when the equivalent normal stress reaches the tensile yield stress, in this case 320 MN/m² (46 ksi). The factor of safety against yielding is therefore

$$\text{Factor of safety (von Mises)} = \frac{320}{145.3} = \underline{2.20}$$

We note that Tresca is some 12% more conservative than von Mises.

EXAMPLE 9.15

A strain rosette with three gages arranged at 120° to each other is located at what is expected to be a critical point of a cast iron press frame. Under test conditions with the press applying a force of 10 kN, the gages register strains of -3.67×10^{-3}, -7.89×10^{-4} and 1.32×10^{-3}. Using the elastic and strength properties given in Appendix A for cast iron, together with the Mohr fracture criterion, determine the press force at which failure at the point concerned would occur. Assume that the system remains linearly elastic.

Since we only need to establish the state of stress at the point of interest in terms of principal stresses, the absolute angular orientations of the gages are not important. With equal relative orientations, the order in which we take the gages is also not important. Figure 9.49 shows the assumed arrangement. Use of program ROSETTE (Section 9.2.6), with elastic properties of 70 GN/m² for Young's modulus and 0.25 for Poisson's ratio, gives the following relevant results based on the measured strains:

Principal stresses $\sigma_1 = 64.3$ MN/m² and $\sigma_2 = -259.6$ MN/m²

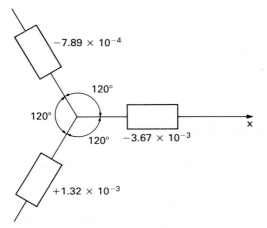

Fig. 9.49.

Now, we are interested in failure when the press force is increased, say by a factor of β. If the system remains linearly elastic, both principal stresses will be increased by the same factor, to 64.3β MN/m² and -259.6β MN/m², respectively. According to the Mohr fracture criterion, Fig. 9.47, with a state of stress lying in the fourth quadrant failure would occur when

$$\frac{\sigma_1}{\sigma_{UT}} - \frac{\sigma_2}{\sigma_{UC}} = 1$$

where $\sigma_{UT} = 200$ MN/m² and $\sigma_{UC} = 700$ MN/m² are the ultimate tensile and compressive strengths, respectively. Hence

$$\frac{64.3\beta}{200} + \frac{259.6\beta}{700} = 1$$

and $\qquad \beta = 1.44$

Therefore, the required press force to cause failure is 1.44×10 kN or <u>14.4 kN</u> (3.2 kip).

EXAMPLE 9.16

A pressure vessel takes the form of a closed-ended thin-walled circular cylinder of diameter D and wall thickness t. The yield stress of the ductile material used is σ_Y. Derive an expression for the internal pressure, p, at which yielding will occur according to the von Mises criterion.

According to equations (2.16) and (2.18), the uniform hoop and axial stresses, σ_θ and σ_z, are given by

$$\sigma_\theta = 2\sigma_z = \frac{pD}{2t}$$

Since these are principal stresses, and the third principal stress is negligibly small, the von Mises equivalent stress is given by equation (9.54) as

$$\sigma_e = \sqrt{(\sigma_\theta^2 + \sigma_z^2 - \sigma_\theta\sigma_z)} = \frac{\sigma_\theta\sqrt{3}}{2} = \frac{pD\sqrt{3}}{4t}$$

Yielding occurs when this is equal to the yield stress, and

$$p = \frac{4\sigma_Y t}{D\sqrt{3}}$$

which is the required expression for internal pressure.

EXAMPLE 9.17

The uniform horizontal steel cantilever shown in Fig. 9.50 is 300 mm (12 in) long, 25 mm (1 in) diameter and carries a torque, T, of 300 Nm (220 lb ft). Find the value of the vertical force, F, at the free end which will just cause yielding according to the Tresca criterion, given that the yield stress of the steel is 350 MN/m² (50 ksi).

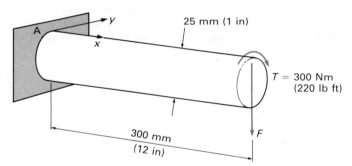

Fig. 9.50.

This is a straightforward example of combined torsion and bending. Although we are not told precisely how the torque and force are applied at the free end, this is not important because the greatest stresses are at the built-in end. Indeed, our first task is to locate the critical state of stress, which can be done by inspection in this case. We know that in torsion the greatest shear stress at a particular cross section occurs at the outer surface of a circular shaft. For a uniform shaft and constant torque, this shear stress remains constant along its length. Due to bending, the greatest normal stress is also at the outer surface, at the point or points furthest from the neutral axis of the cross section (in this case the horizontal diameter). Bending also creates shear stresses, but as we saw in Section 5.3.2, for all but very short beams these stresses are relatively small in magnitude, and are zero where the normal bending stresses are greatest.

With constant shear stress at the outer surface due to the applied torque, the critical state of stress will be where the normal bending stress is greatest, at either the highest or lowest point of the cross section at the built-in end of the cantilever. Using equation (7.11), we find the shear stress there as

$$\tau = \frac{D}{2}\frac{T}{J}$$

where D is the diameter and the polar second moment of area is

SI units

$$J = \frac{\pi D^4}{32} = \frac{\pi \times 0.025^4}{32} = 3.835 \times 10^{-8}\ \text{m}^4$$

US customary units

$$J = \frac{\pi D^4}{32} = \frac{\pi \times 1^4}{32} = 0.098\,17\ \text{in}^4$$

Therefore

$$\tau = \frac{0.025}{2} \times \frac{300}{3.835 \times 10^{-8}}\ \text{N/m}^2 = 97.8\ \text{MN/m}^2$$

Therefore

$$\tau = \frac{1}{2} \times \frac{220 \times 12}{0.098\,17}\ \text{psi} = 13.45\ \text{ksi}$$

The point marked A in Fig. 9.50 is one of the critical points (the other is on the underside of the cantilever where the bending stress is compressive), and the state of plane stress there is shown in Fig. 9.51. Choosing the x and y coordinates as shown, we know the shear stress τ_{xy} and the normal stress σ_y (which is zero), and we wish to find σ_x such that yielding just occurs at this point. From equation (9.16), the principal stresses are

$$\sigma_{1,2} = \frac{\sigma_x}{2} \pm \tfrac{1}{2}\sqrt{(\sigma_x^2 + 4\tau_{xy}^2)}$$

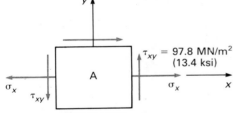

Fig. 9.51.

and must be of opposite signs. Therefore, from equation (9.28), the absolute maximum shear stress is

$$\tau_{\text{max}} = \tfrac{1}{2}(\sigma_1 - \sigma_2) = \tfrac{1}{2}\sqrt{(\sigma_x^2 + 4\tau_{xy}^2)}$$

According to the Tresca criterion, equation (9.47), at yield this is equal to half the tensile yield stress, and

$$\sigma_x^2 + 4\tau_{xy}^2 = \sigma_Y^2$$

SI units
Hence

$$\sigma_x = \sqrt{(350^2 - 4 \times 97.8^2)} = 290.2\ \text{MN/m}^2$$

US customary units
Hence

$$\sigma_x = \sqrt{(50^2 - 4 \times 13.45^2)} = 42.15\ \text{ksi}$$

SI units

The bending moment necessary to create this bending stress is

$$M = \frac{\sigma_x}{D/2} I$$

where the bending second moment of area is $I = J/2 = 1.917 \times 10^{-8}$ m^4. Hence

$$M = \frac{290.2 \times 10^6 \times 1.917 \times 10^{-8}}{0.0125} = 445.1 \text{ Nm}$$

and, since the bending moment at the built-in end of the cantilever is the product of its length and the force F, the required value of F is

$$F = \frac{445.1}{0.300} \text{ N} = \underline{1.48 \text{ kN}}$$

US customary units

The bending moment necessary to create this bending stress is

$$M = \frac{\sigma_x}{D/2} I$$

where the bending second moment of area is $I = J/2 = 0.04908$ in^4. Hence

$$M = \frac{42.15 \times 0.049\,08}{0.5} = 4.137 \text{ kip-in}$$

and, since the bending moment at the built-in end of the cantilever is the product of its length and the force F, the required value of F is

$$F = \frac{4.137}{12} \text{ kip} = \underline{345 \text{ lb}}$$

We note that, had the von Mises criterion been specified, the form of the algebraic equation for σ_x would have been much more difficult to solve.

Problems

SI METRIC UNITS – STRESSES AND STRAINS AT A POINT

9.1 Show that the complementary shear stresses $\tau_{r\theta}$ and $\tau_{\theta r}$ in the cylindrical polar coordinate system are equal in magnitude.

9.2 For the state of plane stress shown in Fig. P9.2, find the normal and shear stresses acting on a plane whose normal is at 70° to the x axis, measured in the counterclockwise direction.

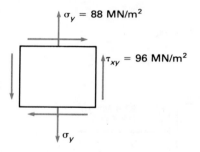

Fig. P9.3.

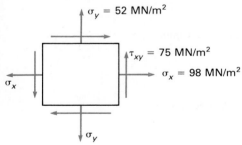

Fig. P9.2.

9.3 For the state of plane stress shown in Fig. P9.3 find the normal and shear stresses acting on a plane whose normal is at 135° to the x axis, measured in the counterclockwise direction.

9.4 and 9.5 For the states of plane stress illustrated in Figs P9.2 and P9.3, find the principal stresses, the directions of the principal planes, the maximum in-plane shear stresses, and the absolute maximum shear stresses.

9.6 and 9.7 Use program ANAL2D to check the results of Problems 9.4 and 9.5.

9.8 and 9.9 Draw Mohr's stress circles for the states of plane stress shown in Figs P9.2 and P9.3, showing the points representing stresses on the planes normal to the x and y axes in each case.

9.10 Draw Mohr's stress circle for a state of biaxial tension where the tensile stresses are of magnitude σ, and hence find the maximum shear stress in the wall of a thin sphere under internal pressure.

9.11 to 9.13 For the states of plane strain shown in the table, find the normal and shear strains associated with planes whose normals are at the angles θ shown to the x axis, measured in the counter-clockwise direction.

Problem	e_x	e_y	γ_{xy}	θ
9.11	1.73×10^{-3}	0.59×10^{-3}	-0.35×10^{-3}	$30°$
9.12	-0.87×10^{-3}	0.43×10^{-3}	0.71×10^{-3}	$-55°$
9.13	$0.$	-1.16×10^{-3}	-0.62×10^{-3}	$125°$

9.14 to 9.16 For the states of plane strain defined in Problems 9.11 to 9.13, find the principal strains, the directions of the principal planes, the maximum in-plane shear strains, and the absolute maximum shear strains.

9.17 to 9.19 Use program ANAL2D to check the results of Problems 9.14 to 9.16.

9.20 to 9.22 Draw Mohr's strain circles for the states of plane strain defined in Problems 9.11 to 9.13, showing the points representing the strains associated with the x and y directions in each case.

9.23 The principal strains e_1 and e_2 are measured at a point on the surface of a solid steel shaft which is subjected to bending and torsion. The values are $e_1 = 1.10 \times 10^{-3}$, $e_2 = -0.60 \times 10^{-3}$, and e_1 is inclined at $20°$ to the axis of the shaft. If the diameter of the shaft is 50 mm, determine the applied torque and the absolute maximum shear stress in the material at the point concerned.

9.24 A solid bar of circular cross section with a diameter of D is subjected to a torque of magnitude T and a bending moment of magnitude M. Find an expression for the absolute maximum shear stress in the shaft.

9.25 The short vertical column shown in Fig. P9.25 is firmly fixed at the base and projects a distance of 1.25 m. The column is of I section, 266 mm deep and 148 mm wide, flanges 13.0 mm thick, web 7.6 mm thick (cross-sectional area 5700 mm², maximum second moment of area 70.8 $\times 10^6$ mm⁴). An inclined force of 110 kN acts at the top of the column, in the center and in the plane containing the center line of the web, the inclination of the line of action of the force being $30°$ to the vertical. Assuming that the shear loading is uniformly distributed over the web, find the principal stresses
1 at the outer faces of the flanges at the base,
2 at the center of the web at the base,
3 at the points in the web 100 mm from the neutral axis at the base.

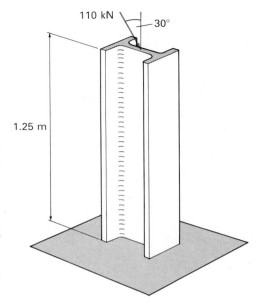

Fig. P9.25.

9.26 A cylindrical gas storage tank has a diameter of 2.5 m and a wall thickness of 30 mm. Strain gages attached to the outer surface of the tank in the hoop and axial directions show strains of 4.59×10^{-4} and 1.10×10^{-4}, respectively, when the tank is pressurized. The value of Young's modulus for the tank material is 200 GN/m². Find the pressure in the tank, the principal stresses and the absolute maximum shear stress in the tank wall.

9.27 Two strain gages are fitted at $\pm 45°$ to the axis of a 75 mm diameter steel shaft, as shown in Fig. P9.27. The shaft is rotating, and in addition to transmitting power it is subjected to an unknown bending moment and axial force. The readings of the gages are recorded, and it is found that the maximum and minimum values for each gage occur at $180°$ intervals of shaft rotation and are -0.60×10^{-3} and 0.30×10^{-3} for the two gages at one instant and -0.50×10^{-3} and 0.40×10^{-3} for the same gages $180°$ of rotation later. Determine

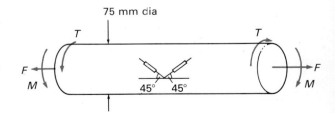

Fig. P9.27.

the transmitted torque and applied bending moment and axial force. Assume all the loads are steady, not varying during each rotation of the shaft.

9.28 A small element of a structural steel member in a two-dimensional stress field is subjected to normal stresses σ_x and σ_y in the x and y directions, together with shear stresses τ_{xy} and τ_{yx}. The shear stress has a magnitude of 46 MN/m², and the measured principal strains are 0.88×10^{-3} and -0.12×10^{-3}. Find the stresses σ_x and σ_y.

9.29 Three strain gages forming a rosette as shown in Fig. P9.29 are attached at a particular point to the surface of a structural member and register the strains shown. Use program ROSETTE to find the magnitudes and directions of the principal strains at the point concerned.

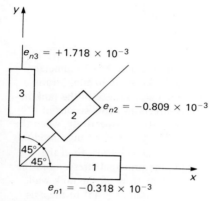

Fig. P9.29.

9.30 Strain measurements are recorded from a 60° strain rosette as shown in Fig. P9.30. Use program ROSETTE to determine the principal strains and the absolute maximum shear strain.

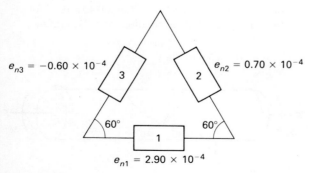

Fig. P9.30.

9.31 Three strain gages A, B and C are fixed to a point on the surface of a structural member at 120° intervals, and the strains recorded are 1.08×10^{-3}, 0.64×10^{-3} and 0.90×10^{-3}, respectively. Determine the principal strains and the inclination of gage A to the direction of the greater principal strain.

9.32 A hollow steel cylinder having diameters 60 mm and 45 mm can be loaded in compression along its axis, torsion about its axis, or any combination of these; it cannot be loaded in bending. To enable the loading at any time to be calculated, a strain rosette is fixed to the outer surface of the cylinder. The rosette has gages A, B and C, arranged as an equilateral triangle, and is fixed so that gage A is parallel to the axis of the cylinder. The strains are measured with two different combinations of load and are found to be as follows:

1 $e_A = -4.83 \times 10^{-4}$, $e_B = e_C = -0.121 \times 10^{-4}$
2 $e_A = -3.52 \times 10^{-4}$, $e_B = 1.82 \times 10^{-4}$,
   $e_C = -2.0 \times 10^{-4}$

Find the corresponding axial forces and torques on the cylinder.

SI METRIC UNITS – YIELD AND FRACTURE CRITERIA

9.33 and 9.34 For the states of plane stress defined in Problems 9.2 and 9.3, determine the following material properties for there to be a safety factor of 2.0 against failure.

1 The tensile yield stress for a ductile material, according to the Tresca criterion.
2 The tensile yield stress for a ductile material, according to the von Mises criterion.
3 The ultimate tensile strength for a brittle material, according to the maximum principal stress criterion.
4 The ultimate tensile strength for a brittle material, which may be taken as one quarter of the ultimate compressive strength, according to the Mohr fracture criterion.

9.35 A 100 mm diameter solid shaft rotating at 600 rev/min transmits 750 kW and carries an end thrust of 180 kN. If the maximum allowable (tensile) stress in the material is 110 MN/m², find whether the shaft is overstressed according to (i) the Tresca yield criterion, and (ii) the von Mises criterion.

9.36 A solid 100 mm diameter shaft is simply supported in bearings 2.5 m apart, and transmits 40 kW at 60 rev/min. A concentrated lateral force of 18 kN is applied to the shaft at a point 0.5 m from one bearing and 2.0 m from the other. Find the safety factor against yielding, according to the Tresca criterion, if the tensile yield stress of the shaft material is 200 MN/m².

9.37 It is proposed to check the safety of a thin-walled cylindrical pressure vessel with closed ends, by measuring the change in diameter as the internal pressure is applied. Derive an expression for the change, δ, in the diameter, D, which would occur before the onset of yielding according to the von Mises criterion, given that the tensile yield stress of the material is σ_Y, Young's modulus E, and the value of Poisson's ratio is 0.29.

9.38 A cylindrical pressure vessel is to be made having a diameter of 280 mm and to work at an internal pressure of 13 MN/m² gage. The material of which it is to be made has a yield stress of 420 MN/m² in uniaxial tension. Assuming that the material obeys the von Mises criterion of yielding, and that there is to be a safety factor against yielding of 3.0, find the necessary wall thickness remote from the ends of the vessel. The walls may be assumed thin compared to the diameter.

9.39 A cylindrical steel pressure vessel has an outside diameter of 500 mm and a wall thickness of 4 mm. It is subjected to an internal pressure of 6 MN/m² gage, and a bending moment of 280 kNm is applied at a particular section remote from the ends or other attachment. Calculate the greatest axial and hoop stresses at that section. Neglecting

the radial stresses and adopting the von Mises criterion of yielding, find the necessary yield stress of the material in uniaxial tension if there is to be a safety factor against yielding of 2.0.

9.40 Figure P9.40 shows a twist drill held vertically in a chuck. The diameter of the drill is 10 mm, and it protrudes 120 mm from the chuck. In order to penetrate the workpiece, an axial force of $F = 20$ kN and a torque of $T = 35$ Nm must be applied. The drill is made from a hard but brittle material which has an ultimate compressive strength of 1000 MN/m², but an ultimate tensile strength of only 200 MN/m². Find the maximum horizontal force, P, which can be applied to the workpiece without breaking the drill. Assume that failure would occur in the solid shank of the drill rather than in the fluted tip.

US CUSTOMARY UNITS – STRESSES AND STRAINS AT A POINT

9.41 Solve Problem 9.1.

9.42 For the state of plane stress shown in Fig. P9.42, find the normal and shear stresses acting on a plane whose normal is at 25° to the x axis, measured in the clockwise direction.

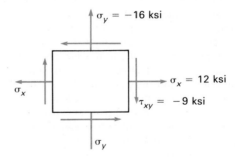

Fig. P9.42.

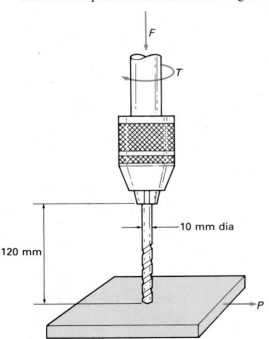

Fig. P9.40.

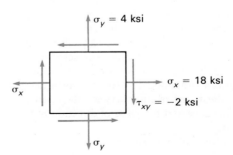

Fig. P9.43.

529

9.43 For the state of plane stress shown in Fig. P9.43, find the normal and shear stresses acting on a plane whose normal is at 45° to the x axis, measured in the counterclockwise direction.

9.44 and 9.45 For the states of plane stress illustrated in Figs P9.42 and P9.43, find the principal stresses, the directions of the principal planes, the maximum in-plane shear stresses, and the absolute maximum shear stresses.

9.46 and 9.47 Use program ANAL2D to check the results of Problems 9.44 and 9.45.

9.48 and 9.49 Draw Mohr's stress circles for the states of plane stress shown in Figs P9.42 and P9.43, showing the points representing stresses on the planes normal to the x and y axes in each case.

9.50 Solve Problem 9.10.

9.51 to 9.62 Solve Problems 9.11 to 9.22.

9.63 Mutually perpendicular planes at a point in a material are subjected to normal stresses of 6 and −3 ksi, together with a shear stress of 4 ksi. Find the values of the shear stresses on all planes that are inclined at an angle of 10° to the principal planes.

9.64 Solve Problem 9.24.

9.65 A strain gage attached to the surface of a closed-ended thin-walled aluminum alloy cylinder at 45° to its axis registers a strain of 0.001 when an internal pressure is applied. If the diameter of the cylinder is 24 in and its thickness is 0.2 in, find the pressure and the change in diameter it produces.

9.66 A shaft of length 6 in and diameter 1 in is fitted with a strain gage, the axis of which is inclined at 45° to the axis of the shaft. When calibrated with the shaft in pure torsion the gage records a strain of 0.80×10^{-3}. In service there is an end load which causes the shaft to extend by 0.0088 in and its diameter to contract by 0.000 48 in. Determine the strain the gage would record in service if subjected to the same torque as before.

9.67 Solve Problem 9.27, taking the diameter of the shaft to be 3 in.

9.68 A uniform solid steel bar with a diameter of 2.2 in has a strain gage fixed to its surface oriented so that it makes an angle of 20° with the axis of the bar. Find the strain reading if the bar is subjected to an axial compressive force of 54 kip. The axial force is then removed and the bar subjected to a torque. Find the value of this torque if the strain gage records a strain of 2.42×10^{-4}. Draw a diagram to indicate the relation between the orientation of the gage and the sense of direction of the applied torque.

9.69 to 9.71 Solve Problems 9.29 to 9.31.

US CUSTOMARY UNITS – YIELD AND
FRACTURE CRITERIA

9.72 and 9.73 For the states of plane stress defined in Problems 9.42 and 9.43, determine the following material properties for there to be a safety factor of 2.0 against failure.
1 The tensile yield stress for a ductile material, according to the Tresca criterion.
2 The tensile yield stress for a ductile material, according to the von Mises criterion.
3 The ultimate tensile strength for a brittle material, according to the maximum principal stress criterion.
4 The ultimate tensile strength for a brittle material, which may be taken as one quarter of the ultimate compressive strength, according to the Mohr fracture criterion.

9.74 A hollow steel shaft 52 in long, 3 in external diameter and 1 in internal diameter is simply supported at its ends and carries a concentrated lateral force at its center. It is also subjected to a torque of 7.4 kip-ft. If the yield stress of the shaft material is 58 ksi, find how great the central force can be without yield occurring in the shaft, according to the maximum shear stress criterion of yielding.

9.75 A solid shaft is subjected simultaneously to a bending moment of 6.6 kip-ft and a torque of 11 kip-ft. The material of the shaft has a tensile yield stress of 36 ksi. Find the minimum outside diameter of the shaft according to (i) the Tresca, (ii) the von Mises, and (iii) the maximum principal stress failure criteria.

9.76 Solve Problem 9.37.

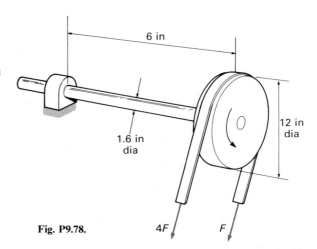

Fig. P9.78.

530

9.77 A hollow steel shaft, 3 in outside diameter and 1 in inside diameter, is subjected to a torque of 2.6 kip-ft and a bending moment of 3.5 kip-ft acting simultaneously. To guard against the danger of fatigue failure, it is desired not to exceed one third of the yield stress, assuming the material obeys the shear strain energy criterion of yielding. Determine the lowest acceptable yield strength in simple tension of the steel to be used.

9.78 A pulley of 12 in diameter is attached to the free end of a 1.6 in diameter solid shaft which overhangs 6 in from the nearest bearing, as in Fig. P9.78. The pulley is driven by a belt from another pulley of the same diameter. The tension on the tight side is four times the tension on the slack side. Find the values of these tensions that would just cause yielding according to the von Mises criterion if the material of the shaft yields at 22 ksi in uniaxial tension.

10

Equilibrium and Compatibility Equations: Applications to Beams and Thick-walled Cylinders

In Chapter 9 we were concerned with the analysis of stresses and strains *at a point* in a solid body. For this purpose, we considered elements of material sufficiently small for stresses and strains to be regarded as constant, and equal to the values at the point concerned. In most situations of practical interest, both stresses and strains vary from point to point. To be able to determine these variations, we must now consider the relationships between stresses and strains *at adjacent points*. In the case of stresses, the approach is similar to that used in Section 5.2.2 to derive relationships between distributed lateral force, shear force and bending moment in a beam (equations (5.10) and (5.11)). In other words, we establish differential equations of equilibrium describing the variations. Similarly, for strain variations we derive differential equations of compatibility.

10.1 Stress equilibrium equations

In Chapter 9 we met states of plane stress and plane strain as two-dimensional approximations to the three-dimensional situations we encounter in practice. In plane stress, there are no stresses in one particular direction; in plane strain the same is true of strains. As we derive stress equilibrium equations describing variations of stresses, we will restrict our attention to these two-dimensional states.

10.1.1 Equilibrium equations in Cartesian coordinates

Figure 10.1 shows a rectangular element ABCD of a solid body, with its sides parallel to Cartesian coordinate axes x and y, and with dimensions of δx and δy in the x and y directions. We can also treat this element as having a uniform thickness of δz in the z direction (which is normal to the plane shown). It is large enough for the stresses to vary significantly in both the x and y directions, but small enough for us to treat these variations as being linear with position. For example, let us take the normal stress component in the x direction at the midpoint E of the left-hand side AD of the element as σ_x. Because the stress variations are assumed to be linear, this is also the average value of the stress over side AD. At the midpoint F of the right-hand side BC it has changed, to

$$\sigma_x + \frac{\partial \sigma_x}{\partial x} \delta x$$

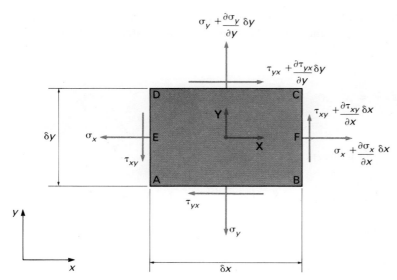

Fig. 10.1.

The rate of change of stress with distance is given by the *partial derivative* of σ_x with respect to x. We need to use the partial derivative, which defines the rate of change with x only, because in general σ_x varies with both x and y. The change in stress is given by the product of this rate of change and the distance, δx, between points E and F. Changes in the other stresses σ_y, τ_{xy} and τ_{yx}, which we can again regard as average values over the relevant element sides, are given by similar products of partial derivatives and element dimensions, as shown in the figure. While there may be a normal stress component σ_z applied to the element in the z direction, this has no effect on its equilibrium in the x and y directions.

In addition to the stresses acting on the sides of the element, it may also be subjected to *body forces*. The most common examples of body forces are the weight of the element and the centrifugal effects of rotation of the body concerned about some axis (as in the case of a thin ring or cylinder which we considered in Section 2.2.3). In Fig. 10.1, the components of body force *per unit volume of material* in the x and y directions are X and Y, respectively, which we can assume act at the center of the element. For example, in the case of gravity acting vertically downward

$$X = 0, \quad Y = -\rho g \tag{10.1}$$

where ρ is the density of the material and g is the acceleration due to gravity.

We can now write equations of equilibrium for the *forces* (products of stresses and areas) acting on the isolated element of material shown in Fig. 10.1, in the x and y directions. In the x direction, we have

$$\left(\sigma_x + \frac{\partial \sigma_x}{\partial x}\,\delta x\right)\delta y\,\delta z - \sigma_x\,\delta y\,\delta z + \left(\tau_{yx} + \frac{\partial \tau_{yx}}{\partial y}\,\delta y\right)\delta x\,\delta z - \tau_{yx}\,\delta x\,\delta z$$

$$+ X\,\delta x\,\delta y\,\delta z = 0$$

which, after simplification, elimination of the common factor of $\delta x\,\delta y\,\delta z$, and replacement of τ_{yx} by the equal complementary shear stress τ_{xy}, becomes

$$\frac{\partial \sigma_x}{\partial x} + \frac{\partial \tau_{xy}}{\partial y} + X = 0 \tag{10.2}$$

Similarly, in the y direction

$$\left(\sigma_y + \frac{\partial \sigma_y}{\partial y}\,\delta y\right)\delta x\,\delta z - \sigma_y \delta x\,\delta z + \left(\tau_{xy} + \frac{\partial \tau_{xy}}{\partial x}\,\delta x\right)\delta y\,\delta z - \tau_{xy}\,\delta y\,\delta z$$

$$+\, Y\,\delta x\,\delta y\,\delta z = 0$$

which becomes

$$\frac{\partial \sigma_y}{\partial y} + \frac{\partial \tau_{xy}}{\partial x} + Y = 0 \tag{10.3}$$

Equations (10.2) and (10.3) are the *stress equilibrium equations* in Cartesian coordinates.

10.1.2 Equilibrium equation in polar coordinates for axial symmetry

Many members of engineering systems are cylindrical in shape: tubes, pipes and shafts are common examples, and the natural coordinate system to use to describe them is the cylindrical polar system. Indeed, we have already used it for thin-walled cylindrical shells (Chapters 2 and 3) and shafts in torsion (Chapter 7). In many cases of practical interest, not only is the geometry of the system symmetrical about the axis of the cylinder, but so is the loading: for instance, internal pressure. Such situations are said to be *axisymmetric*, for which we need a stress equilibrium equation in polar coordinates.

Figure 10.2 shows a sector-shaped element PQRS of an axisymmetric solid body, with its sides lying along lines of constant radius, r, or angle, θ, in the polar coordinate system (r, θ) which has its origin on the axis of symmetry. The dimensions of the element in the r and θ directions are δr and $\delta \theta$. We can also treat this element as having a uniform thickness of δz in the z direction (which is normal to the plane shown). While there may be a normal stress component σ_z applied to the element in the z direction, this has no effect on its equilibrium in the r and θ directions. The in-plane stresses acting on the element are as shown. Due to the axial symmetry, there are no variations with angular position, and the hoop stress, σ_θ, has the same magnitude on sides PQ and RS. Also, since axial symmetry implies mirror symmetry about any plane through the axis, there are no shear stresses, $\tau_{r\theta}$ and $\tau_{\theta r}$. This is consistent with the principle we

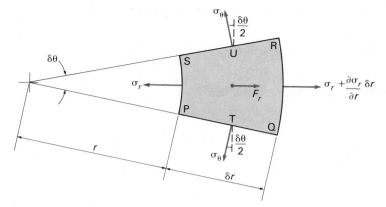

Fig. 10.2.

first met in Section 2.2.1, that there are no shear stresses on a plane of mirror symmetry through a body. Both PQ and RS represent such planes, on which $\tau_{\theta r}=0$. Because complementary shear stresses are equal in magnitude, $\tau_{r\theta}$, which acts on sides PS and QR, is also zero. Both the hoop and radial stress components may vary with radius, r. Assuming the element is small enough for these variations to be treated as linear with position, we can take the hoop stress acting at the midpoints T and U of sides PQ and RS to be the average value over these sides. Also, the radial stress, which we can take as σ_r acting over side PS, changes to

$$\sigma_r + \frac{\partial \sigma_r}{\partial r} \delta r$$

over side QR.

In addition to the stresses acting on the sides of the element, it may also be subject to a body force in the radial direction (axial symmetry prevents there being a body force in the hoop direction). We can take this force as F_r per unit volume in the positive r direction (radially outward), acting at the center of the element as shown. The most common example of such a body force is the centrifugal effect of rotation about the axis of symmetry, a situation which is considered in Section 10.4.3.

We can now write an equation of equilibrium of forces acting on the element of material in the radial direction (equilibrium in the hoop direction is already assured) as

$$\left(\sigma_r + \frac{\partial \sigma_r}{\partial r} \delta r\right)(r+\delta r)\delta\theta\,\delta z - \sigma_r r\delta\theta\,\delta z - 2\sigma_\theta \delta r\,\delta z \sin\frac{\delta\theta}{2}$$

$$+ F_r \delta r\left(r + \frac{\delta r}{2}\right)\delta\theta\,\delta z = 0$$

Replacing the sine of the small angle $\delta\theta/2$ by $\delta\theta/2$, this becomes

$$\frac{\partial \sigma_r}{\partial r}\delta r(r+\delta r)\delta\theta\,\delta z + (\sigma_r - \sigma_\theta)\delta r\,\delta\theta\,\delta z + F_r\delta r\left(r+\frac{\delta r}{2}\right)\delta\theta\,\delta z = 0$$

Dividing through by $r\,\delta r\,\delta\theta\,\delta z$, and taking $\delta r \to 0$, we obtain

$$\frac{\mathrm{d}\sigma_r}{\mathrm{d}r} + \frac{\sigma_r - \sigma_\theta}{r} + F_r = 0 \qquad (10.4)$$

This is the stress equilibrium equation in the radial direction of the plane polar coordinate system (r, θ) under conditions of axial symmetry. Note that, with only one independent variable, r, the partial derivative (of σ_r with respect to r) becomes a total derivative.

10.2 Strain compatibility equations

In earlier chapters we have defined strains in terms of changes of dimensions or shapes of finite pieces of material subjected to simple states of normal or shear strain. In Section 1.3.5 we defined a normal strain as the ratio between the change in length and the original length of a bar in simple tension. We also defined shear strain in Section 1.3.5, and rather more generally in Section 9.2.1, as the change in angle at a corner of a rectangular element of material subjected to pure shear. These forms of definition are useful in that they provide clear physical interpretations of strains. In all cases, however, we were concerned with pieces of material over which the strains were constant.

We now wish to study variations of strains in solid bodies, and to do this we must first find more mathematical ways to define strains, in terms of local changes of displacements. The strain components within a solid body are not free to vary independently. Links between them can be expressed mathematically, using the fact that in the absence of cracks or holes in the body the displacement components are continuous functions of position. These links take the form of differential equations known as the strain *compatibility equations*. We have already met in many applications involving constant stresses and strains the principle of compatibility of strains or geometry of deformation helping to determine the solution to a problem.

We proceed as we did for the stress equilibrium equations, confining our attention to the two-dimensional states of plane stress and plane strain, with possibly a normal strain (but no shear strains) perpendicular to the plane of interest.

10.2.1 Definitions of strains in terms of displacements

Figure 10.3 shows a rectangular element ABCD of a solid body, with its sides parallel to Cartesian coordinate axes x and y, and with dimensions of δx and δy

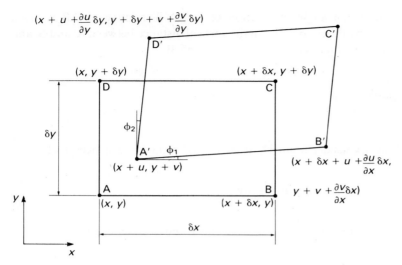

Fig. 10.3.

in the x and y directions. When the body is loaded, the element may be both displaced and deformed into the parallelogram A'B'C'D'. We take the symbols u and v to represent components of displacement in the x and y directions, respectively (also w in the z direction, normal to the x, y plane). In particular, we take the displacements of point A to be u and v. In other words, if the coordinates of point A are (x, y), then those of point A' are $(x + u,\ y + v)$. Similarly, point B, which is at $(x + \delta x, y)$, is displaced to B' at

$$\left(x + \delta x + u + \frac{\partial u}{\partial x}\, \delta x,\ y + v + \frac{\partial v}{\partial x}\, \delta x \right)$$

and the distances between A' and B' in the x and y directions are

$$\delta x + \frac{\partial u}{\partial x}\, \delta x \quad \text{and} \quad \frac{\partial v}{\partial x}\, \delta x$$

respectively. In writing these expressions, we are treating u and v as functions of both x and y: hence the need for partial derivatives with respect to x to define the rates of change of the displacements with distance in the x direction. The expressions are approximate, to the extent that we are effectively assuming that the displacements vary only linearly with x, ignoring the contributions of second and higher derivatives of displacement. The normal strain in the x direction is the ratio between the change in distance and the original horizontal distance between points A and B

$$e_x = \frac{\left(\delta x + \dfrac{\partial u}{\partial x}\, \delta x \right) - \delta x}{\delta x} = \frac{\partial u}{\partial x}$$

Similarly, the small angle of rotation of line A′B′ relative to AB is given approximately by the ratio between the vertical distance between A′ and B′ and the original horizontal distance between A and B

$$\phi_1 = \frac{\frac{\partial v}{\partial x} \delta x}{\delta x} = \frac{\partial v}{\partial x}$$

Let us now consider point D, at $(x, y + \delta y)$, which is displaced to D′ at

$$\left(x + u + \frac{\partial u}{\partial y} \delta y, \; y + \delta y + v + \frac{\partial v}{\partial y} \delta y \right)$$

and the distances between A′ and D′ in the x and y directions are

$$\frac{\partial u}{\partial y} \delta y \quad \text{and} \quad \delta y + \frac{\partial v}{\partial y} \delta y$$

respectively. The normal strain in the y direction is the ratio between the change in distance and the original vertical distance between points A and D

$$e_y = \frac{\left(\delta y + \frac{\partial v}{\partial y} \delta y \right) - \delta y}{\delta y} = \frac{\partial v}{\partial y}$$

Also, the small angle of rotation of line A′D′ relative to AD is given approximately by the ratio between the horizontal distance between A′ and D′ and the original vertical distance between A and D

$$\phi_2 = \frac{\frac{\partial u}{\partial y} \delta y}{\delta y} = \frac{\partial u}{\partial y}$$

According to equation (9.29), the total shear strain is the sum of angles ϕ_1 and ϕ_2, and we can gather together the definitions of the three in-plane strain components as

$$e_x = \frac{\partial u}{\partial x}, \quad e_y = \frac{\partial v}{\partial y}, \quad \gamma_{xy} = \frac{\partial u}{\partial y} + \frac{\partial v}{\partial x} \tag{10.5}$$

We also need mathematical definitions for strains in polar coordinates for axial symmetry. Since there are no shear stresses in this case, there are also no shear strains, and we seek only definitions for the radial and hoop normal strains. These must be expressed in terms of the relevant displacement components, which we can define as u_r and u_θ in the radial and hoop directions, respectively, as shown in Fig. 10.4. Due to the axial symmetry, however, u_θ can be treated as zero. Actually, in addition to any radial displacement, the whole body could rotate about its axis without deforming in the hoop direction, in which case u_θ would be independent of θ. Following equations (10.5) for the Cartesian coordinate case, we might anticipate that the radial and hoop strains are the derivatives of radial displacement with respect to radial direction, and hoop displacement with respect to hoop direction, respectively. Indeed, this is true for the radial strain, but not the hoop: u_θ is independent of θ. In fact, we have already established a definition for hoop strain in an axisymmetric situation, in Section 3.4.2 (equation (3.24)) as the ratio between the change in circumference and the original circumference. With a radial displacement of u_r at a radius r, this is

$$e_\theta = \frac{2\pi(r + u_r) - 2\pi r}{2\pi r} = \frac{u_r}{r}$$

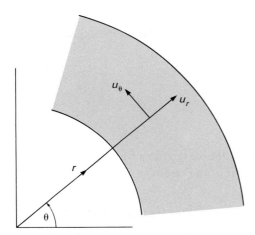

Fig. 10.4.

The required radial and hoop strain components are therefore

$$e_r = \frac{du_r}{dr}, \quad e_\theta = \frac{u_r}{r} \tag{10.6}$$

It must be emphasized that these definitions only apply to axisymmetric problems. Note that the radial strain is the total rather than the partial derivative of u_r with respect to r: there are no variations in the hoop direction, and u_r is a function only of r.

10.2.2 Compatibility equation in Cartesian coordinates

Equations (10.5) define the *three* strain components in the (x, y) plane of a Cartesian coordinate system, in terms of only *two* displacement components. This disparity between the number of strains and the number of displacements implies that not all the strain components are independent, and that one of them can be expressed in terms of the other two. Now, in the absence of cracks and holes within the body, the displacement components are continuous functions of position. The same is true of derivatives of displacements with respect to position. A mathematical consequence of continuity of derivatives is that, for a partial derivative of displacement involving differentiation with respect to both independent variables, x and y, the order of differentiation (with respect to x first, and then y, or vice versa) is not important.

We can therefore establish a relationship between the three strains by differentiation. Differentiating the expression for shear strain γ_{xy} in equations (10.5) with respect to first x and then y, we obtain

$$\frac{\partial}{\partial y}\left(\frac{\partial \gamma_{xy}}{\partial x}\right) = \frac{\partial}{\partial y}\left(\frac{\partial^2 u}{\partial x \partial y}\right) + \frac{\partial}{\partial y}\left(\frac{\partial^2 v}{\partial x^2}\right)$$

which, ignoring the order of differentiation and introducing the expressions for normal strains e_x and e_y, becomes

$$\frac{\partial^2 \gamma_{xy}}{\partial x \partial y} = \frac{\partial^2 e_x}{\partial y^2} + \frac{\partial^2 e_y}{\partial x^2} \tag{10.7}$$

This is the single compatibility equation linking the three strain components for the two-dimensional states of plane stress and strain.

10.2.3 Compatibility equation in polar coordinates for axial symmetry

Equations (10.6) define the two strain components in polar coordinates for axial symmetry in terms of the one displacement component. We can again establish a relationship between the strains by differentiation as

$$e_r = \frac{\mathrm{d}u_r}{\mathrm{d}r} = \frac{\mathrm{d}}{\mathrm{d}r}(re_\theta) \tag{10.8}$$

which is the strain compatibility equation.

10.3 Application to beam bending

In Section 1.5 we laid down the three physical principles governing the mechanics of deformable solids as the equilibrium of forces, compatibility of strains (geometry of deformation), and the stress–strain characteristics of the material concerned. For plane problems in which stresses and strains vary continuously, we have now established the relevant mathematical equations in Cartesian coordinates as (10.2) and (10.3) for equilibrium, (10.7) for strain compatibility, and (3.16) and (9.30) for linear elastic stress–strain relationships. What we need in addition, however, to solve particular problems are *boundary conditions* defining the stresses or displacements on the edges of the solid body concerned. These form part of the equilibrium conditions or geometry of deformation conditions imposed on the problem.

In only a small proportion of problems of practical interest is it possible to obtain exact analytical (algebraic) solutions which satisfy both the governing mathematical equations and the boundary conditions. The derivation of such solutions is the province of the *theory of elasticity*. Even when problems cannot be solved analytically, however, numerical computer methods can be employed, one of the most widely used being the finite element method. We have already met this technique in connection with the analysis of pin-jointed structures and beams, and it can be adapted to solve continuum problems in two (and three) dimensions.

Let us now consider some types of problems which are amenable to exact analytical solution. These involve the bending of beams, and allow us to check the validity and accuracy of the simple beam theory we developed in Chapters 5 and 6.

10.3.1 Pure bending

Consider the simply supported uniform beam of negligible weight shown in Fig. 10.5, which is of length L and has a rectangular cross section of depth d and breadth b. The beam is subjected to externally applied moments of magnitude M_0 in the sagging sense at its ends. This is the problem of pure bending we considered in Examples 5.3 and 6.1, where we showed in Fig. 5.20 that the shear force is zero everywhere along the beam, and the bending moment is equal to M_0 everywhere.

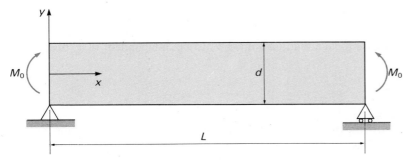

Fig. 10.5.

Since the weight of the beam is negligible, there are no body forces, and $X = Y = 0$ in equations (10.2) and (10.3). It is also reasonable to assume that the normal stress σ_z in the direction perpendicular to the plane shown in Fig. 10.5 is zero, so that the beam is in a state of plane stress. To start the analysis, we make an assumption about how the bending stress, σ_x, varies in the x direction along the beam. Since the bending moment is independent of x, we assume that σ_x is also independent of x. When we have obtained the solution, we can examine the implications of this assumption.

With σ_x independent of x, stress equilibrium equation (10.2) becomes

$$\frac{\partial \tau_{xy}}{\partial y} = 0$$

and shear stress τ_{xy} does not vary with position y through the depth of the beam. But, this shear stress (or, more precisely, the equal complementary shear stress τ_{yx}) is zero on the top and bottom surfaces of the beam, at $y = \pm d/2$. Consequently, $\tau_{xy} = 0$ everywhere, which is consistent with the fact that there is no resultant shear force at any beam cross section. With $\tau_{xy} = 0$, equilibrium equation (10.3) becomes

$$\frac{\partial \sigma_y}{\partial y} = 0$$

and σ_y does not vary with y. But, since σ_y is zero on the top and bottom surfaces of the beam, at $y = \pm d/2$, it must be zero everywhere.

In order to use compatibility equation (10.7), we need expressions for the strain components. For a state of plane stress in which $\tau_{xy} = \sigma_y = 0$ and with no temperature variations, these are given by stress–strain equations (3.16) and (9.30) as

$$e_x = \frac{\sigma_x}{E}, \quad e_y = -\frac{v\sigma_x}{E}, \quad \gamma_{xy} = 0$$

Substituting these expressions into equation (10.7), we obtain

$$0 = \frac{1}{E}\frac{\partial^2 \sigma_x}{\partial y^2} - \frac{v}{E}\frac{\partial^2 \sigma_x}{\partial x^2}$$

which, since we have already assumed σ_x to be independent of x, becomes

$$\frac{d^2 \sigma_x}{dy^2} = 0$$

which is now an ordinary differential equation. Its general solution for the bending stress is

$$\sigma_x = Ay + B$$

where A and B are constants of integration.

To find these constants, we can proceed much as we did in Section 5.3.1, using the fact that there is no resultant force acting on the beam in the x direction

$$\int_{-d/2}^{+d/2} \sigma_x b \, dy = 0 \tag{10.9}$$

also that the resultant moment about the neutral axis is equal to the bending moment at every beam cross section

$$\int_{-d/2}^{+d/2} -\sigma_x b y \, dy = M(x) \tag{10.10}$$

From equation (10.9) we obtain $B = 0$, and from (10.10)

$$-A \int_{-d/2}^{+d/2} b y^2 \, dy = M_0$$

or

$$A = -\frac{M_0}{I}$$

where I is the second moment of area of the beam cross section about its neutral axis, and

$$\sigma_x = \frac{-yM_0}{I} \tag{10.11}$$

This is identical to equation (5.34), which we derived using simple beam theory.

Now we can review the implications of our initial assumption that σ_x is independent of x. Since equation (10.11) applies at all beam cross sections, it must apply at the ends of the beam. In other words, the end moments are applied to the beam by means of linear stress distributions of the form shown in Fig. 10.6. If the end loadings are not of this form then the present analysis cannot give the stress distributions close to the ends, although it would be satisfactory for the rest of the beam. A detailed treatment of the end regions would require a more sophisticated form of analysis, which would almost certainly have to be numerical rather than analytical.

Having found the stress distributions, we can now find the displacements. We start with the definition of the normal strain along the beam, and its relationship to the bending stress

$$e_x = \frac{\partial u}{\partial x} = \frac{\sigma_x}{E} = \frac{-yM_0}{EI}$$

Fig. 10.6.

Integrating partially with respect to x

$$u = -\frac{xyM_0}{EI} + f_1(y) \tag{10.12}$$

where $f_1(y)$ is a *function of integration*: its partial derivative with respect to x is zero. Similarly, from the definition of shear strain

$$\gamma_{xy} = \frac{\partial u}{\partial y} + \frac{\partial v}{\partial x} = 0$$

we find

$$\frac{\partial v}{\partial x} = -\frac{\partial u}{\partial y} = \frac{xM_0}{EI} - \frac{df_1}{dy}$$

which can be integrated to give

$$v = \frac{x^2 M_0}{2EI} - x\frac{df_1}{dy} + f_2(y) \tag{10.13}$$

where $f_2(y)$ is another function of integration.

In order to compare with the results of simple beam theory, we are interested in the deflection of the neutral surface, at $y = 0$. Therefore, let us assume that the beam is supported in such a way that, at its ends, there is no deflection of the neutral surface. Therefore, at the origin of the coordinates, $v = 0$, which in equation (10.13) gives $f_2(0) = 0$. Applying the same zero deflection condition at $x = L$, $y = 0$

$$0 = \frac{L^2 M_0}{2EI} - L\left(\frac{df_1}{dy}\right)_{y=0}$$

from which

$$\left(\frac{df_1}{dy}\right)_{y=0} = \frac{LM_0}{2EI} \tag{10.14}$$

and equation (10.13) for the deflected shape of the neutral surface, along the line $y = 0$, becomes

$$v = \frac{M_0}{2EI}(x^2 - xL) \tag{10.15}$$

This is exactly the same result as we obtained in Example 6.1, equation (6.11), using simple beam theory.

In order to find the displacement component in the axial direction from equation (10.12), we need the full form of function $f_1(y)$: so far we only have its gradient at $y=0$, from equation (10.14). Another condition we can use, however, is that the deformation of the beam is symmetrical about its center, where $x = L/2$. Therefore, at this cross section, we can take the displacement component u as zero, for all values of y. Hence

$$f_1 = \frac{yLM_0}{2EI}$$

and

$$u = \frac{yM_0}{EI}\left(\frac{L}{2} - x\right) \tag{10.16}$$

We note that u is directly proportional to y, the distance from the neutral axis, which confirms the simple beam theory assumption that plane cross sections remain plane.

The results of this more thorough analysis show that, for pure bending, simple bending theory gives exact solutions for stresses and beam deflections. This is because all the assumptions on which the theory is based are valid in this case.

EXAMPLE 10.1

The beam shown in Fig. 10.5 is simply supported at its ends, not at the neutral surface but at its lower surface. Find the correction which must be made to equation (10.15) to allow for this fact, and express this correction as a proportion of the maximum deflection of the beam.

The required correction is the vertical displacement of the neutral surface of the beam relative to its lower surface, at either of the supports. Using equation (10.13) to define displacement v, we can express the normal strain in the y direction as

$$e_y = \frac{\partial v}{\partial y} = -x\frac{d^2 f_1}{dy^2} + \frac{df_2}{dy} = -\frac{v\sigma_x}{E}$$

Considering the support at $x=0$, and using equation (10.11) to define σ_x, this becomes

$$\frac{df_2}{dy} = \frac{yvM_0}{EI}$$

from which

$$f_2 = \frac{y^2 vM_0}{2EI} + D$$

where D is an integration constant. At $x = 0$, the vertical displacement of the neutral surface relative to the support is

$$v' = f_2(0) - f_2\left(-\frac{d}{2}\right) = -\frac{vM_0 d^2}{8EI}$$

Equation (10.15) (also equation (6.12)) gives the maximum beam deflection, at $x = L/2$, as

$$v_{max} = -\frac{M_0 L^2}{8EI}$$

and the ratio between the correction and this maximum deflection is

$$\frac{v'}{v_{max}} = v\left(\frac{d}{L}\right)^2$$

For typical beams, with L/d ratios of 10 or more, this is a very small quantity, and the correction is negligibly small. It is caused by the normal strain in the y direction in the beam produced by the axial bending stress via the Poisson's ratio effect.

10.3.2 Cantilever with a concentrated end force

Consider the uniform cantilever beam of negligible weight shown in Fig. 10.7, which is of length L and has a rectangular cross section of depth d and breadth b. The beam is subjected to a concentrated lateral force of magnitude F at its free end. We previously examined this problem in Example 6.2, and Fig. 5.21 shows that the shear force in the beam is constant and given by $V = F$, and the bending moment distribution is given by

$$M(x) = -F(L-x) \tag{10.17}$$

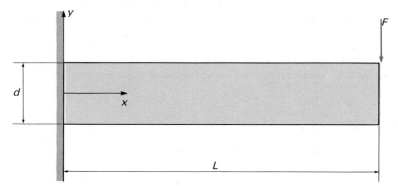

Fig. 10.7.

Therefore, using equation (10.10)

$$\int_{-d/2}^{+d/2} \sigma_x b y \, dy = F(L-x)$$

and from equation (10.9)

$$\int_{-d/2}^{+d/2} \sigma_x \, dy = 0$$

We start by assuming that σ_x takes the form

$$\sigma_x = f_3(y) F(L - x) + f_4(y) \tag{10.18}$$

where the functions $f_3(y)$ and $f_4(y)$ satisfy the following integral conditions

$$\int_{-d/2}^{+d/2} f_3(y) by \, dy = 1 \tag{10.19}$$

$$\int_{-d/2}^{+d/2} f_3(y) \, dy = 0 \tag{10.20}$$

$$\int_{-d/2}^{+d/2} f_4(y) y \, dy = 0 \tag{10.21}$$

and

$$\int_{-d/2}^{+d/2} f_4(y) \, dy = 0 \tag{10.22}$$

This assumption of the form of σ_x defined by equation (10.18) is equivalent to the one we made in Section 10.3.1 for a simply supported beam in pure bending that σ_x is independent of x. As in that case, we can examine the implications of the assumption when we have obtained the solution.

Since the weight of the beam is negligible, there are no body forces. We also assume that the beam is in a state of plane stress. With σ_x given by equation (10.18), stress equilibrium equation (10.2) becomes

$$\frac{\partial \tau_{xy}}{\partial y} = -\frac{\partial \sigma_x}{\partial x} = f_3(y) F$$

Now, since the shear stress is zero at the bottom surface of the beam, at $y = -d/2$, its distribution is given by

$$\tau_{xy} = F \int_{-d/2}^{y} f_3(y) \, dy \tag{10.23}$$

We note that τ_{xy} is proportional to F and independent of x, which is consistent with the fact that there is a constant shear force of magnitude F at every beam cross section.

With τ_{xy} given by equation (10.23), equilibrium equation (10.3) becomes

$$\frac{\partial \sigma_y}{\partial y} = -\frac{\partial \tau_{xy}}{\partial x} = 0$$

and σ_y does not vary with y. But, since σ_y is zero on the top and bottom surfaces of the beam, at $y = \pm d/2$, it must be zero everywhere.

In order to use compatibility equation (10.7), we need expressions for the strain components. For a state of plane stress in which $\sigma_y = 0$, these are given by

$$e_x = \frac{\sigma_x}{E}, \quad e_y = -\frac{v\sigma_x}{E}, \quad \gamma_{xy} = \frac{\tau_{xy}}{G}$$

where τ_{xy} is a function only of y. Substituting these expressions into equation (10.7), we obtain

$$0 = \frac{1}{E} \frac{\partial^2 \sigma_x}{\partial y^2} - \frac{v}{E} \frac{\partial^2 \sigma_x}{\partial x^2}$$

which, with σ_x defined by equation (10.18), becomes

$$\frac{d^2 f_3}{dy^2} F(L-x) + \frac{d^2 f_4}{dy^2} = 0$$

Since this condition must hold for all values of x, we conclude that

$$\frac{d^2 f_3}{dy^2} = \frac{d^2 f_4}{dy^2} = 0$$

and

$$f_3(y) = A_3 y + B_3, \quad f_4(y) = A_4 y + B_4$$

where A_3, A_4, B_3 and B_4 are integration constants. Substituting these forms of the functions into equations (10.19) to (10.22), we find the values of the constants as

$$A_3 = \frac{12}{bd^3} = \frac{1}{I}, \quad B_3 = A_4 = B_4 = 0$$

In this particular case, function f_4 is zero, and equation (10.18) for σ_x becomes

$$\sigma_x = \frac{y}{I} F(L-x) = -\frac{yM(x)}{I} \tag{10.24}$$

This is identical to equation (5.34), which we derived using simple beam theory. Also, equation (10.23) for the shear stress becomes

$$\tau_{xy} = F \int_{-d/2}^{y} \frac{y}{I} \, dy = \frac{F}{2I} \left[y^2 - \left(\frac{d}{2}\right)^2 \right] \tag{10.25}$$

which is equivalent to the result obtained in equation (5.45), again using simple beam theory (bearing in mind that $F = V$, the shear force).

Now we can review the implications of our initial assumption that σ_x takes the form of equation (10.18). Since equation (10.25) applies at all beam cross sections, it must apply at the free end of the beam. In other words, the force F is applied to the end face of the beam by means of a parabolic stress distribution of the form shown in Fig. 10.8. If the real loading is not of this form, then the present analysis cannot give the stress distributions close to the free end, although it would be satisfactory for the rest of the beam. We should note that the concentrated force at the end of the beam shown in Fig. 10.7, although a useful model for simple beam theory, cannot be realized in practice: it would result in an infinite stress at the point of contact.

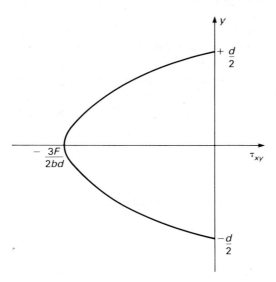

Fig. 10.8.

Having found the stress distributions, we can now find the displacements. We start with the definition of the normal strain along the beam

$$e_x = \frac{\partial u}{\partial x} = \frac{\sigma_x}{E} = \frac{F}{EI}(L-x)y$$

Integrating partially with respect to x

$$u = \frac{F}{EI}\left(Lx - \frac{x^2}{2}\right)y + f_5(y) \tag{10.26}$$

where $f_5(y)$ is a function of integration. Similarly, from the definition of shear strain

$$\gamma_{xy} = \frac{\partial u}{\partial y} + \frac{\partial v}{\partial x} = \frac{\tau_{xy}}{G} = \frac{F}{2GI}\left[y^2 - \left(\frac{d}{2}\right)^2\right]$$

we find

$$\frac{\partial v}{\partial x} = \frac{F}{2GI}\left[y^2 - \left(\frac{d}{2}\right)^2\right] - \frac{F}{EI}\left(Lx - \frac{x^2}{2}\right) - \frac{df_5}{dy}$$

which can be integrated to give

$$v = \frac{F}{2GI}\left[y^2 - \left(\frac{d}{2}\right)^2\right]x - \frac{F}{EI}\left(\frac{Lx^2}{2} - \frac{x^3}{6}\right) - x\frac{df_5}{dy} + f_6(y) \qquad (10.27)$$

where $f_6(y)$ is another function of integration.

In order to compare with the results of simple beam theory, we are interested in the deflection of the neutral surface, at $y = 0$. Therefore, let us assume that the built-in support is such that there is no deflection of the neutral surface there. Hence, at the origin of the coordinates, $v = 0$, which in equation (10.27) gives $f_6(0) = 0$. There are several alternatives for further boundary conditions at the built-in end. For example, if the slope of the neutral surface (the partial derivative of v with respect to x along $y = 0$) is zero at the origin, then

$$\left(\frac{df_5}{dy}\right)_{y=0} = -\frac{Fd^2}{8GI}$$

Equation (10.27) for the deflected shape of the neutral surface, along $y = 0$, then becomes

$$v = -\frac{F}{EI}\left(\frac{Lx^2}{2} - \frac{x^3}{6}\right) \qquad (10.28)$$

giving a maximum deflection at the free end of

$$v = -\frac{FL^3}{3EI}$$

This is exactly the same result as we obtained in Example 6.2, equation (6.14), using simple beam theory. But, the displacement in the x direction at the built-in end of the beam is given by equation (10.26) as $u = f_5(y)$, and the gradient of this function at the neutral axis is not zero. The beam cross section is distorted out of the vertical plane, contravening the simple beam theory assumption of plane cross sections remaining plane. This is due to the presence of the shear stresses and strains, which cause the neutral surface and beam cross section to be no longer mutually perpendicular.

An alternative form of boundary condition at the built-in end of the beam is that of zero distortion of the cross section, that is $u = 0$ for all y, at $x = 0$. According to equation (10.26), this requires that the function $f_5(y) = 0$, and equation (10.27) gives the displacement in the y direction as

$$v = \frac{F}{2GI}\left[y^2 - \left(\frac{d}{2}\right)^2\right]x - \frac{F}{EI}\left(\frac{Lx^2}{2} - \frac{x^3}{6}\right)$$

The deflection of the neutral surface at the free end of the beam, at $x = L$, $y = 0$, is therefore

$$v = -\frac{Fd^2L}{8GI} - \frac{FL^3}{3EI}$$

The two terms making up this expression can be identified as the deflections due to shearing and bending, respectively, v_s and v_b, and the ratio of their magnitudes is

$$\frac{v_s}{v_b} = \frac{3}{8}\frac{E}{G}\left(\frac{d}{L}\right)^2 \tag{10.29}$$

It is interesting to compare this result with the approximation obtained in Example 6.2, equation (6.15), which was of the same form but with a constant of 1/4 in place of the 3/8: the shear deflection according to the present solution is 50% greater than the estimated value, but both are small relative to the bending deflection.

In practice, neither of the two displacement constraint conditions at the built-in end, zero neutral axis slope and zero axial displacement, is entirely satisfactory. The beam itself must continue into the structure to which it is attached, and this structure is not perfectly rigid. A more realistic analysis would have to take into account both of these facts.

The results of this more exact treatment of a cantilever with a concentrated end force show that simple bending theory still gives accurate solutions for stresses. Beam deflections may be slightly less accurately predicted, due to the presence of shearing effects.

10.3.3 Simply supported beam with a distributed force

Consider the uniform simply supported beam of negligible weight shown in Fig. 10.9, which is of length L and has a rectangular cross section of depth d and breadth b. The beam is subjected to a uniformly distributed lateral force of

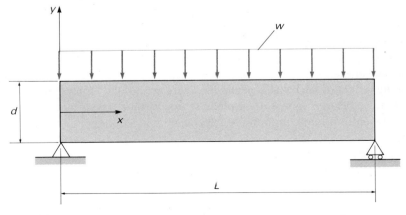

Fig. 10.9.

intensity w per unit length. We previously examined this problem in Example 6.8, and Fig. 5.19 shows that the shear force in the beam is given by

$$V(x) = w\left(\frac{L}{2} - x\right) \tag{10.30}$$

and the bending moment distribution by

$$M(x) = \frac{w}{2}(Lx - x^2) \tag{10.31}$$

Therefore, using equation (10.10)

$$\int_{-d/2}^{+d/2} \sigma_x\, by\, dy = -\frac{w}{2}(Lx - x^2)$$

and from equation (10.9)

$$\int_{-d/2}^{+d/2} \sigma_x\, dy = 0$$

We start by assuming that σ_x takes the form

$$\sigma_x = -\frac{y}{I}\frac{w}{2}(Lx - x^2) + f_7(y) \tag{10.32}$$

where the first term corresponds to the simple beam theory solution, and function $f_7(y)$ satisfies the integral conditions

$$\int_{-d/2}^{+d/2} f_7(y)\, dy = 0 \quad \text{and} \quad \int_{-d/2}^{+d/2} f_7(y)\, y\, dy = 0 \tag{10.33}$$

In fact we could make a more general assumption for the form of σ_x, by writing an unknown function of y in place of y/I, much as we did in equation (10.18) for the cantilever beam problem. This makes the analysis significantly more complicated, and the same final result is obtained. We can justify the particular form we have assumed for σ_x by examining the implications when we have obtained the solution.

Since the weight of the beam is negligible, there are no body forces. We also assume that the beam is in a state of plane stress. With σ_x given by equation (10.32), stress equilibrium equation (10.2) becomes

$$\frac{\partial \tau_{xy}}{\partial y} = -\frac{\partial \sigma_x}{\partial x} = \frac{y}{I}\frac{w}{2}(L - 2x)$$

Now, since the shear stress is zero at the bottom surface of the beam, at $y = -d/2$, its distribution is given by

$$\tau_{xy} = \frac{w}{2I}(L-2x)\int_{-d/2}^{y} y\,dy = \frac{V(x)}{2I}\left[y^2 - \left(\frac{d}{2}\right)^2\right] \tag{10.34}$$

which is the result we would have obtained from simple beam theory. Therefore, equilibrium equation (10.3) becomes

$$\frac{\partial \sigma_y}{\partial y} = -\frac{\partial \tau_{xy}}{\partial x} = \frac{w}{2I}\left[y^2 - \left(\frac{d}{2}\right)^2\right]$$

and in this problem σ_y is not zero, and varies with y. Integrating partially with respect to y, we obtain

$$\sigma_y = \frac{w}{2I}\left[\frac{y^3}{3} - y\left(\frac{d}{2}\right)^2\right] + f_8(x)$$

where $f_8(x)$ is a function of integration. One of the boundary conditions for this normal stress is $\sigma_y = 0$ at the bottom surface of the beam, $y = -d/2$, for all x. Therefore

$$0 = \frac{w}{2I}\left[-\frac{d^3}{24} + \frac{d^3}{8}\right] + f_8(x)$$

from which

$$f_8(x) = -\frac{wd^3}{24I}$$

and the function is a constant. At the top surface of the beam, $y = +d/2$ and

$$\sigma_y = \frac{w}{2I}\left[\frac{d^3}{24} - \frac{d^3}{8}\right] - \frac{wd^3}{24I} = -\frac{wd^3}{12I} = -\frac{w}{b}$$

which is the required condition: a compressive *force per unit length* of beam on this surface of intensity w implies a compressive *stress* of magnitude w/b, where b is the breadth of the beam. The general expression for stress σ_y is

$$\sigma_y = \frac{w}{24I}(4y^3 - 3yd^2 - d^3) \tag{10.35}$$

In order to use compatibility equation (10.7), we need expressions for the strain components. For a state of plane stress, these are given by

$$e_x = \frac{1}{E}(\sigma_x - \nu\sigma_y), \qquad e_y = \frac{1}{E}(\sigma_y - \nu\sigma_x), \qquad \gamma_{xy} = \frac{\sigma_{xy}}{G}$$

Therefore, using equations (10.32) for σ_x, (10.34) for τ_{xy} and (10.35) for σ_y

$$\frac{\partial^2 e_x}{\partial y^2} = \frac{1}{E}\left(\frac{\partial^2 \sigma_x}{\partial y^2} - v\frac{\partial^2 \sigma_y}{\partial y^2}\right) = \frac{1}{E}\left(\frac{d^2 f_7}{dy^2} - v\frac{wy}{I}\right)$$

$$\frac{\partial^2 e_y}{\partial x^2} = \frac{1}{E}\left(\frac{\partial^2 \sigma_y}{\partial x^2} - v\frac{\partial^2 \sigma_x}{\partial x^2}\right) = \frac{1}{E}\left(-v\frac{wy}{I}\right)$$

$$\frac{\partial^2 \gamma_{xy}}{\partial x\,\partial y} = \frac{1}{G}\frac{\partial^2 \tau_{xy}}{\partial x\,\partial y} = -\frac{wy}{GI}$$

Substituting these expressions into equation (10.7), we obtain

$$-\frac{2(1+v)\,wy}{EI} = \frac{1}{E}\left(\frac{d^2 f_7}{dy^2} - v\frac{wy}{I}\right) - v\frac{wy}{EI}$$

from which

$$\frac{d^2 f_7}{dy^2} = -\frac{2\,wy}{I}$$

Integration of this differential equation yields

$$f_7(y) = -\frac{wy^3}{3I} + Ay + B$$

where A and B are integration constants. This function must satisfy equations (10.33), the first of which gives $B=0$ and the second gives

$$\int_{-d/2}^{+d/2}\left(-\frac{wy^4}{3I} + Ay^2\right)dy = 0$$

$$-\frac{2w}{15I}\left(\frac{d}{2}\right)^5 + \frac{2A}{3}\left(\frac{d}{2}\right)^3 = 0$$

$$A = \frac{w}{5I}\left(\frac{d}{2}\right)^2$$

Equation (10.32) for σ_x therefore becomes

$$\sigma_x = -\frac{y}{I}\frac{w}{2}(Lx-x^2) + \frac{w}{I}\left(\frac{d^2 y}{20} - \frac{y^3}{3}\right) \tag{10.36}$$

Now we can review the implications of our initial assumption that σ_x takes the general form of equation (10.32), which satisfies the integral conditions of applied bending moment at any beam cross section, and zero axial force, and results in the particular form of equation (10.36). The corresponding shear stress distribution is given by equation (10.34), which satisfies the conditions of zero shear stress at the top and bottom surfaces of the beam. Similarly, the distribution of σ_y is defined by equation (10.35), which gives zero stress at the bottom surface of the beam and the applied compressive stress at the top surface. So, at least in the central region of the beam, the solution satisfies all the requirements: equilibrium, compatibility, stress–strain relationships and the boundary conditions. At the ends of the beam, however, there are some imperfections. According to equation (10.36), and in particular the second term in the definition, σ_x is not zero for all values of y (with $x=0$ or L) as it should be, although the integral conditions of zero axial force and bending moment there are satisfied. Also, equation (10.34) for the shear stress distribution implies that the support reaction forces are applied to the end faces of the beam by means of parabolic shear stress distributions, of the form shown in Fig. 10.8. If the real loading is not of this form, then the present analysis cannot give the exact stress distributions close to the ends, although it would be satisfactory for the rest of the beam.

Having found the stress distributions, we can now find the displacements. After a rather lengthy analysis, similar in form to that given in Sections 10.3.1 and 10.3.2, and assuming that the neutral surface of the beam does not deflect at the ends of the beam, we arrive at the following expression for the deflection of the neutral surface at the center of the beam

$$v\left(\frac{L}{2}\right) = -\frac{5}{384}\frac{wL^4}{EI}\left[1 + \frac{12}{5}\left(\frac{d}{L}\right)^2\left(\frac{4}{5} + \frac{v}{2}\right)\right] \tag{10.37}$$

The first term within the square brackets gives the deflection according to simple beam theory, while the second is due to the presence of shearing stresses and the compressive stress normal to the beam axis.

The results of this more thorough treatment for the central region of a simply supported beam with a distributed lateral force show that simple beam theory now only gives an exact representation of the shear stress distribution. The expression for bending stress distribution has an extra term which is not predicted by the simple theory. Also, due to the distributed applied force, there are normal stresses perpendicular to the axis of the beam, which are again not predicted by the simple theory. The present analysis also shows how, as problems become more complex (and the present one is still straightforward compared to most which are of practical interest), exact analytical solutions become very lengthy and difficult (if not impossible) to obtain.

EXAMPLE 10.2

The simply supported steel beam shown in Fig. 10.9 has a length-to-depth ratio of $L/d = 10$. Find the percentage error involved in using simple beam theory to determine the maximum bending stress in the beam, rather than the more accurate equation (10.36). Find also the percentage error involved in using the simple theory to predict the maximum beam deflection, as compared to the result given by equation (10.37).

Equation (10.36) gives the maximum (tensile) bending stress, at $x = L/2$, $y = -d/2$, as

$$\sigma_x = \frac{wdL^2}{16I} + \frac{wd^3}{60I}$$

where the first term is the simple beam theory solution. Therefore the relative error involved in using the simple theory is approximately

$$\frac{wd^3}{60I}\frac{16I}{wdL^2} = \frac{16}{60}\left(\frac{d}{L}\right)^2 = \frac{16}{60}\left(\frac{1}{10}\right)^2 = \underline{0.27\%}$$

Equation (10.37) for the maximum beam deflection gives the relative error involved in using simple beam theory as

$$\frac{12}{5}\left(\frac{d}{L}\right)^2\left(\frac{4}{5} + \frac{v}{2}\right) = \frac{12}{5}\left(\frac{1}{10}\right)^2\left(\frac{4}{5} + \frac{0.3}{2}\right) = \underline{2.28\%}$$

So, while the error in the maximum bending stress is negligibly small for practical purposes, the error in the maximum deflection is rather more significant.

10.4 Application to thick-walled cylinders and disks

Thick-walled cylinders form a practically important class of axisymmetric engineering components. They are widely used as containment vessels in, for example, chemical plant, when internal pressures are high. In Section 2.2 we considered thin-walled vessels used for the same purpose, though for relatively low pressures. In a thick-walled vessel the internal pressure may be of the same order as the maximum allowable stress in the vessel material. We can make the distinction between thin and thick walls when the wall thickness is about a tenth of the cylinder radius – an assertion we will be able to confirm when we have developed an analysis for the thick-walled case.

The relevant stress equilibrium and strain compatibility equations in plane polar coordinates for axisymmetric problems are equations (10.4) and (10.8), with strain components defined by equations (10.6). These are repeated here for convenience

$$\frac{d\sigma_r}{dr} + \frac{\sigma_r - \sigma_\theta}{r} + F_r = 0 \tag{10.4}$$

$$e_r = \frac{du_r}{dr}, \qquad e_\theta = \frac{u_r}{r} \tag{10.6}$$

$$e_r = \frac{du_r}{dr} = \frac{d}{dr}(re_\theta) \tag{10.8}$$

Hooke's Law gives the stress–strain equations (in the absence of temperature variations) as

$$e_r = \frac{1}{E}\left[\sigma_r - v(\sigma_\theta + \sigma_z)\right] \qquad (10.38)$$

$$e_\theta = \frac{1}{E}\left[\sigma_\theta - v(\sigma_z + \sigma_r)\right] \qquad (10.39)$$

$$e_z = \frac{1}{E}\left[\sigma_z - v(\sigma_r + \sigma_\theta)\right] \qquad (10.40)$$

While in the beam problems considered in Section 10.3, stresses and strains were functions of two independent variables (the Cartesian coordinates x and y), now there is only one independent variable, namely the radial coordinate r. The derivatives in the above equations are therefore total rather than partial derivatives, and consequently the equations themselves rather easier to solve analytically. We start with the case of an internally pressurized cylinder.

10.4.1 Internally pressurized thick-walled cylinder

Figure 10.10 shows the cross section of a long thick-walled cylinder of internal radius R_1 and external radius R_2, subjected to a hydrostatic pressure of p at its inner surface. We wish to find the variations of radial and hoop stresses with radius through the thickness of the cylinder wall.

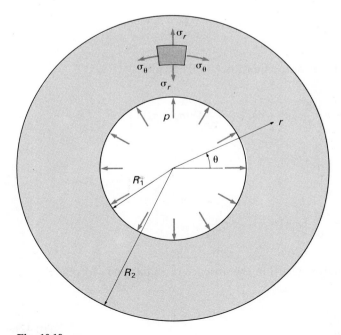

Fig. 10.10.

Starting from compatibility equation (10.8), we can use equations (10.38) and (10.39) to define the radial and hoop strain components in terms of stresses, giving

$$\sigma_r - v(\sigma_\theta + \sigma_z) = \frac{d}{dr}\{r[\sigma_\theta - v(\sigma_z + \sigma_r)]\}$$

Carrying out the differentiation on the right-hand side of this equation, we obtain after some rearrangement

$$(\sigma_r - \sigma_\theta)(1 + v) = r\frac{d}{dr}[\sigma_\theta - v(\sigma_z + \sigma_r)] \tag{10.41}$$

Now, we would like to be able to eliminate the axial stress, leaving this equation and equilibrium equation (10.4) in terms of σ_r and σ_θ only. To do this, we make an assumption about either the normal stress or normal strain in the axial direction. One possibility is to take σ_z as zero, but this is only appropriate if the axial length of the cylinder is small compared to its diameter so that a state of plane stress exists in the wall: the cylinder would have to be a thin circular disk. An alternative is to assume that plane cross sections of the cylinder remain plane, which implies that the axial strain is independent of radius, and which is appropriate if the length of the cylinder is large compared to its diameter: a form of plane strain. Differentiating equation (10.40) with respect to radius and setting the derivative of e_z to zero, we obtain

$$\frac{d\sigma_z}{dr} = \frac{d}{dr}[v(\sigma_r + \sigma_\theta)]$$

which may be substituted into equation (10.41) to give

$$(\sigma_r - \sigma_\theta)(1 + v) = r\left\{\frac{d}{dr}[\sigma_\theta - v\sigma_r] - v\frac{d}{dr}[v(\sigma_r + \sigma_\theta)]\right\}$$

$$= r\frac{d}{dr}[\sigma_\theta(1 - v^2) - v\sigma_r(1 + v)]$$

Therefore

$$(\sigma_r - \sigma_\theta) = r\frac{d}{dr}[\sigma_\theta(1 - v) - v\sigma_r]$$

But, equilibrium equation (10.4) in the absence of radial body forces ($F_r = 0$) gives

$$(\sigma_r - \sigma_\theta) = -r\frac{d\sigma_r}{dr} \tag{10.42}$$

Comparing these two expressions for the difference between the radial and hoop stresses, we deduce that

$$-\sigma_r = \sigma_\theta(1 - v) - v\sigma_r + \text{constant}$$

therefore

$$(\sigma_r + \sigma_\theta)(1 - v) = \text{constant}$$

and

$$\sigma_r + \sigma_\theta = \text{constant} = 2A \qquad (10.43)$$

The reason for choosing the constant in this equation as $2A$, rather than, say, just A, should become clear later. The fact that the sum of the two normal stress components σ_r and σ_θ is constant has an important consequence. From equation (10.40), bearing in mind that we have assumed the axial strain to be constant (independent of radius), we can conclude that the axial stress is also constant.

Using equation (10.43), we can substitute for σ_θ in equation (10.42), to give

$$2A - 2\sigma_r = r\frac{d\sigma_r}{dr}$$

$$2Ar = \frac{d}{dr}(r^2\sigma_r)$$

$$Ar^2 = r^2\sigma_r + B$$

where B is an integration constant. Hence

$$\sigma_r = A - \frac{B}{r^2} \qquad (10.44)$$

and, from equation (10.43)

$$\sigma_\theta = A + \frac{B}{r^2} \qquad (10.45)$$

We now see the reason for taking the constant in equation (10.43) to be $2A$: the forms of expression for the radial and hoop stresses involve A.

We have derived equations (10.44) and (10.45) without reference to the boundary conditions for a particular problem. The integration constants A and

B must now be found from these conditions, which for the situation illustrated in Fig. 10.10 are

$$\sigma_r = -p \text{ at } r = R_1 \quad \text{and} \quad \sigma_r = 0 \text{ at } r = R_2 \tag{10.46}$$

Note that the hydrostatic pressure within the cylinder applies a *compressive* radial stress to its inner surface. For equation (10.44) to satisfy these boundary conditions

$$A - \frac{B}{R_1^2} = -p \quad \text{and} \quad A - \frac{B}{R_2^2} = 0$$

from which

$$A = \frac{p}{K^2 - 1} \quad \text{and} \quad B = \frac{pR_2^2}{K^2 - 1}$$

where $K = R_2/R_1$ is the ratio of outer to inner radius of the cylinder. The stress distributions are therefore given by

$$\sigma_r = \frac{p}{K^2 - 1}\left(1 - \frac{R_2^2}{r^2}\right) \tag{10.47}$$

and

$$\sigma_\theta = \frac{p}{K^2 - 1}\left(1 + \frac{R_2^2}{r^2}\right) \tag{10.48}$$

Figure 10.11 shows these plotted for the case of $K = 2$.

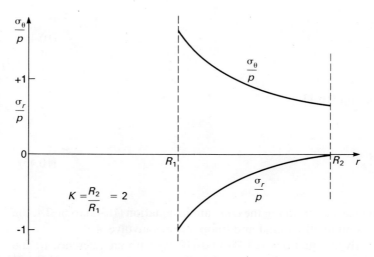

Fig. 10.11.

The greatest absolute values of radial and hoop stress occur at the inner surface of the cylinder, and are given by

$$\sigma_r = -p, \qquad \sigma_\theta = p\frac{K^2+1}{K^2-1} \tag{10.49}$$

The magnitude of the axial stress depends on the end conditions imposed on the cylinder, and is related to the axial strain by equation (10.40), which now becomes

$$e_z = \frac{1}{E}\left(\sigma_z - \frac{2vp}{K^2-1}\right) \tag{10.50}$$

Just as with the thin-walled cylinders we considered in Section 2.2.1, the two most commonly occurring types of end conditions are open-ended and closed-ended.

Open-ended cylinder

When none of the pressure load on the end closures is transferred to the cylinder walls, as in Fig. 10.12a, there is no axial stress in the cylinder, $\sigma_z = 0$ and, from equation (10.50), the axial strain is

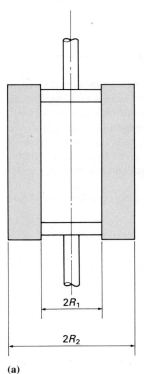

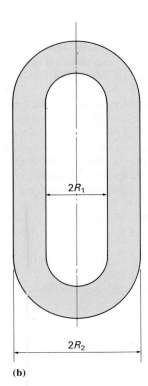

(a)

(b)

Fig. 10.12.

$$e_z = -\frac{2vp}{E(K^2 - 1)} \tag{10.51}$$

We note that this normal strain is negative: under internal pressure the length of the cylinder decreases.

Closed-ended cylinder

When all of the pressure load on each end of the cylinder is transmitted to the walls, as in Fig. 10.12b, this load must be balanced by the axial stress in the wall, which is uniform in the central region of the cylinder away from its ends

$$p\pi R_1^2 = \sigma_z \pi (R_2^2 - R_1^2)$$

from which

$$\sigma_z = \frac{p}{K^2 - 1} \quad \text{and} \quad e_z = \frac{(1 - 2v)p}{E(K^2 - 1)} \tag{10.52}$$

EXAMPLE 10.3

A thick-walled cylinder with an internal diameter of 100 mm (4 in)* and an external diameter of 220 mm (8.8 in) is subjected to an internal pressure of 120 MN/m² (17 ksi) and an external pressure of 6 MN/m² (900 psi). Find the greatest hoop stress in the cylinder wall.

This is not quite the situation for which equations (10.47) and (10.48) were derived, in that we now have pressures applied to both the internal and external surfaces. We can anticipate that the loading can be treated as a general hydrostatic pressure of 6 MN/m², giving rise to a uniform compressive stress of the same magnitude throughout the cylinder, plus a pressure difference between the internal and external surfaces of 114 MN/m², the corresponding greatest hoop stress being given by equation (10.49). This can be confirmed by applying the appropriate boundary conditions to equations (10.44) and (10.45).

If we take the pressures at the inner and outer surfaces of the cylinder to be p_1 and p_2, respectively, the boundary conditions are

$$\sigma_r = -p_1 \text{ at } r = R_1 \quad \text{and} \quad \sigma_r = -p_2 \text{ at } r = R_2$$

which applied to equation (10.44) give

$$A - \frac{B}{R_1^2} = -p_1 \quad \text{and} \quad A - \frac{B}{R_2^2} = -p_2$$

from which

$$A = \frac{p_1 - p_2}{K^2 - 1} - p_2 \quad \text{and} \quad B = \frac{(p_1 - p_2)R_2^2}{K^2 - 1}$$

and the radial and hoop stress distributions are given by

$$\sigma_r = \frac{(p_1 - p_2)}{K^2 - 1}\left(1 - \frac{R_2^2}{r^2}\right) - p_2 \tag{10.53}$$

and

$$\sigma_\theta = \frac{(p_1 - p_2)}{K^2 - 1}\left(1 + \frac{R_2^2}{r^2}\right) - p_2 \tag{10.54}$$

As is to be expected, if we set p_2 to zero and p_1 to p we obtain equations (10.47) and (10.48). These results also confirm that we could have superimposed a uniform compressive stress of magnitude p_2 on the distributions due to a pressure difference of $(p_1 - p_2)$ between the inner and outer surfaces.

The greatest hoop stress is therefore

$$\sigma_\theta = (p_1 - p_2)\frac{(K^2 + 1)}{(K^2 - 1)} - p_2 \tag{10.55}$$

Substituting in the numerical values of $K = 220/100 = 2.2$ (the ratio of either the radii or the diameters of the cylinder), $p_1 = 120 \text{ MN/m}^2$ and $p_2 = 6 \text{ MN/m}^2$, we obtain the required greatest hoop stress as

$$\sigma_\theta = 114 \times \frac{(2.2^2 + 1)}{(2.2^2 - 1)} - 6 = \underline{167 \text{ MN/m}^2} \ (24 \text{ ksi})$$

EXAMPLE 10.4

A solid rod of circular cross section 50 mm (2 in) in diameter is immersed in a fluid at a pressure of 15 MN/m² (2.2 ksi). Find the distributions of radial and hoop stresses in the rod.

While we can anticipate that uniform hydrostatic external loading on the rod will give rise to uniform hydrostatic internal stresses, it should be possible to arrive at this result from equations (10.44) and (10.45): a solid rod is only a special case of a hollow cylinder in which the internal radius is zero.

At the axis of the rod, we use not an applied pressure boundary condition but the fact that the radial and hoop directions are indistinguishable there, and therefore the radial and hoop stresses are equal. This can only occur in equations (10.44) and (10.45) if the constant B is zero, which incidentally has the effect of preventing the stresses becoming infinite at $r = 0$. Therefore, the radial and hoop stresses are equal everywhere to the constant A, and equal to the applied radial stress at the surface of the rod, in this case -15 MN/m^2 (-2.2 ksi).

EXAMPLE 10.5

A steel pressure vessel takes the form of a closed-ended cylinder with an internal diameter of 250 mm (10 in) and an external diameter of 350 mm (14 in). The material follows the von Mises yield criterion, and its yield stress in simple tension is 500 MN/m² (73 ksi). Assuming that the cylindrical walls of the cylinder are more severely stressed than the end closures, find the internal gage pressure at which yielding just occurs.

The ratio of external to internal cylinder radius (or diameter) in this case is $K = 350/250 = 1.4$. Equation (10.49) gives the greatest radial and hoop stresses at the inner surface in terms of the internal pressure p as

$$\sigma_r = -p, \qquad \sigma_\theta = p\frac{1.4^2 + 1}{1.4^2 - 1} = +3.083p$$

while equation (10.52) gives the uniform axial stress as

$$\sigma_z = \frac{p}{1.4^2 - 1} = +1.042p$$

Figure 10.13 shows these three normal stress components acting on a small element of material at the inner surface of the cylinder. We note that, because there are no shear stresses acting on the faces of the element, the radial, hoop and axial stresses are principal stresses. For present purposes, the ordering of these stresses is not important, but if we take $\sigma_1 \geqslant \sigma_2 \geqslant \sigma_3$ (as in equation (9.27)) then $\sigma_1 \equiv \sigma_\theta$, $\sigma_2 \equiv \sigma_z$ and $\sigma_3 \equiv \sigma_r$. Using equation (9.52) to express the von Mises yield criterion in terms of the three principal stresses, we obtain

$$p^2(3.083 - 1.042)^2 + p^2(1.042 + 1)^2 + p^2(-1 - 3.083)^2 = 2 \times 500^2$$

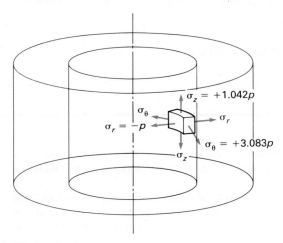

Fig. 10.13.

from which

$$p = \underline{141 \text{ MN/m}^2} \ (20 \text{ ksi})$$

This is the pressure required to just cause yielding in the cylinder wall, at its inner surface.

In Chapter 2, Section 2.2.1, we derived expressions for the stresses in an internally pressurized cylinder on the assumption that the hoop and axial stresses are constant through the wall thickness. We are now in a position to check the accuracy of these results against the more exact formulae derived without making this assumption for thick-walled cylinders.

It is the hoop stress which is of particular interest: axial stress is assumed to be uniform in both thin and thick-walled cylinders, and equations (2.17) and (10.52) give identical values of stress. The extent to which equation (2.16) underestimates the greatest hoop stress is of practical importance. It can be expressed in terms of the variables used in the present thick-walled cylinder analysis as

$$\sigma_\theta = \frac{pR_1}{R_2 - R_1} = \frac{p}{K - 1} \qquad (10.56)$$

This is the same as the *average* of the hoop stress through the wall given by equation (10.48). The ratio between the maximum hoop stress in a cylinder given by equation (10.49), and the average value given by equation (10.56) is

$$S = \frac{K^2 + 1}{K + 1} \qquad (10.57)$$

When $K = 1$, and the cylinder is infinitely thin, the maximum and average hoop stresses are equal. Values of S for larger values of K may be calculated as follows

$K =$	1.01	1.05	1.10	1.20	1.50	2.0
$S =$	1.005	1.026	1.052	1.109	1.300	1.667

We see from these figures that, as we would expect, for K ratios close to 1.0 the thin-walled cylinder analysis gives accurate values of maximum hoop stress, but that this accuracy is lost as the relative thickness of the cylinder wall increases. Depending on how accurately we wish to determine the greatest hoop stresses, we could set a maximum K ratio for use of the simple analysis of about 1.1, where the error involved is some 5%. In other words, the thin-walled cylinder analysis is useful for cylinders whose wall thicknesses are up to about 10% of their radii (inner or outer).

10.4.2 Compound cylinders

In some practical applications two or more concentric cylinders are arranged to fit together. There are two main reasons for doing this. Firstly, it may be necessary to provide a thick-walled cylinder with a lining of a different material, which has particular physical or chemical properties, such as improved wear or corrosion resistance for use in chemical reaction vessels. In this case, the cylinders may be made to fit each other as closely as possible, or an interference fit may be specified. Secondly, two (or occasionally more) cylinders (often of the same material), which are assembled with interference fits between them, can be made to withstand higher internal pressures than a single cylinder with the same overall dimensions. This use of compound cylinders can be very beneficial when very high pressures, of the same order of magnitude as the maximum allowable stress, have to be contained.

Figure 10.11 shows the stress distributions in a typical thick-walled cylinder. While material close to the internal surface of the cylinder is relatively highly stressed, particularly in the hoop direction, towards the outer surface the stresses are relatively low. This effect becomes more pronounced as the K ratio of the cylinder is increased. In other words, the use of a massive single cylinder to contain a very high pressure represents an inefficient use of material. By using a compound cylinder, however, it is possible to arrange for the material to be more uniformly loaded. When a pair of cylinders is first assembled, and before the internal pressure is applied, the interference between them causes the inner and outer cylinders to be in compression and tension, respectively, in the hoop direction. When the tensile hoop stresses due to the internal pressure are superimposed, the resulting distribution is relatively uniform. This effect is demonstrated in Example 10.7 below.

Let us first develop the analysis for two compounded cylinders. Consider the two concentric cylinders shown in Fig. 10.14. The nominal internal and external radii of the cylinders are R_1, R_2 and R_2, R_3, respectively, and there is a small difference (interference) of δ between the interfacial radii before assembly. The Young's modulus and Poisson's ratio of the inner cylinder are E_1 and v_1, while those of the outer cylinder are E_2 and v_2. In practice, the arrangement would be assembled by heating the outer cylinder or cooling the inner cylinder, or both, before bringing them together. Once assembled, and with both cylinders at ambient temperature, the radial stresses in the two cylinders at the interface between them must be the same. Let this interfacial pressure be p_i, giving boundary conditions of

$$(\sigma_r)_1 = (\sigma_r)_2 = -p_i \quad \text{at} \quad r = R_2$$

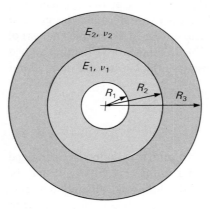

Fig. 10.14.

where $(\sigma_r)_1$ and $(\sigma_r)_2$ are the radial stress components in the inner and outer cylinders, respectively. Also, the geometry of deformation requires that the interfacial surfaces of the cylinders, which initially overlapped due to the interference, must be coincident after assembly. Figure 10.15 shows in very exaggerated form the geometry changes involved. If $(u_r)_1$ and $(u_r)_2$ are the outward radial displacements in the two cylinders, then at the interface the

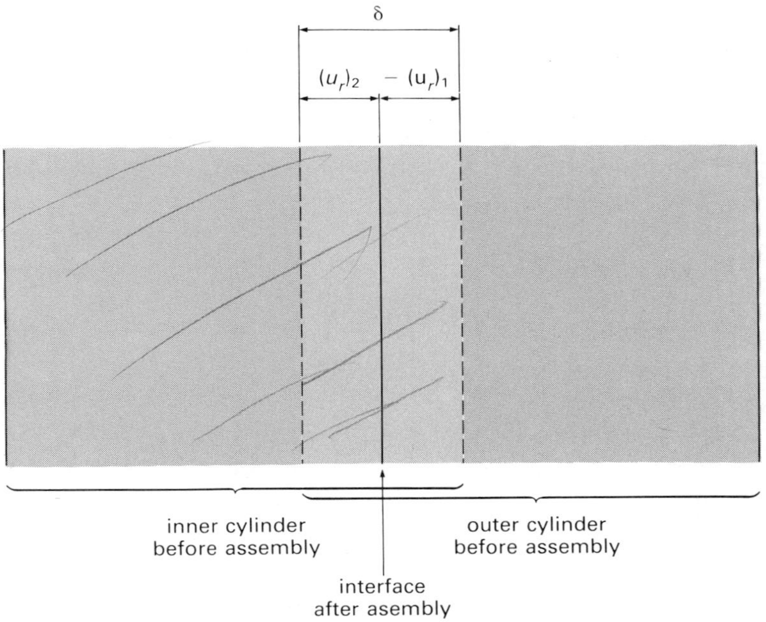

Fig. 10.15.

decrease in radius of the inner cylinder, $-(u_r)_1$, plus the increase in radius of the outer cylinder, $(u_r)_2$, must be equal to the initial interference

$$-(u_r)_1 + (u_r)_2 = \delta \quad \text{at} \quad r = R_2 \tag{10.59}$$

In other words, the radial expansion of the outer cylinder must exceed that of the inner one by an amount equal to the interference if the cylinders are to fit together. Dividing equation (10.59) through by R_2, and using equations (10.6) to define hoop strain, we have

$$-(e_\theta)_1 + (e_\theta)_2 = \frac{\delta}{R_2} \quad \text{at} \quad r = R_2 \tag{10.60}$$

Now, according to equations (10.44) and (10.45), the stresses in the inner cylinder are given by

$$\sigma_r = A_1 - \frac{B_1}{r^2}, \qquad \sigma_\theta = A_1 + \frac{B_1}{r^2} \tag{10.61}$$

while those in the outer cylinder are

$$\sigma_r = A_2 - \frac{B_2}{r^2}, \qquad \sigma_\theta = A_2 + \frac{B_2}{r^2} \tag{10.62}$$

where A_1 and B_1 are constants for the inner cylinder, and A_2 and B_2 are independent constants for the outer cylinder. The stress boundary conditions, which are those of zero radial stress at the inner and outer surfaces of the compound cylinder, and equations (10.58) at the interface, can therefore be expressed as

$$A_1 - \frac{B_1}{R_1^2} = 0 \tag{10.63}$$

$$A_1 - \frac{B_1}{R_2^2} = -p_i \tag{10.64}$$

$$A_2 - \frac{B_2}{R_2^2} = -p_i \tag{10.65}$$

$$A_2 - \frac{B_2}{R_3^2} = 0 \tag{10.66}$$

Assuming that there are no axial stresses in the cylinders due to compounding, hoop strains can be found from radial and hoop stresses as

$$e_\theta = \frac{1}{E}(\sigma_\theta - v\sigma_r) \tag{10.67}$$

where E and v are the appropriate values of Young's modulus and Poisson's ratio. For example, the hoop strain in the inner cylinder at $r = R_2$ is

$$(e_\theta)_1 = \frac{1}{E_1}\left[(1-v_1)A_1 + (1+v_1)\frac{B_1}{R_2^2}\right]$$

and equation (10.60) becomes

$$-\frac{(1-v_1)}{E_1}A_1 - \frac{(1+v_1)}{E_1}\frac{B_1}{R_2^2} + \frac{(1-v_2)}{E_2}A_2 + \frac{(1+v_2)}{E_2}\frac{B_2}{R_2^2} = \frac{\delta}{R_2} \tag{10.68}$$

Equations (10.63) to (10.66) and equation (10.68) provide five equations for the five unknowns A_1, B_1, A_2, B_2 and p_i. While they can be solved analytically by algebraic manipulation, it is often more convenient to be able to solve them numerically with the aid of a computer, particularly if they have to be solved several times with different data. This we can do with the aid of program CYLIND, which is described below.

One simplification of the equations is very straightforward to make: the unknown p_i can be eliminated from the equations by subtracting equation (10.65) from equation (10.64). The resulting set of four linear equations for the unknowns A_1, B_1, A_2 and B_2 can be expressed in matrix form as

$$\begin{bmatrix} 1 & -\dfrac{1}{R_1^2} & 0 & 0 \\[2ex] 1 & -\dfrac{1}{R_2^2} & -1 & \dfrac{1}{R_2^2} \\[2ex] 0 & 0 & 1 & -\dfrac{1}{R_3^2} \\[2ex] -\dfrac{(1-\nu_1)}{E_1} & -\dfrac{(1+\nu_1)}{E_1 R_2^2} & \dfrac{(1-\nu_2)}{E_2} & \dfrac{(1+\nu_2)}{E_2 R_2^2} \end{bmatrix} \begin{bmatrix} A_1 \\[2ex] B_1 \\[2ex] A_2 \\[2ex] B_2 \end{bmatrix} = \begin{bmatrix} 0 \\[2ex] 0 \\[2ex] 0 \\[2ex] \dfrac{\delta}{R_2} \end{bmatrix} \qquad (10.69)$$

When this set of equations has been solved for the four constants, we are able to find the distributions of radial and hoop stresses in the two cylinders with the aid of equations (10.61) and (10.62).

So far we have only considered the effects of compounding. Now we can superimpose those due to an internal pressure applied to the compound cylinder. This we do by replacing the zero on the right-hand side of equation (10.63) by the internal pressure, $-p$, say. Provided the compound cylinder remains open-ended when the internal pressure is applied, such that no axial stresses are generated, then the only modification necessary to equations (10.69) is to replace the zero at the top of the right-hand side vector by $-p$, before solving for the new constants defining the stress distributions produced by the combined effect of compounding and internal pressure.

If the compound cylinder is closed-ended and the axial stress uniformly distributed throughout both individual cylinders, then from equations (10.52)

$$\sigma_z = \frac{p}{\left(\dfrac{R_3}{R_1}\right)^2 - 1} \qquad (10.70)$$

and equation (10.67) must be modified to

$$e_\theta = \frac{1}{E} (\sigma_\theta - \nu \sigma_r) - \frac{\nu}{E} \sigma_z$$

Consequently, the last coefficient in the right-hand side vector in equations (10.69) is modified from δ/R_2 to

$$\frac{\delta}{R_2} + \sigma_z \left(\frac{\nu_2}{E_2} - \frac{\nu_1}{E_1} \right) \qquad (10.71)$$

An alternative to assuming the same uniform axial stress to exist in both cylinders is to assume the same uniform axial strain, with both cylinders extending by the same amount. This would result in somewhat different modifications to the equations. In practice, however, the form of axial loading or deformation assumed does not have a very significant effect on the most severe states of stress in the system.

Program CYLIND provides a numerical method for solving equations (10.69), using subroutine SOLVE which is described in detail in Appendix B, and then finding the stresses and strains in the cylinders.

```
      PROGRAM  CYLIND
C
C  PROGRAM TO FIND THE STRESSES (AND STRAINS) IN TWO COMPOUNDED
C  THICK-WALLED CYLINDERS SUBJECT TO INTERNAL PRESSURE.
C
      DIMENSION  COEFF(4,5),SCOEFF(4,5),AB(4)
      CHARACTER*4  ENDED
      REAL   NU1,NU2
      OPEN(5,FILE='DATA')
      OPEN(6,FILE='RESULTS')
C
C  INPUT THE NOMINAL CYLINDER RADII.
      READ(5,*) R1,R2,R3
      WRITE(6,61) R1,R2,R3
   61 FORMAT('ANALYSIS OF TWO COMPOUNDED THICK-WALLED CYLINDERS'//
     1         'INNER RADIUS OF THE INNER CYLINDER = ',E12.4 /
     2         'NOMINAL RADIUS AT THE INTERFACE = ',E12.4 /
     3         'OUTER RADIUS OF THE OUTER CYLINDER = ',E12.4)
C
C  INPUT THE RADIAL INTERFERENCE AND INTERNAL PRESSURE.
      READ(5,*) DELTA,P
      WRITE(6,62) DELTA,P
   62 FORMAT(/ 'RADIAL INTERFERENCE BETWEEN THE CYLINDERS = ',E12.4 /
     1         'INTERNAL PRESSURE = ',E12.4)
      IF(DELTA.LT.0.) THEN
         WRITE(6,63)
   63    FORMAT(/ 'A NEGATIVE INTERFERENCE IS NOT ACCEPTABLE - STOP')
         STOP
      END IF
C
C  INPUT THE ELASTIC PROPERTIES OF THE CYLINDERS.
      READ(5,*) E1,NU1,E2,NU2
      WRITE(6,64) E1,NU1,E2,NU2
   64 FORMAT(/ 'YOUNGS MODULUS OF THE INNER CYLINDER = ',E12.4 /
     1         'POISSONS RATIO OF THE INNER CYLINDER = ',F6.3  /
     2         'YOUNGS MODULUS OF THE OUTER CYLINDER = ',E12.4 /
     3         'POISSONS RATIO OF THE OUTER CYLINDER = ',F6.3)
C
C  INPUT THE END CONDITION - ASSUMED TO BE CLOSED-ENDED UNLESS THE
C  WORD 'OPEN' IS SUPPLIED AS DATA TO MAKE IT OPEN-ENDED.
      READ(5,*) ENDED
      IF(ENDED.EQ.'OPEN') WRITE(6,65)
   65 FORMAT(/ 'THE COMPOUND CYLINDER IS ASSUMED TO BE OPEN-ENDED')
      IF(ENDED.NE.'OPEN') WRITE(6,66)
   66 FORMAT(/ 'THE COMPOUND CYLINDER IS ASSUMED TO BE CLOSED-ENDED')
C
C  ANALYZE THE STRESSES AND STRAINS DUE TO COMPOUNDING.
C
C  FIRST SET UP THE EQUATIONS FOR THE STRESS DISTRIBUTION CONSTANTS.
      DO 1 I=1,4
      DO 1 J=1,5
   1  COEFF(I,J)=0.
      COEFF(1,1)=1.
      COEFF(1,2)=-1./R1**2
      COEFF(2,1)=1.
      COEFF(2,2)=-1./R2**2
      COEFF(2,3)=-1.
      COEFF(2,4)=-COEFF(2,2)
      COEFF(3,3)=1.
      COEFF(3,4)=-1./R3**2
      COEFF(4,1)=-(1.-NU1)/E1
      COEFF(4,2)=-(1.+NU1)*COEFF(2,4)/E1
      COEFF(4,3)=(1.-NU2)/E2
      COEFF(4,4)=(1.+NU2)*COEFF(2,4)/E2
      COEFF(4,5)=DELTA/R2
C
C  STORE THESE COEFFICIENTS.
      DO 2 I=1,4
      DO 2 J=1,5
```

```
    2     SCOEFF(I,J)=COEFF(I,J)
C
C  AVOID COMPUTING ZERO STRESSES IF THERE IS ZERO INTERFERENCE.
      IF(DELTA.EQ.0.) GO TO 3
C
C  SOLVE THE EQUATIONS.
      CALL  SOLVE(COEFF,AB,4,4,5,IFLAG)
C
C  A UNIT VALUE OF THE ILL-CONDITIONING FLAG IMPLIES DATA ERROR.
      IF(IFLAG.EQ.1) THEN
        WRITE(6,67)
   67   FORMAT(/ 'DATA ERROR CAUSING EQUATION ILL-CONDITIONING - STOP')
        STOP
      END IF
C
C  CALCULATE AND OUTPUT THE STRESSES AND STRAINS.
      WRITE(6,68)
   68   FORMAT(/ '***** DUE TO COMPOUNDING ALONE *****')
      CALL  STRESS(AB,E1,NU1,E2,NU2,0.,R1,R2,R3)
C
C  NOW ANALYZE THE STRESSES AND STRAINS DUE TO COMPOUNDING AND INTERNAL
C  PRESSURE - PROVIDED THE PRESSURE IS POSITIVE.
C
    3 IF(P.LE.0.) STOP
C
C  RECOVER AND MODIFY THE EQUATION COEFFICIENTS.
      DO 4 I=1,4
      DO 4 J=1,5
    4   COEFF(I,J)=SCOEFF(I,J)
      COEFF(1,5)=-P
      RATIO=R3/R1
      SIGZ=0.
      IF(ENDED.NE.'OPEN') SIGZ=P/(RATIO**2-1.)
      COEFF(4,5)=COEFF(4,5)+SIGZ*(NU2/E2-NU1/E1)
C
C  SOLVE THE EQUATIONS.
      CALL  SOLVE(COEFF,AB,4,4,5,IFLAG)
C
C  A UNIT VALUE OF THE ILL-CONDITIONING FLAG IMPLIES DATA ERROR.
      IF(IFLAG.EQ.1) THEN
        WRITE(6,67)
        STOP
      END IF
C
C  CALCULATE AND OUTPUT THE STRESSES AND STRAINS.
      WRITE(6,69)
   69   FORMAT(/ '***** DUE TO COMPOUNDING AND INTERNAL PRESSURE *****')
      CALL  STRESS(AB,E1,NU1,E2,NU2,SIGZ,R1,R2,R3)
      STOP
      END


      SUBROUTINE  STRESS(AB,E1,NU1,E2,NU2,SIGZ,R1,R2,R3)
C
C  SUBROUTINE TO EVALUATE THE STRESSES (AND STRAINS) IN COMPOUND
C  THICK-WALLED CYLINDERS FROM THE STRESS DISTRIBUTION CONSTANTS.
C
      DIMENSION AB(4)
      REAL  NU1,NU2
C
C  OUTPUT THE AXIAL STRESS.
      WRITE(6,61) SIGZ
   61   FORMAT(/ 'MEAN AXIAL STRESS = ',E12.4)
C
C  EXTRACT THE STRESS DISTRIBUTION CONSTANTS FROM THE ARRAY AB.
      A1=AB(1)
      B1=AB(2)
      A2=AB(3)
      B2=AB(4)
C
C  HEADING FOR TABLE OF STRESSES.
      WRITE(6,62)
   62   FORMAT(/ 'CYL SURFACE      RADIAL        HOOP        EQUIVALENT'
      1          '   MAX. SHEAR        HOOP' /
      2          'NO.              STRESS      STRESS        STRESS
      3          '   STRESS          STRAIN')
```

```
      C
      C  STRESSES AT INNER SURFACE OF INNER CYLINDER.
            SIGR=A1-B1/R1**2
            SIGH=A1+B1/R1**2
            CALL  OUTPUT(1,'INNER',SIGR,SIGH,SIGZ,E1,NU1)
      C
      C  STRESSES AT OUTER SURFACE OF INNER CYLINDER.
            SIGR=A1-B1/R2**2
            SIGH=A1+B1/R2**2
            CALL  OUTPUT(1,'OUTER',SIGR,SIGH,SIGZ,E1,NU1)
      C  STRESSES AT INNER SURFACE OF OUTER CYLINDER.
            SIGR=A2-B2/R2**2
            SIGH=A2+B2/R2**2
            CALL  OUTPUT(2,'INNER',SIGR,SIGH,SIGZ,E2,NU2)
      C
      C  STRESSES AT OUTER SURFACE OF OUTER CYLINDER.
            SIGR=A2-B2/R3**2
            SIGH=A2+B2/R3**2
            CALL  OUTPUT(2,'OUTER',SIGR,SIGH,SIGZ,E2,NU2)
            RETURN
            END


            SUBROUTINE  OUTPUT(NCYL,SURF,SIGR,SIGH,SIGZ,E,NU)
      C
      C  SUBROUTINE TO OUTPUT STRESSES AND STRAINS.
      C
            CHARACTER*5  SURF
            REAL  NU
      C
      C  DEFINE THE VON MISES EQUIVALENT STRESS AND ABSOLUTE
      C  MAXIMUM SHEAR STRESS.
            SIGE=SQRT(((SIGR-SIGH)**2+(SIGH-SIGZ)**2+(SIGZ-SIGR)**2)/2.)
            ASHMAX=MAX(ABS(SIGR-SIGH),ABS(SIGH-SIGZ),ABS(SIGZ-SIGR))/2.
      C
      C  DEFINE THE HOOP STRAIN.
            EH=(SIGH-NU*(SIGZ+SIGR))/E
      C
      C  OUTPUT STRESSES AND STRAINS.
            WRITE(6,61) NCYL,SURF,SIGR,SIGH,SIGE,ASHMAX,EH
      61    FORMAT(I2,3X,A5,5E13.4)
            RETURN
            END
```

The following list provides the definitions of the FORTRAN variables used, arranged in alphabetical order.

Variable	Type	Definition
AB	real	Array storing the constants A_1, B_1, A_2 and B_2
ASHMAX	real	Absolute maximum shear stress
A1	real	The constant A_1
A2	real	The constant A_2
B1	real	The constant B_1
B2	real	The constant B_2
COEFF	real	Array storing the coefficients of linear equations (10.69)
DELTA	real	Radial interference between the two cylinders before assembly, δ
EH	real	Hoop strain, e_θ
ENDED	character	Label for the type of end condition (taken to be closed-ended unless ENDED is 'OPEN')
E1	real	Young's modulus of the material of the inner cylinder, E_1
E2	real	Young's modulus of the material of the outer cylinder, E_2
I	integer	Row counter
IFLAG	integer	Flag for a singular or very ill-conditioned coefficient matrix (normally returned as zero by SOLVE, one if the ill-conditioning test is satisfied)
J	integer	Column counter
NCYL	integer	Cylinder number (1 for inner and 2 for outer)

Variable	Type	Definition
NU1	real	Poisson's ratio of the material of the inner cylinder, v_1
NU2	real	Poisson's ratio of the material of the outer cylinder, v_2
P	real	Internal pressure, p
RATIO	real	Ratio of radii R_3 and R_1
R1	real	Inner radius of the inner cylinder, R_1
R2	real	Nominal outer radius of the inner cylinder and inner radius of the outer cylinder, R_2
R3	real	Outer radius of the outer cylinder, R_3
SCOEFF	real	Array forming temporary store for array COEFF
SIGE	real	Von Mises equivalent stress, σ_e
SIGH	real	Hoop stress, σ_θ
SIGR	real	Radial stress, σ_r
SIGZ	real	Mean axial stress, σ_z
SURF	character	Label for inner or outer surface of a cylinder

The program reads information from a file named DATA, and writes onto a file named RESULTS. All READ statements call for free format input data. After reading in the three nominal radii defining the geometry of the two cylinders, the program writes these out, together with a heading. It then reads in and writes out the radial interference and internal pressure to be applied after compounding. Execution is terminated if a negative value of interference is detected. This would imply an initial clearance between the cylinders: a situation which could be analyzed, but not by the present program. The elastic properties of the cylinders are then read in and written out, followed by the type of end condition. The end condition is only relevant when pressure is applied, when both cylinders are assumed to be closed-ended (and to carry the same axial stress) unless the word 'OPEN' is supplied as data, when they are taken to be open-ended.

The effect of compounding alone is considered first, the program calculating and storing in array COEFF the coefficients of the linear equations (10.69), in preparation for solution by subroutine SOLVE. Before solving the equations (which has the effect of modifying their coefficients), however, the program stores the coefficients in array SCOEFF for later use. Also, this first stage of the analysis is abandoned if there is no interference, which implies no stresses due to compounding alone. Subroutine SOLVE is only likely to detect an ill-conditioned set of equations if there is an error in the data. Having found the values of the constants A_1, B_1, A_2 and B_2, the program then writes out a heading to indicate results due to compounding only, and calls subroutine STRESS to calculate and output stresses and strains. The reason for putting this part of the program in a subroutine is that it can then be used again after the effects of internal pressure have been added.

A zero or negative value of internal pressure halts execution of the program. Otherwise, the coefficients of equations (10.69) are recovered from array SCOEFF and modified to take into account the nonzero radial stress at the innermost surface. Also, if the compound cylinder is closed-ended, the mean axial stress is calculated using equation (10.70), and the last coefficient in the right-hand side vector is modified according to (10.71). Subroutine SOLVE is then called to solve the modified equations, a heading for the results written out and subroutine STRESS called again to calculate and output stresses and strains.

In subroutine STRESS, the mean axial stress is first written out, before the constants A_1, B_1, A_2 and B_2 are recovered from array AB, and a heading for a table of stresses and strains written out. The radial and hoop stresses are then calculated using equations (10.61) and (10.62) at each of the four cylinder surfaces in turn. In each of these cases, the subroutine calls a further subroutine OUTPUT to perform the final calculations and output the results. The first two arguments of this subroutine are the cylinder number (1 for inner, 2 for outer) and either the label 'INNER' or 'OUTER' according to the surface involved, a label which is used in the subroutine to present results.

In subroutine OUTPUT, the von Mises equivalent stress is first found using equation (9.52), the absolute maximum shear stress with the aid of (9.25), and the hoop strain from equation (10.67). The radial and axial strains could also be found, but are much less useful. From the hoop strain at any cylinder surface, we can find the amount by which its diameter there changes, which is a useful practical measure of cylinder deformation. Finally, the stresses and strain are written out, together with labels indicating the cylinder number and surface concerned.

EXAMPLE 10.6

Use program CYLIND to check the results of Example 10.5, and find the change in the outer diameter of the cylinder when yielding just occurs.

Although CYLIND is intended for the analysis of pairs of compounded cylinders, it can also be applied to single cylinders. To do this, we define an imaginary interface at some radial position within the wall which divides the cylinder into two, with the same material properties and no interference. In this case, the internal and external diameters of the closed-ended steel cylinder are 250 and 350 mm, and we can define the radii of two cylinders as $R_1 = 125$ mm, $R_2 = 150$ mm (say), and $R_3 = 175$ mm. Figure 10.16 shows the file of input data defining this problem for program CYLIND, annotated to indicate what each of the lines of data represents. Meter and N/m² units are used for lengths and stresses, respectively. Note that a value of unity for internal pressure (that is, 1 N/m²) is supplied: it is the pressure we wish to find, given a maximum allowable von Mises equivalent stress of 500 MN/m².

The file RESULTS produced by the program is shown in Fig. 10.17. While hoop strain is dimensionless, the stresses are in N/m² (for an internal pressure of 1 N/m²). Looking at the tabulated stresses and strains, we see that, at the imaginary interface, the values on either side of it are identical, justifying the treatment of the single cylinder as two. Also, the radial, hoop and axial stresses at the inner surface of the cylinder are identical to the values we calculated in Example 10.5. The greatest value of von Mises equivalent stress is 3.536 times the internal pressure, occurring at the inner surface of the cylinder. With a yield stress of 500 MN/m², this means that the pressure required to just cause yielding is $p = 500/3.536 = 141$ MN/m², which is precisely the result we obtained in Example 10.5.

```
0.125   0.15   0.175                    (Cylinder radii)
0.   1.                      (Interference and internal pressure)
207.0E9   0.3   207.0E9   0.3   (Young's moduli and Poisson's ratios)
CLOSED                                   (End condition)
```

Fig. 10.16.

```
INNER RADIUS OF THE INNER CYLINDER =    0.1250E+00
NOMINAL RADIUS AT THE INTERFACE =    0.1500E+00
OUTER RADIUS OF THE OUTER CYLINDER =    0.1750E+00

RADIAL INTERFERENCE BETWEEN THE CYLINDERS =    0.0000E+00
INTERNAL PRESSURE =    0.1000E+01

YOUNGS MODULUS OF THE INNER CYLINDER =    0.2070E+12
POISSONS RATIO OF THE INNER CYLINDER =    .300
YOUNGS MODULUS OF THE OUTER CYLINDER =    0.2070E+12
POISSONS RATIO OF THE OUTER CYLINDER =    .300

THE COMPOUND CYLINDER IS ASSUMED TO BE CLOSED-ENDED

***** DUE TO COMPOUNDING AND INTERNAL PRESSURE *****

MEAN AXIAL STRESS =    0.1042E+01
```

CYL NO.	SURFACE	RADIAL STRESS	HOOP STRESS	EQUIVALENT STRESS	MAX. SHEAR STRESS	HOOP STRAIN
1	INNER	-0.1000E+01	0.3083E+01	0.3536E+01	0.2042E+01	0.1483E-10
1	OUTER	-0.3762E+00	0.2459E+01	0.2456E+01	0.1418E+01	0.1092E-10
2	INNER	-0.3762E+00	0.2459E+01	0.2456E+01	0.1418E+01	0.1092E-10
2	OUTER	-0.1192E-06	0.2083E+01	0.1804E+01	0.1042E+01	0.8555E-11

Fig. 10.17.

To find the change in the outer diameter, we use the hoop strain there. For an internal pressure of 1 N/m^2, this is computed as 0.8555×10^{-11}, and for the pressure to just cause yielding this becomes

$$e_\theta = 0.8555 \times 10^{-11} \times 141 \times 10^6 = 1.21 \times 10^{-3}$$

With an outer diameter of 350 mm, the change in this diameter is therefore $1.21 \times 10^{-3} \times 350 = \underline{0.423 \text{ mm}}$ (0.017 in).

EXAMPLE 10.7

An open-ended cylinder for a high-pressure compressor having an internal diameter of 25 mm (1 in) is to be made from a steel for which the maximum allowable normal stress according to the von Mises criterion is 400 MN/m^2 (58 ksi). If the internal pressure is to be 300 MN/m^2 (44 ksi), show that a single cylinder would not be satisfactory. Show that a pair of compounded cylinders, each with a radius ratio of 2, with a radial interference between them of 0.04 mm (0.0016 in) would meet the requirements.

　　If a single cylinder is used, the greatest stresses occur at the inner surface. The radial stress there would be -300 MN/m^2. Equations (10.49) show that, even for an infinitely large outer cylinder radius, the hoop stress at its inner surface cannot be less than $+300$ MN/m^2. Therefore, with no axial stress, equation (9.54) gives the corresponding von Mises equivalent stress as

$$\sigma_e = \sqrt{(300^2 + 300^2 + 300^2)} = 520 \text{ MN/m}^2 \text{ (75 ksi)}$$

which is greater than the maximum allowable stress. From this we conclude that the pressure cannot be contained by a single cylinder.

　　Figure 10.18 shows the file of input data defining the compound cylinder problem for program CYLIND. Meter and N/m^2 units are used for lengths and stresses, respectively. With radius ratios of 2 for both cylinders, their radii are 12.5 mm and

```
0.0125   0.025   0.050
0.040E-3   300.0E6
207.0E9   0.3   207.0E9   0.3
OPEN
```

Fig. 10.18.

```
ANALYSIS OF TWO COMPOUNDED THICK-WALLED CYLINDERS

INNER RADIUS OF THE INNER CYLINDER =    0.1250E-01
NOMINAL RADIUS AT THE INTERFACE =    0.2500E-01
OUTER RADIUS OF THE OUTER CYLINDER =    0.5000E-01

RADIAL INTERFERENCE BETWEEN THE CYLINDERS =    0.4000E-04
INTERNAL PRESSURE =    0.3000E+09

YOUNGS MODULUS OF THE INNER CYLINDER =    0.2070E+12
POISSONS RATIO OF THE INNER CYLINDER =    .300
YOUNGS MODULUS OF THE OUTER CYLINDER =    0.2070E+12
POISSONS RATIO OF THE OUTER CYLINDER =    .300

THE COMPOUND CYLINDER IS ASSUMED TO BE OPEN-ENDED

***** DUE TO COMPOUNDING ALONE *****

MEAN AXIAL STRESS =    0.0000E+00

CYL  SURFACE    RADIAL       HOOP      EQUIVALENT   MAX. SHEAR     HOOP
NO.              STRESS      STRESS       STRESS       STRESS      STRAIN
 1    INNER    0.8000E+01  -0.2650E+09  0.2650E+09  0.1325E+09  -0.1280E-02
 1    OUTER   -0.9936E+08  -0.1656E+09  0.1444E+09  0.8280E+08  -0.6560E-03
 2    INNER   -0.9936E+08   0.1656E+09  0.2318E+09  0.1325E+09   0.9440E-03
 2    OUTER    0.0000E+00   0.6624E+08  0.6624E+08  0.3312E+08   0.3200E-03

***** DUE TO COMPOUNDING AND INTERNAL PRESSURE *****

MEAN AXIAL STRESS =    0.0000E+00

CYL  SURFACE    RADIAL       HOOP      EQUIVALENT   MAX. SHEAR     HOOP
NO.              STRESS      STRESS       STRESS       STRESS      STRAIN
 1    INNER   -0.3000E+09   0.7504E+08  0.3437E+09  0.1875E+09   0.7973E-03
 1    OUTER   -0.1594E+09  -0.6560E+08  0.1387E+09  0.7968E+08  -0.8595E-04
 2    INNER   -0.1594E+09   0.2656E+09  0.3718E+09  0.2125E+09   0.1514E-02
 2    OUTER    0.0000E+00   0.1062E+09  0.1062E+09  0.5312E+08   0.5132E-03
```

Fig. 10.19.

25 mm (nominal) for the inner, and 25 mm (nominal) and 50 mm for the outer. The file RESULTS produced by the program is shown in Fig. 10.19, with stresses given in N/m².

Figure 10.20 shows the plotted distributions of hoop stresses in the two cylinders, both after compounding alone and with the internal pressure added. Compounding alone creates very substantial compressive hoop stresses in the inner cylinder, while the outer cylinder is in tension. When the internal pressure is applied, the effect of which is to make all the hoop stresses more tensile, the inner cylinder goes partially into tension, and the tensile hoop stresses in the outer cylinder increase. Since the radial stresses are also very substantial, we must use the equivalent stresses to assess whether the material is anywhere overstressed. Due to compounding alone, the greatest equivalent stress is 265 MN/m² (38 ksi) at the innermost surface, although a value of 232 MN/m² (34 ksi), which is almost as high, occurs at the inner surface of the outer cylinder. Due to the combined effect of compounding and internal pressure, the greatest equivalent stress is 372 MN/m² (54 ksi) at the inner surface of the outer cylinder (the value at the innermost surface of 344 MN/m² (50 ksi) is almost as high). With all of these values being less than the maximum allowable stress of 400 MN/m² (58 ksi), the design is satisfactory.

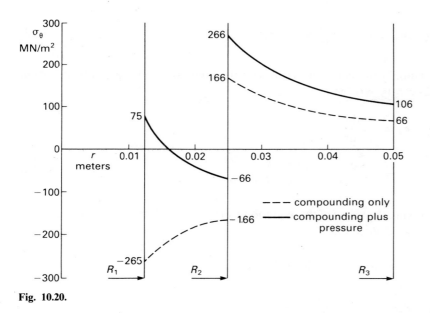

Fig. 10.20.

10.4.3 Rotating disks

Disks rotating at high speed can also be regarded as thick-walled cylinders. In this case however, the loading is due to centrifugal body forces rather than pressure. Practical examples include the rotors of steam and gas turbines, flywheels and other parts of high speed machinery such as wheels and pulleys. Indeed, we have already considered in Section 2.2.3 some examples of rotating rings and cylinders which are thin in the radial direction.

Figure 10.21 shows a flat circular disk of uniform thickness h, inner radius R_1 and outer radius R_2, which is rotating at an angular speed ω (in units of radians per second) about its axis. At a radial distance of r from this axis, the centripetal acceleration is $\omega^2 r$, and the centrifugal body force per unit volume at this radius is

$$F_r = \rho \omega^2 r \qquad (10.72)$$

The analysis of stresses in the disk is similar to that given in Section 10.4.1 for pressurized cylinders. In this case, however, because the disk is thin relative to its radial dimensions we assume that the disk is in a state of plane stress, with axial stress $\sigma_z = 0$. Starting from compatibility equation (10.8), we can use equations (10.38) and (10.39) to define the strains in terms of stresses, giving

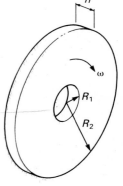

Fig. 10.21.

$$\sigma_r - v\sigma_\theta = \frac{d}{dr}\left[r(\sigma_\theta - v\sigma_r)\right]$$

Carrying out the differentiation on the right-hand side of this equation, we obtain after some rearrangement

$$(\sigma_r - \sigma_\theta)(1 + v) = r\frac{d}{dr}(\sigma_\theta - v\sigma_r) \tag{10.73}$$

Also, equilibrium equation (10.4) with the body force term given by equation (10.72) can be written as

$$\sigma_r - \sigma_\theta = -r\frac{d\sigma_r}{dr} - \rho\omega^2 r^2$$

$$= -r\frac{d}{dr}\left(\sigma_r + \frac{\rho\omega^2 r^2}{2}\right) \tag{10.74}$$

Multiplying the terms of this equation by $(1 + v)$ and comparing with equation (10.73), we deduce that

$$-(1 + v)\left(\sigma_r + \frac{\rho\omega^2 r^2}{2}\right) = \sigma_\theta - v\sigma_r + \text{constant}$$

from which

$$\sigma_r + \sigma_\theta = -\frac{\rho\omega^2 r^2}{2}(1 + v) + 2A \tag{10.75}$$

where A is a constant. This result is equivalent to equation (10.43) for a pressurized cylinder.

Using equation (10.75), we can substitute for σ_θ in equation (10.74) to give

$$2A - 2\sigma_r = r\frac{d\sigma_r}{dr} + \frac{(3 + v)}{2}\rho\omega^2 r^2$$

$$2Ar = \frac{d}{dr}(r^2\sigma_r) + \frac{(3 + v)}{2}\rho\omega^2 r^3$$

$$Ar^2 = r^2\sigma_r + \frac{(3 + v)}{8}\rho\omega^2 r^4 + B$$

where B is an integration constant. Hence

$$\sigma_r = A - \frac{B}{r^2} - \frac{(3 + v)}{8}\rho\omega^2 r^2 \tag{10.76}$$

and, from equation (10.75)

$$\sigma_\theta = A + \frac{B}{r^2} - \frac{(1 + 3v)}{8}\rho\omega^2 r^2 \tag{10.77}$$

We note that these expressions for the stresses do not depend on the thickness, h, of the disk.

We have derived equations (10.76) and (10.77) without reference to the boundary conditions for a particular problem. The integration constants A and B must now be found from these conditions, which for the situation illustrated in Fig. 10.21 are

$$\sigma_r = 0 \text{ at } r = R_1 \quad \text{and} \quad \sigma_r = 0 \text{ at } r = R_2 \tag{10.78}$$

from which

$$A = \frac{(3+v)}{8} \rho\omega^2 (R_1^2 + R_2^2) \qquad B = \frac{(3+v)}{8} \rho\omega^2 R_1^2 R_2^2$$

and

$$\sigma_r = \frac{(3+v)}{8} \rho\omega^2 \left(R_1^2 + R_2^2 - \frac{R_1^2 R_2^2}{r^2} - r^2 \right) \tag{10.79}$$

$$\sigma_\theta = \frac{(3+v)}{8} \rho\omega^2 \left(R_1^2 + R_2^2 + \frac{R_1^2 R_2^2}{r^2} - \frac{1+3v}{3+v} r^2 \right) \tag{10.80}$$

Since the radial stress is zero at both of the boundaries, we can expect it to take a maximum or minimum value somewhere between these limits. The derivative of σ_r with respect to r is zero when

$$\frac{2R_1^2 R_2^2}{r^3} - 2r = 0$$

or

$$r = \sqrt{(R_1 R_2)} \tag{10.81}$$

at which position the maximum radial stress is

$$\sigma_r = \frac{(3+v)}{8} \rho\omega^2 (R_2 - R_1)^2 \tag{10.82}$$

Equation (10.80) for the hoop stress can be compared with equation (2.22) which we obtained for a thin rotating ring. If $r \approx R_1 \approx R_2 \approx R_m$, where R_m is the mean radius, then equation (10.80) becomes

$$\sigma_\theta = \frac{(3+v)}{8} \rho\omega^2 \left(3R_m^2 - \frac{1+3v}{3+v} R_m^2 \right) = \rho\omega^2 R_m^2$$

which is identical to equation (2.22).

Another case which is of interest in practice is that of a solid rotating disk, that is with $R_1 = 0$ in Fig. 10.21. As in Example 10.4, the boundary condition we use at the axis of the disk, $r = 0$, is that the hoop and radial directions are indistinguishable, and $\sigma_r = \sigma_\theta$. Applying this condition to equations (10.76) and (10.77), we conclude that the constant $B = 0$. For the radial stress to be zero at the outer radius, R_2, equation (10.76) gives

$$A = \frac{(3+v)}{8} \rho\omega^2 R_2^2$$

and

$$\sigma_r = \frac{(3+v)}{8} \rho\omega^2 (R_2^2 - r^2) \tag{10.83}$$

$$\sigma_\theta = \frac{(3+v)}{8} \rho\omega^2 \left(R_2^2 - \frac{1+3v}{3+v} r^2 \right) \tag{10.84}$$

EXAMPLE 10.8

A turbine rotor disk has an external diameter of 1.2 m (48 in) and has a 0.1 m (4 in) diameter hole bored along its axis. The rotor is to operate at 4000 rev/min and is made of steel. Find the distributions of the radial and hoop stresses in the disk, and the greatest values of these stresses.

The required values of density and Poisson's ratio for steel are given in Appendix A as $\rho = 7850 \text{ kg/m}^3$ and $v = 0.3$, respectively. The angular velocity of the disk is 4000 rev/min, or $\omega = 418.9$ rad/s. Therefore, the common factor in equations (10.79) and (10.80) is

$$\frac{(3+v)}{8} \rho\omega^2 = \frac{(3+0.3)}{8} \times 7850 \times 418.9^2 \text{ N/m}^4 = 568.2 \text{ MN/m}^4$$

With disk radii $R_1 = 0.05$ m and $R_2 = 0.6$ m, the radial and hoop stress distributions are given by

$$\sigma_r = 568.2 \times \left(0.3625 - \frac{9 \times 10^{-4}}{r^2} - r^2 \right) \text{MN/m}^2$$

$$\sigma_\theta = 568.2 \times \left(0.3625 + \frac{9 \times 10^{-4}}{r^2} - 0.5758\, r^2 \right) \text{MN/m}^2$$

where radius r is in meters. These are plotted in Fig. 10.22. Note the maximum in the radial stress distribution, also that the hoop stress is greater than the radial stress at all radial positions, and is greatest at the inner radius of the disk. According to equation (10.81), the maximum radial stress occurs at $r = \sqrt{(R_1 R_2)} = 0.173$ m, and equation (10.82) gives its value as

$$\sigma_r = 568.2 \times (0.6 - 0.05)^2 = \underline{172 \text{ MN/m}^2} \text{ (25 ksi)}$$

The greatest hoop stress, which occurs at $r = 0.05$ m, is

$$\sigma_\theta = 568.2 \times \left(0.3625 + \frac{9 \times 10^{-4}}{0.05^2} - 0.5758 \times 0.05^2 \right) = \underline{410 \text{ MN/m}^2} \text{ (59 ksi)}$$

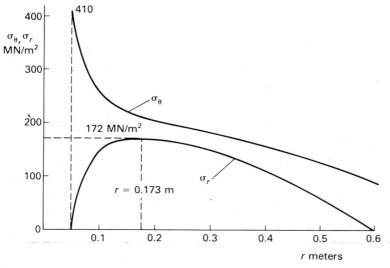

Fig. 10.22.

EXAMPLE 10.9

Repeat Example 10.8 assuming there is no central hole in the disk, and compare the
results.

For a solid disk, equations (10.83) and (10.84) give the distributions of radial and
hoop stress as

$$\sigma_r = 568.2 \times (0.36 - r^2) \ \text{MN/m}^2$$

$$\sigma_\theta = 568.2 \times (0.36 - 0.5758 \, r^2) \ \text{MN/m}^2$$

where radius r is in meters. These are plotted in Fig. 10.23. Once again, the hoop stress is

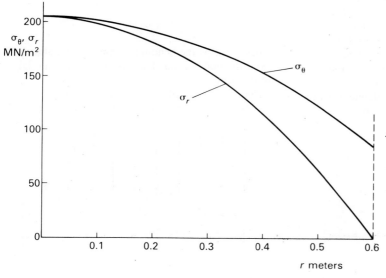

Fig. 10.23.

at least equal in magnitude to the radial stress at all radial positions. Both stresses take their greatest values at the axis of the disk, where $r = 0$ and

$$\sigma_r = \sigma_\theta = 568.2 \times 0.36 = \underline{205 \text{ MN/m}^2} \text{ (30 ksi)}$$

The effect of removing the central hole in the disk is to remove the hoop stress concentration there, thereby halving the greatest hoop stress. The greatest radial stress is, however, somewhat increased, and occurs at the same position as the greatest hoop stress. In order to compare the worst states of stress in the two cases, we should really compare, say, von Mises equivalent stresses. For the disk with the central hole, the most severe state of stress is one of uniaxial tension in the hoop direction at the inner radius, with an equivalent stress of 410 MN/m². For the solid disk, the worst state of stress is equal biaxial tension at its axis, and from equation (9.54) the equivalent stress is equal to the common value of 205 MN/m² of the radial and hoop stresses. In these examples the greatest equivalent stresses happen to be equal in magnitude to the greatest hoop stresses.

Problems

SI METRIC UNITS – BEAMS

10.1 A simply supported uniform steel beam of length L and rectangular cross section of depth d carries a uniformly distributed applied force, and is of negligible weight. The maximum deflection at the center can be regarded as the sum of two contributions, one obtainable from simple beam theory, the other being due to shearing and other effects in the beam not allowed for in the simple theory. Determine the ratios of length to depth of the beam which cause the latter shear deflections to be (i) 1% and (ii) 10% of the total central deflection.

10.2 For the beam defined in Problem 10.1, find an expression for the ratio between the maximum lateral normal stress and the maximum axial bending stress. Show that in practice this ratio is small enough for the simple beam theory assumption of no lateral stress to be a reasonable one.

10.3 The uniform cantilever beam of negligible weight shown in Fig. P10.3 is of length L and has a

rectangular cross section of depth d and breadth b. It is subjected to a uniform shear stress of magnitude τ applied to its lower surface. Assuming that the normal stress σ_x takes the form

$$\sigma_x = -f_1(y)M(x) + f_2(x)$$

where $M(x)$ is the bending moment at a distance x from the built-in end, find expressions for the stress distributions in the beam.

10.4 If the shear stress applied to the cantilever beam in Fig. P10.3 is removed and the weight of the beam is no longer negligible, find the stresses due to this weight. Assume that the normal stress σ_x takes the form

$$\sigma_x = -f_3(y)M(x) + f_4(y)$$

where $M(x)$ is the bending moment at a distance x from the built-in end. The density of the material of the beam is ρ.

10.5 The uniform cantilever beam of negligible weight shown in Fig. P10.5 is of length L and has a

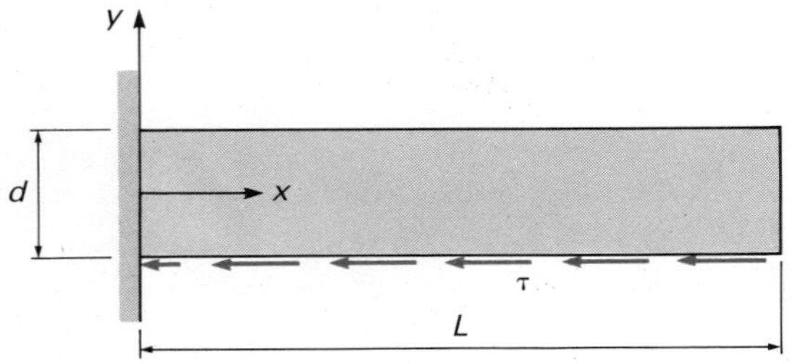

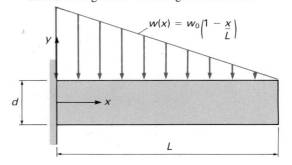

Fig. P10.3.

Fig. P10.5.

rectangular cross section of depth d and breadth b. It is subjected to a distributed lateral force whose intensity decreases linearly with distance along the beam, from w_0 at the built-in end to zero at the free end. Assuming that the normal stress σ_x takes the form

$$\sigma_x = -\frac{yM(x)}{I} + f_5(y)(L - x)$$

where $M(x)$ is the bending moment at a distance x from the built-in end, find expressions for the stress distributions in the beam.

SI METRIC UNITS – CYLINDERS AND DISKS

10.6 A thick-walled cylinder has a ratio between its inner and outer diameters of K, and is subjected to an internal gage pressure of p. Find expressions for the greatest von Mises equivalent stress, and the greatest absolute maximum shear stress in the wall of the cylinder when it is (i) open-ended, and (ii) closed-ended.

10.7 The thick-walled cylindrical steel pressure vessel shown in Fig. P10.7 is subjected to an internal gage pressure of 25 MN/m². Find the greatest values of the radial, hoop, axial, equivalent and absolute maximum shear stresses in the walls of the vessel remote from its ends. Find also the change in the internal diameter when the pressure is applied.

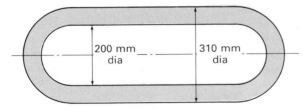

200 mm dia   310 mm dia

Fig. P10.7.

10.8 A closed-ended cylindrical pressure vessel is to be designed to have an internal diameter of 125 mm and to be subjected to an internal gage pressure of 10 MN/m². Find a suitable external diameter for the cylinder: (i) if the hoop stress is not to exceed 100 MN/m², (ii) if the maximum shear stress is not to exceed 60 MN/m², and (iii) if the von Mises equivalent stress is not to exceed 100 MN/m².

10.9 A closed-ended thick-walled cylinder has a ratio between its inner and outer diameters of K, and is subjected to an internal gage pressure of p. Find expressions for the radial, hoop and axial strains at the inner and outer surfaces.

10.10 A closed-ended thick-walled cylinder with internal and external diameters of 150 mm and 200 mm, respectively, is subjected to internal and external pressures of 50 MN/m² and 12 MN/m², respectively. Find the greatest values of the radial, hoop, axial, equivalent and maximum shear stresses in the walls of the cylinder remote from its ends.

10.11 A solid steel rod 70 mm in diameter and 120 mm long is immersed in a fluid at a pressure of 25 MN/m². Find the greatest values of the radial, hoop, axial, equivalent and absolute maximum shear stresses in the rod, and the amounts by which its diameter and length change.

10.12 A cylindrical tube with inner and outer diameters of 80 mm and 160 mm, respectively, is internally pressurized, and the axial and hoop strains at its outer surface are measured by means of strain gages as 3.52×10^{-4} and 1.41×10^{-3}. Find the Poisson's ratio of the tube material.

10.13 A cylindrical steel sleeve is pressed onto a 35 mm diameter solid steel shaft and exerts a uniform compressive radial stress of 30 MN/m². A tensile force of 80 kN is then applied to the shaft in the axial direction, reducing the radial stress between them to 20 MN/m². Find the change in diameter of the interface between the sleeve and shaft due to the application of this force.

10.14 Use program CYLIND to solve Problem 10.7.

10.15 Show that the compound cylinder design analyzed in Example 10.7 remains satisfactory if the cylinder becomes effectively closed-ended.

10.16 A compound steel cylinder is formed by shrinking a tube of 300 mm external diameter and 225 mm internal diameter onto another tube of 175 mm internal diameter. If the difference between the diameters of the contacting surfaces before assembly is 0.10 mm, find the von Mises equivalent stresses at the inner and outer surfaces of each tube due to compounding.

10.17 If the maximum permissible hoop stress for the open-ended compound cylinder defined in Problem 10.16 is 200 MN/m², find the greatest internal pressure which can be applied. Compare this to the corresponding pressure which can be applied to a single cylinder with the same overall dimensions and the same limiting value of hoop stress.

10.18 Repeat Problem 10.17 assuming that the limiting value of 200 MN/m² is applied not to the hoop stresses but to the von Mises equivalent stresses in the cylinder.

10.19 A closed-ended compound cylinder is formed by shrinking a steel tube of 320 mm external diameter and 250 mm internal diameter onto a brass tube of

200 mm internal diameter. If the difference between the diameters of the contacting surfaces before assembly is 0.10 mm, find the von Mises equivalent stresses at the inner and outer surfaces of each tube due to compounding, and the effect on these stresses of an internal pressure of 50 MN/m².

10.20 A solid steel flywheel has a diameter of 500 mm. If the material yields according to the von Mises criterion, and at 400 MN/m² in simple tension, find the maximum speed at which the flywheel can rotate without yielding.

10.21 A circular disk of uniform thickness with an external radius of R_2 and a central hole with a radius of R_1 is press-fitted onto a shaft. When the assembly is rotated at an angular speed of ω, the pressure at the interface between the shaft and disk is p. Derive expressions for the hoop stress in the disk at its inner and outer edges.

10.22 A thin steel disk has internal and external diameters of 175 and 375 mm, respectively. Find the maximum speed at which the disk can be rotated if the hoop stress is not to exceed 100 MN/m², also the corresponding changes in the diameters of the disk.

10.23 A thin solid steel disk with a nominal diameter of 700 mm has a steel ring of 900 mm external diameter and the same thickness shrunk onto it. If the difference between the diameters of the contacting surfaces of the two components before assembly is 0.1 mm, find the rotational speed at which the compressive radial stress at the interface is reduced to zero.

US CUSTOMARY UNITS – BEAMS

10.24 to 10.28 Solve Problems 10.1 to 10.5.

US CUSTOMARY UNITS –
CYLINDERS AND DISKS

10.29 Solve Problem 10.6.

10.30 A steel cylinder of a hydraulic press is shown in Fig. P10.30. A force of 200 kip is applied to the frictionless piston which closes one end of the cylinder. Find the greatest values of the radial, hoop, axial, equivalent and absolute maximum shear stresses in the walls of the vessel remote from its ends. Find also the change in the internal diameter when the pressure is applied.

10.31 A thick-walled cast iron pipe carrying a liquid at a pressure of 220 psi has internal and external diameters of 8 in and 10 in, respectively. Find the greatest hoop stress and the change in the external diameter of the pipe due to the pressure. Compare

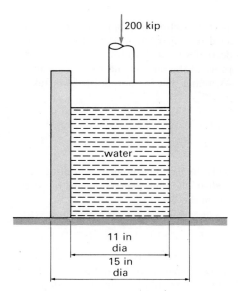

Fig. P10.30

these results with the values obtained assuming the pipe to be thin-walled.

10.32 Solve Problem 10.9.

10.33 A closed-ended thick-walled cylinder has internal and external diameters of 6 in and 8 in, respectively. If the yield stress in simple tension of the material is 36 ksi, find the gage pressure within the cylinder at which yielding just occurs according to (i) the Tresca criterion, and (ii) the von Mises criterion.

10.34 A solid steel rod of 1.5 in diameter and 30 in long is immersed in a fluid at a pressure of 4 ksi. Find the greatest values of the radial, hoop, axial, equivalent and absolute maximum shear stresses in the rod, and the amounts by which its diameter and length change.

10.35 Solve Problem 10.12, taking the inner and outer diameters of the tube to be 3.2 in and 6.4 in, respectively.

10.36 and 10.37 Use program CYLIND to solve ·Problems 10.30 and 10.33.

10.38 A gun barrel is constructed by shrinking a tube of 10 in external diameter and 7.2 in internal diameter over another tube 7.22 in external diameter and 5 in internal diameter. Find the von Mises equivalent stresses at the inner and outer surfaces of each tube.

10.39 A compound steel cylinder is formed by shrinking a tube of 12 in external diameter and 9 in internal diameter onto another tube of 7 in internal diameter. If the difference between the diameters of the contacting surfaces before assembly is 0.004 in, find the von Mises equivalent stresses at the inner and outer surfaces of each tube due to compounding.

10.40 If the maximum permissible hoop stress for the open-ended compound cylinder defined in Problem 10.39 is 30 ksi, find the greatest internal pressure which can be applied. Compare this to the corresponding pressure which can be applied to a single cylinder with the same overall dimensions and the same limiting value of hoop stress.

10.41 Repeat Problem 10.40 assuming that the limiting value of 30 ksi is applied not to the hoop stresses but to the von Mises equivalent stresses in the cylinder.

10.42 A compound cylinder is formed by shrinking an aluminum alloy tube of 3 in external diameter and 2.4 in internal diameter onto a copper tube of 1.8 in internal diameter. If the difference between the diameters of the contacting surfaces before assembly is 0.0016 in, find the greatest von Mises equivalent stresses in each of the tubes due to compounding, and the changes in the inner diameter of the copper tube and the outer diameter of the aluminum alloy tube.

10.43 A solid steel flywheel has a diameter of 20 in. If the material yields according to the von Mises criterion, and at 60 ksi in simple tension, find the maximum speed at which the flywheel can rotate without yielding.

10.44 Solve Problem 10.21.

10.45 A thin steel disk has internal and external diameters of 7 and 15 in, respectively. Find the maximum speed at which the disk can be rotated if the hoop stress is not to exceed 15 ksi, also the corresponding changes in the diameters of the disk.

Appendices

A: PROPERTIES OF MATERIALS

The following tables display typical values at ambient temperature for the properties of engineering materials needed for solid mechanics calculations. These values are used in the worked Examples presented in the book, and are needed for many of the Problems provided at the ends of the chapters. Properties are tabulated in both SI metric and US customary (Imperial) units. Conversion factors between the two systems are provided in Table A.3.

Density, elastic and thermal properties

The density, elastic constants and thermal expansion coefficients for many commonly used engineering materials do not vary very much with changes in chemical composition, heat treatment and mechanical working. For present purposes, it is therefore reasonable to work with single values of the properties which are typical of those met in practice.

The symbols used to represent these properties are as follows:

density	ρ
Young's modulus of elasticity	E
shear modulus of elasticity	G
bulk modulus of elasticity	K
Poisson's ratio	v
coefficient of linear thermal expansion	α
coefficient of volumetric thermal expansion	β

These properties are not all independent of each other. In particular, the elastic properties are connected by the relationships

$$E = 2G(1 + v) \quad \text{and} \quad E = 3K(1 - 2v)$$

Also, the coefficients of thermal expansion are related by $\beta = 3\alpha$, although it is usual to quote only one value for a material: α for a solid, and β for a fluid.

Table A.1 (SI metric units)

Material	ρ (kg/m^3)	E (GN/m^2)	G (GN/m^2)	K (GN/m^2)	v	α (°C^{-1})	β (°C^{-1})
Steel	7850	207.	79.6	172.0	0.30	11×10^{-6}	
Cast iron (gray)	7200	70.0	28.0	46.7	0.25	12×10^{-6}	
Aluminum	2700	70.0	26.0	75.0	0.34	25×10^{-6}	
Aluminum alloy	2720	68.9	26.5	57.5	0.30	23×10^{-6}	
Copper	8930	120.	45.0	120.0	0.33	16×10^{-6}	
Brass	8410	103.	38.3	115.0	0.35	19×10^{-6}	
Phosphor bronze	8800	110.	40.7	122.0	0.35	18×10^{-6}	
Concrete	2400	13.8			0.10	11×10^{-6}	
Timber	550	12.0				4×10^{-6}	
Glass	2200	65.0	28.0	31.0	0.15	8×10^{-6}	
Rubber	900				0.49	160×10^{-6}	
Water	1000			2.3			2.1×10^{-4}

Table A.1 (US customary units)

Material	ρ (lb/in^3)	E (10^6 psi)	G (10^6 psi)	K (10^6 psi)	v	α (°F^{-1})	β (°F^{-1})
Steel	0.284	30	11.5	25.0	0.30	6.5×10^{-6}	
Cast iron (gray)	0.260	10	4.0	6.7	0.25	6.7×10^{-6}	
Aluminum	0.098	10	3.7	10.4	0.34	14×10^{-6}	
Aluminum alloy	0.098	10	3.8	8.3	0.30	13×10^{-6}	
Copper	0.323	17	6.4	17.0	0.33	8.9×10^{-6}	
Brass	0.305	15	5.6	16.7	0.35	11×10^{-6}	
Phosphor bronze	0.318	16	5.9	17.8	0.35	10×10^{-6}	
Concrete	0.086	2.0			0.10	6.5×10^{-6}	
Timber	0.019	1.7				2.2×10^{-6}	
Glass	0.079	9.4	4.1	4.5	0.15	4.0×10^{-6}	
Rubber	0.033				0.49	90×10^{-6}	
Water	0.036			0.33			1.2×10^{-4}

Strength properties

The strength properties, particularly of materials such as steels, depend very
much on chemical composition, heat treatment and mechanical working. The
values given in the following Table A.2 are therefore only broadly represen-
tative of the ranges met in practice. Where necessary, more precise data are
given in the Examples and Problems. The abbreviations used are as follows:

UTS   ultimate tensile strength
UCS   ultimate compressive strength
USS   ultimate shear strength

Where no value is shown for UCS (for relatively ductile materials), it may be
assumed to be equal to the UTS. Brittle materials such as cast iron and concrete
are much stronger in compression than in tension. Indeed, concrete is usually
assumed to have zero strength in tension.

Table A.2 (SI metric units)

Material	Yield stress (MN/m^2)	UTS (MN/m^2)	UCS (MN/m^2)	USS (MN/m^2)
Steel	400	650		240
Cast iron (gray)		200	700	250
Aluminum alloy	350	400		220
Brass (annealed)	100	300		200
Concrete		0	30	
Glass			50	
Rubber		15		

Table A.2 (US customary units)

Material	Yield stress (ksi)	UTS (ksi)	UCS (ksi)	USS (ksi)
Steel	60	95		35
Cast iron (gray)		30	100	36
Aluminum alloy	50	60		32
Brass (annealed)	15	40		30
Concrete		0	5	
Glass			7	
Rubber		2.2		

Conversion factors between SI metric and US customary units

Table A.3

Quantity	US customary unit	SI equivalent
Mass	lb (mass)	0.4536 kg
Length	in	25.4 mm
	ft	0.3048 m
Density	lb/in^3	27.68×10^3 kg/m^3
	lb/ft^3	16.02 kg/m^3
Force	lb (force)	4.448 N
	kip (10^3 lb)	4.448 kN
Pressure or stress	lb/in^2 (psi)	6.895 kN/m^2
	ksi (10^3 psi)	6.895 MN/m^2
Moment or torque	lb in	0.1130 Nm
	lb ft	1.356 Nm
Power	ft lb/s	1.356 W
	hp(550 ft lb/s)	745.7 W
Temperature	°F	0.5556 °C

B: SOLUTION OF LINEAR ALGEBRAIC EQUATIONS BY GAUSSIAN ELIMINATION

Simultaneous linear algebraic equations arise in methods for analyzing many different problems in solid mechanics, and indeed other branches of engineering science. In this book alone, we meet examples in the analysis of both statically determinate and statically indeterminate pin-jointed structures, the finite element analysis of beams, the analysis of strain gage measurements, and the determination of stresses in thick-walled cylinders. Gaussian elimination is a direct method for solving such equations by successive elimination of the unknowns. Let us consider first an example involving just three equations

$$2x_1 + x_2 - x_3 = 1$$
$$x_1 + 3x_2 + 2x_3 = 13 \qquad \text{(B.1)}$$
$$x_1 - x_2 + 4x_3 = 11$$

where x_1, x_2 and x_3 are the unknowns to be found. We can use the first equation to eliminate x_1 from the other two equations. To do this, we divide the first equation through by 2 and subtract from the second and third equations to give

$$2x_1 + x_2 - x_3 = 1$$
$$2.5x_2 + 2.5x_3 = 12.5 \qquad \text{(B.2)}$$
$$-1.5x_2 + 4.5x_3 = 10.5$$

Similarly, the second of these is multiplied through by 1.5/2.5 and added to the third equation to eliminate x_2

$$2x_1 + x_2 - x_3 = 1$$
$$2.5x_2 + 2.5x_3 = 12.5 \qquad \text{(B.3)}$$
$$6x_3 = 18$$

Solving these equations in reverse order, a process known as *back substitution*, we obtain: $x_3 = 3$, then $x_2 = (12.5 - 2.5 \times 3)/2.5 = 2$, and finally $x_1 = (1 - 2 + 3)/2 = 1$.

Let us now try to generalize this process to any set of n linear equations, which we may express in the general form

$$a_{11}x_1 + a_{12}x_2 + \ldots + a_{1n}x_n = b_1$$
$$a_{21}x_1 + a_{22}x_2 + \ldots + a_{2n}x_n = b_2$$
$$\ldots \ldots \ldots \ldots \ldots \ldots \ldots \ldots \ldots \ldots \qquad \text{(B.4)}$$
$$a_{n1}x_1 + a_{n2}x_2 + \ldots + a_{nn}x_n = b_n$$

where the coefficients a_{ij} and b_i are all known constants. We may also write these equations in *matrix* form as

$$\begin{bmatrix} a_{11} & a_{12} & \cdots & a_{1n} \\ a_{21} & a_{22} & \cdots & a_{2n} \\ \cdots & \cdots & \cdots & \cdots \\ a_{n1} & a_{n2} & \cdots & a_{nn} \end{bmatrix} \begin{bmatrix} x_1 \\ x_2 \\ \cdots \\ x_n \end{bmatrix} = \begin{bmatrix} b_1 \\ b_2 \\ \cdots \\ b_n \end{bmatrix} \qquad \text{(B.5)}$$

$$\text{or} \qquad [A][X] = [B] \qquad \text{(B.6)}$$

where $[A]$ is an $n \times n$ square matrix, and $[X]$ and $[B]$ are *column vectors* or $n \times 1$ matrices. We may represent the general coefficients in these matrices as a_{ij}, x_i and b_i, where subscript i specifies the row number, and j defines the column number.

In order to define the elimination process, let us give the initial coefficients of $[A]$ and $[B]$ the notation $a_{ij}^{(1)}$ and $b_i^{(1)}$. After the first elimination, of x_1 from all equations from the 2nd to the nth, the modified coefficients in the corresponding rows of $[A]$ and $[B]$ are

$$a_{ij}^{(2)} = a_{ij}^{(1)} - f a_{1j}^{(1)}$$
$$b_i^{(2)} = b_i^{(1)} - f b_1^{(1)} \qquad \text{(B.7)}$$

for $i = 2, 3, \ldots, n$ and $j = 1, 2, \ldots, n$, where the factor f is defined by

$$f = a_{i1}^{(1)} / a_{11}^{(1)}$$

Similarly, after the kth elimination

$$a_{ij}^{(k+1)} = a_{ij}^{(k)} - f a_{kj}^{(k)}$$
$$b_i^{(k+1)} = b_i^{(k)} - f b_k^{(k)} \qquad \text{(B.8)}$$

for $i = k+1, k+2, \ldots, n$ and $j = k, k+1, \ldots, n$, where f is now defined by

$$f = a_{ik}^{(k)} / a_{kk}^{(k)}$$

It is important to notice that vector $[B]$ is treated just like a column of $[A]$, and we can take advantage of this fact to simplify the programming of the elimination process, by treating $[B]$ as the $(n+1)$th column of $[A]$. The final set of equations, after $(n-1)$ eliminations have been performed, is

$$a_{11}^{(1)} x_1 + a_{12}^{(1)} x_2 + \ldots + a_{1n}^{(1)} x_n = b_1^{(1)}$$
$$a_{22}^{(2)} x_2 + \ldots + a_{2n}^{(2)} x_n = b_2^{(2)} \qquad \text{(B.9)}$$
$$\cdots\cdots\cdots\cdots\cdots\cdots\cdots\cdots\cdots$$
$$a_{nn}^{(n)} x_n = b_n^{(n)}$$

The solutions we require for the x_i may now be obtained by back substitution

$$x_n = b_n^{(n)} / a_{nn}^{(n)} \qquad \text{(B.10)}$$

$$x_i = \left[b_i^{(i)} - \sum_{j=i+1}^{n} a_{ij}^{(i)} x_j \right] \Big/ a_{ii}^{(i)} \qquad \text{(B.11)}$$

for $i = n - 1, n - 2, \ldots, 1$.

A great many arithmetic operations are involved in solving large sets of equations by elimination. Any errors introduced, such as *roundoff errors* caused by representing numbers with only a finite number of significant figures, tend to be magnified and may become unacceptably large. Equations (B.8) show that the elimination process involves many multiplications by the factors f. In order to minimize the effects of roundoff errors, we should make these factors as small as possible, and certainly less than one. Therefore, the *pivotal* coefficient $a_{kk}^{(k)}$ should be the largest coefficient in the leading column of the remaining submatrix

$$|a_{kk}^{(k)}| > |a_{ik}^{(k)}| \quad \text{for } i = k+1, k+2, \ldots, n \qquad \text{(B.12)}$$

Satisfying this condition also helps to avoid division by zero in equations (B.8), and we can achieve it by a process known as *partial pivoting*. Immediately before each elimination, the leading column is searched for the largest coefficient. By interchanging equations, this can be made the pivotal coefficient to satisfy equation (B.12). This process can be illustrated with the following equations

$$-3x_1 + x_2 - 3x_3 = -10$$

$$3x_1 + 2x_2 + 2x_3 = 15 \qquad \text{(B.13)}$$

$$6x_1 + 4x_2 + x_3 = 27$$

The largest coefficient of x_1 occurs in the third equation, which is interchanged with the first before the first elimination is performed to give

$$6x_1 + 4x_2 + x_3 = 27$$

$$1.5x_3 = 1.5 \qquad \text{(B.14)}$$

$$3x_2 - 2.5x_3 = 3.5$$

Now, we could not use the second of these equations to eliminate x_2 from the third equation, because it does not involve x_2, the coefficient of x_2 having been reduced to zero by the first elimination. But, we can overcome this difficulty by interchanging the second and third equations to satisfy equation (B.12). In fact, no further eliminations are then necessary, and we may obtain the solutions by back substitution as $x_3 = 1$, then $x_2 = (3.5 + 2.5 \times 1)/3 = 2$, and finally $x_1 = (27 - 4 \times 2 - 1)/6 = 3$.

We could extend the idea of partial pivoting to searching the whole of the remaining submatrix for the largest coefficient. Such *complete pivoting* involves interchanging both rows and columns, and is more difficult to program. Since it offers only modest advantages in terms of accuracy over partial pivoting, it is rarely used.

Another refinement which helps to minimize the effects of roundoff errors is to scale the equations to make their coefficients similar in magnitude. One way we can do this is to normalize each equation so that the largest coefficient in each row of $[A]$ is of magnitude one. Scaling can be particularly important when corresponding coefficients in different equations differ by several orders of magnitude.

Let us consider now another example, which presents a further type of difficulty

$$12x_1 + 2x_2 - x_3 = 13$$

$$6x_1 + 8x_2 + 4x_3 = 18 \qquad\qquad (B.15)$$

$$3x_1 + 4x_2 + 2x_3 = 9$$

After the first elimination, of x_1 from the second and third equations, we obtain

$$12x_1 + 2x_2 - x_3 = 13$$

$$7x_2 + 4.5x_3 = 11.5 \qquad\qquad (B.16)$$

$$3.5x_2 + 2.25x_3 = 5.75$$

and after the second elimination, of x_2 from the third equation

$$12x_1 + 2x_2 - x_3 = 13$$

$$7x_2 + 4.5x_3 = 11.5 \qquad\qquad (B.17)$$

$$0x_3 = 0$$

The coefficient of x_3 is now zero, which makes it impossible to solve the equations by back substitution. In fact, x_3 could take any numerical value, including zero. The reason for this is that, in equations (B.15), the third equation is merely the second equation divided through by a common factor of 2, and therefore does not provide any new information about the unknowns. Alternatively, if the third equation had a numerical value on the right-hand side other than 9, then the second and third equations would have provided conflicting information, and the set of equations would have been *inconsistent*. In either case, the matrix of coefficients $[A]$ on the left-hand sides of the equations is said to be *singular*, a condition which is due to the *determinant* of matrix $[A]$ being zero. For our present purposes, what we need to know is that such a condition can be detected during partial pivoting if it is impossible to find a nonzero pivotal value at some stage of the elimination process or at the start of the back substitution process. What is more difficult to detect, however, is when a set of equations is nearly singular or *ill-conditioned*. This arises when, for example, two equations provide very nearly the same information about the unknowns. Alternatively, the effect of roundoff errors may be to make what is a singular set of equations apparently ill-conditioned, in that rather than a zero pivotal coefficient, a very small value is obtained. Indeed, we may use the detection of a very small pivotal coefficient as a convenient test for either a singular or very ill-conditioned set of equations, but without being able to distinguish between them. By very small pivotal coefficient, we mean a value very small in relation to the magnitudes of the coefficients of the matrix at the start of the elimination process, which are of order unity if we have previously applied scaling.

Subroutine SOLVE provides a FORTRAN implementation of Gaussian elimination.

```
      SUBROUTINE SOLVE(A,X,NEQN,NROW,NCOL,IFLAG)
C
C  SUBROUTINE FOR SOLVING SIMULTANEOUS LINEAR EQUATIONS BY GAUSSIAN
C  ELIMINATION WITH PARTIAL PIVOTING.
C
      DIMENSION   A(NROW,NCOL),X(NROW)
      IF(NEQN.GT.NROW.OR.NEQN.GE.NCOL) THEN
        WRITE(6,'DIMENSION ERROR IN SOLVE - STOP')
        STOP
      END IF
C
C  INITIALIZE ILL-CONDITIONING FLAG.
      IFLAG=0
C
C  SCALE EACH EQUATION TO HAVE A MAXIMUM COEFFICIENT MAGNITUDE OF UNITY.
      JMAX=NEQN+1
      DO 2 I=1,NEQN
      AMAX=0.
      DO 1 J=1,NEQN
      ABSA=ABS(A(I,J))
    1 IF(ABSA.GT.AMAX) AMAX=ABSA
      DO 2 J=1,JMAX
    2 A(I,J)=A(I,J)/AMAX
C
C  COMMENCE ELIMINATION PROCESS.
      DO 5 K=1,NEQN-1
C
C  SEARCH LEADING COLUMN OF THE COEFFICIENT MATRIX FROM THE DIAGONAL
C  DOWNWARDS FOR THE LARGEST VALUE.
      IMAX=K
      DO 3 I=K+1,NEQN
    3 IF(ABS(A(I,K)).GT.ABS(A(IMAX,K))) IMAX=I
C
C  IF NECESSARY, INTERCHANGE EQUATIONS TO MAKE THE LARGEST COEFFICIENT
C  BECOME THE PIVOTAL COEFFICIENT.
      IF(IMAX.NE.K) THEN
        DO 4 J=K,JMAX
        ATEMP=A(K,J)
        A(K,J)=A(IMAX,J)
    4   A(IMAX,J)=ATEMP
      END IF
C
C  ELIMINATE X(K) FROM EQUATIONS (K+1) TO NEQN, FIRST TESTING FOR
C  EXCESSIVELY SMALL PIVOTAL COEFFICIENT (ASSOCIATED WITH A SINGULAR
C  OR VERY ILL-CONDITIONED MATRIX).
      IF(ABS(A(K,K)).LT.1.E-5) THEN
        IFLAG=1
        RETURN
      END IF
      DO 5 I=K+1,NEQN
      FACT=A(I,K)/A(K,K)
      DO 5 J=K,JMAX
    5 A(I,J)=A(I,J)-FACT*A(K,J)
```

```
C
C   SOLVE THE EQUATIONS BY BACK SUBSTITUTION, FIRST TESTING
C   FOR AN EXCESSIVELY SMALL LAST DIAGONAL COEFFICIENT.
        IF(ABS(A(NEQN,NEQN)).LT.1.E-5) THEN
          IFLAG=1
          RETURN
        END IF
        X(NEQN)=A(NEQN,JMAX)/A(NEQN,NEQN)
        DO 7 I=NEQN-1,1,-1
        SUM=A(I,JMAX)
        DO 6 J=I+1,NEQN
6       SUM=SUM-A(I,J)*X(J)
7       X(I)=SUM/A(I,I)
        RETURN
        END
```

The following list provides the definitions of the FORTRAN variables used, arranged in alphabetical order.

Variable	Type	Definition
A	real	Coefficient matrix $[A]$ extended to include vector $[B]$ as its last column
ABSA	real	Absolute value of coefficient a_{ij}
AMAX	real	Coefficient of maximum magnitude in current row of $[A]$
ATEMP	real	Temporary store for coefficients during equation interchange
FACT	real	Factor f
I	integer	Row number in array A
IMAX	integer	Value of row number I corresponding to the largest coefficient in the leading column of $[A]$
IFLAG	integer	Flag for a singular or very ill-conditioned coefficient matrix (normally returned as zero to the calling program, one if the ill-conditioning test is satisfied)
J	integer	Column number in array A
JMAX	integer	Maximum value of the column number, J (equal to $n+1$)
K	integer	Elimination counter, k
NCOL	integer	Number of columns in the DIMENSION for array A
NEQN	integer	Number of equations to be solved, n.
NROW	integer	Number of rows in the DIMENSION for array A
SUM	real	Temporary store for the summation term in equation (B.11)
X	real	Solution vector $[X]$

The arguments of subroutine SOLVE include the array of equation coefficients, A, which incorporates the vector of constants $[B]$ as its last column, and the solution vector, X. The variables NROW and NCOL, which enter the maximum numbers of rows and columns permitted in A and X, are used for dimensioning purposes. Normally, NCOL = NROW + 1, but the main consideration is that the dimensions of A and X should be compatible with those in the calling program. The argument NEQN enters the number of equations to be solved, while IFLAG returns an integer flag value to the calling program to warn of a singular or very ill-conditioned coefficient matrix.

The first action of the program is to test the acceptability of the number of equations to be solved, in relation to the maximum array sizes. Then IFLAG is set to zero, before the equations are scaled to give a maximum coefficient magnitude of unity in each row of $[A]$. The elimination process is then started: each equation, with the exception of the last, is used in turn to eliminate the corresponding unknown from the subsequent equations. The current elimination number is given by K, and is equivalent to k in equation (B.8). Before performing the necessary eliminations with a particular equation, however, a search is made down the leading column of the remaining submatrix to find the coefficient of greatest magnitude, as defined by equation (B.12). The search technique locates the row number of the largest coefficient, IMAX, by first assuming that it corresponds to the coefficient on the diagonal, and only changing this if a larger coefficient is found. If the pivotal coefficient is not the largest, the relevant equations are interchanged by interchanging all their coefficients. In computing terms, this movement of data between storage locations is inefficient. While it could be avoided by keeping a record of the revised order in which the equations are to be considered, this makes the program significantly more difficult to understand. For present purposes, we sacrifice computational efficiency in the interests of clarity, in the knowledge that for the relatively small sets of equations we wish to solve, the extra computing time involved is very small.

Despite the search for the largest coefficient, it is still possible for the pivotal coefficient to be extremely small or zero if the coefficient matrix is singular or very ill-conditioned. Bearing in mind that the equations were scaled initially, an appropriate test of magnitude on a computer working to 32 bit precision for real variables is

$$|a_{kk}| < 10^{-5} \tag{B.18}$$

If this condition is satisfied at any stage of the elimination process, the problem is rejected. Rejection is indicated by setting the value of IFLAG to one, which may be detected by the calling program. If the partial pivoting is successful, however, the eliminations defined by equations (B.8) are performed, with the variable FACT being used to store the values of the factor f.

After testing the magnitude of the last diagonal coefficient ($a_{nn}^{(n)}$ in equations (B.9)), the back substitutions defined in equations (B.10) and (B.11) are performed to find the required solutions. The variable SUM is used to accumulate the results of the summations indicated in equations (B.11).

C: MOMENTS OF AREA

C.1 First moment of area

Consider the region of total area A shown in Fig. C.1, a typical element of which is the small shaded piece δA. We define the first moment of area of this piece about the axis RS as $p\delta A$, where p is the perpendicular distance of δA from the axis. The first moment of area of the whole region about this axis is found by summing over all the elements which make up the region

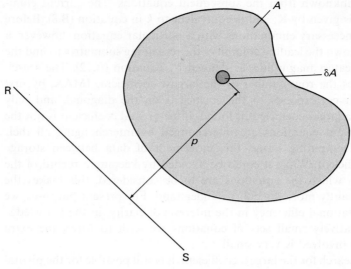

Fig. C.1.

$$Q = \sum p\delta A = \int_A p\, \mathrm{d}A \tag{C.1}$$

Working from this general definition, if we now consider the region to lie in the x, y plane as in Fig. C.2, with δA being a small rectangle with dimensions δx by δy and with its center at the point (x, y), then we can define the first moment of area of the region about the x axis as

$$Q_x = \int_A y\, \mathrm{d}A = \int_A y\, \mathrm{d}x\, \mathrm{d}y \tag{C.2}$$

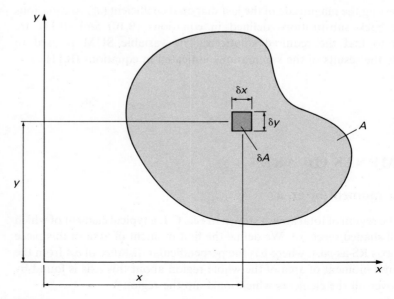

Fig. C.2.

Similarly, the first moment of area about the y axis is

$$Q_y = \int_A x \, dA = \int_A x \, dx \, dy \tag{C.3}$$

Now, if we consider Q_x in a little more detail, since the function we wish to integrate with respect to x and y, namely y, is independent of x, the first integration is trivial. If $b(y)$ is the breadth of the region at a height y above the x axis, as shown in Fig. C.3, then we may express Q_x as

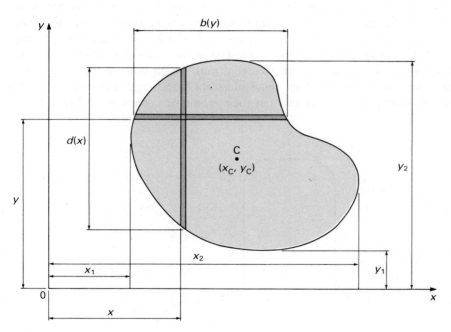

Fig. C.3.

$$Q_x = \int_{y_1}^{y_2} yb(y) \, dy \tag{C.4}$$

Similarly, if $d(x)$ is the depth of the region at a horizontal distance x from the y axis, the first moment of area about this axis is

$$Q_y = \int_{x_1}^{x_2} x d(x) \, dx \tag{C.5}$$

The center of area or *centroid* of the region A is defined as the point at which the entire area would have to be concentrated in order to have the same first moment of area about any axis as the actual region A. In other words, if C is the centroid, with coordinates (x_C, y_C), as shown in Fig. C.3, then

$$Q_x = Ay_C \quad \text{and} \quad Q_y = Ax_C \tag{C.6}$$

This means that if, for example, the x axis passes through the centroid of the region, then y_C is zero and

$$Q_x = \int_{y_1}^{y_2} y b(y)\,dy = 0 \tag{C.7}$$

If the first moment of area about the x axis is also zero, then the centroid and the origin of the coordinates coincide.

If a region has as *axis of symmetry*, the first moment of area about this axis is zero. In Fig. C.4, for example, the y axis is the axis of symmetry, and for every element of area δA to the right of this axis there is an identical element at the same distance to the left of the axis, and their first moments are equal, giving $Q_y = 0$. Consequently, the centroid must lie on the axis of symmetry. If a region has two axes of symmetry, the centroid must lie on both of these axes, and therefore at the point where they intersect. A rectangle and circle provide simple examples, with centroids at their geometric centers, as in Fig. C.5.

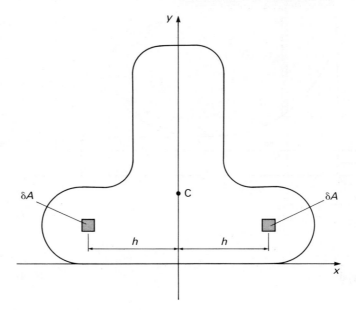

Fig. C.4.

The main applications for first moments of area in this book are to finding shear stresses in beams (Section 5.3.2) and determining the positions of the neutral axes of beam cross sections (Sections 5.3.1 and 5.3.3), which pass through their centroids. In many cases, these applications involve finding first moments of area of composite regions made up of rectangles, circles, triangles and other relatively simple shapes. Consider, for example, the inverted T-shaped region in Fig. C.6. We may divide this into two rectangular subregions with areas of $A_1 = b_1 d_1$ and $A_2 = b_2 d_2$. Because the region is symmetrical about the y axis, the centroid must lie on this axis, and $x_C = 0$. Then, to find coordinate y_C of the centroid, and hence the position of the neutral axis, we can

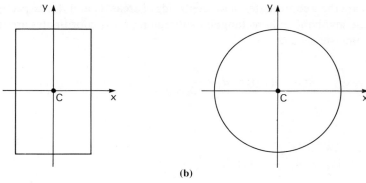

(a)   **(b)**

Fig. C.5.

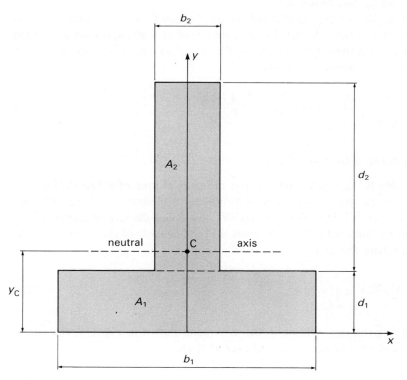

Fig. C.6.

take moments of area about the lower edge of the cross section, which coincides
with the x axis, and use equation (C.6)

$$(A_1 + A_2)y_C = Q_x = \int_0^{d_1 + d_2} y b(y)\, dy$$

Now, while the integral on the right-hand side of this equation is straightforward to evaluate, we note that it represents the sum of the first moments of
areas A_1 and A_2 about the x axis, which are $A_1(y_C)_1$ and $A_2(y_C)_2$ where $(y_C)_1$

and $(y_C)_2$ are the y coordinates of the centroids of areas A_1 and A_2, respectively. In this case involving only rectangular subregions, these coordinates are those of the geometric centers

$$(y_C)_1 = \frac{d_1}{2} \quad \text{and} \quad (y_C)_2 = d_1 + \frac{d_2}{2}$$

Hence

$$(A_1 + A_2)y_C = A_1\frac{d_1}{2} + A_2\left(d_1 + \frac{d_2}{2}\right)$$

from which y_C can be found.

This result can be generalized for any number of subregions. If the ith subregion has an area A_i, and the coordinates of its centroid (in any convenient Cartesian coordinate system) are $(x_C)_i$ and $(y_C)_i$, then the coordinates of the centroid of the whole region are

$$x_C = \frac{\sum A_i(x_C)_i}{\sum A_i} \quad \text{and} \quad y_C = \frac{\sum A_i(y_C)_i}{\sum A_i} \tag{C.8}$$

C.2 Second moment of area

In equation (C.1), we defined the first moment of area of a region (Fig. C.1) about a given axis as the sum of products of elements of area and their perpendicular distances from the axis. We now define the *second moment of area* as a similar sum of products of areas and the *squares* of their perpendicular distances from the axis

$$I = \sum p^2 \delta A = \int_A p^2 \, \mathrm{d}A \tag{C.9}$$

Considering the region lying in the x, y plane shown in Fig. C.2, the second moments of area about the coordinate axes are

$$I_x = \int_A y^2 \, \mathrm{d}x \, \mathrm{d}y \quad \text{and} \quad I_y = \int_A x^2 \, \mathrm{d}x \, \mathrm{d}y \tag{C.10}$$

and, with the breadth $b(y)$ and depth $d(x)$ defined as in Fig. C.3, these become

$$I_x = \int_{y_1}^{y_2} y^2 b(y) \, \mathrm{d}y \quad \text{and} \quad I_y = \int_{x_1}^{x_2} x^2 d(x) \, \mathrm{d}x \tag{C.11}$$

In this book, the main application for such second moments of area is to finding bending stresses in beams (Sections 5.3.1 and 5.3.3), for which purpose we need second moments of area about the neutral axes of the beam cross sections. To find these for the types of composite regions we meet in practical beam cross sections, we need the *parallel axis theorem*, which is presented below.

Another important practical application, for a somewhat different type of second moment of area, is to torsion (Chapter 7). Here, we are interested not in moments about axes lying in the plane of the region concerned, but about an axis *perpendicular* to that plane. Such a second moment is referred to as a *polar second moment of area*, and is given the symbol J rather than I. Consider the region of total area A shown in Fig. C.7, a typical element of which is the small shaded piece δA, whose center is at a distance r from the origin of coordinates x, y. The second polar moment of area of this piece about the z axis, which is perpendicular to the x, y plane and passes through the origin, is $r^2 \delta A$, and of the whole region about the same axis is

$$J_z = \sum r^2 \delta A = \int_A r^2 \, dA \tag{C.12}$$

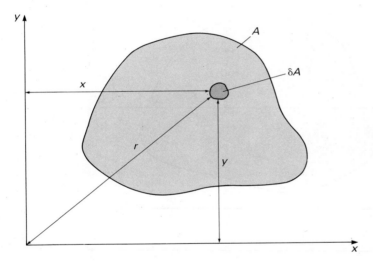

Fig. C.7.

There is a simple relationship between second moments of area about axes x, y and z, known as the *perpendicular axis theorem*, which is presented below.

There is an additional quantity which is sometimes introduced in connection with second moments of area, namely the *radius of gyration*, a term which reflects the close links between second moments of area and moments of inertia. We may define the second moments of area of an area A about, say, the x and y axes as

$$I_x = r_x^2 A \quad \text{and} \quad I_y = r_y^2 A \tag{C.13}$$

where r_x and r_y are the radii of gyration about these axes.

C.3 Parallel axis theorem

Given the second moment of area of a region about a particular axis, it is often necessary to find the second moment about some other axis, parallel to the first. The relationship we use to do this is the *parallel axis theorem*, which avoids the need to reevaluate the second moment from first principles. This is especially useful for finding second moments of area of composite regions. Suppose that, for the region of area A shown in Fig. C.8, we already know the second moment of area, I, about the horizontal axis GG′, which passes through the centroid, C, of the region. Now we wish to find the second moment, I', about the parallel axis HH′ which is at a distance h from GG′. By definition

$$I = \int_A (y - y_C)^2 \, dA \quad \text{and} \quad I' = \int_A (y - y_C + h)^2 \, dA$$

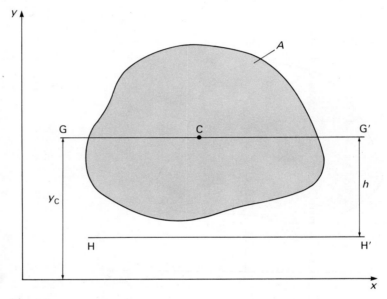

Fig. C.8

and, by expanding the bracketed term, the latter expression can be rearranged as

$$I' = \int_A (y - y_C)^2 \, dA + 2h \int_A (y - y_C) \, dA + h^2 \int_A dA$$

The second of these integrals is the first moment of area of the region about GG′, and because this axis passes through the centroid the integral is zero. We may therefore express I' as

$$I' = I + Ah^2 \tag{C.14}$$

which involves adding the area of the region multiplied by the square of the distance between the axes to the known second moment of area. This is the parallel axis theorem, which can be applied to any shape of region. The only restriction is that *I must be about an axis through the centroid of the region*. If it is not, and we are given the second moment about an axis not through the centroid, we can still use the theorem to first find *I*.

We can now use the parallel axis theorem to find the second moment of area about the neutral axis of a composite region such as that shown in Fig. C.6. In equation (5.38), we found the second moment of area of a rectangle, of breadth *b* and depth *d*, *about its own neutral axis* (passing horizontally through its geometric center) to be $bd^3/12$. The neutral axis of the first rectangle of area A_1 in Fig. C.6 is at a height of $d_1/2$ above the *x* axis, and the second moment of area of this subregion about the neutral axis of the whole section is therefore

$$I_1 = \frac{b_1 d_1^3}{12} + b_1 d_1 \left(y_C - \frac{d_1}{2} \right)^2$$

where y_C is the previously determined height of the neutral axis above the *x* axis. Similarly, the second moment of area of the second rectangle about the same neutral axis is

$$I_2 = \frac{b_2 d_2^3}{12} + b_2 d_2 \left(d_1 + \frac{d_2}{2} - y_C \right)^2$$

and the second moment of area of the whole section is found by adding these two contributions

$$I = I_1 + I_2$$

C.4 Perpendicular axis theorem

In equation (C.12), we obtained an expression for the polar second moment of area about the *z* axis of a region lying in the *x*, *y* plane, in terms of *r*, the distance from the origin of points in the region. Since this distance can be defined in terms of point coordinates *x* and *y* as

$$r^2 = x^2 + y^2$$

equation (C.12) can be expressed as

$$J_z = \int_A x^2 \, dA + \int_A y^2 \, dA$$

or

$$J_z = I_x + I_y \tag{C.15}$$

This is the *perpendicular axis theorem*, which defines the polar second moment of area of a region about any axis perpendicular to its plane as the sum of the second moments of area about any pair of mutually perpendicular axes in its plane which intersect on the first axis.

In this book, the main application of this theorem is to relating the second moments of area used in the bending of beams to the polar second moments of area used in the torsion of shafts. Consider the circular region of radius D shown in Fig. C.9, and let the origin of coordinates x and y be at the center of the circle. In this case, the polar second moment of area about the z axis is simpler to evaluate than those about the x and y axes. The narrow shaded ring, of width δr and area $2\pi r \delta r$, which is at a mean distance of r from the z axis, has a second moment of area about this axis of $2\pi r^3 \delta r$. The polar second moment of area of the whole circular region is therefore given by

$$J_z = \int_0^{D/2} 2\pi r^3 \, \mathrm{d}r = 2\pi \left[\frac{r^4}{4} \right]_0^{D/2} = \frac{\pi D^4}{32} \tag{C.16}$$

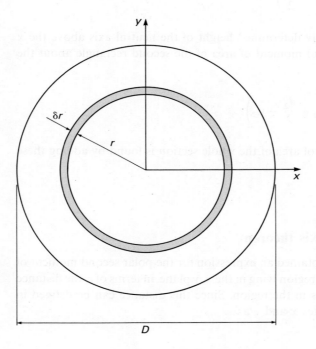

Fig. C.9.

For this highly symmetrical region, there is no difference between taking second moments of area about the x and y axes, and we can therefore use the perpendicular axis theorem to give

$$I_x = I_y = \frac{J_z}{2} = \frac{\pi D^4}{64} \tag{C.17}$$

C.5 Properties of some simple shapes

Table C.1 summarizes the main geometric properties of some simple shapes which may be met in mechanics of solids problems, particularly those involving bending or torsion. These properties are the position of the centroid, the area, and the second moments of area.

Table C.1.

Shape	(x_C, y_C)	Area	I_x	I_y	J_z
Rectangle	$(0, 0)$	bh	$\dfrac{bh^3}{12}$	$\dfrac{hb^3}{12}$	$\dfrac{bh(b^2 + h^2)}{12}$
Circle	$(0, 0)$	$\dfrac{\pi D^2}{4}$	$\dfrac{\pi D^4}{64}$	$\dfrac{\pi D^4}{64}$	$\dfrac{\pi D^4}{32}$
Triangle	$\left(0, \dfrac{h}{3}\right)$	$\dfrac{bh}{2}$	$\dfrac{bh^3}{12}$	$\dfrac{hb^3}{48}$	$\dfrac{bh}{12}\dfrac{(h^2 + b^2)}{4}$
Ellipse	$(0, 0)$	πab	$\dfrac{\pi ab^3}{4}$	$\dfrac{\pi a^3 b}{4}$	$\dfrac{\pi ab(a^2 + b^2)}{4}$
Semicircle	$\left(0, \dfrac{4r}{3\pi}\right)$	$\dfrac{\pi r^2}{2}$	$\dfrac{\pi r^4}{8}$	$\dfrac{\pi r^4}{8}$	$\dfrac{\pi r^4}{4}$
Quarter circle	$\left(\dfrac{4r}{3\pi}, \dfrac{4r}{3\pi}\right)$	$\dfrac{\pi r^2}{4}$	$\dfrac{\pi r^4}{16}$	$\dfrac{\pi r^4}{16}$	$\dfrac{\pi r^4}{8}$

D: DEFLECTIONS AND SLOPES FOR SOME COMMON CASES OF THE BENDING OF UNIFORM BEAMS

D.1 Cantilevers

Case	End deflection	End slope	Comment
M_0	$\dfrac{M_0 L^2}{2EI}$	$\dfrac{M_0 L}{EI}$	
F	$-\dfrac{FL^3}{3EI}$	$-\dfrac{FL^2}{2EI}$	see Example 6.2
w	$-\dfrac{wL^4}{8EI}$	$-\dfrac{wL^3}{6EI}$	see Example 6.3
$w(x) = w_0\left(1 - \dfrac{x}{L}\right)$	$-\dfrac{w_0 L^4}{30EI}$	$-\dfrac{w_0 L^3}{24EI}$	
$w(x) = w_0\left(1 - \dfrac{x}{L}\right)^2$	$-\dfrac{w_0 L^4}{72EI}$	$-\dfrac{w_0 L^3}{60EI}$	

D.2 Simply supported beams

Case	Central deflection	End slope	Comment
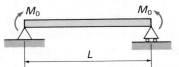	$-\dfrac{M_0 L^2}{8EI}$	$\pm\dfrac{M_0 L}{2EI}$	see Example 6.1
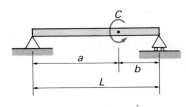		see Example 6.7 (especially Fig. 6.16)	
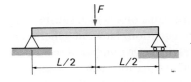	$-\dfrac{FL^3}{48EI}$	$\pm\dfrac{FL^2}{16EI}$	see Example 6.5
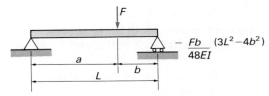	$-\dfrac{Fb}{48EI}(3L^2-4b^2)$	$-\dfrac{Fb}{6EIL}(L^2-b^2)$ $+\dfrac{Fa}{6EIL}(L^2-a^2)$	see Example 6.6 (especially Fig. 6.14)
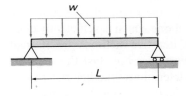	$-\dfrac{5wL^4}{384EI}$	$\pm\dfrac{wL^3}{24EI}$	see Example 6.4

D.3 Built-in beams

Case	Central deflection	End moment	Comment
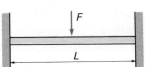	$-\dfrac{FL^3}{192EI}$	$\dfrac{FL}{8}$	see Example 6.13 (including F at any position along beam)
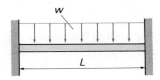	$-\dfrac{wL^4}{384EI}$	$\dfrac{wL^2}{12}$	see Fig. 6.36 for end moment

E: SOLUTION OF NONLINEAR ALGEBRAIC EQUATIONS

Nonlinear algebraic equations arise in the analysis of many different problems in engineering. In this book we meet examples of polynomial equations in the treatment of beam deflection problems, and equations involving trigonometric functions in the analysis of buckling problems. Any nonlinear algebraic equation with a single variable can be expressed in the form

$$f(X) = 0 \tag{E.1}$$

where $f(X)$ is some nonlinear function of the variable X. Consider, for example, the case arising in Example 6.8, equation (6.51)

$$0.12X^2 - \tfrac{1}{6}(X - 0.2)^3 - 0.024 = 0 \tag{E.2}$$

The object of solving such an equation is to find a value of X which satisfies the equation. This value is said to be a *root* of the equation. Now, in general an equation in the form of (E.1) may have an arbitrarily large number of real roots (real as opposed to imaginary or complex): a cubic equation such as (E.2), may have as many as three real roots. In practice, however, we are usually interested in only one of the roots, which we can often identify from the nature of the physical problem from which the equation is derived. In the case of equation (E.2), for instance, we seek a root in the range $0.2 < X < 0.6$, and we expect there to be only one root in that range.

Many different methods have been developed for solving nonlinear algebraic equations, and are described in detail in numerical analysis textbooks. All are *iterative* in nature: we first guess a value for the required root and then repeatedly obtain successively better values until one which is sufficiently close to the true root is achieved. Two properties of iterative methods are of great importance: the *rate of convergence*, which determines how fast the required root is approached, and the *stability*, which determines whether the root is approached at all, or whether successive estimates diverge from the root. Unfortunately, those methods which show good rates of convergence under favorable conditions also tend to be the most unstable under adverse conditions. The best choice of method therefore depends on the conditions under which it is to be used. If an equation is to be solved manually, then calculations are expensive and the rate of convergence is important. Any tendency towards unstable behavior can be rapidly detected and action taken. On the other hand, if an equation is to be solved using a computer, calculations are cheap but more difficult to monitor and control: stability of the method is then more important than its rate of convergence.

For present computational purposes we choose to use a technique often referred to as the *bisection method*, which is very simple, relatively slow, but highly stable. Indeed, provided we can first identify a range of the variable, say between $X = a$ and $X = b$, in which there is at least one root of equation (E.1), and provided that the function $f(X)$ is continuous and single-valued over this range, then the bisection method will find a root.

Figure E.1 shows a typical plot of a function $f(X)$ against X, which cuts the X axis at the root X_r, somewhere between $X = a$ and $X = b$. We test for whether the root lies between a and b by whether $f(a)$ and $f(b)$ are of different sign, that is

$$f(a)f(b) \leqslant 0 \tag{E.3}$$

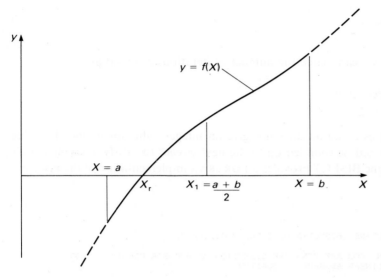

Fig. E.1.

If the product is exactly zero, then either a or b, or both, are roots of the equation. The way we proceed is to bisect the given range and to take the average of a and b as the first estimate of X_r

$$X_1 = \frac{a+b}{2} \tag{E.4}$$

We then test (by the same function product method) whether the root lies in the left-hand or right-hand half of the original range, and adjust either the upper or lower limit of the range accordingly. For example, if the product $f(a)f(X_1)$ is negative then X_1 replaces b as the upper limit. The method is then applied repeatedly, each time halving the range of X in which the root must lie, thereby slowly but surely converging to this root.

Given a set of estimates of the required root, $X_1, X_2, X_3 \ldots$ and so on, we must decide when we are sufficiently close to the true value so that the iterative process can be stopped. To do this, we test the difference between successive estimates, $X_i - X_{i-1}$, and require that this be less than some small tolerance, say ε. Under certain circumstances, we might prefer to test the magnitude of the difference relative to the magnitude of the root, say $(X_i - X_{i-1})/X_i$, but this is potentially hazardous as X_i might be zero, and computers normally reject division by zero. Because of the very simple nature of the bisection method, we are able to predict in advance how many iterations will be required to achieve convergence to a given tolerance. Taking the initial estimate for the root as either a or b, the first iteration, defined by equation (E.4), gives a new estimate which differs by $(b - a)/2$ from the initial one. After the second iteration, the difference is halved at $(b - a)/2^2$. Every further iteration halves the difference, until after n iterations it is $(b - a)/2^n$, and in order to achieve convergence to the given tolerance we require that

$$\frac{|b - a|}{2^n} < \varepsilon \tag{E.5}$$

from which we can define the number of iterations required as

$$n = \frac{\ln |(b - a)/\varepsilon|}{\ln 2} \tag{E.6}$$

In general this equation does not give an integer value for n, and the value obtained should be rounded up to the next integer to satisfy inequality (E.5).

Subroutine BISECT provides a FORTRAN implementation of the bisection method.

```
      SUBROUTINE   BISECT(A,B,TOLER,XROOT,IFLAG)
C
C SUBROUTINE FOR APPLYING THE BISECTION METHOD FOR FINDING A ROOT
C OF A NONLINEAR ALGEBRAIC EQUATION.
C
C DEFINE THE MINIMUM AND MAXIMUM POSSIBLE VALUES FOR X, INITIALLY
C AS THE LIMITS OF THE GIVEN RANGE.
      XMIN=A
      XMAX=B
C
C FIND THE CORRESPONDING VALUES OF THE FUNCTION.
      FXMIN=FN(A,0)
      FXMAX=FN(B,0)
C
C TEST FOR A ROOT WITHIN THE GIVEN RANGE.
      PROD=FXMIN*FXMAX
      IF(PROD.GT.0.) THEN
        IFLAG=1
        RETURN
      END IF
      IFLAG=0
C
C TEST FOR A ROOT AT EITHER END OF THE RANGE.
      IF(PROD.EQ.0.) THEN
        IF(FXMIN.EQ.0) XROOT=XMIN
        IF(FXMAX.EQ.0) XROOT=XMAX
        RETURN
      END IF
C
C FIND THE NUMBER OF ITERATIONS REQUIRED.
      RATIO=ABS((B-A)/TOLER)
      NITER=INT(ALOG(RATIO)/ALOG(2.))+1
C
C SET UP ITERATION LOOP.
      DO 1 ITER=1,NITER
C
C FIND NEW ESTIMATE OF THE ROOT AS THE AVERAGE OF THE MINIMUM AND
C MAXIMUM VALUES.
      XNEW=(XMIN+XMAX)/2.
C
C FIND THE CORRESPONDING VALUE OF THE FUNCTION.
      FXNEW=FN(XNEW,0)
```

```
C
C   TEST FOR NEW ESTIMATE BEING THE REQUIRED ROOT.
        IF(FXNEW.EQ.0.) THEN
            XROOT=XNEW
            RETURN
        END IF
C
C   USE NEW ESTIMATE TO UPDATE EITHER THE LOWER OR UPPER LIMIT.
        IF(FXNEW*FXMIN.LT.0.) THEN
            XMAX=XNEW
            FXMAX=FXNEW
        ELSE
            XMIN=XNEW
            FXMIN=FXNEW
        END IF
    1   CONTINUE
C
C   AFTER ITERATION, TAKE THE ROOT AS THE BEST AVAILABLE ESTIMATE.
        XROOT=XNEW
        RETURN
        END
```

The following list provides the definitions of the FORTRAN variables used, arranged in alphabetical order.

Variable	Type	Definition
A	real	Lower limit, a, of the given range of variable X
B	real	Upper limit, b, of the given range of variable X
FN	real	Function subprogram to define the particular form of nonlinear algebraic function, f
FXMAX	real	Value of the nonlinear function f corresponding to value XMAX of variable X
FXMIN	real	Value of the nonlinear function f corresponding to value XMIN of variable X
FXNEW	real	Value of the nonlinear function f corresponding to value XNEW of variable X
IFLAG	integer	Flag for failure to find a root within the given range of X (set to one if test shows no root, otherwise zero)
ITER	integer	Iteration counter
NITER	integer	Number of iterations to be performed
PROD	real	Product of function values
RATIO	real	Absolute value of the ratio $(b - a)/\varepsilon$
TOLER	real	Convergence tolerance, ε
XMAX	real	Current maximum possible value of X
XMIN	real	Current minimum possible value of X
XNEW	real	New estimate of the root, found by the bisection process
XROOT	real	Value of the required root of the equation

The arguments of subroutine BISECT are the given limits, A and B, of the possible range for the root, the convergence tolerance, TOLER, the computed value of the root, XROOT, and the variable IFLAG. While the first three of these provide data to the subroutine, the last two provide the results, the last one returning an integer flag value to the calling program to warn of no root in the given range.

The first action of the subroutine is to compute the values of the nonlinear function, f in equation (E.1), corresponding to the given limits of the possible range for X. This is done using a further function subprogram, FN, whose form must be suited to the particular equation being solved. The reason for having two arguments for this subprogram, the value of X and another integer quantity (here set to zero), is explained below. We test for a root within the given range by testing the product of the two function values. A positive value of the product indicates that there is not a single root in the range (although for a continuous function there could be an even number of roots), and a zero value indicates that either *a* or *b* (or both) must be a root.

The number of iterations is then computed with the aid of equation (E.6) (note that, since FORTRAN function INT truncates its argument to the integer immediately below it, one is added to obtain the integer immediately above it). An iteration loop is then set up to bisect repeatedly the possible range for the root. If one of the resulting estimates for the root should happen to exactly satisfy the nonlinear equation, this value is returned as the root to the calling program. Otherwise, the value obtained from the last iteration is taken as the root.

Now, subroutine BISECT is designed as a piece of coding for applying the bisection method which is independent of both the particular nonlinear equation to be solved and the program which requires the solution. It can therefore be used in any situation where such an equation has to be solved, including, for example, program SIBEAM which is described in Section 6.4.2. Similarly, it can be used to solve the cubic equation derived in Example 6.8, which is repeated here in equation (E.2). To do this, we must prepare a small main program, and an appropriate form of function subprogram FN.

Program ROOT provides the main program for entering the data for solving a nonlinear algebraic equation. After writing out an appropriate title onto an

```
      PROGRAM  ROOT
C
C  PROGRAM TO FIND A ROOT OF A NONLINEAR ALGEBRAIC EQUATION.
C
      OPEN(5,FILE='DATA')
      OPEN(6,FILE='RESULTS')
C
C  OUTPUT A TITLE AND THE FORM OF FUNCTION, BY CALLING THE
C  DEFINING FUNCTION SUBPROGRAM.
      WRITE(6,61)
   61 FORMAT('SOLUTION OF A NONLINEAR ALGEBRAIC EQUATION BY THE
     1         'BISECTION METHOD' /)
      F=FN(0.,1)
C
C  INPUT AND OUTPUT THE LOWER AND UPPER LIMITS OF THE RANGE OF THE
C  VARIABLE, ALSO THE CONVERGENCE TOLERANCE.
C
      READ(5,*) A,B,TOLER
      WRITE(6,62) A,B,TOLER
   62 FORMAT('LOWER LIMIT OF POSSIBLE RANGE OF VARIABLE X = ',E12.4//
     1         'UPPER LIMIT OF POSSIBLE RANGE OF VARIABLE X = ',E12.4//
     2         'MAXIMUM DIFFERENCE BETWEEN SUCCESSIVE ESTIMATES OF '
     3         'THE ROOT = ',E12.4 /)
C
C  CALL SUBROUTINE TO APPLY THE BISECTION METHOD.
      CALL   BISECT(A,B,TOLER,XROOT,IFLAG)
C
C  UNIT VALUE OF IFLAG INDICATES NO SOLUTION WITHIN THE GIVEN RANGE.
      IF(IFLAG.EQ.1) THEN
         WRITE(6,63)
   63    FORMAT('NO ROOT WITHIN THE GIVEN RANGE - STOP')
         STOP
      END IF
C
C  OUTPUT THE COMPUTED ROOT.
      WRITE(6,64) XROOT
   64 FORMAT('COMPUTED ROOT = ',E14.6)
      STOP
      END
```

output file named RESULTS, ROOT calls function subprogram FN to also write out the form of nonlinear equation being solved. The values of a, b and ε are then read in from an input file named DATA, and immediately written out, and subroutine BISECT is called to solve the equation. The computed results are written out, either in the form of the computed root or a warning message that there is no single solution within the given range.

```
      FUNCTION  FN(X,IOTEST)
C
C  FUNCTION SUBPROGRAM TO DEFINE THE NONLINEAR ALGEBRAIC FUNCTION
C  WHICH IS REQUIRED TO BE ZERO.
C
C  THIS VERSION DEFINES THE CUBIC FUNCTION DERIVED IN EXAMPLE 6.8
C
C  IF IOTEST IS UNITY, ONLY OUTPUT THE FORM OF FUNCTION.
      IF(IOTEST.EQ.1) THEN
        WRITE(6,61)
 61     FORMAT('FUNCTION IS   0.12*X**2 - (X-0.2)**3/6 - 0.024 = 0'/)
        RETURN
      END IF
      FN=0.12*X**2-(X-0.2)**3/6.-0.024
      RETURN
      END
```

Fig. E.2.

Function subprogram FN displayed in Fig. E.2 provides the definition of the cubic function defined in equation (E.2). Note that the function has two arguments, X for the current value of the variable X, and IOTEST, which is an integer parameter for controlling the output of a message defining the form of function. A unit value of IOTEST must be entered to obtain this output, in which case the value of the function is not computed and the value of argument X is irrelevant. In the context of the program which comprises ROOT, BISECT and FN, FN is only called once with IOTEST set to unity, this being from main program ROOT when the form of function message is required. All calls from BISECT, where function values are required, are made with IOTEST set to zero. The purpose of these arrangements is to ensure that details of the particular function under consideration, both for computation and output messages, are confined to the single subprogram. This makes the analysis of any other form of function straightforward to implement, as demonstrated below.

Figure E.3 shows the results obtained, the computed root being $X = 0.4801$.

```
SOLUTION OF A NONLINEAR ALGEBRAIC EQUATION BY THE BISECTION METHOD

FUNCTION IS   0.12*X**2 - (X-0.2)**3/6 - 0.024 = 0

LOWER LIMIT OF POSSIBLE RANGE OF VARIABLE X =    0.2000E+00

UPPER LIMIT OF POSSIBLE RANGE OF VARIABLE X =    0.6000E+00

MAXIMUM DIFFERENCE BETWEEN SUCCESSIVE ESTIMATES OF THE ROOT =    0.1000E-05

COMPUTED ROOT =    0.480140E+00
```

Fig. E.3.

Another illustration of the use of program ROOT is provided by the nonlinear equation (8.20) derived in Example 8.3, which is

$$X\left[1 + 1.6 \sec\left(\frac{\pi}{2}\sqrt{X}\right)\right] - 5.014 = 0 \tag{E.7}$$

with a root expected in the range $0 < X < 1$. The subprogram displayed in Fig. E.4 provides the definition of this function, and Fig. E.5 shows the results obtained. Note that the upper limit of the possible range of X is set not at 1.0 exactly (which would cause computer overflow due to an infinite value of the secant function), but 0.99. The computed root is $X = 0.6956$.

```
      FUNCTION  FN(X,IOTEST)
C
C     FUNCTION SUBPROGRAM TO DEFINE THE NONLINEAR ALGEBRAIC FUNCTION
C     WHICH IS REQUIRED TO BE ZERO.
C
C     THIS VERSION DEFINES THE FUNCTION DERIVED IN EXAMPLE 8.3
C
C     IF IOTEST IS UNITY, ONLY OUTPUT THE FORM OF FUNCTION.
      IF(IOTEST.EQ.1) THEN
        WRITE(6,61)
   61   FORMAT('FUNCTION IS
     1          'X(1 + 1.6*SEC(0.5*PI*SQRT(X))) - 5.014 = 0'/)
        RETURN
      END IF
      PI=4.*ATAN(1.)
      ANG=0.5*PI*SQRT(X)
      SEC=1./COS(ANG)
      FN=X*(1.+1.6*SEC)-5.014
      RETURN
      END
```

Fig. E.4.

```
SOLUTION OF A NONLINEAR ALGEBRAIC EQUATION BY THE BISECTION METHOD

FUNCTION IS    X(1 + 1.6*SEC(0.5*PI*SQRT(X))) - 5.014 = 0

LOWER LIMIT OF POSSIBLE RANGE OF VARIABLE X =    0.0000E+00

UPPER LIMIT OF POSSIBLE RANGE OF VARIABLE X =    0.9900E+00

MAXIMUM DIFFERENCE BETWEEN SUCCESSIVE ESTIMATES OF THE ROOT =    0.1000E-05

COMPUTED ROOT =    0.695632E+00
```

Fig. E.5.

```
      FUNCTION  FN(X,IOTEST)
C
C     FUNCTION SUBPROGRAM TO DEFINE THE NONLINEAR ALGEBRAIC FUNCTION
C     WHICH IS REQUIRED TO BE ZERO.
C
C     THIS VERSION DEFINES THE FUNCTION DERIVED FROM EQUATION (8.43)
C
C     IF IOTEST IS UNITY, ONLY OUTPUT THE FORM OF FUNCTION.
      IF(IOTEST.EQ.1) THEN
        WRITE(6,61)
   61   FORMAT('FUNCTION IS    TAN(X) - X = 0'/)
        RETURN
      END IF
      FN=TAN(X)-X
      RETURN
      END
```

Fig. E.6.

A further example of the use of program ROOT is provided by the nonlinear equation (8.43), which can be written as

$$\tan X - X = 0 \tag{E.8}$$

with a root expected in the range $4.0 < X < 4.7$. The subprogram displayed in Fig. E.6 provides the definition of this function, and Fig. E.7 shows the results obtained. The computed root is $X = 4.493$.

```
SOLUTION OF A NONLINEAR ALGEBRAIC EQUATION BY THE BISECTION METHOD

FUNCTION IS     TAN(X) - X = 0

LOWER LIMIT OF POSSIBLE RANGE OF VARIABLE X =    0.4000E+01

UPPER LIMIT OF POSSIBLE RANGE OF VARIABLE X =    0.4700E+01

MAXIMUM DIFFERENCE BETWEEN SUCCESSIVE ESTIMATES OF THE ROOT =    0.1000E-05

COMPUTED ROOT =    0.449341E+01
```

Fig. E.7

Answers to Even-Numbered Problems

Chapter 1

1.2	797 N at foot, 123 N at wall
1.4	Statically indeterminate (see Chapter 4)
1.6	$F_1 = F_2 = 0.03\,W$
1.8	376 N, 504 N
1.10	Statically indeterminate (see Chapter 6)
1.12	90.7 mm
1.14	25.5 MN/m^2 in A, 102 MN/m^2 in B
1.16	Diameter ratio $= 1$
1.18	393 kN, -200 MN/m^2
1.20	2.29 mm
1.22	1.414
1.24	192 lb at foot, 29.5 lb at wall
1.26	$F_1 = F_2 = 0.03\,W$
1.28	19.4 kip, 20.6 kip
1.30	3.63 in
1.32	(i) 3.33 ksi (ii) 4.44 ksi (iii) 5.66 ksi (iv) 6.67 ksi
1.34	(i) 5.61 (ii) 1.87
1.36	0.0112° counterclockwise
1.38	52 lb at A, 178 lb (resultant) at B, 1.32 ksi

Chapter 2

2.4	$-6.333P$ in AB, $-5.667P$ in BC, $+4.908P$ in CD and DE, $+5.485P$ in EA, 0 in BD, $+1.155P$ in BE
2.6	-176.8 kN in AB, DE, CF and CH; -125.0 kN in BC and CD; $+125.0$ kN in EF, AH, BH and DF; $+250.0$ kN in FG, GH and CG
2.8	2.5F at A and E; $-3F$ in AB and DE; $-2.5F$ in BC and CD; $+0.8660F$ in CJ, CK, JH and KG; 0 in BI, DF, BJ and KD; $-0.8660F$ in BH and DG; $+2.598F$ in AI, IH, GF and FE; $+1.732F$ in HG
2.12	$+24.04$ kN in AB and DE; $+26.81$ kN in BC and CD; -32.84 kN in AF and FE; -19.63 kN in BF and FD; -26.82 kN in CF
2.14	-111.5 kN in CD (load at H), $+55.77$ kN in DH (load at H), $+70.71$ kN in GH (load at H)
2.16	Mechanism
2.18	Statically determinate
2.20	Mechanism

2.28	$+31.83$ kN in KN, $+2.445$ kN in FL, -29.75 kN in FU
2.30	-695.3 kN in DE, -571.7 kN in DM, $+937.5$ kN in NM
2.32	$+9.33$ MN/m^2 (hoop), -3.80 MN/m^2 (axial)
2.34	408 m/s, no optimum diameter
2.36	1590 kN and 198 kN at A, 1590 kN and 252 kN at B, 1610 kN
2.38	16.0 kN
2.46	$+26.8$ kip in BC, -19.6 kip in BF
2.48	-83.65 kip in AB, BC, DE and EF; -111.5 kip in CD; $+22.42$ kip in FG; $+70.71$ kip in GH; $+63.24$ kip in HI; -31.35 kip in IA; $+83.65$ kip in BI and GE; -55.77 kip in IC and DG; $+55.77$ kip in CH and HD
2.62	-63.45 kip in HI
2.64	16 ksi (hoop), 8 ksi (axial)
2.66	7.80 ksi (hoop), -1.83 ksi (axial)
2.68	49.1 kip at A
2.70	(i) 0.7L from C, 0.0411 L (ii) 6.39 W (iii) 8.97 W

Chapter 3

3.2	0.582 mm
3.4	3.92 mm
3.6	0.452 mm
3.8	$\sigma_x = 22.3$ MN/m^2, $\sigma_y = 28.4$ MN/m^2
3.10	$\sigma_x = 82.5$ MN/m^2, $\sigma_y = 88.6$ MN/m^2
3.12	0.0217 mm, $-0.002\,54$ mm
3.14	

$$\sigma_x = \frac{Ee_x}{(1+v)} + \frac{vE}{(1+v)(1-2v)}(e_x + e_y + e_z) - \frac{E\alpha\Delta T}{(1-2v)}$$

$$\sigma_y = \frac{Ee_y}{(1+v)} + \frac{vE}{(1+v)(1-2v)}(e_x + e_y + e_z) - \frac{E\alpha\Delta T}{(1-2v)}$$

$$\sigma_z = \frac{Ee_z}{(1+v)} + \frac{vE}{(1+v)(1-2v)}(e_x + e_y + e_z) - \frac{E\alpha\Delta T}{(1-2v)}$$

3.16	19.8 kN, -0.00568 mm
3.18	4.69 mm, -2.71 mm
3.20	-0.0183 mm, -0.00102 mm
3.22	-0.0165 mm, $+0.00848$ mm, -5.64×10^{-5} mm
3.24	$$\frac{2p\pi}{E}(1-v)\left(\frac{R_1^4}{t_1}+\frac{R_2^4}{t_2}\right)$$
3.26	0.132 in
3.28	0.0153 in
3.30	$\sigma_x = 3.25$ ksi, $\sigma_y = 4.14$ ksi
3.32	$\sigma_x = 12.4$ ksi, $\sigma_y = 13.3$ ksi
3.34	-0.0016 in, -0.0032 in, -0.0096 in, -23.0 in^3
3.36	415 kip
3.38	5.06 kip, -0.246×10^{-3} in
3.40	(i) $+100\,°$F (ii) $-154\,°$F (iii) $+286\,°$F
3.42	0.706
3.44	0.0525 in, 74.2 in^3
3.46	80.8 °F

Chapter 4

4.2	(i) 10 kN (ii) 16.62 kN
4.4	$-6.333\,P$ in AB, $-5.667\,P$ in BC, $+4.908\,P$ in CD and DE, $+5.485P$ in AE, 0 in BD, $+1.155P$ in BE
4.6	-15.17 mm at G
4.8	$+22.11$ kN in AB and AD, -27.89 kN in BC, -31.27 kN in AC, $+39.44$ kN in BD
4.12	-1.974 mm
4.14	$+6.408$ kN in AD, $+0.9384$ kN in BD, -7.735 kN in CD; 0.09066 mm (horizontal), -0.8380 mm (vertical)
4.16	-1.393 mm
4.18	37.32 mm
4.20	-155.3 mm
4.22	-22.72 mm at N
4.24	-0.838×10^{-3}, -1.33×10^{-3}
4.26	18.9 MN/m
4.28	11.5 kN, 17.2 MN/m^2 (aluminum); 11.5 kN, 50.9 MN/m^2 (steel); 7.0 kN, 2.5 MN/m^2 (thermoplastic)
4.30	$$\tfrac{2}{3}\pi p D^3\left(\frac{D}{Et}+\frac{1}{K}\right)$$
4.32	8.5 kN, 12.6 MN/m^2 (aluminum); 21.5 kN, 95.2 MN/m^2 (steel)
4.34	$(\beta - 3\alpha)\varDelta T$
4.36	6.32 MN/m, 19.5 MN/m^2 (steel), -29.2 MN/m^2 (aluminum)
4.38	-961 kN, $+0.64 \times 10^{-4}$
4.40	(i) 2.34 kip (ii) 3.76 kip
4.44	$+12$ kN in AB, $+16.97$ kN in BC, -12 kN in CD and DE, 0 in BD, -4.971 kN in BE, $+7.029$ kN in BF
4.48	-27.24 ksi, -0.9558 in
4.52	2.06 ksi (AG and GD), 2.40 ksi (AC, BD, AE and FD), 0.00823 in
4.54	1.44
4.56	132.6 lb in AF, 229.6 lb in BF, 341.7 lb in CF, 299.8 lb in DF, 157.1 lb in EF
4.58	2.53 kip, 2.53 ksi (aluminum); 2.75 kip, 7.59 ksi (steel); 1.72 kip, 0.38 ksi (thermoplastic)
4.60	-1.49 ksi, -2.46 ksi (cylinder); $+6.84$ ksi, $+20.8$ ksi (rods)
4.66	296 psi
4.68	9.68 ksi (brass), -6.05 ksi (steel)
4.70	10.9 ksi (steel), -4.83 ksi (aluminum)

Chapter 5

5.2	3.5 kN, 2.63 kNm
5.4	7 kN, 10.5 kNm
5.6	$$wa \text{ or } w\left(\frac{L}{2}-a\right),\ \frac{wa^2}{2} \text{ or } \frac{wL}{2}\left(\frac{L}{4}-a\right)$$
5.8	wL, $3\,wL^2/8$
5.10	$3\,wL$, $5\,wL^2/2$
5.12	50 kN, 0 and -140 kNm
5.14	22.9 kN, $+52.2$ kNm and 0
5.16	23 kN, $+5.06$ kNm and -43 kNm
5.26	0.91 m
5.30	101.7 kN, $+166.2$ kNm and 0
5.32	877 N/m
5.34	(i) 17 kN (ii) 17 kNm, both at right-hand support (iii) 94.6 MN/m^2 (iv) 7.08 MN/m^2
5.36	0.375 MN/m^2
5.38	2.70 MNm, 1.53 MN
5.40	88.9 mm below top edge, 96.3×10^{-6} m^4, 86.7 kNm
5.42	Neutral axis through center, 34.1×10^{-6} m^4, 22.7 kNm
5.44	8.33 kN
5.46	Two pairs of plates: 1st pair 7.44 m long, 2nd pair 5.38 m
5.48	203 mm
5.50	0.350 MN/m^2
5.52	5.50 mm upward
5.54	14.8 mm
5.56	$+20.6$ MN/m^2, -23.6 MN/m^2
5.58	1 kip, 3 kip-ft
5.60	2 kip, 12 kip-ft
5.68	13 kip, 0 and -28 kip-ft
5.70	8 kip, $+6.61$ kip-ft and -16 kip-ft
5.72	8.3 kip, 0 and -32.8 kip-ft
5.74	1.332 kip, $+0.3750$ kip-ft and -2.911 kip-ft

5.82	800 lb, 1500 lb ft		**6.36**	-0.3231 mm
5.84	$L/4$, $2WL/3$		**6.38**	-74.65 mm
5.86	62.5 in		**6.44**	11.34 kN, $+1.841$ kNm and -9 kNm, 0 and
5.88	869 lb			-7.034 mm

5.82 800 lb, 1500 lb ft

5.84 $L/4$, $2WL/3$

5.86 62.5 in

5.88 869 lb

5.90 75 psi

5.92 110 psi

5.94 310.7 in^4, 90.7 kip-ft

5.96 51.3 kip-ft

5.98 Steel rods with a total area of 1.25 in^2 to have their centers about 12 in below top surface of beam

5.100 $+18$ ksi, -14 ksi

5.102 $+19.4$ ksi and -6.01 ksi (axial), $+13.5$ ksi (hoop)

Chapter 6

6.2 -5.57 mm

6.4 -5.53 mm

6.6
$$-\frac{w_0 L^4}{30EI}, \quad -\frac{w_0 L^3}{24EI}$$

6.8
$$-\frac{w_0 L^3}{45EI}$$

6.10
$$-\frac{w}{24EI}\left(\frac{L}{2}-a\right)^2\left(\tfrac{5}{4}L^2 - 5aL - a^2\right)$$

6.12
$$0.144\,\frac{WL^3}{EI}$$

6.14
$$-0.009\,82\,\frac{wL^4}{EI}$$

6.16
$$-\frac{FL^3}{EI_0} \quad \text{where} \quad I_0 = \frac{\pi d_0^4}{64}$$

6.18
$$\frac{wL^2}{12}, \quad -\frac{wL^4}{384\,EI}$$

6.20 $19F/8$

6.22
$$-\frac{27\,FL^3}{256\,EI}$$

6.24 5.71 MN/m^2 at end nearer valve

6.26
$$-\frac{CL^2}{216EI}$$

6.28 -1.42 mm at 139 mm from A

6.30 $-wL^2/14$

6.32 -31.83 mm

6.34 -19.64 mm

6.36 -0.3231 mm

6.38 -74.65 mm

6.44 11.34 kN, $+1.841$ kNm and -9 kNm, 0 and -7.034 mm

6.46 10.8%, 21.7%, 49.6%, 15.8%, 2.1% (from left to right)

6.48 -0.331 in

6.50 -0.353 in

6.78 -3.560 in

6.80 $+2.850$ in

6.82 -0.6217 in

6.84 $+0.038\,18$ in

6.88 8.32 ksi at 23.72 in from left-hand end

Chapter 7

7.2 6.68 kNm, 42.4 MN/m^2

7.4 9.60 revolutions

7.6 110 mm, 184 mm

7.8 49.8%

7.10 624 Nm, 74.1 MN/m^2

7.12 215 Nm, 3.81°

7.14 102 MN/m^2 (shaft 1), 138 MN/m^2 (shaft 2), 5.58 revolutions

7.16 60 mm outer diameter, 4 mm wall thickness, 1.76° per m

7.18
$$\frac{(3.6+3\beta)}{(1+\beta)}T, \quad \frac{(0.84+1.2\beta)}{(1+\beta)}\frac{TL}{GJ}, \quad \text{where } \beta = \frac{GJ}{\mu L}$$

7.20 33.5 mm, 52.8 MN/m^2

7.22 2.03 kip-ft

7.24 13.8 kip-ft, 6.73 ksi

7.26 1.56 kip-ft, 0.602° per foot

7.32 294 lb ft, 5.33 ksi

7.34 4.07 ksi (shaft 1), 4.16 ksi (shaft 2), 2.15°

7.36 66.0 lb ft

7.40
$$-\frac{FL^3}{EI}(5+2v)$$

Chapter 8

8.2 $kL/2$

8.4 474 kN, 245 MN/m^2

8.6 10.5 kN

8.8 8.74×10^{-5} m^4

8.10
$$54.7°, \quad \frac{3\sqrt{3}\,H^2 W}{2\pi^2 E}$$

8.14 85.3 kN, 36.4 kN

8.16 $L/3$

8.18 Effective length = actual length

8.20

$$P_c = \frac{4\pi^2 EI}{L^2}, \quad v(x) = \frac{v_0(x)}{\left(\dfrac{P_c}{P} - 1\right)}$$

8.22 $2L$

8.24

$$\sigma_{max} = -\frac{P}{A}\left[1 + \frac{ce}{r^2}\sec\left(\frac{L}{r}\sqrt{\frac{P}{EA}}\right)\right]$$

where A is the cross-sectional area of the column, and r the minimum radius of gyration

8.26 Deflected shape is that of a half sine wave ($L_e = L$), with zero slope at each end

8.28 $\lambda \to 0$, $P = \pi^2 EI/L^2$ (pin-ended); $\lambda \to \infty$, $P = 4\pi^2 EI/L^2$ (built-in)

8.30

$$\frac{L}{D} = \frac{\pi}{8}\sqrt{\frac{E}{\sigma_Y}}, \quad 9.55$$

8.36 3.76×3.76 in

8.38 0.120 (structure would buckle: members should be hollow rather than solid)

8.56 103 kip

8.60 197 kip

Chapter 9

9.2 106 MN/m², 72.2 MN/m²

9.4 153.4 MN/m² at 36.5°, − 3.45 MN/m² at 126.5°, 78.45 MN/m², 78.45 MN/m²

9.10 0 (in the plane of the wall)

9.12 − 0.331 × 10⁻³, 1.46 × 10⁻³

9.14 + 1.756 × 10⁻³ at 171.5°, + 0.564 × 10⁻³ at 81.5°, 1.193 × 10⁻³, 1.756 × 10⁻³

9.16 + 0.078 × 10⁻³ at 165.9°, − 1.238 × 10⁻³ at 75.9°, 1.315 × 10⁻³, 1.315 × 10⁻³

9.24

$$\frac{16}{\pi D^3}\sqrt{M^2 + T^2}$$

9.26 2.59 MN/m², 108 MN/m² and 53.9 MN/m², 53.9 MN/m²

9.28 $\sigma_x = 177$ MN/m², $\sigma_y = 47.4$ MN/m²

9.30 3.043 × 10⁻⁴, − 1.043 × 10⁻⁴, − 0.857 × 10⁻⁴ (out of plane), 4.086 × 10⁻⁴

9.32 (1) − 124 kN, 0 (2) −90.2 kN, 1.02 kNm

9.34 (1) 422 MN/m² (2) 376 MN/m² (3) 299 MN/m² (4) 330 MN/m²

9.36 2.04

9.38 11.3 mm

9.40 178 N

9.42 13.9 ksi, − 4.94 ksi

9.44 14.64 ksi at 163.6°, − 18.64 ksi at 73.6°, 16.64 ksi, 16.64 ksi

9.66 1.29 × 10⁻³

9.68 − 4.02 × 10⁻⁴, 18.1 kip-in

9.72 (1) 66.6 ksi (2) 57.8 ksi (3) 37.3 ksi (4) 38.6 ksi

9.74 9.47 kip

9.78 262 lb, 1050 lb

Chapter 10

10.2

$$\left[0.75\left(\frac{L}{d}\right)^2 + 0.2\right]^{-1}$$

10.4

$$\sigma_x = -\frac{yM(x)}{I} + \frac{\rho gbd}{I}\left(\frac{d^2 y}{20} - \frac{y^3}{3}\right)$$

$$\sigma_y = \frac{\rho gbd}{2I}\left(\frac{y^3}{3} - \frac{yd^2}{4}\right) + \rho gy$$

$$\tau_{xy} = \frac{\rho gbd(L-x)}{2I}\left(y^2 - \frac{d^2}{4}\right)$$

10.6

(i) $\dfrac{p\sqrt{3K^4 + 1}}{K^2 - 1}$, $\dfrac{pK^2}{K^2 - 1}$ (ii) $\dfrac{p\sqrt{3}K^2}{K^2 - 1}$, $\dfrac{pK^2}{K^2 - 1}$

10.8 (i) 138.3 mm (ii) 136.9 mm (iii) 137.5 mm

10.10 − 50 MN/m², + 124 MN/m², + 36.9 MN/m², 150 MN/m², 86.9 MN/m²

10.12 0.286

10.16 61.0 MN/m² and 44.2 MN/m² (inner tube), 50.2 MN/m² and 31.0 MN/m² (outer tube)

10.18 89.7 MN/m²

10.20 13 400 rev/min

10.22 6190 rev/min, + 0.0845 mm (inner), + 0.0744 mm (outer)

10.30 − 2.10 ksi, + 7.00 ksi, 0, 8.26 ksi, 4.55 ksi, + 0.002 80 in

10.32

$$e_r = -\frac{p}{E}\left[1 + \frac{v(K^2 + 2)}{K^2 - 1}\right] \quad \text{and} \quad -\frac{3vp}{E(K^2 - 1)}$$

$$e_\theta = \frac{p}{E(K^2 - 1)}[(K^2 + 1) + v(K^2 - 2)]$$

$$\text{and} \quad \frac{p(2 - v)}{E(K^2 - 1)}$$

$$e_z = \frac{p(1 - 2v)}{E(K^2 - 1)}$$

10.34 -4 ksi, -4 ksi, -4 ksi, 0, 0, -8×10^{-5} in, -1.6×10^{-3} in

10.38 53.51 ksi and 34.86 ksi (inner), 52.00 ksi and 29.82 ksi (outer)

10.40 16.6 ksi, 14.8 ksi

10.42 4.507 ksi (inner), 5.057 ksi (outer), -0.477×10^{-3} in, $+1.05 \times 10^{-3}$ in

10.44

$$\sigma_\theta = p \frac{K^2 + 1}{K^2 - 1} + \frac{\rho\omega^2}{4} [R_1^2 + 3R_2^2 + v(R_2^2 - R_1^2)]$$

(inner edge)

$$\sigma_\theta = \frac{2p}{K^2 - 1} + \frac{\rho\omega^2}{4} [3R_1^2 + R_2^2 + v(R_1^2 - R_2^2)]$$

(outer edge)

Index

623